Applied Materials Characterization

MATERIALS RESEARCH SOCIETY SYMPOSIA PROCEEDINGS

ISSN 0272 - 9172

Volume 1—Laser and Electron-Beam Solid Interactions and Materials Processing, J. F. Gibbons, L. D. Hess, T. W. Sigmon, 1981

Volume 2—Defects in Semiconductors, J. Narayan, T. Y. Tan, 1981

Volume 3—Nuclear and Electron Resonance Spectroscopies Applied to Materials Science, E. N. Kaufmann, G. K. Shenoy, 1981

Volume 4—Laser and Electron-Beam Interactions with Solids, B. R. Appleton, G. K. Celler, 1982

Volume 5—Grain Boundaries in Semiconductors, H. J. Leamy, G. E. Pike, C. H. Seager, 1982

Volume 6—Scientific Basis for Nuclear Waste Management, S. V. Topp, 1982

Volume 7—Metastable Materials Formation by Ion Implantation, S. T. Picraux, W. J. Choyke, 1982

Volume 8—Rapidly Solidified Amorphous and Crystalline Alloys, B. H. Kear, B. C. Giessen, M. Cohen, 1982

Volume 9—Materials Processing in the Reduced Gravity Environment of Space, G. E. Rindone, 1982

Volume 10—Thin Films and Interfaces, P. S. Ho, K.-N. Tu, 1982

Volume 11—Scientific Basis for Nuclear Waste Management V, W. Lutze, 1982

Volume 12—In Situ Composites IV, F. D. Lemkey, H. E. Cline, M. McLean, 1982

Volume 13—Laser Solid Interactions and Transient Thermal Processing of Materials, J. Narayan, W. L. Brown, R. A. Lemons, 1983

Volume 14—Defects in Semiconductors II, S. Mahajan, J. W. Corbett, 1983

Volume 15—Scientific Basis for Nuclear Waste Management VI, D. G. Brookins, 1983

Volume 16—Nuclear Radiation Detector Materials, E. E. Haller, H. W. Kraner, W. A. Higinbotham, 1983

Volume 17—Laser Diagnostics and Photochemical Processing for Semiconductor Devices, R. M. Osgood, S. R. J. Brueck, H. R. Schlossberg, 1983

Volume 18—Interfaces and Contacts, R. Ludeke, K. Rose, 1983

Volume 19—Alloy Phase Diagrams, L. H. Bennett, T. B. Massalski, B. C. Giessen, 1983

Volume 20—Intercalated Graphite, M. S. Dresselhaus, G. Dresselhaus, J. E. Fischer, M. J. Moran, 1983

Volume 21—Phase Transformations in Solids, T. Tsakalakos, 1984

Volume 22—High Pressure in Science and Technology, C. Homan, R. K. MacCrone, E. Whalley, 1984

Volume 23—Energy Beam-Solid Interactions and Transient Thermal Processing, J. C. C. Fan, N. M. Johnson, 1984

Volume 24—Defect Properties and Processing of High-Technology Nonmetallic Materials, J. H. Crawford, Jr., Y. Chen, W. A. Sibley, 1984

MATERIALS RESEARCH SOCIETY SYMPOSIA PROCEEDINGS

Volume 25—Thin Films and Interfaces II, J. E. E. Baglin, D. R. Campbell, W. K. Chu, 1984

Volume 26—Scientific Basis for Nuclear Waste Management VII, G. L. McVay, 1984

Volume 27—Ion Implantation and Ion Beam Processing of Materials, G. K. Hubler, O. W. Holland, C. R. Clayton, C. W. White, 1984

Volume 28—Rapidly Solidified Metastable Materials, B. H. Kear, B. C. Giessen, 1984

Volume 29—Laser-Controlled Chemical Processing of Surfaces, A. W. Johnson, D. J. Ehrlich, H. R. Schlossberg, 1984

Volume 30—Plasma Processing and Synthesis of Materials, J. Szekely, D. Apelian, 1984

Volume 31—Electron Microscopy of Materials, W. Krakow, D. Smith, L. W. Hobbs, 1984

Volume 32—Better Ceramics Through Chemistry, C. J. Brinker, D. E. Clark, D. R. Ulrich, 1984

Volume 33—Comparison of Thin Film Transistor and SOI Technologies, H. W. Lam, M. J. Thompson, 1984

Volume 34—Physical Metallurgy of Cast Iron, H. Fredriksson, M. Hillerts, 1985

Volume 35—Energy Beam-Solid Interactions and Transient Thermal Processing/1984, D. K. Biegelsen, G. Rozgonyi, C. Shank, 1985

Volume 36—Impurity Diffusion and Gettering in Silicon, R. B. Fair, C. W. Pearce, J. Washburn, 1985

Volume 37—Layered Structures, Epitaxy and Interfaces, J. M. Gibson, L. R. Dawson, 1985

Volume 38—Plasma Synthesis and Etching of Electronic Materials, R. P. H. Chang, B. Abeles, 1985

Volume 39—High-Temperature Ordered Intermetallic Alloys, C. C. Koch, C. T. Liu, N. S. Stoloff, 1985

Volume 40—Electronic Packaging Materials Science, E. A. Giess, K.-N. Tu, D. R. Uhlmann, 1985

Volume 41— Advanced Photon and Particle Techniques for the Characterization of Defects in Solids, J. B. Roberto, R. W. Carpenter, M. C. Wittels, 1985

Volume 42—Very High Strength Cement-Based Materials, J. F. Young, 1985

Volume 43—Coal Combustion and Conversion Wastes: Characterization, Utilization, and Disposal, G. J. McCarthy, R. J. Lauf, 1985

Volume 44—Scientific Basis for Nuclear Waste Management VIII, C. M. Jantzen, J. A. Stone, R. C. Ewing, 1985

Volume 45—Ion Beam Processes in Advanced Electronic Materials and Device Technology, F. H. Eisen, T. W. Sigmon, B. R. Appleton, 1985

Volume 46—Microscopic Identification of Electronic Defects in Semiconductors, N. M. Johnson, S. G. Bishop, G. D. Watkins, 1985

MATERIALS RESEARCH SOCIETY SYMPOSIA PROCEEDINGS

Volume 47—Thin Films: The Relationship of Structure to Properties, C. R. Aita, K. S. SreeHarsha, 1985

Volume 48—Applied Material Characterization, W. Katz, P. Williams, 1985

Volume 49—Materials Issues in Applications of Amorphous Silicon Technology, D. Adler, A. Madan, M. J. Thompson, 1985

Volume 50—Scientific Basis for Nuclear Waste Management IX, L. O. Werme, 1985

MATERIALS RESEARCH SOCIETY SYMPOSIA PROCEEDINGS VOLUME 48

Applied Materials Characterization

Symposium held April 15-18, 1985, San Francisco, California, U.S.A.

EDITORS:

W. Katz
General Electric Company, Schenectady, New York, U.S.A.

P. Williams
Arizona State University, Tempe, Arizona, U.S.A.

MATERIALS RESEARCH SOCIETY
Pittsburgh, Pennsylvania

Copyright 1985 by Materials Research Society.
All rights reserved.

This book has been registered with Copyright Clearance Center, Inc. For further information, please contact the Copyright Clearance Center, Salem, Massachusetts.

Published by:

Materials Research Society
9800 McKnight Road, Suite 327
Pittsburgh, Pennsylvania 15237
telephone (412) 367-3003

Library of Congress Cataloging in Publication Data

Library of Congress Cataloging-in-Publication Data

Main entry under title:

Applied materials characterization.

(Materials Research Society symposia proceedings, ISSN 0272-9172; v. 48)

Bibliography: p
Includes index.
1. Materials—Testing—Congresses.   2. Materials—Analysis—Congresses.
  I. Katz, W. (William), 1953-   II. Williams, P. (Peter), 1942-   III. Series.
TA410.A67 1985   620.1'10287   85-25936
ISBN 0-931837-13-8

Printed in the United States of America

Manufactured by Publishers Choice Book Mfg. Co.
Mars, Pennsylvania 16046

Contents

Preface   xi

Acknowledgments   xiii

PART I: ATOMIC ORDERING OF MATERIALS AND CHEMICAL BONDING ANALYSIS MATERIALS

THE USE OF LEED FOR THE CHARACTERIZATION OF SURFACE DAMAGE FROM PULSED-LASER IRRADIATION
    Aubrey L. Helms, Jr., Chih-Chen Cho, Steven L. Bernasek, and Clifton W. Draper   3

REFRACTORY METALS GROWTH ON MBE GaAs
    J. Bloch and M. Heiblum   13

ELECTRON SPECTROSCOPIC STUDIES OF SUBSTOICHIOMETRIC TANTALUM CARBIDE
    G.R. Gruzalski, D,M, Zehner, and G.W. Ownby   19

CATION SOLUTE SEGREGATION TO SURFACES OF MgO and α-Al_2O_3
    Robert C. McCune   27

PROPERTIES OF SINGLE-CRYSTAL SILICON FILMS ON AMORPHOUS SiO_2 ON SINGLE-CRYSTAL CUBIC ZIRCONIA SUBSTRATES
    I. Golecki, R.L. Maddox, H.L. Glass, A.L. Lin, and H.M. Manasevit   37

ATOMIC INTERACTIONS IN SILICON-METAL COMPLEXES ON W(110)
    John D. Wrigley and Gert Ehrlich   47

THE THICKNESS EFFECT ON THE MICROSTRUCTURE OF SPUTTERED FILMS STUDIED BY A NEW X-RAY DIFFRACTION METHOD
    M. Hecq   55

SURFACE ELECTRONIC FUNCTION PROPERTIES FROM DEFECT HETEROGENEITY DOMINATED SPECULAR/GLANCING/GRAZING VERSUS BULK TRANSMISSION SMALL-ANGLE SCATTERING (SAS) DIFFRACTION-PATTERN VIA THE STATIC SYNERGETICS ALGORITHM/EXPERIMENTAL MODEL
    Edward Siegel   63

CORE-LEVEL ELECTRON BINDING-ENERGY CHANGE OF EVAPORATED Pd
    Shigemi Kohiki   71

DEPTH OF PENETRATION OF THE PLASMA FLUORINATION REACTION INTO VARIOUS POLYMERS
    Eve A. Wildi, Gerald J. Scilla, and Alan DeLuca   79

SURFACE COMPOSITION OF CARBURIZED TUNGSTEN TRIOXIDE AND ITS CATALYTIC ACTIVITY
    Masatoshi Nakazawa and H. Okamoto   85

INSTANTANEOUS IMPEDANCE OF ALUMINIUM IN ANODIC
POLARIZATION STATUS
    Zhu Yingyang, Wang Kuang, Zhu Rizhang, and
    Zhang Wengi   91

PART II: MICROSTRUCTURES OF MATERIALS AND ELEMENTAL ANALYSIS

PROGRESS AND PROSPECTS OF MATERIALS CHARACTERIZATION AT
SUBNANOMETER SPATIAL RESOLUTION USING FINELY FOCUSSED
ELECTRON BEAMS
    Michael Issacson   107

TRANSMISSION ELECTRON MICROSCOPE INVESTIGATION OF SPUTTERED
Co-Pt THIN FILMS
    P. Alexopoulos, R.H. Geiss, and M. Schlenker   117

CHARACTERIZATION OF THE Ni/NiO INTERFACE REGION IN
OXIDIZED HIGH-PURITY NICKEL BY TRANSMISSION ELECTRON
MICROSCOPY
    Howard T. Sawhill and Linn W. Hobbs   127

INTERFACE STUDY OF Mo/GaAs
    Peiching Ling, Jyh-Kao Chang, Min-Shyong Lin, and
    Jen-Chung Lou   137

DETERMINATION OF THE COMPOSITION AND THICKNESS OF THIN
POTASSIUM POLYPHOSPHIDE FILMS
    Klara Kiss and Paul M. Figura   145

DEFECTS IN PLATINUM SILICIDE FORMATION
    Michael J. Warburton   159

AUGER ELECTRON ANALYSIS OF OXIDES GROWN ON A DILUTE
ZIRCONIUM/NICKEL ALLOY
    R.A. Ploc, R.D. Davidson, and J.A. Roy   169

AUGER SPUTTER DEPTH PROFILING APPLIED TO ADVANCED
SEMICONDUCTOR DEVICE STRUCTURES
    D.K. Skinner, C. Hill, and M.W. Jones   179

SURFACE SEGREGATION OF Ni-Cr ALLOY
    N.Q. Chen, Q.J. Zhang, and Z.Y. Hua   185

THE EFFECT OF OXYGEN ON DIFFUSION AND COMPOUNDING
AT Ni-GaAs (100) INTERFACES
    J.S. Solomon, D.R. Thomas, and S.R. Smith   191

THE CHARACTERIZATION OF ALLOYED NiGeAuAgAu OHMIC CONTACTS
TO AlInAs/GaInAs HETEROSTRUCTURE BY AUGER ELECTRON
SPECTROSCOPY AND WAVE LENGTH DISPERSIVE X-RAY ANALYSIS
    P.M. Capani, S.D. Mukherjee, L. Rathbun, H.T. Griem,
    G.W. Wicks, L.F. Eastman, and J. Hunt   203

EXPERIMENTAL INVESTIGATION OF GaAs SURFACE OXIDATION
    S. Matteson and R.A. Bowling   215

PART III: MATERIALS CHARACTERIZATION
WITH ION BEAMS

ON THE USE OF SECONDARY ION MASS SPECTROMETRY IN
SEMICONDUCTOR DEVICE MATERIALS AND PROCESS DEVELOPMENT
    Charles W. Magee and Ephraim M. Botnick   229

SIMS CHARACTERIZATION OF THIN THERMAL OXIDE LAYERS
ON POLYCRYSTALLINE ALUMINIUM
    F. Degreve and J.M. Lang   241

SIMS, SAM, AND RBS STUDY OF HIGH-DOSE OXYGEN IMPLANTATION
INTO SILICON
    W.M. Lau, P. Ratnam, and C.A.T. Salama   263

IN SITU ION IMPLANTATION FOR QUANTITATIVE SIMS ANAYLSIS
    Richard T. Lareau and Peter Williams   273

SECONDARY ION MASS SPECTROSCOPY OF CERAMICS
    Jenifer A.T. Taylor, Paul F. Johnson, and
    Vasantha R.W. Amarakoon   281

SIMS ANALYSIS OF PURE AND HYDRATED CEMENTS
    Erich Naegele and Ulrich Schneider   289

APPLICATION OF SIMS DEPTH PROFILING TO CERAMIC
MATERIALS
    Jenifer A.T. Taylor, Paul F. Johnson, and
    Vasantha R.W. Amarakoon   299

ULTRASENSITIVE ELEMENTAL ANALYSIS OF MATERIALS USING
SPUTTER-INITIATED, RESONANCE-IONIZATION SPECTROSCOPY
    J.E. Parks, D.W. Beekman, H.W. Schmitt, and M.T. Spaar   309

MICROFOCUSSED ION BEAMS FOR SURFACE ANALYSIS AND
DEPTH PROFILING
    David R. Kinghan, P. Vohralik, D. Fathers, A.R. Waugh,
    and A.R. Bayly   319

SOME APPLICATIONS OF SIMS AND SSMS IN MATERIALS
CHARACTERIZATION
    J. Verlinden, R. Vlaeminck, F. Adams, and R. Gijbels   331

CHARACTERIZATION OF THIN METAL FILMS BY SIMS, AUGER,
AND TEM
    B.K. Furman, J.P. Benedict, K.L. Granato,
    R.M. Prestipino, and D.Y. Shih   341

TITANIUM SILICIDE FORMATION ON HEAVILY DOPED
ARSENIC-IMPLANTED SILICON
    S.L. Dowben, D.W. Marsh, G.A. Smith, N. Lewis,
    T.P. Chow, and W. Katz   355

THE CHARACTERIZATION OF INTENTIONAL DOPANTS IN
HgCdTe USING SIMS, HALL-EFFECT, AND C-V MEASUREMENTS
    L.E. Lapides, R.L. Whitney, and C.A. Crosson   365

PART IV: HIGH ENERGY METHODS AND MATERIALS CHARACTERIZATION USING PHOTON BEAMS

HYDROGEN MEASUREMENT OF THIN FILM SILICON: HYDROGEN ALLOY FILMS TECHNIQUE COMPARISON
  Gary A. Pollock 379

HYDROGENATION DURING THERMAL NITRIDATION OF SiO_2
  A.E.T. Kuiper, F.H.P.M. Habraken, and James T. Chen 387

GROWTH AND COMPOSITION OF LPCVD SILICON OXYNITRIDE FILMS
  F.H.P.M. Habraken and A.E.T. Kuiper 395

MeV HELIUM MICROBEAM ANALYSIS: APPLICATIONS TO SEMICONDUCTOR STRUCTURES
  R.A. Brown, J.C. McCallum, C.D. McKenzie, and
  J.S. Williams 403

CONCENTRATION MEASUREMENTS AND DEPTH PROFILING OF PHOSPHORUS AND BORON BY MEANS OF (p,) RESONANT REACTIONS
  M.J.M. Pruppers, F. Zijderhand, F.H.P.M. Habraken,
  and W.F. van der Weg 409

MATERIALS CHARACTERIZATION WITH INTENSE POSITRON BEAMS
  I.J. Rosenberg, R.H. Howell, M.J. Fluss, and P. Meyer 419

EFFECTS OF AMBIENT GAS ON THE OUT-DIFFUSION OF NICKEL AND COPPER THROUGH THIN GOLD FILMS
  R.K. Lewis, S.K. Ray, K. Seshan 425

ELASTIC RECOIL ANALYSIS OF HYDROGEN IN ION-IMPLANTED MAGNETIC BUBBLE GARNETS USING 44 MeV CHLORINE IONS
  A. Leiberich, B. Flaugher, and R. Wolfe 431

APPLICATIONS OF SURFACE ANALYSIS BY LASER IONIZATION (SALI) TO INSULATORS AND II-VI COMPOUNDS
  C.H. Becker, C.M. Stahle, and D.J. Thomson 447

INFLUENCE OF THE DENSITY OF OXIDE PARTICLES ON THE DIFFUSIONAL BEHAVIOR OF OXYGEN IN INTERNALLY OXIDIZED, SILVER-BASED ALLOYS
  F.H. Sanchez, R.C. Mercader, A.F. Pasquevich,
  A.G. Bibiloni, and A. Lopez-Garcia 455

HIGH-TEMPERATURE RAMAN STUDIES OF PHASE TRANSITIONS IN THIN-FILM DIELECTRICS
  Gregory J. Exarhos 461

USE OF ENERGY-LOSS STRUCTURES IN XPS CHARACTERISATION OF SURFACES
  J.E. Castle, I. Abu-Talib, and S.A. Richardson 471

AUTHOR INDEX 481

SUBJECT INDEX 483

Preface

This book contains papers presented at the symposium on "Applied Materials Characterization" held in San Francisco, CA, April 15-18, 1985. This symposium, which was part of the Materials Research Society meeting, was the first ever to address the topic of materials characterization. The symposium provided an international forum consisting of eight oral sessions in which eight invited and 55 contributed papers were presented.

The symposium dealt with both the development of advanced characterization techniques and the application of characterization technology to novel materials. Papers addressed understanding materials from an atomic level as well as bulk chemical and physical properties. Sessions dealt with attempting to unravel materials on a fundamental level using many of the more common electron and ion spectroscopies. In addition, frontier areas in new characterization techniques were also covered - novel methods such as Raman spectroscopy for studies of high-temperature phase transitions, conversion-electron Mossbauer spectroscopy, and positron annihilation studies were shown to be promising surface analytical tools.

Symposium Co-Chairmen

W. Katz　　　　　　　　　　　　　　　　　　　　　　　　　　　　　　　P. Williams

Acknowledgments

We would like to thank all the conference participants, particularly the invited speakers who provided excellent reviews of their topics. They are:

W. Gibson, R.J. Hitzman, M. Isaacson, M. Lagally, C.W. Magee, R.P. Messmer, J. Spence, N. Winograd

We are also grateful to the session chairmen who directed the sessions and guided discussion:

J. Burkstrand, V.R. Deline, B.K. Furman, C.W. Magee, R.P. Messmer, G.A. Smith

It is our pleasure to acknowledge with gratitude the financial support provided by the following companies:

A.G. Associates, Cameca Instruments, Charles Evans and associates General Ionex, Instruments, S.A., JOEL, Leybold-Heraeus, Perkin-Elmer-Physical Electronics Division, Philips, Surface Science Instruments, V.G. Instruments

PART I

Atomic Ordering of Materials and Chemical Bonding Analysis of Materials

THE USE OF LEED FOR THE CHARACTERIZATION OF
SURFACE DAMAGE FROM PULSED LASER IRRADIATION

AUBREY L. HELMS, JR.,[*] CHIN-CHEN CHO[*], STEVEN L. BERNASEK,[*] AND
CLIFTON W. DRAPER[**]

[*]Department of Chemistry, Frick Chemical Laboratory, Princeton University,
Princeton N.J. 08544

[**]AT & T Technologies Engineering Research Center, P.O. Box 900,
Princeton N.J. 08540

ABSTRACT

Low Energy Electron Diffraction (LEED)-Spot Profile Analysis and Auger
Electron Spectroscopy (AES) have been used to study the response of Mo(100)
single crystal surfaces to Q-switched, frequency doubled Nd:YAG laser pulses.
The experiments were conducted in a special ultra-high vacuum (UHV) system
which allowed the surfaces to be irradiated under controlled conditions.
Laser fluences both above and below the melt threshold were employed. For
the melted surfaces, good epitaxial regrowth was observed. The spot profile
analysis indicates the formation of random islands on the surfaces. Surfaces
which had been previously disordered by 3 KeV Ar^+ implantation were laser
surface melted and observed to regrow epitaxially as has been observed in the
case of ion implanted silicon. The formation of the islands and stepped
structures is explained by considering the activation of dislocation sources
by the induced thermal stresses resulting in slip.

INTRODUCTION

The interaction between high power laser pulses and materials has been
the subject of active research for many years. In contrast to the good
epitaxial regrowth observed in semiconductors, metals have been shown to
exhibit liquid phase epitaxial regrowth with a marked increase in disorder
after pulsed laser melting [1-8]. The quality of the epitaxy has been shown
to be dependent upon crystallographic orientation, and the damage scales with
laser power.

The responses of the metals have taken many forms. Transmission
electron microscopy has shown the formation of an extended dislocation network in laser melted nickel [2]. Other modes of damage include slip [5],
pitting, dislocation motion and multiplication [7], increased vacancy concentration [8], and cratering [6]. The various forms of damage have been
found to generally have different laser fluence thresholds.

Paralleling the experimental studies has been an equally intensive
theoretical effort [9-13]. The laser/solid interaction for large spot -
short pulse experiments can be modeled as a semi-infinite slab with one-
dimensional heat flow [9]. By choosing appropriate boundary conditions, the
spatial and temporal profiles of the temperature excursions in the solid may
be calculated. Using such a model and concentrating on the near surface
region, Musal has calculated the minimum temperature rise at the surface
required for the onset of plastic deformation [9]. This temperature rise is
suprisingly low, being only 20 K for copper [9]. In the proposed model, the
deformation takes place through thermally activated slip, resulting in steps
at the surface.

The development of spot profile analysis applied to LEED over the past decade has made the study of surface defect structures possible [14-20]. The angular profile of the diffraction spot in reciprocal space yields information about the atomic morphology of the surface [19]. The behavior of the profile as a function of incident electron energy can also be used to make a qualitative assessment of the defect structure. LEED is very surface sensitive, probing only the top 3-4 atomic layers and spot profile analysis allows the distinction between random point defects, ordered terraces, ordered islands, random islands, and facets [16].

We report here the use of LEED-spot profile analysis, AES, and Nomarski Interference Contrast (NIC) microscopy to study the response of Mo(100) surfaces to pulsed laser irradiation at laser fluences both above and below the threshold for melting. The use of LEED has been shown to be very sensitive to the introduction of disorder. The results indicate that the surface regrows epitaxially with the formation of random islands on the surface after pulsed laser melting. Similar structures are observed for fluences below the melt threshold. These islands coalesce back toward the flat surface after long periods of annealing at 1000°C. Samples which had been disordered by Ar^+ ion implantation were annealed by pulsed laser irradiation in a manner similar to the laser annealing of silicon [21]. The LEED analysis also indicates the formation of random islands in the irradiated region of these samples.

EXPERIMENTAL

The samples were cut from a single crystal boule which had been oriented to within 1° of the (100) face of molybdenum. The orientation was obtained using the back reflecting Laue technique. The samples were mechanically polished to a mirror finish using standard metallographic techniques. The samples were spotwelded to thin tantalum foils which were mounted on a specially designed transfer block attached to the sample manipulator. The resistively heated tantalum foil was used to heat the sample to temperatures in excess of 1000°C for cleaning and annealing purposes. The sample temperature was monitored by means of an optical pyrometer focussed on the sample through an 8" viewport at the front of the vacuum system.

The experiments were conducted in a conventional ion-pumped UHV chamber. Additional pumping speed was supplied by a titanium sublimation pump and a liquid nitrogen cryotrap. The base pressure of the system was less than 2×10^{-10} torr. The chamber was equipped with standard 4-grid LEED optics, a cylindrical mirror analyzer (CMA) for AES, a quadrupole mass spectrometer for residual gas analysis, a 3 KeV ion gun for Ar^+ sputtering, a high precision manipulator with X,Y,Z, rotary and tilt motions, and a variable leak valve.

The samples were transferred out of the main chamber by long rods on the end of a magnetically coupled drive system built in the departmental machine shop. The transfer system coupled the main chamber with a custom built laser chamber (Fig. 1). The laser chamber was independently pumped by a 20 l/s ion pump and could be isolated by means of two high vacuum valves. A complete description of the design and capabilities of the transfer block, transfer system, and laser chamber will appear elsewhere [22].

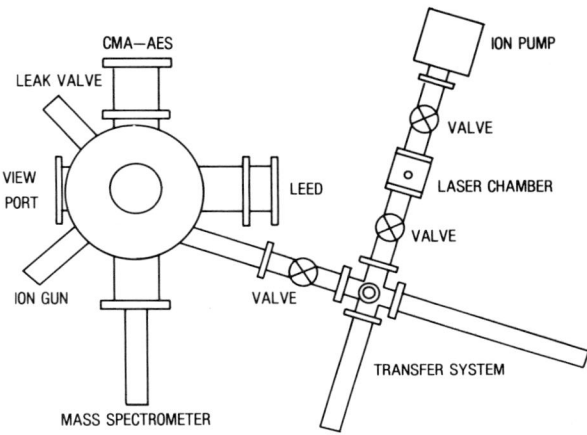

Figure 1. Schematic of the experimental apparatus showing the main chamber, transfer system, and laser chamber.

The samples were cleaned and characterized in the main chamber and then transferred to the laser chamber. The high vacuum valves were closed and the laser chamber decoupled from the main system. The chamber was then transported to the laser facilities at the AT & T Technologies Engineering Research Center (ERC). The samples were irradiated under UHV conditions and returned to Princeton. The sample was transferred back to the main chamber for analysis. UHV conditions were maintained throughout the experiment.

The laser system employed for these experiments is a Q-switched, frequency doubled, Nd:YAG that is part of a laser trimming test system. The repetition rate was 5 KHz with a pulse width of 140 ns FWHM. The melted spot size was nominally 25 μm. To ensure complete coverage, the beam was rastered across the surface by means of optics coupled to computer controlled linear induction motors. The table speed, displacement, and coordinates of the spot pattern were chosen to give good spot-to-spot overlap. One half of the sample was irradiated five times to ensure uniform modification, leaving the other half to serve as an internal standard.

Samples were irradiated under three different laser conditions. One set of samples were irradiated at 15 MW/cm^2 which is below the threshold for melting. These samples will be designated as 'Type A' in the following discussion. A second set of samples were irradiated at laser fluences between 25 and 90 MW/cm^2 which were above the threshold for melting. These samples will be designated as 'Type B'. The third set were samples that had been Ar$^+$ bombarded at 3 KeV in 1x10^{-6} torr. Ar for one hour at normal incidence. This bombardment produced a disordered surface as evidenced by the abscence of a LEED pattern. These samples were irradiated at a laser fluence of 75 MW/cm^2 and will carry the designation 'Type C'.

The LEED data were collected by photographing the LEED patterns with a 35 mm camera mounted in front of the 8" viewport and focussed on the LEED screen. The intensity profiles were evaluated from the negatives by means of a computer interfaced Vidicon camera system described previously [24]. The information collected using this system include the angular spot profile, spot-to-spot distance, and relative intensity between the spot and the background. This technique provides a permanent record of the experiment while allowing the accumulation of a large quantity of data in a very short time.

The composition of the surface was monitored using AES. The conditions and instrument parameters were carefully controlled so that all of the spectra were acquired under the same conditions. The surface concentrations of carbon, sulfur, and oxygen were calculated by evaluating the ratio of the peak-to-peak heights of those elements to the peak-to-peak height of the Mo-220 eV transition. The ratios could be compared to literature values to give surface concentrations in units of fractional monolayers. Additional calibrations were obtained in the cases where an ordered overlayer structure appeared in the LEED pattern.

The data accumulation involved characterizing the surface using AES and LEED followed by transfer and laser irradiation. After the sample was returned to the main chamber, the compositional differences between the virgin and irradiated surfaces were determined using AES. The sample was flashed to 1000°C for five minutes to desorb any contamination adsorbed from the background during the transfer process. Photographic LEED data were collected as a function of incident electron energy. The sample was annealed at 1000°C with AES and LEED data collected following 15,30,45,60,90,120, and 180 minutes of heating. The AES and LEED data were collected with the sample at room temperature.

The defect nature of the surface was deduced by comparing the spot profiles of the virgin and irradiated surfaces with trends expected from the work of Lagally, Henzler, and others [14-20]. By restricting the study to the spot profiles and neglecting the use of integral intensities, they have calculated the expected diffraction profiles for a variety of surface defect structures within the kinematical approximation. Figure 2 summarizes the profiles expected from a variety of defect surfaces. The results described in the following section are derived from qualitative comparisons of their calculated spot profiles and the observed spot profiles under different irradiation conditions.

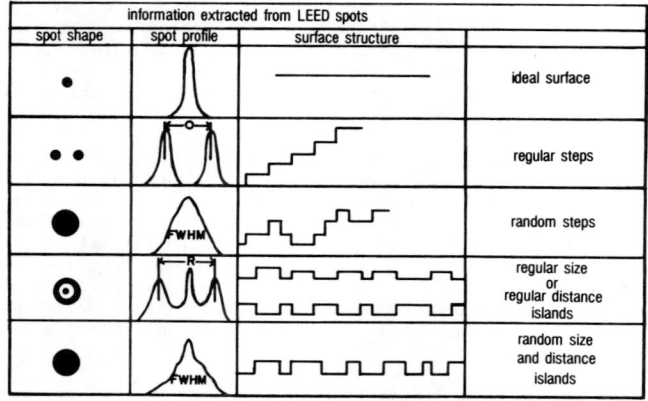

Figure 2. Table from Gronwald and Henzler (19) summarizing the information available from LEED - Spot Profile Analysis.

RESULTS

The results for types A and B were similar and will be described simultaneously. For both types, the surface concentrations of carbon and oxygen were observed to increase after returning to the main chamber. This was probably due to the adsorption of background gases during the transportation between Princeton and ERC. In each case the levels of contamination were lower on the irradiated surface indicating that laser stimulated desorption occurred during the irradiation phase. This observation correlated well with a slight pressure rise registered on the ion pump controller during the laser processing. The contaminants were easily removed by flashing the sample to 1000°C after transferring to the main chamber.

A clear LEED pattern was observed for both the irradiated and virgin regions of both types of samples. The geometry was consistent with that expected for the (100) surface. The spots for the virgin surface were visually sharper than the spots for the irradiated area. This clearly indicates the introduction of disorder into the surface. The appearance of a LEED pattern for the irradiated region of samples of Type B is an indication of liquid phase epitaxial regrowth during the resolidification of the melt.

A comparison of the angular profiles of spots taken from each region indicates that in each case the half-widths of the spots from the irradiated region were greater than those corresponding to spots from the virgin region (Fig. 3). A plot of the relative half-widths versus incident electron energy for first order spots from the irradiated region indicated an oscillatory behavior indicative of the formation of stepped structures on the surface (Fig 4). The amplitude and frequency of these oscillations, as well as the amplitude of the relative half-width, were observed to decrease as a function of accumulated annealing time, suggesting a coalescence of the

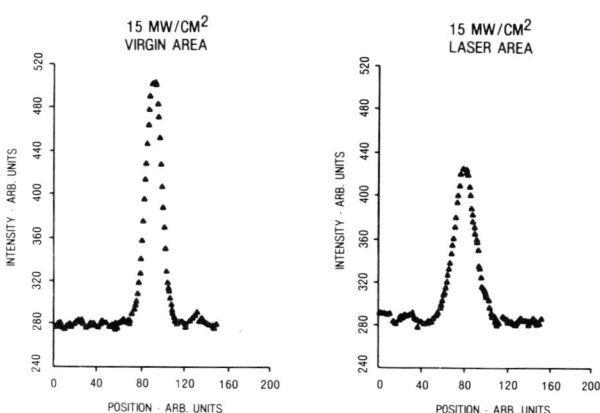

Figure 3. A plot of intensity versus position for typical spot profiles for both the virgin and irradiated regions from a sample of Type A.

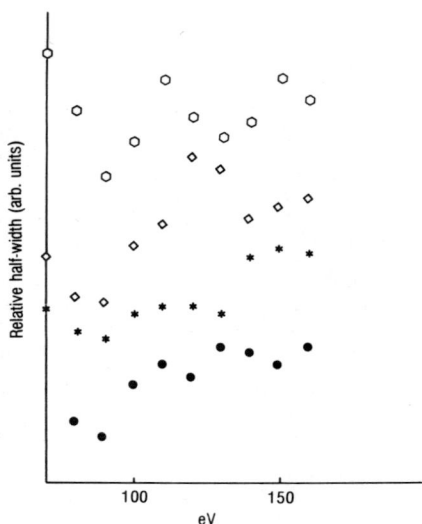

Figure 4. A plot of relative half-width (half-width divided by the (0,0) - (0,1) spot-to-spot distance) versus incident electron energy for several heating times. (○) 5 minutes, (◇) 20 minutes, (✱) 35 minutes, (●) 80 minutes.

stepped surface back toward a flat surface. The effects were observed for both types A and B, with the magnitudes of the features being larger for Type B.

Careful analysis of the shape of the spots indicated that in some instances the spots from the irradiated region contained 'shoulders' aligned along the <010> and <001> directions (Fig 5). These shoulders indicate the orientation of the terrace edges. As with the half-width, the prominence of the shoulders was observed to decrease with accumulated annealing time indicating the coalescence of the steps.

The LEED patterns in the irradiated regions were slightly expanded relative to the patterns for the virgin areas as evidenced by an increase in the spot-to-spot distance. The expansion was 3% for Type A and 7% for Type B. The expansion was isotropic and was determined not to be an experimental artifact due to errors in sample positioning or changes in the incident electron energy. This expansion indicates that the irradiated region is slightly contracted in the plane of the surface in real space.

The AES results for Type C followed the same trends as types A and B. A distinct LEED pattern was observed for the irradiated regions of samples of type C. This indicates that the melted region extended beyond the range of damage due to the Ar^+ implantation to the single crystal substrate below and underwent liquid phase epitaxial regrowth resulting in a net annealing of the

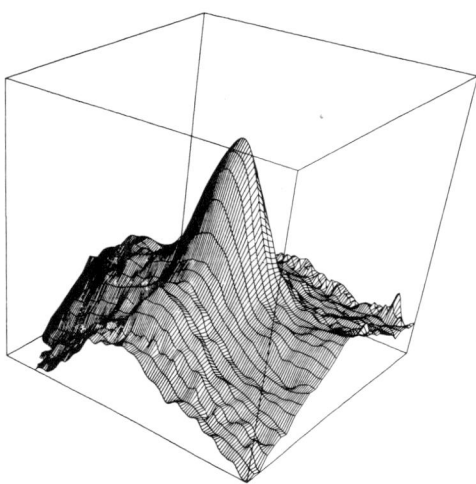

Figure 5. A plot of intensity versus position for a spot in the irradiated region of a sample of Type B showing the 'shoulders'.

surface. A very poor LEED pattern was observed for the virgin region after a heating time of several minutes, which remained very diffuse relative to the sharp spots observed for the irradiated region throughout the study. For this type, the half-widths of the spots from the irradiated region were smaller than from the virgin region indicating the extent of the annealing induced by the laser. The irradiated region still showed an increase in disorder relative to the well annealed surface. As with samples of types A and B, the spot profile analysis indicated some spots with 'shoulders'. However, unlike the other two types, the shoulders for this type of sample were circularly symmetric. The half-widths of spots from both regions decreased as a function of accumulated heating time indicating the previously mentioned annealing behavior observed for samples of types A and B.

DISCUSSION

The melt puddles observed in the irradiated region of samples of types B and C using NIC are clear indications that the surface had melted. The appearance of a LEED pattern in these areas indicates that the surface has regrown via the process of liquid phase epitaxy. This is further evidenced by the fact that the geometry of the LEED pattern in these areas is consistent with the original (100) surface. The irradiated surface of samples of Type A was visually indistinguishable from the virgin area indicating that melting had not occurred. The damage in these regions can only be due to thermal effects in the solid phase.

The angular profiles of the spots in reciprocal space yield information about the structure of the surface. The observation of a broader half-width that oscillated as a function of incident electron energy in the irradiated regions indicates the formation of multi-tiered structures on the surface. These trends were observed for both types A and B implying that the features are not simply due to the macroscopic roughness of the laser melted region of samples of type B.

The observation of shoulders on the spots indicates the formation of islands with a random size distribution and a random distribution of distances between their centers. The orientation of the shoulders indicates a preferred orientation of the island edges. The circularly symmetric shoulders observed for samples of Type C indicate that the preferential orientation of the island edges in these samples is not as strong as in the other two types.

It has been shown previously that the damage introduced into metals by pulsed laser irradiation takes the form of dislocation networks [2] and an increase in the number of vacancies [8]. This is consistent with the thermomechanical model proposed by Musal [9]. The thermal gradients introduced during the pulse cause the heated volume to expand. During the heating phase, this results in compressive stresses accumulated in the heated volume. If the surface melts (types B and C) these stresses are relieved in the liquid phase. After resolidification, the melted volume cools and again stresses are induced by the thermal gradients. If this stress exceeds the critical resolved shear stress (CRSS) for dislocation activation on one of the slip planes of the material, the solid will be plastically deformed through the mechanism of dislocation movement/multiplication. It is well known that the CRSS decreases with increasing temperature, so the hot, resolidified volume is expected to deform before the cooler surroundings. The motion/multiplication of a dislocation can result in the formation of an atomic step at the surface.

The observed LEED results are consistent with the aforementioned model. The major slip planes for the bcc crystal structure are of the $\{211\}$ and $\{110\}$ families [25]. Clear slip patterns are rarely observed in bcc metals because of the cross slip and interaction of these two systems. The edges of the islands may be due to the activation of dislocation sources in one or both of these slip systems. The preferential orientation of the edges seems to indicate that the $\{110\}$ system is slightly favored over the $\{211\}$ system. This conclusion is derived by noticing that the edges lie along the intersecting lines of the $\{110\}$ planes with the (100) surface.

CONCLUSIONS

The response of Mo(100) surfaces to pulsed laser irradiation was studied using the technique of LEED-spot profile analysis. The method has been shown to be very sensitive, indicating damage even at laser fluences below the melt threshold. The behavior of the relative half-width as a function of incident electron energy and the angular profiles of the spots after irradiation suggest the formation of random islands on the surface with edges aligned along the $\{110\}$ slip planes. The LEED pattern for the irradiated region of the samples which were laser surface melted indicates liquid phase epitaxial regrowth during resolidification. Futhermore, the appearance of a LEED pattern for the irradiated region of samples that had been disordered by 3 KeV Ar^+ implantation implies that the surface was ordered by epitaxial regrowth from a melt which extended beyond the damaged region to the underlying single crystal. The formation of the islands can be explained using a thermomechanical model in which the metal plastically deforms in response to the introduced thermal stress by the activation of dislocation sources. The movement/multiplication of these dislocations can result in slip and the formation of step edges and multi-tiered structures at the surface.

ACKNOWLEDGEMENTS

The expertise and patience of Ms. Lisa Kennedy '86 of Princeton University is gratefully acknowledged in the preparation of high quality single crystal molybdenum surfaces. We also gratefully acknowledge the help and guidance of Mr. Randy Crisci of AT & T Technologies in the maintenance of the laser system. Special thanks are given to Dr. Everett J. Canning, Jr. of AT & T Technologies for helpful discussions on dislocation theory. This work was partially supported by the National Science Foundation, Division of Materials Research, under grant DMR-80-19772.

REFERENCES

1. Helms, Jr., A.L., Draper, C.W., Jacobson, D.C., Poate, J.M., and Bernasek, S.L., in ENERGY BEAM-SOLID INTERACTIONS AND TRANSIENT THERMAL PROCESSING, Biegelsen, Rozgonyi, and Shank eds., (Elsevier North Holland, New York, 1985) in Press.
2. Buene, L., Jacobson, D.C., Nakahara, S., Poate, J.M., Draper, C.W., and Hirvonen, J.K., in LASER AND ELECTRON-BEAM SOLID INTERACTIONS AND MATERIALS PROCESSING, Gibbons, Hess, and Sigmon eds., (Elsevier North Holland, New York, 1981), pp. 583-590.
3. Buene, L., Kaufmann, E.N., Preece, C.M., and Draper, C.W., in LASER AND ELECTRON-BEAM SOLID INTERACTIONS AND MATERIALSPROCESSING, Gibbons, Hess, and Sigmon eds., (Elsevier North Holland, New York,1981), pp. 591-597.
4. Porteus, J.O., Decker, D.L., Jernigan, J.L., Faith, W.N., and Bass, M., IEEE J. of Quantum Electronics, **QE-14**, 776 (1978).
5. Porteus, J.O., Soileau, M.J., and Fountain, C.W., Appl. Phys. Lett., **29**, 156 (1976). 6. Chun, M.K., and Rose, K., J. of Appl. Phys., **41**, 614 (1970).
7. Haessner, F., and Seitz, W., J. of Mat. Sci., **6**, 16 (1971).
8. Metz, S.A., and Smidt, Jr., F.A., Appl. Phys. Lett., **19**, 207 (1971).
9. Musal, Jr., H.N., Symp. on Optical materials for High Power Lasers, Boulder, (1979), pp. 159.
10. Bechtel, J.H., J. of Appl. Phys., **46**, 1585 (1975).
11. Lax, M., J. of Appl. Phys., **48**, 3919 (1977).
12. L.R., Tucker, T.R., Schriempf, J.T., Stegman, R.L., and Metz, S.A., J. of Appl. Phys., **47**, 1415 (1976).
13. Rimini, E., in SURFACE MODIFICATION AND ALLOYING BY LASER, ION AND ELECTRON BEAMS, Poate, Foti, and Jacobson (Plenum, New York, 1981), ch. 2.
14. Lagally, M.G., Appl. of Surf. Sci., **13**, 260 (1982).
15. Lu, T.M., and Lagally, M.G., Surf. Sci., **120**, 47 (1982).
16. Henzler, M., in ELECTRON SPECTROSCOPY FOR SURFACE ANALYSIS, Ibach ed., (Springer, Berlin, 1977), ch. 4.
17. Henzler, M., Surf. Sci., **73**, 240 (1978).
18. Houston, J.E., and Park, R.L., Surf. Sci., **21**, 209 (1970).
19. Gronwald, K.D., and Henzler, M., Surf. Sci., **117**, 180 (1982).
20. Henzler, M., Appl. of Surf. Sci., **11/12**, 450 (1982).
21. Foti, G., and Rimini, E., in LASER ANNEALING OF SEMICONDUCTORS, Poate and Mayer eds., (Academic, New York, 1982), ch. 7.
22. Helms, Jr., A.L., Schiedt, W.A., Biwer, B.M., and Bernasek, S.L., to be published.
23. Tommet, T.N., Olszewski, G.B., Chadwick, P.A., and Bernasek, S.L., Rev. Sci. Instrum., **50**, 147, (1979).
24. Salmeron, M., Somorjai, G.A., and Chianelli, R.R., Surf. Sci., **127**, 526 (1983).
25. Hull, D., Byron, J.F., and Noble, F.W., Can. J. of Phys., **45**, 1091 (1967).

REFRACTORY METALS GROWTH ON MBE GaAs

J. BLOCH[*] AND M. HEIBLUM
IBM Thomas J. Watson Research Center, Yorktown Heights, NY 10598,
*Permanent address: Nuclear Research Center-Negev, POB 9001,
Beer-Sheva, ISRAEL 84190.

ABSTRACT

Molybenum and tungsten metals have been grown on an MBE grown (100)GaAs at various substrate temperatures. RHEED technique was used in situ to analyse the crystalline structure and the growth mechanism of the mechanism of the thin metal films. It was found that Mo epilayers can be grown on the (100) GaAs at temperatures between 150-400°C. The epitaxal arrangement is (111)Mo ‖ (100)GaAs with [011]Mo ‖ [011]GaAs. Tungsten films are not growing epitaxially under similar experimental conditions.

INTRODUCTION

Properties of metal thin films on GaAs are important for understanding the behaviour of ohmic contracts as well as Schottky barrier devises. The growth of several metals on GaAs and other III-V compounds in molecular beam epitaxy (MBE) systems has been studied recently [1]. Some complications are involved in depositing refractory metals under ultra-high vacuum (UHV) conditions, due to the relatively high temperatures required for the evaporation. An UHV-compatible electron-gun which has been developed recently [2] was mounted into a RIBER 1000-1 MBE system and used for evaporation of Mo and W films on MBE grown (100)GaAs. The growth process was studied in situ using a reflection high-energy electron diffraction (RHEED) technique.

RESULTS AND DISCUSSION

MBE grown (100)GaAs substrates were brought to the desired growth temperature. Thin films of Mo (up to a thickness of 150Å) were then evaporated onto the substrates. Figure 1 shows the RHEED patterns of the Mo films at various substrate temperatures. Three different temperature regions can be distinguished;
(1) The low temperatures range, between room temperature and 150°C, at which the Mo films are growing in a polycrystalline structure as evident of the characteristic rings shown in the RHEED pattern. Starting at 100°C for Mo (250°C for W), a tendency towards preferred orientation is shown by the appearance of high intensity regions along the rings.
(2) The epitaxial growth range; between 150 and 450°C the Mo film RHEED patterns consist of circular dots or wide streaks indicating some extent of epitaxy. Increasing the substrate temperature above 300°C results in changes in the Mo film electron diffraction. Initially a faint spike located at the (200) diffraction spot is observed in the RHEED pattern. At 370°C it becomes more spread and at 420°C it is actually divided into two close together spots.

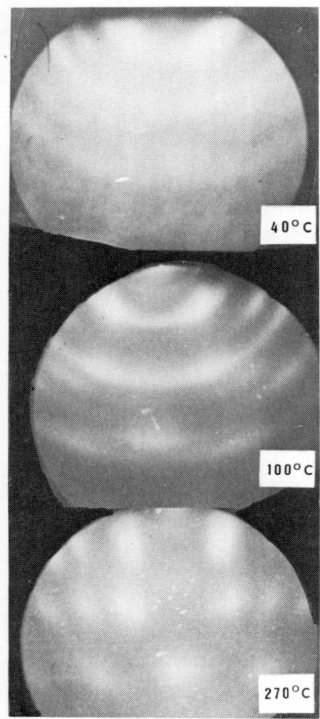 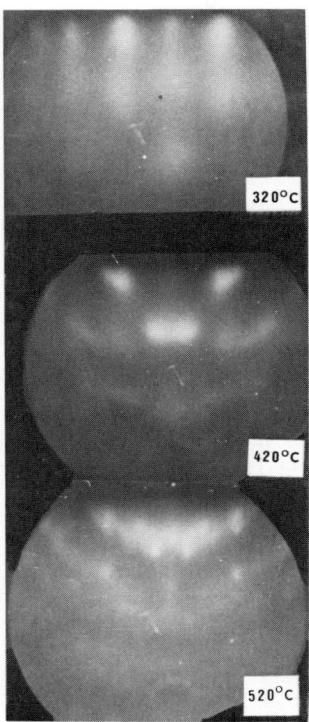

Fig.1. RHEED patterns of molybdinum thin films (100-150Å thick) grown on (100)GaAs at various substrate temperatures.

(3) The high-temperature range. At 520°C the RHEED patterns is considerably changed indicating the presence of a new phase. This phase is identified as Mo_5As_4 [3], the product of a high-temperature solid-state reaction. Note that this new phase appears also to have some preferred orientation structure. For W, no epitaxial growth is observed; For substrate temperatures around 300°C, a rather complex RHEED pattern is developed. Though the growth is not completely polycrystalline, it is evident that more than single plane of the W lattice can match the (100)GaAs. A new phase appears at 520°C as a result of a high-temperature solid-state reaction.

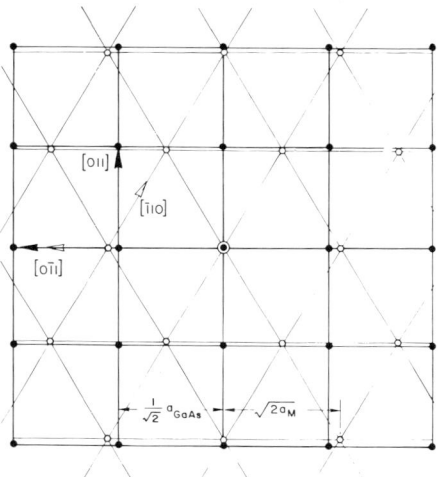

Fig.2. Atomic arrangements of (111)Mo (open circles) on (100)GaAs (solid circles) with [011]Mo ‖ [011]GaAs. Solid and open arrows show directions in the GaAs and the metal lattices, respectively.

The epitaxial arrangement of the Mo bcc lattice on top of the (100)GaAs zincblende lattice is (111)Mo ‖ (100)GaAs with the [011]Mo ‖ [011]GaAs as shown in the model in Figure 2. Note that the smallest nearest neighbour linear misfit (along [01̄1]) is 11.5% while the second neighbour misfit (along [011]) is -3.5%. Rotating the (111)Mo by 30° yields an equivalent arrangement. Four seperate such arrangements are possible. Apparently all of them are present as can be seen in Figure 3. At the bottom, the RHEED pattern of a Mo film on a substrate at 310°C, is shown. The diffraction was taken along the [011] azimuth of the (100)GaAs, but it does not change considerably with rotating the sample because identical planes will be diffracted at angles 60° to each other. The pattern represents two types of grain orientation related to each other by twining about the [112̄][4]. The other two grain orientations which are parallel to the [101] are somewhat difficult to be observed. They should appear as two dots, one to each side of the (222) dot.

The evolution of the RHEED patterns during the first stages of the deposition in the epitaxial range can provide a great deal of information concerning the growth mechanism. Figure 4 represents some stages in the growth of Mo film on substrate at 200-300°C. The following steps are observed:

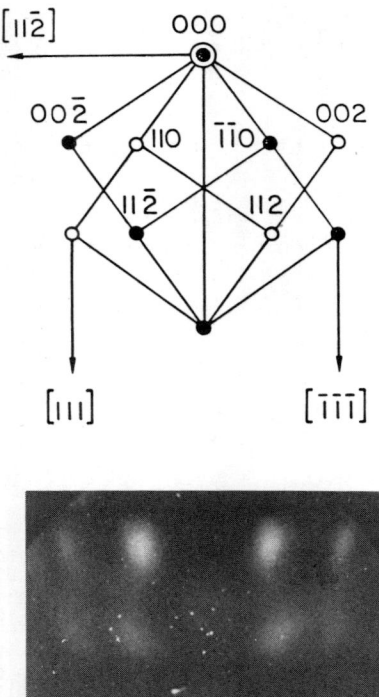

Fig.3. Mo film of 150Å thickness deposited on (100)GaAs at 310°C: bottom: the RHEED pattern. top: indexing of (111)Mo pattern (open circles) and its twin pattern (solid circles).

(1) For very low coverage, typically around 1Å, the surface reconstruction of the (100)GaAs (Fig.4a) is completely gone. The GaAs bulk lines are weakened and disappear completely around 5Å of Mo coverage.
(2) Up to 20-30Å of Mo coverage no pattern is observed. The background level, however, is relatively high. A very faint and diffuse ring is sometimes observed.
(3) At 20-30Å thickness, a line pattern of the Mo diffraction appears rather abruptly (Fig 4b), indicating two-dimensional nucleation [1].

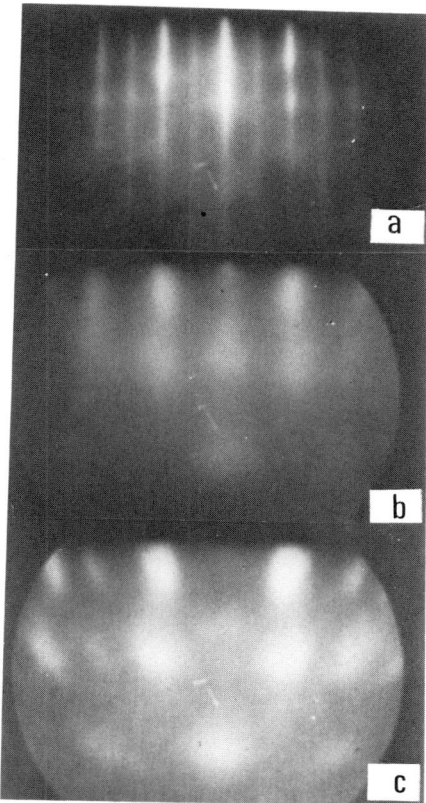

Fig.4. The evolution of the RHEED pattern of Mo grown on (100)GaAs at 270°C. (a) (100)GaAs (2x4) reconstruction along [011] azimuthal direction. (b) the pattern of Mo film of ~ 50Å thickness. (c) the pattern of the Mo film after a growth of ~ 150Å.

(4). As the Mo film is thickening, a transfer into three-dimensional structure is taking place (Fig 4c). Above about 50Å the film exhibits a stable structure which does not change up to 150Å thickness (at which stage the deposition was typically over).

In summary we showed that the use of electron-gun evaporators for refractory metlas in MBE systems can be highly beneficial. It makes possible the study of the important systems of refractory metals on semiconductor substrates and enhance our understanding of their film structures and growth mechanisms.

REFERENCES

1. R. Ludeke, J. Vac. Sci. Technol. $\underline{B2}$, 400 (1984).

2. M. Heiblum, J. Bloch and J.J. O'Sullivan, unpublished.

3. K. Suh, H.K. Park and K.L. Moazed, J. Vac. Sci. Technol. $\underline{B1}$, 365 (1983).

4. J.E. O'Neal and B.B. Rath, Thin Solid Films $\underline{23}$, 363 (1974).

ELECTRON SPECTROSCOPIC STUDIES OF SUBSTOICHIOMETRIC TANTALUM CARBIDE*

G. R. GRUZALSKI, D. M. ZEHNER, and G. W. OWNBY
Solid State Division, Oak Ridge National Laboratory,
Oak Ridge, Tennessee 37831

ABSTRACT

XPS was used to determine core-level binding energies and valence-band structure for TaC_x over the range $0.5 \lesssim x \lesssim 1.0$. As x decreased, the carbon-1s binding energy (BE) changed very little, the carbon-2s BE shifted toward the Fermi level, the position of the p-d valence-band peak shifted toward the Fermi level more, and the tantalum-4d and -4f BE's shifted toward the Fermi level even more, about 0.16 eV for a change in x of 0.1. In addition, the valence-band spectra exhibited structure between about 1 and 2 eV BE, and this structure increased as x decreased. These observations are explicable in terms of charge transfer and the formation of occupied defect states associated with carbon vacancies.

INTRODUCTION

Understanding the electronic structure of group IVB and VB transition-metal monocarbides has been a long-standing objective, and efforts to this end have been relatively successful for the stoichiometric monocarbides. Calculations have shown that the electronic structures of all the stoichiometric monocarbides are similar: each consists of a low-lying band primarily derived from carbon-2s states and, separated from this band by an energy gap, several overlapping bands primarily derived from carbon-2p and metal-5d states. The density of states (DOS) of these p-d bands are made up of two parts separated by a minimum, near which lies the Fermi energy E_F; the lower- and higher-energy parts are mainly p- and d-like, respectively. In many cases, this description of the electronic structure has been borne out by experiment [1].
The search for a satisfactory understanding of the electronic structure of the substoichiometric carbides (i.e., those deficient in carbon) has been relatively recent. Systems that have been studied theoretically include carbides of titanium [2-4], niobium [5-9], tantalum [7], and hafnium [7]. Although two of these studies [2,7] have suggested that removing carbon has little effect upon the shape of the DOS, the others have suggested that new "defect" states are formed, that these states are associated with carbon vacancies, and that their ground-state binding energies (BE) lie near the minimum in the DOS of the p-d bands. For substoichiometric titanium carbide (denoted as TiC_x), these defect states lie near or above E_F [3,4]; for NbC_x they lie between 1.5 and 2.5 eV below E_F [5,6,8,9].
Few experimental investigations have supplied information relevant to the question of whether or not these defect states exist; those that have, however, are consistent with the theoretical studies suggesting that the defect states do form. Regarding the group IVB carbides, there has been the work of Pflüger et al. [10] and Johnson et al. [11] on TiC_x. Pflüger et al. have reported that E_F remained constant (presumably with respect to the carbon-1s level) as the carbon-to-titanium ratio x was varied in this material. From this observation they have inferred that new states were

---

*Research sponsored by the Division of Materials Sciences, U.S. Department of Energy under contract DE-AC05-84OR21400 with Martin Marietta Energy Systems, Inc.

created as carbon was removed. (Because of the charge transfer in these materials, removing carbon would result in charge being transferred back to the transition metal in such a way that would cause E_F to increase, were new states not created at or below E_F; see Discussion below.) Since Johansson et al. [11] did not observe structure in TiC_x valence-band spectra attributable to defect states, it appears that the defect states were located very near E_F, in agreement with theoretical predictions [3,4]. Regarding the group VB carbides, there has been the work of Höchst et al. [12], who have observed structure near 1.9 eV BE in the valence-band spectrum of $NbC_{0.85}$. This structure was attributed to defect states associated with carbon vacancies, in qualitative agreement with theory [5,6,8,9]. Moreover, similar results have been obtained in related experimental studies [13-15] on group IVB nitrides, which are believed to be structurally and electronically similar to the group VB carbides. Hence, the emerging experimental picture is that defect states associated with carbon vacancies do indeed form and that their ground states lie near E_F for the group IVB carbides but below E_F for the group VB carbides. Of course, additional studies of other substoichiometric carbides are required to verify this induction.

In a previous study [16] we have shown that the valence- and conduction-band spectra of single-crystalline nearly-stoichiometric TaC are consistent with the DOS obtained from APW calculations [7]. In the course of that study it was found that near-surface TaC_x regions of various compositions (and degrees of order) could be prepared. This was accomplished by first sputtering with argon ions (which left the near-surface region carbon deficient) and then annealing (which not only allowed the damaged near-surface region to reorder but also allowed carbon to diffuse to the surface region from the bulk). Having done this, XPS was used to determine core-level binding energies and valence-band structure for several of these near-surface regions, which varied in composition from $x \simeq 0.5$ to $x \simeq 1.0$. In this paper our observations are summarized and they are explained in terms of charge transfer and the idea formulated above: in substoichiometric group VB transition-metal carbides, defect states associated with carbon vacancies form below the Fermi level.

EXPERIMENTAL INFORMATION

Measurements were made in an ion-pumped UHV chamber having a base pressure of less than 10^{-8} Pa and equipped with a double-pass CMA containing a coaxial electron gun, an unmonochromatic aluminum x-ray source, LEED optics, and a quadrupole mass spectrometer. The XPS data were obtained with the CMA in the retarding mode (15-eV pass energy). The energy scale of the spectrometer was calibrated by setting the measured gold $4f_{7/2}$ BE equal to 84.1 eV with respect to E_F and by using procedures similar to those described elsewhere [17]. The precision of the spectrometer was estimated as ±0.1 eV.

The tantalum carbide crystal used in this study was polished to within 0.1° of the (100) plane. It was supported in the UHV chamber by a thermally insulated loop of tungsten wire (0.010-inch diameter); two slots were spark cut into opposite edges of the specimen disk to accommodate the wire. An infrared pyrometer was used to monitor the surface temperature of the crystal, which was heated by bombarding its back surface with 800-eV electrons. The surfaces investigated were prepared by argon ion bombardment (1000 V, 10 μA, 10 min) followed by 10-min anneals at temperatures up to 2600°C. The as-sputtered surfaces did not exhibit well-resolved LEED patterns until they were annealed to near or above 700°C for 10 min; very sharp LEED patterns were seen subsequent to 10-min anneals near or above 1600°C.

Oxygen and argon were evidenced immediately after sputtering; no other contaminants were detected. The amount of oxygen and argon present decreased with increasing annealing temperature. If the as-sputtered sample remained at room temperature, however, the amount of oxygen present increased with time: as much as 0.1 monolayer may have been present within 50 h after sputtering. Unfortunately, it was not possible experimentally to separate or sort out the effect of all factors that changed as a result of an anneal (e.g., the carbon-to-tantalum ratio, the degree of ordering, the amount of contaminant present). Yet, by comparing XPS valence-band and core-level spectra from as-sputtered surfaces with those from surfaces exposed to oxygen (but otherwise similarly prepared), we were able to demonstrate that the oxygen contaminant was an adsorbate and that its presence probably did not affect the results and conclusions presented herein.

Atomic sensitivity factors taken from ref. [18] and areas under XPS carbon-1s and tantalum-4f lines were used to determine x, the average carbon-to-tantalum ratio in the near-surface region. The relative precision of x was only dependent on the statistical precision of the areas and on how precisely they could be determined. The largest values of x were within a few percent of unity, and these were obtained for well-annealed well-ordered surfaces. This value of x was also inferred from impact-collision ion-scattering spectra from similarly prepared (100) surfaces of TaC [19]. Hence, although this XPS method of determining composition often is not accurate, it appears to have been so in the present study. The bulk composition of the tantalum carbide crystal was about $TaC_{0.99}$.

RESULTS

In Fig. 1 is shown tantalum-4f XPS spectra from (100) surfaces of tantalum carbide having different carbon-to-tantalum ratios x. The spectra are normalized in that the intensities of the $4f_{7/2}$ peaks are equal. As x decreases the spectra broaden somewhat and the valley regions between peaks fill in. This lineshape change probably is due to there being more than one distinguishable tantalum site in those near-surface regions deficient in carbon.

The spectra also shift toward the Fermi level as x decreases. This shift is displayed better in Fig. 2(a), which is a plot of x vs tantalum $4f_{7/2}$ BE. A similar plot is shown in Fig. 2(b) for tantalum $4d_{5/2}$. The slope of the solid line in Fig. 2(a) is the same as that in Fig. 2(b). That is, essentially identical BE shifts are exhibited by these two core levels, and they are approximately linear in x (though in a direction opposite to that reported earlier by Ramqvist et al. [20]).

In Fig. 2(c) is shown a similar plot for carbon 1s. Note that the overall change in BE (about 0.4 eV) is more than a factor of two smaller than that for the tantalum core levels. Note too that for x > 0.6, the BE appears to be constant (282.86 ± 0.05 eV); if so, these data would be consistent with those obtained by others who found that the metalloid core-level BE was independent of x for TaC_x [20] and for ZrN_x and TiN_x [13]. Because of the scatter in the present data, however, we cannot rule out some monotonic relationship between the carbon-1s BE and x over the entire composition range investigated. In addition, since the shift in BE is small, it is possible that it may not have been resolvable in the earlier studies [13,20], for which the composition varied over only a relatively narrow range. We also note that the present values of the carbon-1s BE significantly differ from those reported previously by Ramqvist et al. [20].

Shown in Fig. 3 are valence-band XPS spectra from surfaces of different x. The spectra are normalized in that the intensities of the peaks near 5 eV BE are equal. As x decreases, this peak, which is due to carbon-2p and tantalum-5d states [16], shifts toward E_F. The magnitude of this shift is about 10% less than that of the shift for the tantalum-4d and -4f levels. The lower-lying carbon-2s peak (not shown) also appears to shift somewhat and in the same direction, but the extent of this shift is difficult to determine because it is small (though not as small as that of the carbon-1s levels) and because the carbon-2s peak overlaps structure owing

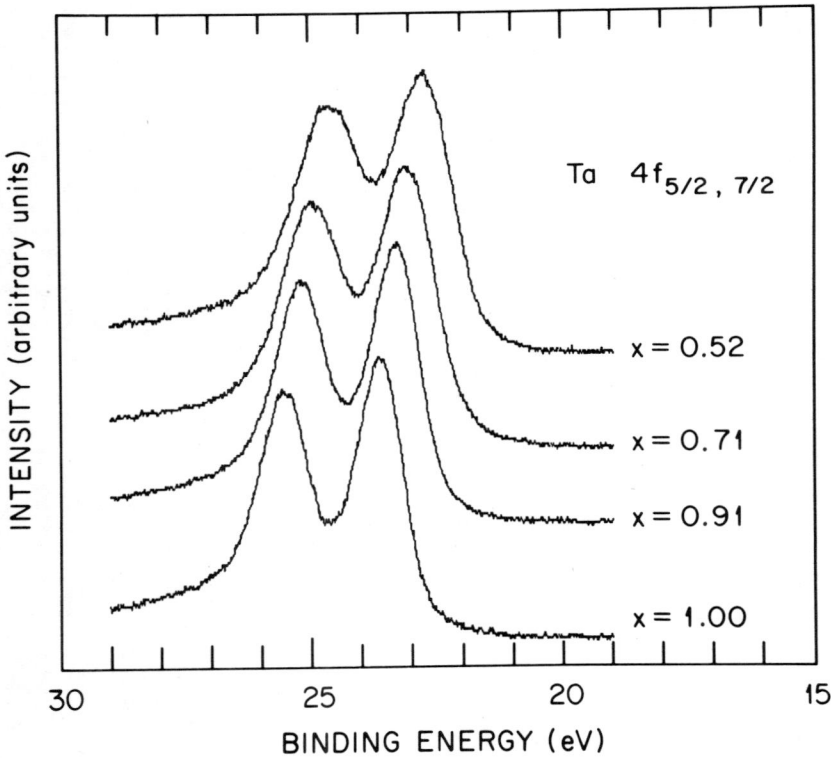

Fig. 1: XPS Ta-4f spectra from surfaces having different carbon-to-metal ratios x. The binding energies are with respect to E_F.

to α_3 and α_4 satellite excitations of tantalum $4f_{7/2}$ electrons. In addition, structure appears in the spectra between about 1 and 2 eV BE, and it increases as x decreases. As discussed above, similar structure has been observed previously for related systems [12-15] and has been attributed to defect states associated with metalloid vacancies. In these earlier studies, however, shifts in the valence-band peaks owing to metalloid-metal hybridization were not observed as x was varied.

DISCUSSION

The results exemplified in Figs. 1-3 can be understood in terms of charge transfer and the formation of occupied defect states associated with carbon vacancies. When tantalum and carbon combine to form tantalum carbide, charge is transferred from the tantalum atoms to the carbon atoms [7]. Hence, when a carbon vacancy is created, the "tantalum" electrons that were previously transferred to the carbon are transferred back toward the tantalum. In a rigid-band sense, these electrons will occupy states just above

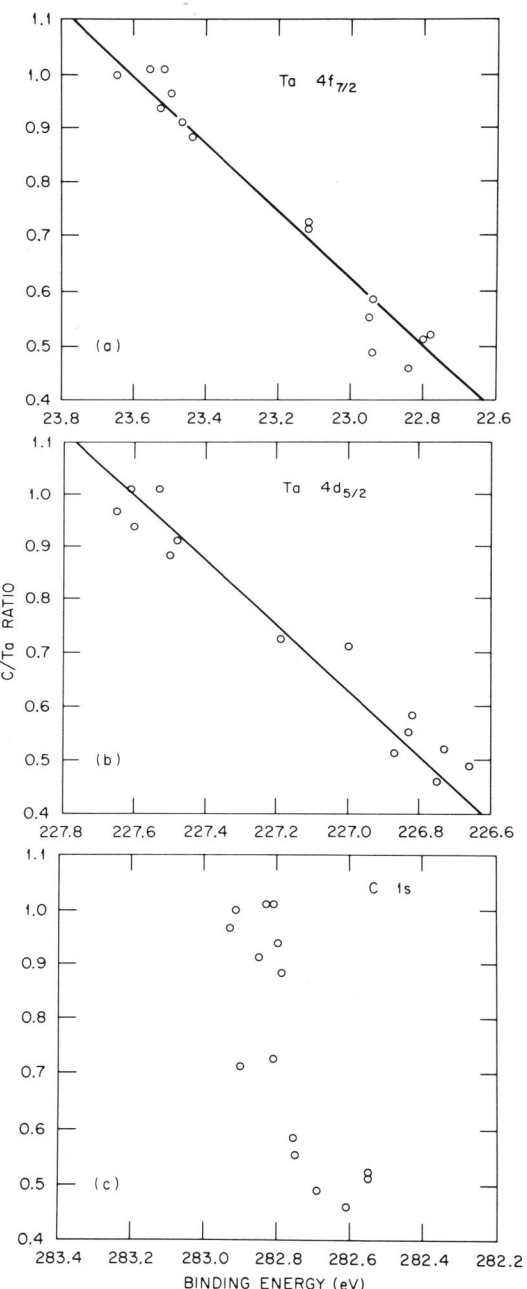

Fig. 2: Carbon-to-tantalum ratio x vs binding energy for (a) Ta $4f_{7/2}$, (b) Ta $4d_{5/2}$, and (c) C 1s. The slope of the solid line in (a) is identical with that in (b): 0.61/eV.

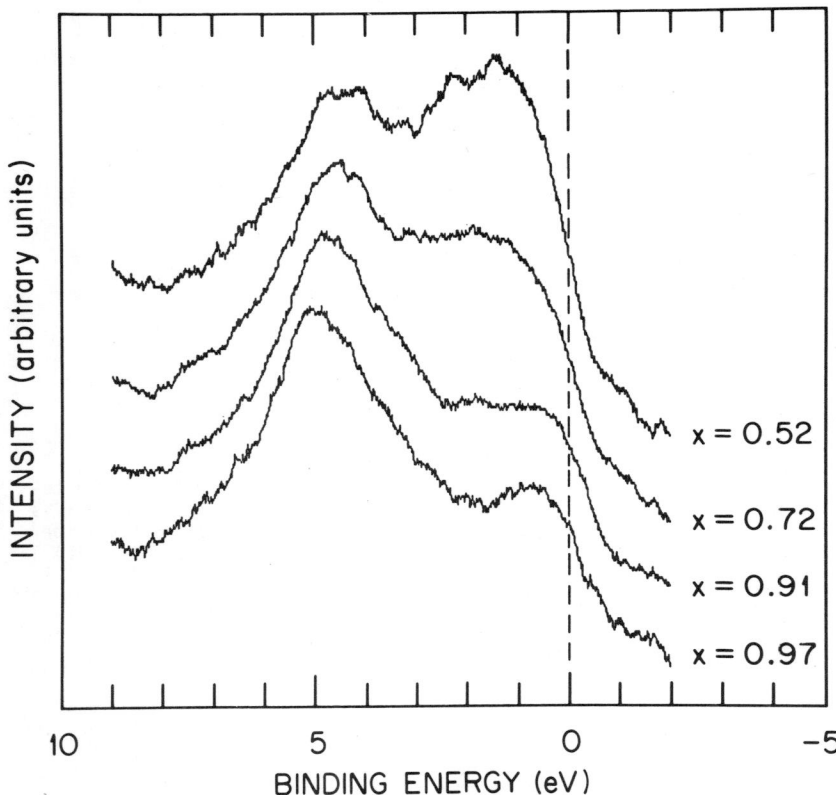

Fig. 3: XPS valence-band spectra from surfaces having different carbon-to-tantalum ratios x. The spectra are shown only to 9 eV below E_F.

the original Fermi level, unless new states (defect states) are created below it. In this picture, E_F could increase, decrease, or remain the same, depending upon the number of defect states created below the original Fermi level; we believe the last possibility best describes TaC_x. In any case, the added negative charge in the vicinity of the tantalum atoms will reduce the binding energy of the tantalum levels. This would be true even if all of the added charge were in the defect states (unless, of course, the defect states were extremely well localized on the vacancy site itself, which we believe is not the case). This reduction in BE of the tantalum levels is exactly what we observed.

In contrast, the local carbon environment changes very little as x is varied in a well-ordered substoichiometric carbide: each carbon atom remains surrounded by six tantalum atoms. It should not be surprising, therefore, that the change in binding energy of the carbon levels was relatively small as x was varied.

Because the local carbon environment in well-ordered TaC_x changes very little with x, the carbon-1s level is probably the best reference level accessible experimentally. That is, a decrease or increase in this level is probably the best available measure of a corresponding decrease or increase in the Fermi level (in a rigid-band sense). As already indicated, it is difficult to determine whether the carbon-1s BE varied with x for x > 0.6. It certainly appears that it did not increase with decreasing x, as was reported by Porte et al. for the nitrogen-1s BE in TiN_x [14]. For variations in x that encompassed x < 0.6, the carbon-1s BE clearly did decrease as x decreased [see Fig. 2(c)]; yet, because the near-surface region for smaller x was more disordered, effects of defects other than carbon vacancies may have been significant. The possibility of large extended defects having been present (i.e., regions of carbon or tantalum) can be eliminated since the appropriate core levels differed significantly from those of carbon or tantalum, even for the as-sputtered surfaces. Nonetheless, and even though carbon vacancies may have been the dominant defect for the as-sputtered material (after all, the material is partially ionic and its vacancies tend to repel one another), we cannot rule out the possibility that other defects were playing an important role when x < 0.6. In any case, we tentatively conclude that carbon vacancies have little if any effect upon the position of the Fermi level (in a rigid-band sense). Moreover, since E_F did not increase with decreasing x, we infer that occupied defect states were created as carbon was removed; we interpret the additional intensity appearing in the valence-band spectra between about 1 and 2 eV BE as being due to photoemission from these states.

It is difficult also to determine the extent of the carbon-2s BE change with x. It appears, however, that this level shifted with x more than the carbon-1s level did. This result is surprising and may suggest that the carbon-2s level is mixing with tantalum levels more strongly than that suggested by APW calculations [7]. (Another explanation would involve changes in the shape of the 2s band; results of coherent-potential-approximation calculations for TaC_x [7], however, do not support this possibility.)

The shift of the p-d valence-band peak toward E_F as x decreased is consistent with the idea that a decrease in x results in added negative charge in the vicinity of the tantalum atoms. That this shift is somewhat smaller than the corresponding shift of the tantalum core levels can be ascribed to hybridization effects. It is puzzling, however, that similar shifts of valence-band peaks owing to metalloid-metal mixing were not observed in previous studies [12-15] on related systems for which metal core-level BE's changed with x [14,20].

Two results presented above (the value of the carbon-1s BE and the direction of the shift of the tantalum-4f BE as x varied) differ from those reported earlier [20]. Although we do not fully understand these differences, we feel that they are attributable to differences in sample preparation and, moreover, to surface contamination in the earlier work; we note, for example, that in the earlier work, carbon in TaC_x gave rise to a 1s peak many times weaker than that resulting from hydrocarbon contamination.

CONCLUDING REMARKS

XPS was used to determine core-level binding energies and valence-band structure for TaC_x over the composition range $0.5 \lesssim x \lesssim 1.0$. The near-surface regions studied were prepared by argon-ion bombardment (which left them carbon deficient) and subsequent anneals. The anneals not only allowed the near-surface regions to order but also allowed x to increase, because of carbon diffusion from the bulk. The degree of ordering varied as x varied, however, and this may have affected our results, which may be summarized as follows.

As x decreased: (i) the carbon-1s BE changed very little if at all for x > 0.6 eV, (ii) the carbon-2s BE shifted toward the Fermi level, although the extent of the shift is small and difficult to determine, (iii) the position of the p-d valence-band peak shifted toward the Fermi level more than the 2s BE did, and (iv) the tantalum 4d and 4f BE's shifted toward the Fermi level even more, about 0.16 eV for a change in x of 0.1. We emphasize that the magnitude of these shifts increased in the order the shifts are listed. In addition, the valence-band spectra exhibited structure between about 1 and 2 eV BE, and this structure increased as x decreased. All of these observations are explicable in terms of charge transfer and the formation of occupied defect states associated with carbon vacancies.

ACKNOWLEDGMENTS

S. H. Liu is thanked for helpful discussions.

REFERENCES

[1] For a recent review of both theoretical and experimental work, see A. Neckel, Intern. J. Quantum Chem. 23, 1317 (1983).
[2] J. Klima, J. Phys. C. 12, 3691 (1979).
[3] L. M. Huisman, A. E. Carlsson, C. D. Gelatt, Jr., and H. Ehrenreich, Phys. Rev. B 22, 991 (1980).
[4] V. A. Gubanov, A. L. Ivanovsky, G. P. Shveikin, and D. E. Ellis, J. Phys. Chem. Solids 45, 719 (1984).
[5] K. Schwarz and R. Rösch, J. Phys. C 9, L433 (1976).
[6] G. Ries and H. Winter, J. Phys. F 10, 1 (1980).
[7] B. M. Klein, D. A. Papaconstantopoulos, and L. L. Boyer, Phys. Rev. B 22, 1946 (1980).
[8] P. Pecheur, G. Toussaint, and E. Kauffer, Phys. Rev. B 29, 6606 (1984).
[9] W. E. Pickett, B. M. Klein, and R. Zeller, Bull. Am. Phys. Soc. 30, 620 (1985).
[10] J. Pflüger, J. Fink, G. Crecelius, K. P. Bohnen, and H. Winter, Solid State Commun. 44, 489 (1982).
[11] L. I. Johansson, A. L. Hagström, B. E. Jacobson, and S. B. M. Hagström, J. Electron Spectrosc. Related Phenomena 10, 259 (1977).
[12] H. Höchst, P. Steiner, S. Hüfner, and C. Politis, Z. Physik B 37, 27 (1980).
[13] H. Höchst, R. D. Bringans, and P. Steiner, Phys. Rev. B 25, 7183 (1982).
[14] L. Porte, L. Roux, and J. Hanus, Phys. Rev. B 28, 3214 (1983).
[15] R. D. Bringans and H. Höchst, Phys. Rev. B 30, 5416 (1984).
[16] G. R. Gruzalski, D. M. Zehner, and G. W. Ownby, Surf. Sci. 157, L395 (1985).
[17] C. J. Powell, N. E. Erikson, and T. Jach, J. Vacuum Sci. Technol 20, 625 (1982).
[18] C. D. Wagner, et al., p. 188 in Handbook of X-Ray Photoelectron Spectroscopy, ed. by G. E. Muilenberg, Perkin-Elmer Corporation, Eden Prairie, MN, 1979.
[19] C. Oshima, R. Soudia, M. Aono, S. Otani, and Y. Ishizawa, Phys. Rev. B 30, 5361 (1984).
[20] L. Ramqvist, K. Hamrin, G. Johansson, U. Gelius, and C. Nordling, J. Phys. Chem. Solids 31, 2669 (1970).

CATION SOLUTE SEGREGATION TO SURFACES OF MgO AND α-Al_2O_3

ROBERT C. McCUNE
Research Staff, Ford Motor Company, Dearborn, MI 48121-2053

ABSTRACT

Low energy ion scattering spectroscopy (LEIS) and Auger electron spectroscopy (AES) were used to measure the extent of impurity cation solute segregation to MgO (100) and various surfaces of single and polycrystalline Al_2O_3, following equilibration anneals at temperatures above $1000^\circ C$ in ultra-high vacuum. Systems studied include Ca/MgO, Ni/MgO, Ca/Al_2O_3 and Y/Al_2O_3. Calcium segregation to MgO (100) is reversible and exhibits monolayer adsorption behavior with an enthalpy of segregation of approximately -55 kJ/mole with maximum occupation of surface cation sites approaching 40%. Nickel segregation to MgO(100) is masked by a preferential segregation of calcium. Calcium segregation to Al_2O_3 surfaces has been found to be transient with a maximum surface cation site occupation of less than 10% as determined by LEIS and estimated enthalpy of segregation in the range -90 to -190 kJ/mole. Yttrium segregation to surfaces of a polycrystalline Al_2O_3 compact was limited by competing calcium segregation at temperatures below $1600^\circ C$. An estimated enthalpy of segregation for yttrium to Al_2O_3 surfaces was in the range -23 to -43 kJ/mole, with maximum cation surface site occupation of about 15%. Practical limitations to free surface measurements of solute segregation in these materials are discussed.

INTRODUCTION

Equilibrium interfacial segregation of solutes in metals and ceramics is known to have pronounced effects on the associated physical and chemical properties of these materials. Corrosion, catalysis of chemical reactions, sintering of compacts, electronic and tribological properties are all influenced by interfacial composition, which, in most practical materials, differs from that of the bulk. While metal alloys have perhaps received the greatest experimental attention [1-4], there is a growing body of knowledge for ceramics indicating a crucial role of solute segregation in the densification, mechanical and electrical properties [5-8]. Experimental approaches to studies of interfacial composition in ceramics usually involve examination of fracture surfaces [9] or microanalytical studies of intact boundaries [10]. In the studies reported here, free surface composition is determined following anneals at elevated temperatures where the solute species can diffuse to the surface. A driving force for segregation of the solute can be assessed from the temperature dependence of solute adsorption in the limit of dilute solutions which exhibit regular solution behavior. This approach has been used in studies of solute segregation at metal surfaces [2], and in studies of calcium segregation to MgO (100) [8,11] and surfaces of Al_2O_3 [12,13]. These earlier studies are summarized and recent findings are reported for the Ni/MgO and Y/Al_2O_3 systems.

THE MODEL

For many dilute metal alloys, it has been shown [2] that the free-surface solute enrichment behavior can be described by a statistical model due to McLean [14] for the case of solute adsorption at grain boundaries. In this approach, interfacial adsorption of the solute at sub-monlayer concentrations is described by an equation of the form:

$$X_1{}^s/X_2{}^s = (X_1{}^b/X_2{}^b) \exp(-\Delta H_s/RT) \qquad (1)$$

where $X_i{}^s$ and $X_i{}^b$ are the surface and bulk mole fractions of the ith species, respectively, and ΔH_s is the enthalpy of segregation. For non-regular solutions, there will be an excess interfacial entropy associated with the segregation, and ΔH_s should more properly be replaced by a free energy of segregation which will reflect the temperature dependence. Similarly, the adsorption model of McLean is predicated on non-interacting adsorption sites, which may not be the case for highly enriched surfaces [15], so that in general ΔH_s will be coverage dependent. In the limiting case of low coverages (i.e. less than a monolayer), it may be possible to extract ΔH_s from the logarithmic plot of surface mole fraction ratio vs. inverse temperature as suggested by Eqn. (1).

In a preliminary study of surface segregation of calcium to MgO (100) [8], this procedure was employed using Auger electron spectroscopy as a means of assessing surface composition. A value for ΔH_s of -75 ± 21 kJ/mole was obtained, and found to be in relatively good agreement with an estimate of -59 kJ/mole from tabulated values of bulk properties [11], and calculated values in the range -40 to -75 kJ/mole obtained from computer simulations using pair potentials [15,16].

Empirical estimates of ΔH_s take into account the solute misfit strain energy and specific surface work difference on segregation of a solute atom to the surface layer [2]. Additionally, for the case of metal alloys, a contribution can be identified resulting from the pairwise bonding between solute and solvent species. For solute or solvent cation species in an ionically bonded ceramic, the nearest neighbors are generally the anionic species which are tightly bound to the cations, and an analog to this quasichemical term, usually derived from the heat of mixing for the solute-solvent couple [2] is not immediately apparent. It could be expected, however, that binary systems which exhibit a tendency for new compound formation (i.e. having large negative enthalpies of mixing) would show a lesser tendency for segregation of the solute than for systems in which the solute species tends to be immiscible and "clusters" or is rejected from the solution to precipitate its parent compound.

In ionically bonded solids it is possible to develop space-charge layers in the vicinity of interfaces [5] and it is expected that this will influence the segregation of aliovalent solute species [17]. In the absence of other driving forces, this electrostatic contribution will act to determine the solute distribution near the interface [18]. In general, the effect of the space-charge layer on the segregation of a particular solute cation is a complex issue since it requires both a knowledge of the charge-compensating defect for a particular solute, as well as the influence of other charged defects. For MgO and Al_2O_3 ceramics, the defect structures are generally dominated by impurity additions in the temperature regimes considered here [19], so that the space-charge effect for a given solute will depend primarily on the possibility for adsorption of its compensating defect, as well the relative magnitude of the other driving forces such as elastic misfit strain [17]. If the solute accumulates in the sub-surface Debye layer [20], a surface sensitive technique such as LEIS may understate its overall enrichment.

EXPERIMENTAL PROCEDURES

Details of the experimental approach have been described in the earlier works [11-13]. In brief, specimens of single or polycrystalline oxides in the form of wafers of size approximately 5x5x0.5 mm are secured to heating strips of either tantalum (MgO samples) or rhenium (Al_2O_3 samples) and the resulting assemblies are heated by alternating current in the vacuum chambers of the spectrometers used. For MgO samples at temperatures below about 1400°C, it was possible to monitor the surface composition by both AES and LEIS in-situ at the temperature of interest. For Al_2O_3 specimens, temperatures usually in excess of 1400°C were required to initiate segregation, and measurements were conducted on samples quenched by cessation of the heating current.

Materials

Earlier studies of calcium segregation to MgO (100) [11,12] were conducted on single crystals of varying overall purity, but with bulk calcium concentrations in the range 180-220 wt. ppm. The nickel-doped MgO crystal used in this work was obtained from W. & C. Spicer, Ltd., Cheltenham, England, and contained the following major cation impurities in wt. ppm as determined by atomic absorption: Ni: 870, Ca: 11, Fe: 150, Al: 70, and As: 80. The specimen was prepared by cleavage from the parent crystal, prior to its mounting on a tantalum heating strip.

A number of single and polycrystalline Al_2O_3 specimens have been studied, and descriptions of these materials are included in the prior work [13]. In general, the calcium levels in these specimens were less than 40 wt ppm. The material used for study of yttrium segregation in this work was a sintered polycrystalline compact obtained from the Coors Porcelain Company, Golden, CO and contained the following observed segregant species as reported in wt. ppm and determined by emission spectroscopy: Y:160, Ca:23, Mg:270 and Fe:83. Uncertainties in these values are on the order of 10%. Coupons of the material were prepared by diamond saw wafering, followed by diamond polishing to a final finish of one micrometer. The wafer was solvent cleaned, followed by etches in hot phosphoric acid and boiling aqua regia. The specimen was rinsed in deionized water and secured to a rhenium heating strip.

Spectroscopy

LEIS spectra were obtained with a 3M Model 525 ion scattering spectrometer, using $^4He^+$ ions, backscattered through 138° into a cylindrical mirror type electrostatic analyser. A beam of size ~ 0.3 mm diameter at a primary beam energy of 500 eV and current of ~ 40 nA was used at normal incidence to the target surface. A number of spectra from the Coors Al_2O_3 were acquired at primary beam energies of 250 eV in hopes of resolving an apparent magnesium enrichment at the surface. Calibrations of surface enrichment for this spectroscopy were made using relative sensitivities established from pure single crystal oxides and compacts of the constituent oxides MgO, Al_2O_3, CaO, Y_2O_3 and NiO. Details of the calibration scheme are reported elsewhere [12].

AES spectra were obtained with a Physical Electronics Industries Model 545 scanning Auger microprobe, with probe size of approximately 10 μm, beam energy of 3 keV, 4 eV modulation amplitude, and beam currents on the order of 1.0 μA. Values reported are based on rastered areas of approximate size 100 × 100 μm. The specimen was held normal to the electron beam. In general, a number of areas could be sampled on a specimen surface, and these values are averaged. Both LEIS and AES measurements of surface solute enrichment indicated some variability with position on the specimen surface. Surface solute enrichments were determined to a first order by comparison of relative sensitivities for the various cationic species relative to oxygen in pure compounds of known stoichiometry including MgO, Al_2O_3, CaO, Y_2O_3, and $Y_3Al_5O_{12}$ (YAG). Possible adjustments of the data to compensate for different electron inelastic mean free paths in the various substances are discussed in the next section.

RESULTS AND DISCUSSION

A summary of the experimental findings for the systems discussed is provided in Table I. Synopses of the solute segregation behavior in the individual systems considered follows.

TABLE I. Summary of solute segregation data for systems studied.

Segregant/ Solvent	Bulk Segregant Concentration (wt ppm)	Maximum Observed Surface Enrichment (X_i^s/X_i^b)	Measured ΔH_s (kJ/mole)	Theoretical ΔH_s (kJ/mole)	Ref.
Ca/MgO (100)*	Ca: 200	2000 (LEIS)	-57	-59 to -63 -51 (avg.) -53	[12] [15] [16]
Ni/MgO (100)*	Ni: 870 Ca: 11	32 (LEIS) 2900 (LEIS)	N.A.	-4	
Ca/Al$_2$O$_3$ (MgO-doped)+	Ca: 39	1900 (AES)	-184	-74 to -159 -134	[12] [8]
Ca/Al$_2$O$_3$ (10$\bar{1}$0)*	Ca: 40	420 (LEIS)	-188 (LEIS)		[13]
Y/Al$_2$O$_3$+	Y: 160 Mg: 270 Ca: 23 Fe: 83	1700 (AES) 1450 (AES)	-23 to -43	-91	[25]

*Single Crystal   +Polycrystal

Ca/MgO

The segregation of calcium to MgO (100), as previously reported [8,11-13], was found to be reversible and exhibits McLean-type adsorption behavior with ΔH_s of -57 + 15 kJ/mole over the temperature range 1100-1450°C. The maximum observed enrichment approaches 40% of the cation sites at approximately 1100°C. The measured values for ΔH_s agree both with theoretical determinations in the range -40 to -75 kJ/mole[15,16], and with empirical estimates of solute strain energy and surface work difference on the order of -60 kJ/mole-[12]. LEIS and AES determinations of surface mole fraction ratio for one experimental run are shown in Figure 1. Included are AES data points adjusted for electron inelastic mean free path (IMFP) in the calibration standards, and projected AES values estimated from the LEIS assessment of top monolayer composition using the IMFP corrected standards and a layer analysis due to Marchut and McMahon [21].

Ni/MgO

The equilibrium phase diagram for the MgO-NiO binary system [22] indicates complete miscibility over the entire compositional range. An estimate of ΔH_s for Ni^{2+} segregated to MgO (100) indicates a driving force of only about -4 kJ/mole. Earlier studies of the MgO-NiO system by photoelectron spectroscopy [23] did not reveal nickel enrichment. Figure 2 shows LEIS spectra taken from the Spicer Ni-doped MgO following an anneal at 1000°C. Calcium, which is present at significantly lower concentrations (~11 ppm wt) is enriched at the outermost surface. Sputter depth profiling with $^4He^+$ showed nickel accumulation in the subsurface region. Indications are that initial enrichment of nickel occurred due to its larger concentration, but that it was displaced by calcium which exhibits a greater driving force for segregation in the MgO lattice.

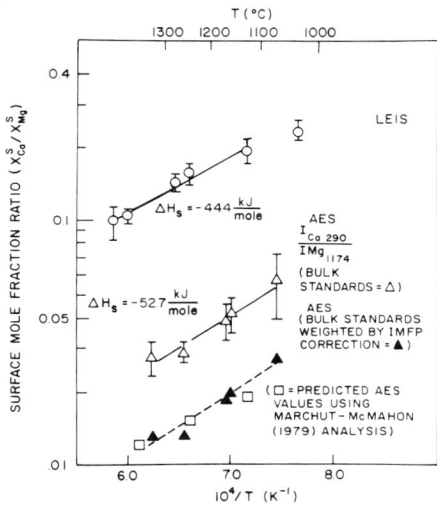

Fig. 1. Comparison of LEIS and AES determinations of calcium surface enrichment on MgO (100) and effects of IMFP correction to calibration.

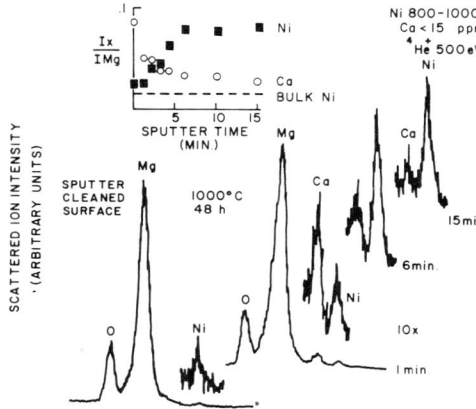

Fig. 2. LEIS spectra from the sputtered and annealed surfaces of a Ni-doped MgO crystal.

Ca/Al$_2$O$_3$

Segregation of calcium at grain boundaries in alumina is a widely observed phenomenon, and a summary of experimental data is included in earlier work [13]. Calcium segregation to free surfaces of single crystal and polycrystalline Al$_2$O$_3$ showed transient behavior, characterized by apparent loss of the segregant by volatilization at temperatures above about 1600°C. At much lower temperatures diffusivity of the calcium in single crystal specimens appears to be limited, although polycrystalline specimens including an MgO-doped alumina, showed calcium segregation at temperatures near 1400°C. Estimates of ΔH_s for calcium to Al$_2$O$_3$ surfaces from the data collected lie in the range -156 to -189 kJ/mole. The solute strain magnitude suggests a value near -134 kJ/mole [8]. Both space-charge effects and quasichemical terms should also contribute to ΔH_s since the solute is aliovalent in Al$_2$O$_3$ and exhibits a propensity for compound formation [24]. A maximum surface cation mole fraction of less than 10% was observed. Comparison of AES and LEIS on a similar specimen of single crystal material did not indicate major concentration differences as observed with segregation of calcium to MgO surfaces. This suggests that the calcium segregation is not confined to the surface monolayer as appears to be the case in MgO.

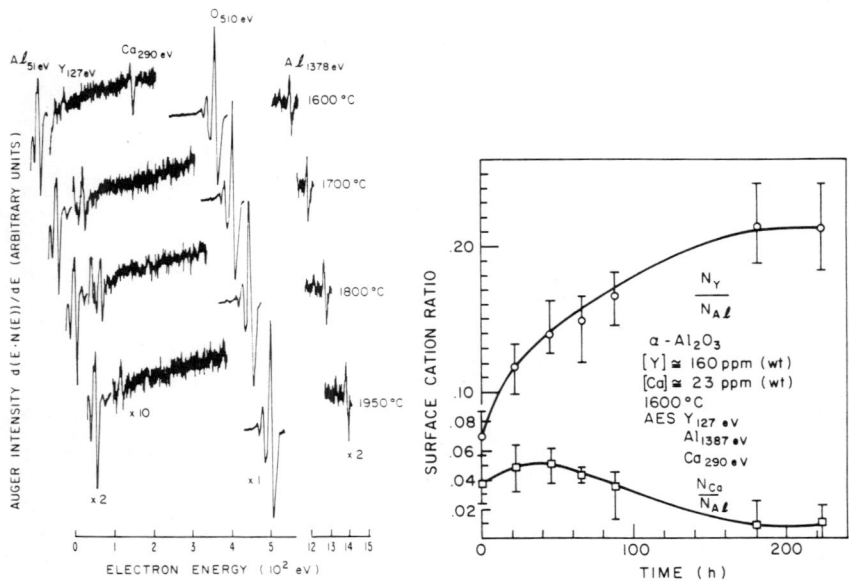

Fig. 3. Typical AES spectra for Y-doped Al$_2$O$_3$ after anneals at temperatures indicated.

Fig. 4. Yttrium and calcium enrichments at the surface of a Y-doped Al$_2$O$_3$ at 1600°C.

Y/Al$_2$O$_3$

Segregation of yttrium to grain boundaries in Al$_2$O$_3$ has been reported for both intact boundaries [10] and for fractured compacts [25,26]. Because Y^{3+} is isovalent with the host cation, space-charge considerations should not be a factor in ΔH_s. The solute strain term, based on reported ionic radii, should be on the order of -92 kJ/mole[25]. The binary system shows a propensity for compound formation [27], suggesting a lowering of the magnitude of ΔH_s. Differences in specific surface work are not well-known. Typical AES spectra for the Coors polycrystal, annealed at temperatures between 1600 and 1950°C are shown in Figure 3. Calcium segregation was observed at temperatures of 1600°C, however, the magnitude of the segregation diminished with time indicating a probable loss due to volatilization, Figure 4 shows the relative Ca/Al and Y/Al atom ratios as a function of time. As calcium is lost from the surface by volatilization, yttrium segregation is enhanced and reaches a limiting level. Calcium was not observed at higher anneal temperatures for this sample.

LEIS spectra from a sputter-cleaned surface of the Coors polycrystal, and that following an anneal at 1400°C are compared in Figure 5. Apparent enrichments in magnesium, calcium, iron and yttrium are observed. Despite its relatively large bulk concentration, magnesium enrichment at the surface was not observed by AES. LEIS, on the other hand, has a large sensitivity to magnesium relative to aluminum in oxide matrices, however it is not easily resolved due to the mass proximity of magnesium and aluminum [28].

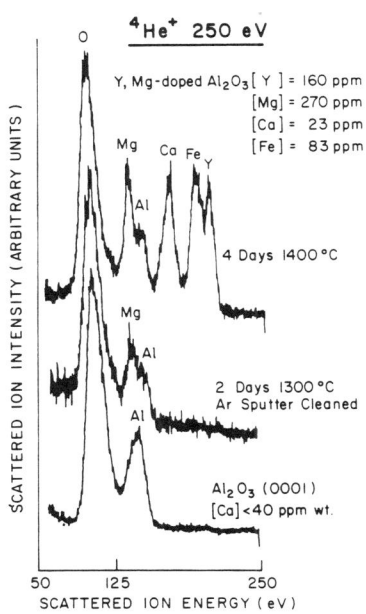

Fig. 5. LEIS spectra of annealed and sputtered Y-doped Al$_2$O$_3$ and comparisons to sapphire (0001) at primary beam energy of 250 eV.

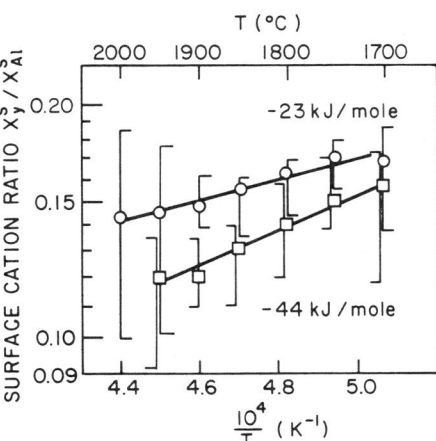

Fig. 6. Estimates of surface enrichment of yttrium and segregation enthalpy from AES measurements using Al 1387 eV peak (circles) and Al 61 eV peak (squares).

Surface mole fraction ratio data obtained by AES for yttrium segregated to the surface of the alumina studied is summarized in Figure 6, where results from two different calibration schemes are compared. Somewhat greater surface enrichments are obtained using the 1387 eV Al KLL Auger peak, and pure oxide standards (Al_2O_3 and Y_2O_3) where the relative Y(127 eV)/Al(1387 eV) sensitivity was determined from the cation to oxygen intensity in the reference materials. No compensation was made for IMFP. A somewhat greater ΔH_s was obtained when the lower energy 61 eV Al LVV Auger peak was considered, and the calibrations from pure oxides obtained in a similar manner. The low energy aluminum Auger line also resulted in a more consistent calibration when relative sensitivites for Y(127 eV)/Al(61 eV) were compared between pure oxides and a specimen of yttrium aluminum garnet (YAG) with composition $Y_3Al_5O_{12}$. The low energy aluminum Auger line, however, has been found to be more prone to distortion and degrading effects of the impinging electron beam used in the analysis. Values for ΔH_s for yttrium to the alumina surface in the temperature range 1700 to 1950°C are in the range -23 to -43 kJ/mole. These values are less than that indicated by solute strain alone and the influence of both quasichemical effects and surface work differences are possible. The maximum observed surface cation site occupation by yttrium is on the order of 15%.

SUMMARY

Measurement of surface solute segregation in MgO and Al_2O_3 ceramics by LEIS and AES, following equilibration at elevated temperatures in ultra-high vacuum was useful in estimating the magnitude of the segregation enthalpy for a given species, as well as the proclivity for various impurity cations to segregate. Such determinations are prone to a number of disturbing effects of the spectroscopic techniques as well as solute volatilization and inherent limitations in depth resolution. Site competition among numerous cationic solute species shows calcium to be a prodominant segregant in the systems studies. Solutes segregated under one set of conditions may apparently be replaced as the anneal time and temperature are changed. Segregation enthalpies obtained from surface analytical measurements are in order of magnitude agreement with theoretical estimates for Ca/MgO, Ca/Al_2O_3 and Y/Al_2O_3.

The author would like to thank R.C. Ku and L. Toth for assistance with data acquisition and W. T. Donlon for reviewing the manuscript.

REFERENCES

1. S. H. Overbury, P. A. Bertrand and G. A. Somorjai, Chem. Reviews 75, 547 (1975).
2. P. Wynblatt and R. C. Ku in: Interfacial Segregation, edited by W. C. Johnson and J. M. Blakely (American Society for Metals, Metals Park, OH 1979) p. 115.
3. F. F. Abraham and C. R. Brundle, J. Vac. Sci. Technol. 18, 506 (1981).
4. A. R. Miedema, Z. Metallkunde 69, 455 (1978).
5. W. D. Kingery, J. Am. Ceram. Soc. 57, pp. 1 and 74 (1974).
6. W. D. Kingery in: Advances in Ceramics, Vol. 1, edited by L. M. Levinson (American Ceramic Society, Columbus, OH 1981) p. 1.
7. W. C. Johnson, Metall. Trans. A. 8A, 1413 (1977).
8. P. Wynblatt and R. C. McCune in: Surfaces and Interfaces in Ceramic and Ceramic-Metal Systems, edited by J. A. Pask and A. G. Evans (Plenum, New York 1981) p. 83.
9. H. L. Marcus and M. E. Fine, J. Am. Ceram. Soc. 55, 568 (1972).
10. B. Bender, D. B. Williams and M. R. Notis, J. Am. Ceram. Soc. 63, 542 (1980).

11. R. C. McCune and P. Wynblatt, J. Am. Ceram. Soc. **66**, 111 (1983).
12. R. C. McCune, PhD. Thesis, University of Michigan, Ann Arbor, MI (1983).
13. R. C. McCune and R. C. Ku in: Advances in Ceramics, Vol. 10, edited by W. D. Kingery (American Ceramic Society, Columbus, OH 1984) p. 217.
14. D. McLean, Grain Boundaries in Metals (University Press, Oxford 1957).
15. E. A. Colbourn, W. C. Mackrodt and P. W. Tasker, J. Mater. Sci. **18**, 1917 (1983.
16. D. Wolf, "Formation Energy of Point Defects in Free Surfaces and Grain Boundaries in MgO," Presented at the 4th Europhysical Topical Conference on Lattice Defects in Ionic Crystals, Dublin, 1982.
17. M. F. Yan, R. M. Cannon and H. K. Bowen, J. Appl. Phys. **54**, 764 (1983).
18. Y. M. Chiang, A. F. Henriksen, W. D. Kingery and D. Finello, J. Am. Ceram.-Soc. **64**, 385 (1981).
19. W. C. Mackrodt in: Advances in Ceramics, Vol 10, edited by W. D. Kingery (American Ceramic Society, Columbus, OH 1984) p. 62.
20. J. M. Blakely and S. Danyluk, Surf. Sci. **40**, 37 (1973).
21. L. Marchut and C. J. McMahon, Jr. in: Electron and Positron Spectroscopies in Materials Science and Engineering, edited by O. Buck, J.K. Tien and H. L. Marcus (Academic Press, New York 1979) p. 183.
22. H. V. Wartenberg and E. Prophet, Z. Anorg. u. Allgem. Chem. **208**, 379 (1932).
23. A. Cimino, B.A. DeAngelis, G. Minelli, T. Persini and P. Scarpino, J. Solid State Chem. **33**, 403 (1980).
24. M. Rolin and P.-H. Thanh, Rev. Hautes Temp. Refractaires **2**, 175 (1965).
25. W. C. Johnson, D. F. Stein and R. W. Rice in: Grain Boundaries in Engineering Materials, edited by J. L. Walter, J. H. Westbrook and D. A. Woodford (Claitor's, Baton Rouge, LA 1975) p. 261.
26. P. Nanni, C.T.H. Stoddart and E. D. Hondros, Mater. Chem. **1**, 297 (1976).
27. J. L. Caslavsky and D. J. Viechnicki, J. Mater. Sci. **15**, 1709 (1980).
28. R. C. McCune, Anal. Chem. **51**, 1249 (1979).

PROPERTIES OF SINGLE-CRYSTAL SILICON FILMS
ON AMORPHOUS SiO$_2$ ON SINGLE-CRYSTAL CUBIC ZIRCONIA SUBSTRATES

I. GOLECKI[1,*], R.L. MADDOX[1], H.L. GLASS[2], A.L. LIN[2] AND H.M. MANASEVIT[1,**]
Rockwell International Corporation, Defense Electronics Operations,
Microelectronics Research and Development Center[1]/Science Center[2], 3370
Miraloma Avenue, Anaheim, CA 92803.

ABSTRACT

A new approach to achieving a large-area silicon-on-insulator technology without pre-patterning is described. (100) Si films are first grown epitaxially on (100) yttria-stabilized cubic zirconia (YSZ) substrates by the pyrolysis of SiH$_4$. The Si side of the <Si>/<YSZ> interface is then oxidized in pyrogenic steam (at 925°C) or dry oxygen (at 1100°C) to form the structure <Si>/amorphous SiO$_2$/<YSZ>. The oxidation occurs by the rapid diffusion of oxidants through the 0.42 mm thick YSZ substrate; e.g., a 0.3 μm SiO$_2$ layer is obtained in 6 h in steam. The samples are analyzed by Rutherford backscattering and channeling spectrometry, X-ray diffraction, infra-red reflectance, Auger electron spectroscopy and sheet resistance measurements. In addition to forming the preferred Si/SiO$_2$ interface, the back-side oxidation eliminates the most defective part of the Si film.

INTRODUCTION

A novel Si-on-insulator (SOI) heteroepitaxial system for integrated circuit applications has recently been developed in our laboratories [1,2]. The system consists of a (100) Si single-crystalline film on a thermally grown SiO$_2$ film on a (100) single-crystalline, yttria-stabilized cubic zirconia (YSZ) bulk substrate. The unique superionic oxygen conduction properties of cubic zirconia readily allow the thermal oxidation of the Si side of the <Si>/<cubic zirconia> interface, in either dry oxygen or pyrogenic steam, to result in the above structure [2]. The principal advantages of <Si>/a-SiO$_2$/<YSZ> over the commercially available Si-on-sapphire (SOS) system are [3,4]: (a) substantially better Si surface and near-surface crystalline quality in 0.4-0.5 μm thick films, e.g. minimum 1.5 MeV ^{4}He$^+$ channeling yield of 0.048 in Si/YSZ vs. 0.12 in SOS, (b) elimination of the most defective region in the Si film by oxidation, resulting in the preferred Si/SiO$_2$ interface, and (c) thirty-fold lower optimum epitaxial growth rate of Si on YSZ (0.08 μm/min) compared to SOS (2.4 μm/min). SOS films, on the other hand, are less compressively strained (-0.42%) than Si/YSZ or Si/SiO$_2$/YSZ (-(0.55-0.58)%), due to the correspondingly smaller mismatch in thermal expansion coefficients between Si and Al$_2$O$_3$, compared to Si and YSZ.
In this paper, we provide a more detailed description of the properties of the <Si>/a-SiO$_2$/<YSZ> system than given earlier [2], in particular concerning the chemical nature of the a-SiO$_2$ and the concentration of the crystallographic defects in the Si films.

EXPERIMENTAL CONDITIONS

The initially 0.5-0.6 μm thick, (100) oriented Si films were grown on 425 μm thick, (100) oriented $(Y_2O_3)_m (ZrO_2)_{0.99-m} (HfO_2)_{0.01}$ substrates, having compositions m = 0.15 and 0.21. The YSZ substrates were oriented,
_____
* Also Visiting Associate, California Institute of Technology, Mail Code 116-81, Pasadena, CA 91125.
** Present Address: TRW, Electro-Optics Research Center, 2525 El Segundo Boulevard, El Segundo, CA 90245.

sliced and polished at Rockwell International from randomly-oriented single-crystal boules grown by the skull melting technique at Singh Industries, Inc., Cedar Knolls, NJ (m = 0.15) and Ceres Corp., North Billerica, MA (m = 0.21). Si epitaxial films were grown by the pyrolysis of SiH_4 at 945 or 1000°C and at deposition rates of 0.11 - 0.56 μm/min. The detailed substrate preparation and film growth procedures have been described [1,3,5-6]. Each sample, 7-10 cm^2 in size, was divided into smaller pieces. In order to protect the top Si surface from subsequent oxidation, a 0.09 or 0.17 μm thick Si_3N_4 layer was deposited by the pyrolysis of $SiCl_2H_2$ and NH_3 on top of the Si surface in a low-pressure reactor at 800°C (45 min) or 820°C (36 min). Next, several oxidation runs at 1 atm were performed on separate samples. The wet oxidation runs at 925°C consisted of the following sequences: 10 min N_2, 100 min (or 360 min) pyrogenic steam, and 30 min N_2. The dry oxidation run at 1100°C lasted 270 min. Bulk (100) Si wafers were included for reference in all runs, and the 360 min, 925°C steam run also included Si_3N_4-capped, 0.6 μm <100 Si>/ 440 μm <1102 Al_2O_3> and 2.3 μm <111 Si>/ 300 μm <111 spinel> samples.

After oxidation, the samples were analyzed by Rutherford backscattering (RBS) without removal of the Si_3N_4, and also by RBS/channeling after the Si_3N_4 layer had been removed in hot H_3PO_4. These measurements were performed using 1.5 - 2.9 MeV $^4He^+$ ions, as appropriate, both at a large scattering angle ($\theta \simeq 150°$) and in a grazing-exit geometry ($\theta \simeq 100°$). The latter arrangement provided a background-free signal for the top half of the Si film, and thus enabled a precise alignment on and an accurate measurement of the surface and near-surface channeling yields of Si to be made [7]. The $^4He^+$ beam spot size was $\simeq 0.5 \times 0.25$ or $\simeq 2 \times 2$ mm^2. Si film thicknesses were also measured on a 10 μm lateral scale with a Nanospec spectrophotometric reflectometer (Nanometrics, Sunnyvale, CA). A Fourier transform infra-red (IR) spectrometer (Digilab Model QS 100) was used to measure differential IR reflectance spectra of the samples, especially in the frequency range 400 - 1500 cm^{-1}, with a resolution of 2 cm^{-1} and a 4 mm diameter sampling spot. A background spectrum was measured with a high-purity, float-zone bulk Si wafer, and a reference spectrum of the appropriate as-deposited Si/YSZ (control) sample was subtracted to obtain the net contribution of the a-SiO_2 layer. The in-plane macrostrain in the Si films was determined by measuring the Si lattice parameter perpendicular to the plane of the film, using a double-crystal X-ray diffractometer [8]. The lattice parameters of the YSZ substrates, needed as reference values, were measured separately [5,6]. The volume concentration of the {221} microtwins in the Si films, relative to the (100) Si matrix, was measured using full-circle X-ray goniometry [9,10]. The area probed by the X-ray spectrometers was $\simeq 1-4$ mm^2. The sheet resistance of the Si films was also measured by means of a high-sensitivity four-point probe (Model 101, Four Dimensions, Inc., Hayward, CA), with a probe-to-probe spacing of 1 mm. Several samples were analyzed by Auger electron spectroscopy (AES) combined with sputter profiling over an area of $\simeq 0.5 \times 0.5$ mm^2.

RESULTS

The RBS/channeling spectra, one example of which is given in Fig.1, showed that following oxidation, an SiO_2 layer had formed at the Si/YSZ interface in all the Si/YSZ samples. The layer was stoichiometric SiO_2, within the experimental accuracy, and had a channeling yield of unity throughout its thickness, indicating that it was not single-crystalline. From the known properties of thermal silicon oxides, we believe that the SiO_2 layer was amorphous, rather than fine-grained polycrystalline. No metallic constituents above $\simeq 1\%$ (the detection limit) were detected by RBS and AES in the SiO_2. The thickness of the SiO_2 layer varied from 650 to 3200 Å, depending on the sample and oxidation conditions (see Table I).

The differential IR reflectance measurements, examples of which are given in Fig.2(a-c), confirmed the presence of chemical SiO_2 by the characteristic peaks [11] at $\simeq 1090$ cm^{-1}, $\simeq 460$ cm^{-1}, and for the thicker SiO_2 layers

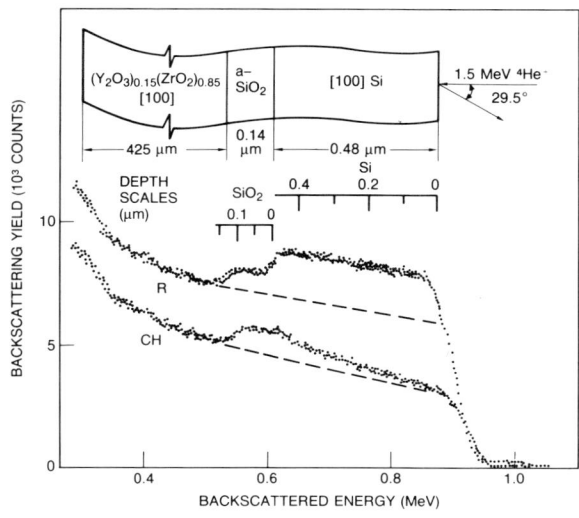

Fig. 1. Rutherford backscattering and axial channeling spectra of (100)Si/a-SiO$_2$/(100)YSZ, in which SiO$_2$ was formed by a 100 min pyrogenic steam oxidation at 925°C. CH=channeled, R=random. The Si RBS yields are obtained by subtracting the signals of Zr, Y and Hf in the YSZ substrate, shown as dashed lines (from Ref.2).

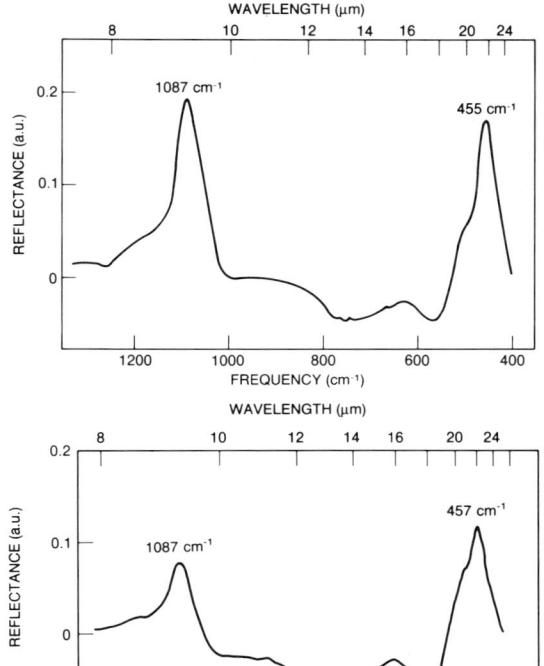

Fig. 2a

Fig. 2a, 2b, 2c. Net infra-red reflectance spectra of <Si>/a-SiO$_2$/<YSZ> samples: (a) 0.48 µm Si/0.14 µm a-SiO$_2$ (steam, 925°C, 100 min)/YSZ (m= 0.15), (b) 0.52 µm Si/ 0.065 µm a-SiO$_2$ (dry O$_2$, 1100°C, 270 min)/YSZ (m= 0.15), (c) 0.44 µm Si/ 0.28 µm a-SiO$_2$ (steam, 925°C, 360 min)/YSZ (m= 0.21).

Fig. 2b

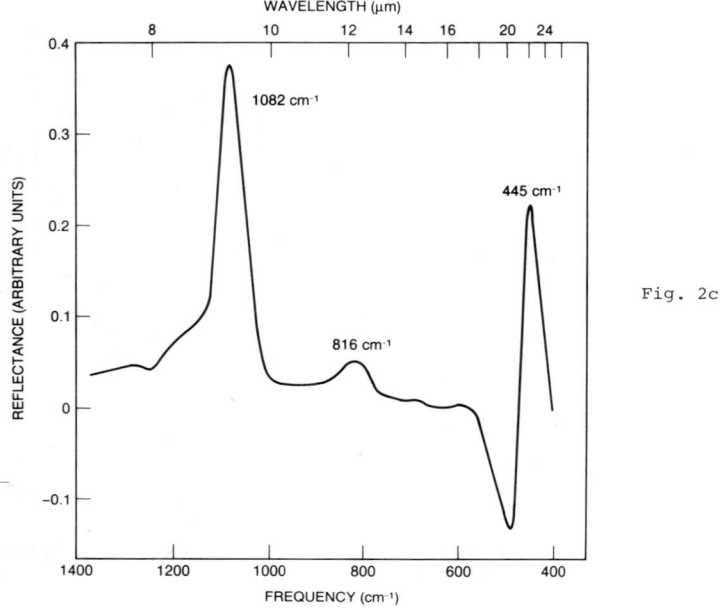

Fig. 2c

also at ≈820 cm^{-1}. The height of the main peak at 1090 cm^{-1} was found to be proportional to the SiO$_2$ thickness, as measured by RBS, except for the 0.3 μm thick SiO$_2$ layers, where the IR readings were lower by ≈15%. The FWHM of this peak was ≈66 cm^{-1}, independent of thickness, for both wet and dry oxidations. The 460 cm^{-1} SiO$_2$ peak, on the other hand, was not useful for quantification, due to residual interference from an intense Si peak at ≈500 cm^{-1}, which was not subtracted out perfectly by the spectrometer; this is seen as a shoulder in Fig.2(a-b) and a dip in Fig.2c. The small peak at ≈630 cm^{-1} in Fig.2(a-b) was apparently also unrelated to the SiO$_2$.

Generally, the oxides formed in pyrogenic steam were thicker and more uniform laterally, as measured by RBS, than those formed in dry oxygen, for oxidation conditions resulting in equal oxide thicknesses on bulk Si. The oxides formed through the zirconia substrates with 21% mol yttria were in most samples slightly thicker than in the 15% mol yttria case, for both wet and dry oxidation conditions; the variations in the thicknesses of the YSZ substrates were in the range 1-3.5%. The SiO$_2$ layers obtained in steam on the (100) bulk Si control wafers under the same experimental conditions were thicker by factors of 2.0-2.5 compared to those formed at the Si/YSZ interface for YSZ with m=0.15; for YSZ with m=0.21 the enhancement factor was 1.7-1.8 for the 1.67 h run but 2.3-2.9 for the 6 h run. In one case (with m=0.21) we found after the 6 h steam oxidation that the Si$_3$N$_4$ layer had developed circular openings, ≈50-150 μm in diameter, and oxidation of the Si from the top had occurred through these openings. In this sample, the ratio between the thicknesses of the top oxide and the interfacial oxide (given in italics in Table I) was 2.0, somewhat lower than for uncapped bulk Si, possibly due to a nucleation time for the formation of the openings in this defective nitride. For the dry oxidation, the enhancement factor was 3.1-4.5 in the m=0.15 case and 2.2-3.7 in the m=0.21 case.

RBS and differential IR reflectance measurements did not detect any interfacial SiO$_2$ layer in the SOS and Si/spinel samples annealed in steam at 925°C for 6 h. It is estimated that the sensitivity of the IR technique was ≤30 Å

Table I. Thicknesses of thermal SiO₂ layers formed at the Si/YSZ interface in Si₃N₄/Si/YSZ samples by diffusion of oxidizing species through the YSZ substrates. The range in values represents lateral variation. SiO₂ thicknesses formed on bulk (100) Si under the same conditions are given for comparison. The values in italics are for a sample where unintentional openings in the top Si₃N₄ layer allowed oxidation of Si from the top as well, resulting in the structure SiO₂/Si/SiO₂/YSZ.

YSZ SUBSTRATE COMPOSITION $m(Y_2O_3)$	SiO_2 THICKNESS (μm)		
	2H₂/O₂, 925°C		O₂, 1100°C
	1.67 h	6 h	4.5 h
0.15	0.115 - 0.149	0.32	0.065 - 0.094
0.21	0.163 - 0.173	0.23 - 0.29 *0.26*	0.079 - 0.134
Bulk Si	0.29	0.66 *0.50*	0.29

SiO₂ under the measurement conditions used.

The X-ray diffraction and 1.5 MeV ^{4}He$^+$ ion channeling results for one set of samples on YSZ with 15% mol yttria, following dry or wet oxidations, are given in Table II. Similar results were obtained for the samples on YSZ with 21% mol yttria. The depth-averaged microtwin concentrations in the Si films decreased significantly as a result of the back-side oxidations: the thicker the interfacial SiO₂ layer, the lower the microtwin concentration. For example, after the 6 h, 925°C steam oxidation, which converted 30% of the initial Si film thickness into SiO₂, the microtwin concentration has been reduced by a factor of 2.4. This result is to be expected, since it is clear from our channeling spectra (see Fig.1 and Ref. 1-3, 5-6) that the concentration of the planar crystallographic defects in heteroepitaxial Si/YSZ films increases monotonically from the Si surface towards the Si/YSZ interface. Thus, the back-side oxidation, which removes the most defective part of the Si film, has a pronounced effect on the defect density. The distribution of microtwins on the four {111} planes in the Si/YSZ films was generally much more uniform than is observed in SOS films [9, 10, 15], where variations of up to a factor of 3 are commonly seen. This difference in defect microstructure between Si/YSZ and SOS may be related to different defect formation mechanisms or nucleation sites during cooldown from the growth temperature of 900 - 1000°C, due to the different crystallographic structures of Al₂O₃ (rhombohedral) and YSZ (cubic fluorite).

In order to compare the {221} microtwin concentration, as measured by X-ray diffraction, with the ion channeling measurements, we used the standard dechanneling analysis method [12], which expresses the channeling yield at depth z, $\chi(z)$, in a crystal containing defects, in terms of the channeling yield in a virgin crystal of the same material (without defects), $\chi_V(z)$, the concentration of defects of type i, $N_i(z)$, and their dechanneling cross-section, $\sigma_i(z)$:

$$1 - \chi(z) = [1 - \chi_V(z)]\exp(-\sum_i \int_0^z \sigma_i(z')N_i(z')dz') \quad (1)$$

In Si/YSZ films the defects causing the observed dechanneling are mainly the microtwins and stacking faults, although the contribution due to dislocations cannot be completely neglected. There are no significant concentrations of defects, such as interstitials, which would cause direct scattering of the analyzing ^{4}He$^+$ beam and, in any case, the cross-section for direct scattering is much smaller than that for dechanneling [12]. In principle, it is possible to develop a detailed model for dechanneling by twins, similar to that of Ref. 16-17, taking into account the 9-fold increase in dechanneling rate, $d\chi/dz$, in the [221] direction, relative to the [100] direction, the possible slight misorientation of the twinned region relative to the matrix, and the contribution of the twin boundaries. However, from a practical standpoint, the information available on the sizes and shapes of the extended defects in Si/YSZ, e.g. from TEM observations [6, 13, 14], is too limited to allow such an exact calculation of their dechanneling cross-sections, σ_i. Therefore, we have listed in Table II the exponent appearing in Eq.1, which is obtained from a plot of $\ln [1 - \chi_v(z)]/[1 - \chi(z)]$ vs. z, where $\chi_v(z)$ and $\chi(z)$ are the measured channeling yields at depth z of virgin (100) Si and Si/YSZ (or Si/SiO$_2$/YSZ), respectively. In all cases, the channeling results give a somewhat higher relative defect density than the X-ray measurements, because the latter measure only the {221} twins, while the former are influenced by all the defects in the Si films. In the case of the dry oxidation, the channeling data even shows a slight increase in the overall weighted defect concentration, while the twin concentration in the film has actually decreased. This discrepancy may be real, and due to additional imperfections generated by the 1100°C, 4.5 h oxidation, or possibly a manifestation of lateral nonuniformities in the sample.

Additional X-ray measurements indicated a ≃30% reduction in the FWHM of the Si rocking curves following either wet or dry oxidation, signifying an overall improvement in crystalline quality in the films. The width of the rocking curve is affected not only by planar defects but also by strain fields associated with dislocations. The depth-averaged in-plane compressive strain increased slightly following oxidation. By comparison, Si/YSZ films where the YSZ thickness was 432 μm (m=0.15) or 734 μm (m=0.24), which were annealed in flowing N$_2$ at 850 or 1000°C for 2 h, also showed 13-23% reductions in the widths of the rocking curves. The in-plane strain in these samples either increased or decreased slightly. RBS measurements showed no chemical reaction at the Si/YSZ interface in these samples [6], but differential IR reflectivity measurements indicated a ≃100 Å thick SiO$_2$ film in the samples with m=0.15, and no SiO$_2$ in the thicker samples with m=0.24.

The depth-averaged resistivities of the Si films, measured after a brief dip in 10% HF, were initially in the range 5 - 20 Ωcm, n-type (probably due to unintentional phosphorus doping during growth [1, 3, 5-6]). There was no significant change in resistivity after the steam oxidations at 925°C; however, a reduction by a factor of ten resulted from the dry oxidation at 1100°C. The reason for this decrease in resistivity was not investigated further.

RBS and AES measurements detected the presence of an SiO$_x$N$_y$ layer, less than 200 Å thick and possibly discontinuous, on the unpolished YSZ side of the samples, following oxidation. This layer may have formed as Si$_3$N$_4$ during nitride deposition on the Si surface, and subsequently been partly oxidized.

DISCUSSION AND SUMMARY

The SiO$_2$ layers were formed at the Si/YSZ interface as a result of the rapid diffusion of oxidizing species through the 425 μm thick YSZ substrates. The driving force for this diffusion is the difference in oxygen chemical potentials between the YSZ/ambient and Si (or SiO$_2$)/YSZ interfaces. Without such a driving force, e.g. in a reducing ambient, no SiO$_2$ can form, since Si cannot completely reduce ZrO$_2$, HfO$_2$, or Y$_2$O$_3$ to form SiO$_2$; the Gibbs free energies of these reduction reactions are all positive and large: 42.5, 45, and 74 kcal/mol O$_2$, respectively, at 925°C, with similar values at 1100°C [18]. The essential lack of any interfacial reaction in the Si/YSZ samples annealed for 2 h in flowing N$_2$ at 850 or 1000°C supports this thermodynamic argument. The 100 Å thick SiO$_2$ layer detected by IR reflectance in some of the samples was most probably due to some water vapor contamination in the furnace. The nature of the diffusing species is probably O^{--} and h^+ (holes) in the dry

Table II. Results of X-ray diffraction and 1.5 MeV ^{4}He$^+$ channeling measurements of Si/YSZ (m=0.15) before and after back-side oxidation to form Si/a-SiO$_2$/YSZ. t_{Si} or t_{SiO_2} = Si or SiO$_2$ film thickness; [twins] = depth-averaged concentration of {221} microtwins in Si, measured by X-ray diffraction; $\Sigma \int \sigma_D N_D dz$ = integrated dechanneling probability; $\varepsilon_{\parallel}$ = in-plane macrostrain; $\Delta\omega$ = rocking-curve width.

QUANTITY	AS-DEPOSITED	2H$_2$/O$_2$, 925°C		O$_2$, 1100°C
		1.67 h	6 h	4.5 h
t_{Si} (μm)	0.55	0.48	0.36	0.52
t_{SiO_2} (μm)	-	0.14	0.32	0.065
[twins] (% vol)	5.1	3.7	2.2	4.2
Relative to as-deposited	*1.00*	*0.72*	*0.42*	*0.82*
$\Sigma \int_0^{t_{Si}} \sigma_D N_D dz$	0.86	0.68	0.46	0.96
Relative to as-deposited	*1.00*	*0.79*	*0.54*	*1.12*
$\varepsilon_{\parallel}$ (10^{-3})	-5.5	-5.8	-5.8	-5.9
$\Delta\omega$ (min of arc)	20.3	15.1	14.2	15.6

(intentional) oxidation, and O^{--} and H$^+$ (protons) in the steam oxidations. The presence of hydroxyl-containing species at the Si/SiO$_2$ interface is necessary to explain the measured SiO$_2$ thicknesses. The diffusion is ambipolar, so that no net transport of charge occurs, and is rate-limited by the hole or proton diffusivity. To our knowledge, no transport data are available in the literature for YSZ with the present yttria contents, especially in the single-crystal form. However, if we assume that diffusion through the YSZ is the rate-limiting step in the back-side oxidation of SiO$_2$ [2], then the thicknesses of our steam-generated SiO$_2$ layers are within order-of-magnitude agreement with published data on the proton diffusivity and H$_2$O permeability of polycrystalline YSZ with m = 0.045 or 0.10 [19]. From our results, the kinetics of back-side steam oxidation of Si are sub-linear. A more detailed discussion of the oxidation mechanism is given in Ref.2. In experiments similar to ours performed recently at Thomson-CSF/University of Paris [14], it was found that the kinetics of dry back-side oxidation at 1150°C were linear up to 40 h, and the SiO$_2$ thickness varied linearly with partial pressure of oxygen in the range 0.1 - 2 atm; by comparison, square-root dependences on time and pressure prevailed for the top SiO$_2$ layer in those uncapped SiO$_2$/<Si>/SiO$_2$/<YSZ> samples. The interfacial SiO$_2$ layer was ≃3 times thinner than the top SiO$_2$ layer, e.g. 1500/4500 Å after 20 h at 1200°C at 0.3 atm O$_2$ for a 0.7 mm thick, m = 0.21 substrate; these results are in agreement with ours. Cross-sectional TEM analysis revealed [14a] that the Si/SiO$_2$ interface was sharp, and no other compounds were present, again in accord with our data.

To summarize, we have developed a new, simple and elegant approach to the achievement of an SOI technology, <Si>/a-SiO_2/<YSZ>, which results in single-crystal Si over amorphous SiO_2, the preferred SOI combination, without any pre-patterning requirements. The two principal processing technologies used, viz. chemical vapor deposition to grow the epitaxial Si films and oxidation to form the back-side SiO_2, are well established, readily available, and applicable to large areas, and the YSZ substrates can also be obtained in larger sizes than used in this study (2.2 in.). The quality of the back-side Si/SiO_2 interface appears to be high, and the blocking action of YSZ against the relatively slow diffusers, such as Zr, Hf, Y and other external metallic impurities, assures the purity of the SiO_2 film. The back-side oxidation rate is sufficiently high (especially in steam) for the process to be practical and compatible with standard device fabrication procedures. In addition, one could, in principle, also form the back-side SiO_2 film electrochemically, close to room temperature. For integrated circuit applications, the low dielectric constant (3.9) of SiO_2 is expected to result, for an appropriately chosen thickness, in reduced parasitic capacitances and thus higher operating speeds. In order to eliminate the microtwins from the remainder of the Si film, the full Si ion implantation, amorphization and solid-phase epitaxial regrowth process [4] can be applied to the <Si>/SiO_2/<YSZ> system, without damaging the substrate or introducing foreign impurities (except oxygen) in the Si film, in contrast to the situation in the Si/Al_2O_3 system. In principle, one could also adjust the ratio of Si/SiO_2 thicknesses, the oxidation temperature and pre/post-oxidation annealing cycles in order to achieve some control over the compressive strain in the Si film. This statement is based on the fact that SiO_2 is a glass and its thermal expansion coefficient is 8 times lower than that of Si (and 25 times lower than that of YSZ). In the present study, although we believe that during the oxidations the SiO_2 behaved essentially as a fluid, with a relaxation time of 3 min or less [20-21], during subsequent cooldown the SiO_2 viscosity increased rapidly and the Si film ended up in compression, as before oxidation. Finally, we have used a combination of complementary analytical techniques (RBS/channeling, X-ray diffraction, IR reflectance, AES, sheet resistance) to obtain a fairly complete picture of the <Si>/SiO_2/<YSZ> system.

ACKNOWLEDGEMENTS

We wish to thank D. Medellin for expert polishing and M.D. Lind for X-ray measurements of the lattice parameters of the YSZ substrates, N. Casey for technical assistance, T.J. Raab for AES measurements, and J.E. Mee for support (all at Rockwell International Corporation). Part of this study was funded by the Electromagnetic Materials Division, Materials Laboratory, AFWAL, Wright-Patterson Air Force Base, OH 45433, under Contract # F33615-81-C-5041 (J.O. Crist and M.C. Ohmer).

REFERENCES

1. I. Golecki, H.M. Manasevit, L.A. Moudy, J.J. Yang, and J.E. Mee, Appl. Phys. Lett. 42, 501 (1983).
2. I. Golecki, R.L. Maddox, H.L. Glass, A.L. Lin, T.J. Raab, and H.M. Manasevit, J. Electron. Mater. 14 (in press, 1985).
3. A.L. Lin and I. Golecki, J. Electrochem. Soc. 132, 239 (1985).
4. I. Golecki, Mat. Res. Soc. Symp. Proc. Vol. 33, 3 (1984).
5. H.M. Manasevit, I. Golecki, L.A. Moudy, J.J. Yang, and J.E. Mee, J. Electrochem. Soc. 130, 1752 (1983).
6. I. Golecki, Final Technical Report # AFWAL-TR-83-4137 (January 1984).
7. I. Golecki, Nucl. Instrum. and Methods in Phys. Res. 218, 63 (1983).
8. I. Golecki, H.L. Glass, G. Kinoshita, and T.J. Magee, Applic. of Surf. Sci. 9, 299 (1981).

9. R.T. Smith and C.E. Weitzel, J. Cryst. Growth 58, 61 (1982).
10. R.A. Kjar, P.E. Haynes, and J. Maurits, Final Technical Report # DELET-TR-80-0311-3 (November 1983).
11. J.E. Dial, R.E. Gong, and J.N. Fordemwalt, J. Electrochem. Soc. 115, 326 (1968).
12. E. Bøgh, Can. J. Phys. 46, 653 (1968).
13. I. Golecki, H.M. Manasevit, L.A. Moudy, J.J. Yang, J.E. Mee, and T.J. Magee, paper # D-8, presented at the 24th Electronic Materials Conference, Fort Collins, CO, June 1982.
14. (a) M. Dupuy, J. Microsc. Spectrosc. Electron. 9, 163 (1984).
    (b) D. Pribat, L.M. Mercandalli, M. Croset, D. Dieumegard, and J. Siejka, Mater. Lett. 2, 524 (1984).
    (c) L.M. Mercandalli, D. Pribat, M. Dupuy, C. Arnodo, D. Rondi, and D. Dieumegard, paper # D.3.12, presented at the Symp. on Layered Structures, Epitaxy and Interfaces, 1984 Fall Meeting of the Materials Research Society, Boston, MA, November 1984 (J.M. Gibson and L.R. Dawson, eds.).
15. H.L. Glass (unpublished).
16. S.U. Campisano, G. Foti, E. Rimini, and S.T. Picraux, Nucl. Instrum. and Methods 149, 371 (1978).
17. G. Foti, L. Csepregi, E.F. Kennedy, J.W. Mayer, P.P. Pronko, and M.D. Rechtin, Phil. Mag. A 37, 591 (1978).
18. T.B. Reed, Free Energy of Formation of Binary Compounds (MIT Press: Cambridge, MA, 1971).
19. C. Wagner, Ber. Bunsenges. Phys. Chem. 72, 778 (1968).
20. J.R. Patel and N. Kato, J. Appl. Phys. 44, 971 (1973).
21. E.P. EerNisse, Appl. Phys. Lett. 30, 290 (1977) and 35, 8 (1979).

ATOMIC INTERACTIONS IN SILICON-METAL COMPLEXES ON W(110)

JOHN D. WRIGLEY AND GERT EHRLICH
Coordinated Science Laboratory, University of Illinois, Urbana, IL 61801

ABSTRACT

With the field-ion microscope one not only can locate individual atoms at a surface, but under some circumstances also can discriminate between chemically different species. These capabilities are illustrated in an examination of the strength of binding in silicon-metal surface clusters and in a detailed analysis of their mobility.

INTRODUCTION

The metal-silicon interface is of considerable interest in solid-state technology, and the structural and electronic properties of the interface have received much attention [1]. However, little is known on the atomic level about the formation and properties of metal-silicon surface complexes. This state of affairs can be attributed to the difficulty of characterizing the chemical constitution and structure of a surface with atomic resolution. In this regard, the field-ion microscope (FIM) has long been recognized for its unrivaled abilities in revealing individual adatoms [2]. Here we illustrate the use of this powerful technique in providing a better understanding of the behavior and properties of metal-silicon clusters on tungsten, not only to characterize atomic arrangement, but also to provide chemical information. These clusters constitute one stage in the process of silicide formation at the surface; we have been specifically interested in the interactions between metal and silicon adatoms responsible for the stability of such complexes, and also in the transport properties of clusters [3].

CLUSTER DISSOCIATION

The strength of the interactions between metal and silicon atoms in a cluster is indicated by the binding energy -- the difference in energy when two adatoms are far apart on the surface, and when the dimer is in its ground-state configuration. One way of ascertaining the strength of binding is through kinetic measurements of the rate of dissociation of a dimer into its constituent atoms. The underlying idea is simple. The rate of dissociation or equivalently the inverse of the average lifetime $<\tau>$ for dissociation, can be written in the usual way as $1/<\tau> = 1/\tau_0 \exp(-E_{Diss}/kT)$ where E_{Diss}, the barrier to the dissociation process, can be derived from the temperature dependence, or from a knowledge of the prefactor $1/\tau_0$. As suggested in Fig. 1, there are at least two contributions to the barrier E_{Diss} -- the binding between adatoms in the cluster and the barrier to the motion of a single adatom over the surface. Even without any interactions, and therefore with no binding between two adatoms on the surface, the atoms will still have to have enough energy to jump from one site on the surface to another in order to move apart. The binding energy is therefore given as the difference between the activation energy for dissociation and for diffusion.

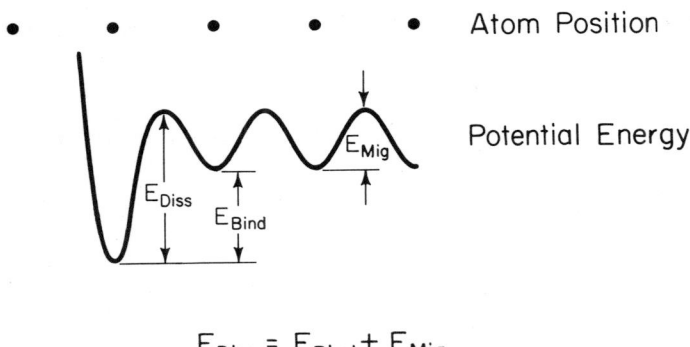

$$E_{Diss} = E_{Bind} + E_{Mig}$$

Fig. 1. Atomic potentials in dimer dissociation at a surface.

The average lifetime $\langle \tau \rangle$ can be determined in a straightforward way using the FIM. A dimer is first formed on the surface by association of the two adatoms. The surface is then heated for a predetermined time interval in the absence of an applied field. Thereafter, under conditions such that the field cannot induce dissociation, an image is taken at $T \sim 20K$ to establish if the dimer still exists. This sequence is repeated until dissociation is observed, to yield one lifetime value. Dissociation measurements must be done on many separately created dimers to generate a statistical sample from which the mean lifetime is derived by fitting to an exponential distribution. Such lifetime measurements for PdSi at 272K on W(110) are shown in Fig. 2. Assuming a normal prefactor, we estimate the dissociation energy for PdSi to be 17.3 kcal/mole.

From observations of single Si atoms as well as of single Pd atoms on W(110) we know that Pd is considerably more mobile than Si on the surface. The PdSi dimers are formed when a Pd atom migrating over the surface attaches itself to a stationary Si adatom. In the reverse process, dissociation, it is also the Pd which migrates away. The activation energy for surface diffusion of Pd amounts to 11.5 kcal/mole; hence we estimate a binding energy of 6 kcal/mole for PdSi on W(110).

Lifetime measurements have also been carried out for WSi in W(110). From their temperature dependence we arrive at an activation energy for dissociation of 19.6 kcal/mole. In this system, the Si adatom is more mobile than the W atom, and dissociation presumably involves diffusion of the former over the surface, with an activation energy of 19.5 kcal/mole. The primary contribution to the dissociation energy now comes from the barrier to diffusion rather than from the binding of W to the Si adatom. Binding is small and well within the limits of error of the measurements. Although only these two systems have been examined quantitatively, it is clear from the available data that there is a pronounced chemical specificity in the binding between metal atoms and silicon.

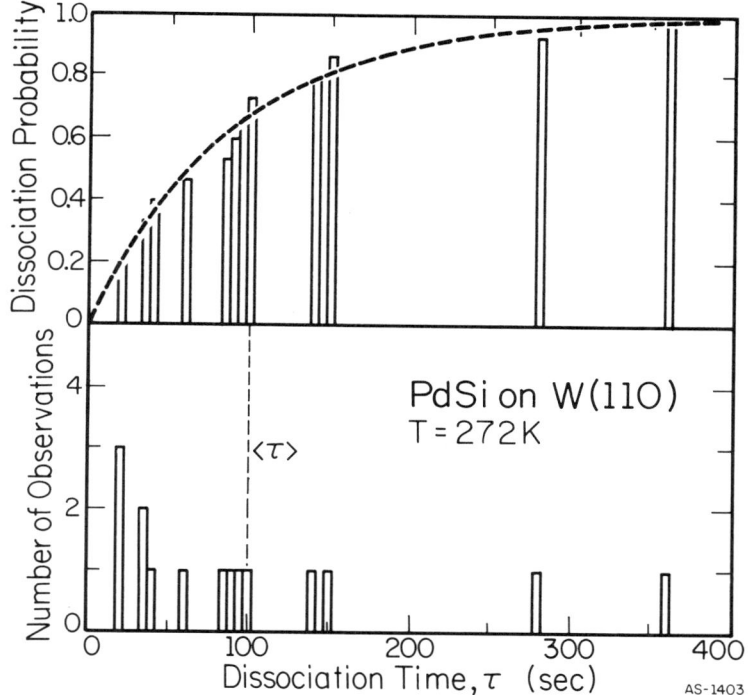

Fig. 2. Dissociation lifetime of PdSi on W(110). Bar graph at top shows experimental cumulative probability of dissociation. Dashed curve at top gives values expected for a Poisson process. Bar graph at bottom shows values of individual lifetimes.

ATOMIC JUMPS IN DIMER DIFFUSION

WSi complexes have considerable mobility on the surface at temperatures below dissociation. In fact, these dimers are more mobile than their constituent atoms. Evidently interactions between metal and silicon atoms are significant enough to affect the forces between adatom and substrate responsible for the diffusion barrier. The diffusion of such complexes is an important part of the overall process of layer growth; it also can provide additional information about the forces between metal and silicon adatoms. We have therefore posed the question -- how do metal-silicon dimers migrate over the surface?

From observations in the FIM it appears that on W(110) individual atoms are held at binding sites which form a grid with the same symmetry and spacing as the (110) plane itself [3,4]. For silicon metal dimers we assume that the two atoms occupy nearest-neighbor sites on this grid. That is, in its ground state, the dimer is oriented along the close-packed ⟨111⟩ direction, as suggested in Fig. 3. It should be noted that in all our observations metal-silicon dimers have indeed been observed in this orientation.

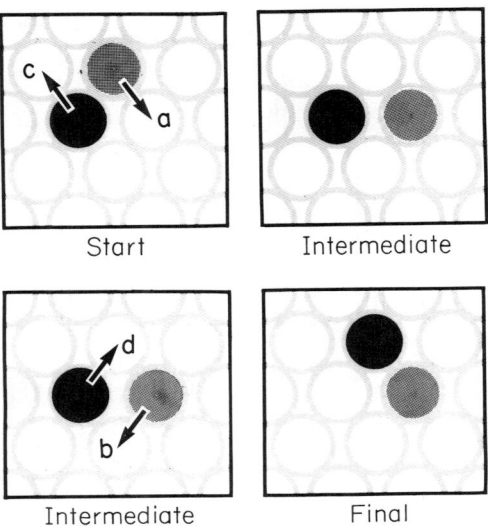

Fig. 3. Dimer states and jump rates on BCC(110). Upper left and lower right are ground states oriented along the close-packed ⟨111⟩ directions. The intermediate state is oriented along the ⟨100⟩.

Presumably motion of the dimer occurs through jumps of individual atoms, just as for metal dimers in one-dimensional motion [5], and involves a sequence of steps: the first jump moves the dimer out of its ground state into an intermediate state of higher energy; this is followed by another jump returning the dimer to the ground-state configuration. On W(110) the obvious intermediate [3] consists of two adatoms in second-nearest-neighbor sites, at a separation $2/\sqrt{3}$ larger than when they are in the ground state. The atomic jumps in such a sequence are illustrated schematically in Fig. 3. The dimer moves into the intermediate position, in which it is oriented along ⟨100⟩, as either a silicon or a metal atom jumps to a second-nearest site; this occurs at rates $\underline{a}$ for the silicon and $\underline{c}$ for the metal. In the next jump the dimer starts in the intermediate state, and the complex will either return to its original site or to a new position; that is, a displacement may occur. This is the model for diffusion of metal-silicon complexes on W(110). Can it be verified by experiment?

One of the postulates is that intermediates are oriented along ⟨100⟩, whereas the stable dimer lies along ⟨111⟩. However, all metal-silicon dimers seen in the FIM lie along ⟨111⟩. Presumably dimers in the intermediate state are of higher energy and have only a transient existence, so they are not seen. Nevertheless, there is a critical consequence of the model just presented that can be checked. Once a dimer is formed on the surface, its configuration -- that is, the orientation of the line between silicon and metal atom with respect to the lattice -- is preordained wherever on the surface the dimer may be. This is illustrated by the schematic in Fig. 4. All that needs to be done is to establish the orientation of the dimer after diffusion, to prove that the orientation conforms to Fig. 4. The chemical identity of the components of a silicon-metal dimer can be differentiated in the FIM. The voltage at which a

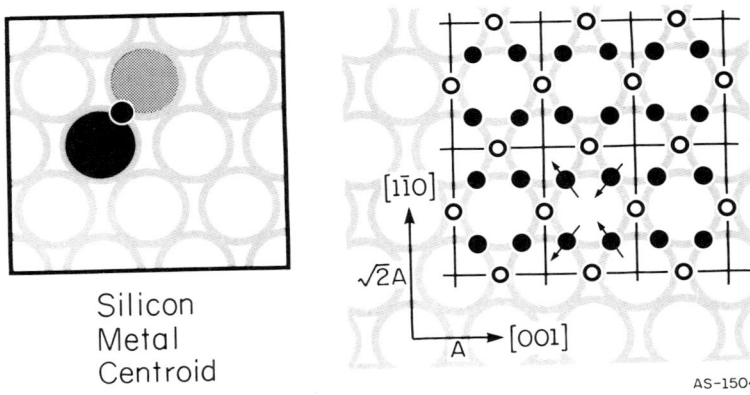

Fig. 4. Center-of-position grid for dimers in the ground state. Arrows indicate possible orientations for diffusion by second-nearest-neighbor intermediates only.

silicon atom is imaged differs from the voltage for imaging a metal atom. By taking two images of a dimer, at the previously established best image voltages for the two different atoms, one can map out the orientation. In all our measurements on WSi, the orientation after diffusion on W(110) is in agreement with an intermediate along <100>, confirming the prediction of the model.

The FIM can also be used to measure quantitatively the individual jump rates important in diffusion. Inasmuch as the intermediates are present in very low concentrations only, it is clear from the principle of microscopic reversibility that the jump rates $\underline{a}$ and $\underline{c}$ in Fig. 3, which lead to formation of the intermediate, are much smaller than the rates $\underline{b}$ and $\underline{d}$ for jumps out of the intermediate state. Rates $\underline{a}$ and $\underline{c}$ limit the speed of dimer diffusion over the surface. The mean-square displacements along the <100> and <110> directions are given [6] in the limit of long times by

$$\frac{\langle \Delta x^2 \rangle}{L^2} = \frac{1}{2}\frac{\langle \Delta y^2 \rangle}{L^2} = \frac{act}{2(a+c)} \qquad (1)$$

where L is the lattice spacing. Measurements of the mean-square displacement in the FIM therefore does not suffice to give all the jump rates important in diffusion of the metal-silicon complexes. To accomplish this we have measured the distribution function governing the probability of finding a dimer at a specified coordinate after a time t, and compared it with the distribution found by Monte Carlo simulations for different jump rates, as in Fig. 5. In this way it has been shown, for example, that in WSi the jump ate of the silicon atom is roughly four times that of the tungsten.

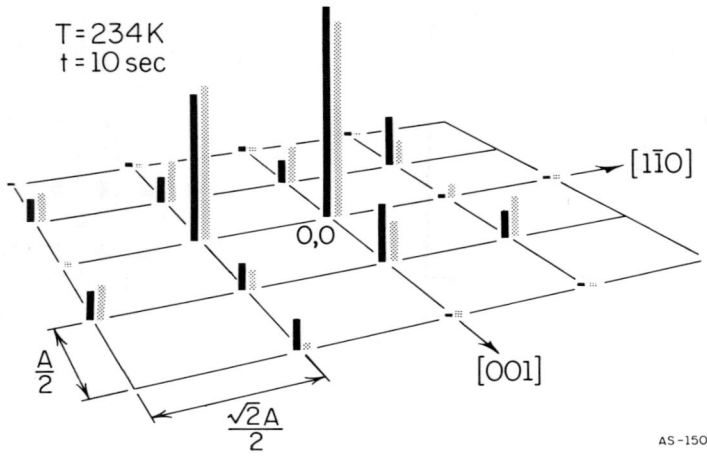

Fig. 5. Position distribution of WSi on W(110) after 10 sec diffusion at 234K. Grey bars are computer simulation. Black bars are experimental data.

SUMMARY

More extensive experiments in progress should make it possible to determine the energetics of the different jumps in WSi, and therefore to attain an insight into the effective potentials operating between silicon and metal atoms. Even at the present stage it is clear that the ability of the FIM to locate individual adatoms, and in some instances to discriminate between chemically different adatoms, provides a powerful capability for the analysis of surface processes on the atomic level.

ACKNOWLEDGMENTS

This work was supported under JSEP Contract N00014-84-C-0149. The award of a Guggenheim Fellowship to one of us (Gert Ehrlich) has made it possible to contribute to this study from the stimulating environment of the Division of Applied Science, Harvard University.

REFERENCES

1. See, for example, P. S. Ho and K. N. Tu (eds.), Thin Films and Interfaces, North-Holland, New York, 1982; also J. Vac. Sci. Technol. B2 (1984) 693-784.

2. The technique has been reviewed recently by J. A. Panitz, J. Phys E, 15 (1982) 1281.

3. Studies on metal clusters are reviewed by D. W. Bassett, in V. T. Binh (ed.), Surface Mobilities on Solid Materials, Plenum, New York, 1983 p. 83.

4. For recent observations, see H.-W. Fink and G. Ehrlich, J. Chem. Phys, 81 (1984) 4657.

5. K. Stolt, W. R. Graham, and G. Ehrlich, J. Chem. Phys., 65 (1976) 3206.

6. J. D. Wrigley and G. Ehrlich, to be published.

THE THICKNESS EFFECT ON THE MICROSTRUCTURE OF SPUTTERED
FILMS STUDIED BY A NEW X-RAY DIFFRACTION METHOD

M. HECQ
Laboratoire de Chimie Inorganique
Université de l'Etat à Mons
Avenue Maistriau, 23
B - 7000 MONS, Belgium

ABSTRACT

The microstructure of sputtered platinum thin films has been investigated by X-ray diffraction analysis. The (111) reflection was analysed by the variance and the Fourier methods. The effective particle size was calculated from the variance slope coefficient and the microstrain from the Fourier cosine coefficients assuming that the effective sizes calculated from both methods are identical. Various thickness thin films have been studied. The effective particle size (D_{ev}) increases with increasing film thickness until a thickness of ~ 1500 Å, then D_{ev} becomes constant and independent on the thickness. The microstrain varies in an inverse way to D_{ev}; it decreases with increasing thickness. Above a thickness of 1500 Å, the microstrain is constant.

1. INTRODUCTION

It has been [1] shown that the microstructure of materials can be studied by X-ray diffraction. Breadths, asymmetry and displacements of diffraction peaks give informations on crystallite size, microstrain, dislocation density, faulting probability and lattice-parameter changes. However, the original application of the variance [2] or the Fourier (WarrenAverbach [1]) methods require at least two orders of a reflection. In many practical cases (supported catalysts [3], sputtered thin films [4], the peaks are very broad and the high orders of a reflection are overlapped by the tails (for example, (311) and (222) in F.C.C. materials). It is why several single line methods have been reported [5], most are based on the Fourier cosine coefficients of the profile. The power distribution per unit length of a Debye-Scherrer line is given by [19]

$$P'(S) = k \int_{-\infty}^{+\infty} A(L) \exp(-2\pi i L(s-s_o)) dL \qquad (1)$$

where $s = 2 \sin \theta/\lambda$, $s_o = 2 \sin \theta_o/\lambda$, θ_o being the peak maximum and L ($=nd_{hkl}$) is a distance normal to the reflecting planes khl of interplanar spacing d_{hkl}.

The Fourier coefficients A(L) are given by

$$A(L) = \frac{\int_{s_1}^{s_2} P'(s) \exp(2\pi i L(s-s_o)) ds}{\int_{s_1}^{s_2} P'(s) ds}$$

where $s_2 = 2 \sin \theta_2/\lambda$ and $s_1 = 2 \sin \theta_1/\lambda$ are the upper and lower integration limits, respectively. The Fourier cosine coefficients A(L) have two components given by

$$A(L) = A_S(L) \cdot A_D(L) \qquad (2)$$

where $A_S(L)$ is the "size coefficient" and $A_D(L)$ the "strain coefficient". These can be separated if two orders of a reflection (e.g. (111) and (222) are available). $A_S(L)$ does not depend on the order of reflection

There is no accurate relation between $A_S(L)$ and L but it has been proved [1] that the initial slope of $A_S(L)$ versus L is given by $-1/D_{eF}$ where D_{eF} is the average crystallite size perpendicular to the reflection planes calculated following the Warren-Averbach procedure. For small values of L where L is such that the number of coherent domains or particles in the sample with this dimension is insignificant, the particle size term can be expanded as [20]

$$A_S(L) = 1 - L/D_{eF} \qquad (3)$$

This equation is valid irrespective of the particle size distribution provided the above criterion is met. From Gangulee [20] eq. (3) is valid up to $L < 0.25 * D_{eF}$.
$A_D(L)$ is reduced to :

$$A_D(L) = 1 - 2 \pi^2 \, 1^2 \, \frac{L^2}{d^2} < \varepsilon_L^2 > \qquad (4)$$

where ε_L denotes the mean strain over a length L within a column and the brackets indicate an averaging over the diffraction volume and l the order of reflection. For cubic crystals, the interplanar spacing (d) is a/h_o, where a is the crystal parameter and $h_o^2 = h^2 + k^2 + l^2$, h, k, l are the Miller indices. Eq. (4) is expressed in the form :

$$A_D(L) = 1 - 2 \pi^2 < \varepsilon_L^2 > \frac{L^2 \, h_o^2}{a^2} \qquad (4')$$

Most of the single line methods [5] which have been reported until now are based on equation (2).
Normally, the $< \varepsilon_L^2 >$ values will increase as L decrease. If dislocations are the principal cause of microstrains then it may be deduced that [5]

$$< \varepsilon_L^2 > = \frac{C}{L} \qquad (5)$$

where C = constant. After substitution of relations (3), (4') and (5) in (2), a polynomial of second degree is obtained and D_{eF} and C can be calculated by a least squares method [6]. However equation (5) is not valid for sputtered films [7]. Another method, using the ratio of the width at half the maximum intensity to the integral breadth has been recently suggested [8]. In this last method it is assumed that the crystallite size profile has a Cauchy distribution and the microstrain profile a gaussian distribution.

In this paper we suggest a new method which is a combination of the variance method and the Fourier method. The variance of line profile about the centroid s_g is given by

$$W(s) = \frac{\int_{s_1}^{s_2} (s-s_g)^2 \, P'(s) ds}{S_\infty} \qquad (6)$$

where $S_\infty = \int_{-\infty}^{+\infty} P'(s) ds$. In practice the line is measured between s_1 and s_2. If the observed intensity of the truncated profile is S, then the true value is [9]

$$S_\infty = S(1-k/\sigma)^{-1} \qquad (7)$$

where $2\sigma = s_2 - s_1 = \Delta 2\theta \cos\theta_0 /\lambda$, $\Delta 2\theta$ being the range of variance.
Wilson [2] showed that the variance of line profile (eq. 6) for large values of σ is given by

$$W(\sigma) = W_o + k\sigma \qquad (8)$$

The intercept W_o and slope k depend on the breadth of the diffraction profile and hence on the imperfections. The slope of (8) is inversely proportional to the harmonic mean of the average crystallite or domain size and the mean distances between mistakes and dislocations (D_{ev}) all measured in a direction perpendicular to the reflecting planes. Thus :

$$k = \frac{1}{\pi^2 D_{ev}} \qquad (9)$$

Strain does not normally affect the slope of the variance range curve. Dislocations introduce a logarithmic term [10] in the range dependent term of the variance but the introduced curvature is normally small for the first order.

2. PROCEDURE

It has been shown by Wilson [2] that the variance method and the Fourier technique of Warren and Averbach must give identical estimates of the effective crystallite or domain size (D_e) ; any difference between the two values of D_e can only be due to inadequate data, incorrect analysis or invalid assumptions. In another paper [4], D_{eV} and D_{eF} have been compared for various Pt film prepared in various experimental conditions. A good correlation between sizes estimated by the two methods has been found. However a systematic difference of about 18 % has been calculated on average with $D_{eV} > D_{eF}$. The discrepancy has been ascribed to a truncation of the 222 line which is overlapped by the tail of the 311 line.

The single line method used in this study is then as follow
(a) The variance slope for the (111) relection is used to obtain the effective particle size D_{eV} ;
(b) $A_S(L)$ is calculated from (3) with $D_{eF} \equiv D_{eV}$;
(c) The rms strain is obtained from (2) and (4).
Equation (2) becomes :

$$A(L) = A_S(L) \left(1 - 2\pi^2 <\varepsilon_L^2> \frac{L^2 h_o^2}{a^2}\right)$$

$<\varepsilon_L^2>$ can be determined as a function of L with no assumption about a particular dependence of $<\varepsilon_L^2>$ on L.

The method will be applied to the study of the thickness effect on the microstructure of sputtered platinum thin films. X-ray studies of evaporated thin films have been investigated by several workers (for example see Sen Gupta [11]) but less work has been done on sputtered thin films [7].

3. EXPERIMENTAL

3.1. Preparation of thin film

Thin films of Pt were prepared by sputtering onto glass microscope slides. The sputtering deposition took place in an ultra high vacuum chamber evacuated by a turbomolecular pumping unit. The platinum target of purity 99.99 %, was a water cooled disk of diameter 4 cm. The voltage during the deposition was 2400 V, the current 6.5 mA and the argon pressure 6×10^{-2} torr. The holding substrate temperature was maintained at 200°C. The thickness of the films was measured by X-ray fluorescence. A calibration curve was obtained from the intensity of the platinum L_α line and the film thickness measured by the Tolansky technique. Other experimental details are given elsewhere [4].

3.2. X-ray measurements

The X-ray diffraction profiles for the thin film were recorded with a Siemens diffractometer having Bragg-Brentano geometry. The aperture of the divergent slit was 1° and that of the receiving slit was 0.10°. Soller slits were inserted in the incident and the diffracted beams. The diffractometer was operated in the step-scanning mode with a step length of 0.05° (2θ). A well annealed platinum foil was used to measure the instrumental broadening. In the case of the Warren-Averbach analysis, the Fourier coefficients were calculated by a least-squares method [12], normalized so that the total intensity under the profile is unity and finally corrected by the Stokes method [1]. In the variance method, the slope and intercept of the variance range function, were obtained using the program of Edwards and Toman [13] with the inclusion of the background correction of Langford. Corrections to the experimental values for the contribution from instrumental broadening, spectral broadening, non additivity, truncation and curvature [11] were made.

4. RESULTS AND DISCUSSION

The effective particle size (D_{ev}) was calculated from the variance slope coefficient for platinum thin films at various thickness. Results are plotted in Fig. 1. With an increasing thickness, D_{ev} increases from 40 Å to 100 Å until a thin film thickness of 1500 Å, then D_{ev} remains constant and independant of the thickness.

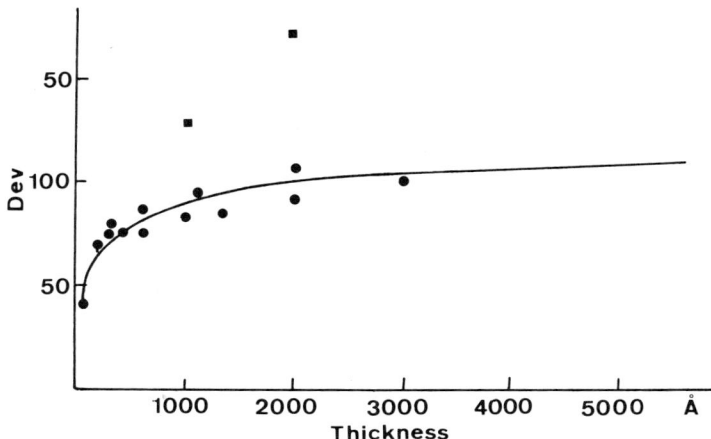

Fig. 1. Effective particle size calculated from the variance slope coefficient as a function of the thin film thickness ( ) as deposited ; ( ) after annealing in air at 400°C.

The rms strain for L = 14 Å was also calculated. If the background level is taken too high (truncature effect) the $A_s(L)$ curve shows a hook at small L and the initial slope has no significance [1]. We note that truncature effects are specially important (see above) in the second order but they are likely weak for the first order and also in our work where the films present an orientation with the (111) plane parallel to the substrate. Nevertheless to avoid any mistake in the first values of A(L), the $< \varepsilon_L^2 >^{1/2}$ with very low values of L were disregarded. As we showed above, L $_{max}$ may not be too large because eq. (3) must be applied so we selected $< \varepsilon_L^2 >^{1/2}$ for L = 14 Å arbitrarily (Fig. 2).

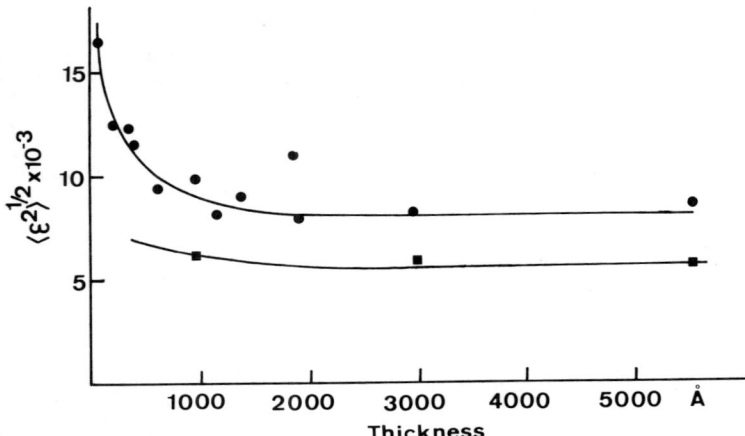

Fig. 2. Rms strain as a function of the thin film thickness ( ) as deposited ; ( ) after annealing in air at 400°C.

The microstrain decreases with an increase of L (Fig. 3). Usually, the large microstrain at short distances are attributed to strain field of the dislocations. However, due to presence of additive inclusions in sputtered films [14] other strain fields are also expected [15]. In addition, strain relief by dislocation movement is also restricted with these inclusions [15]. Microstrains in sputtered film should be larger than in vapor deposited thin films [7].

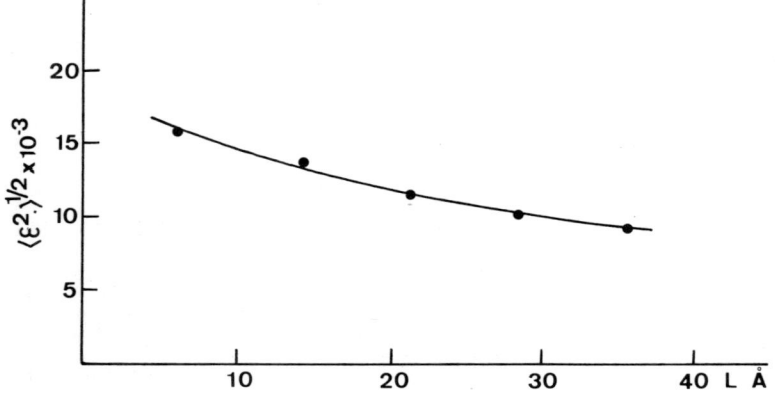

Fig. 3. Rms strain against L where L is a length in the direction perpendicular to the reflecting planes. Thin film thickness ; 400 Å.

Microstrain and effective particle size show a dependence on the film thickness. Contrary to the D_{ev} variation, the microstrain is rather high at low thickness, it decreases with an increasing thickness until around 1500 Å.
Above 1500 Å, the microstrain like D_{ev} is constant. After the samples were annealed for 24 h. at 400°C in air, the rms strain is reduced (Fig. 2) and D_{ev} is increased (Fig. 1).

These results show below 1500 Å that the microstructure of thin films depends strongly on the thickness. In this thickness range, high strain and low effective particle size can be explained by the lattice mismatch of platinum films on glass substrates and also by substrate induced thermalstrain. In sputtering deposition, the surface temperature of the substrate is determinated by three factors :
(1) the temperature of the substrate holder ;
(2) the temperature of the condensing atoms ;
(3) the temperature produced by plasma particle bombardment (electrons and ions).

Only factor (1) is known so a quantitative estimate of the substrate induced thermalstrain is not possible. The interfacial region between the films and the substrate is strained differently from the rest of the film. The strain is defined by :

$$\varepsilon = \frac{d_n - d_o}{d_o}$$

where d_n is the interplanar spacing under stress and d_o is the spacing of the same planes in the absence of stress. It is to point out that in our measurements the strains refer always in a direction perpendicular to the diffraction planes and also perpendicular to the substrate due to the geometry of our diffractometer. Corresponding to ε_L, the interplanar spacing is the average over L of all interplanar spacing d_i^L in all the crystallites. Each d_i is dependent on bulk and interfacial strain. Since the surface strain is assumed to decrease rapidly as one goes away from the surface, the interplanar spacing must change with the thickness of the crystallite.
Strain can be relieved by dislocation movements but the efficiency decreases with decreasing thickness [15]. Thinner films lead to larger residual rms microstrain. The thickness effect is expected until 1000 Å [16]. Dislocations are also found to appear and to grow with the thickness in the coalescence stage [17]. They appear when slightly misoriented adjacent islands join up [18].
Below a thickness of 1500 Å, stress and small crystal size are mainly due to the substrate influence and to the coalescence mode. Above 1500 Å, the microstructure does not depend on the thickness, the high microstrain level is likely stabilized by impurities in the film.

In conclusion, the influence of thickness on the microstructure of sputtered platinum thin films has been studied by an X-ray method. The method is justified by the fact that the effective particle size calculated from the variance slope coefficient is identical to the effective particle size calculated from the Warren-Averbach method. However the use of this method is restricted to an L range for which one can assume that the relation (3) holds.

5. BIBLIOGRAPHY

1. B.E. WARREN, X-Ray Diffraction (Addision-Wesley, Reading, Massachussetts, 1969).
2. A.J.C. WILSON, Mathematical Theory of X-Ray Powder Diffractometry (Centrex, Eidhoven, 1).
3. W.L. SMITH, J. Appl. Cryst. $\underline{5}$ (1972) 127.
4. M. HECQ, A. HECQ and J.I. LANGFORD, J. Appl. Phys. $\underline{53}$ (1982) 421.
5. R. DELHEZ, T.H. de KEIJSER and E.J. MITTEMEIJER, Proceedings of Symposium on Accuracy in Powder Diffraction, N.B.S. Special Publication $\underline{567}$ (1980) 213.
6. J. MIGNOT and D. RONDOT, Acta Met. $\underline{23}$ (1975) 1321.
7. T. ADLER and C.R. HOUSKA, J. Appl. Phys. $\underline{50}$ (1979) 3288.
8. Th. h. de KEIJSER, J.I. LANGFORD, E.J. MITTEMIEJER and A.B.P. VOGELS, J. Appl. Cryst. $\underline{15}$ (1982) 308.
9. J.I. LANGFORD, J. Appl. Cryst. $\underline{15}$ (1982) 315.
10. A.J.C. WILSON, Proc. Phys. Soc. London $\underline{82}$ (1963) 986.
11. S.K. HALDER, SUCHITRA SEN and S.P. SEN GUPTA, J. Phys. D. Appl. Phys. $\underline{9}$ (1976) 1867.
12. A. KIDRON and R.J. DE ANGELIS, Acta Cryst. $\underline{A27}$ (1971) 596.
13. H.J. EDWARDS and K. TOMAN, J. Appl. Cryst. $\underline{4}$ (1971) 323.
14. J.J. CUOMO and R.J. GAMBINO, J. Vac. Sci. Technol. $\underline{14}$ (1977) 152.
15. A. GANGULEE, J. Appl. Phys. $\underline{43}$ (1972) 867.
16. R.W. VOOK in "Epitaxial Growth" ed. J.W. Mattews (Ac. Press. New-York 1975).
17. C.A. NEUGEBAUER in "Handbook of Thin Film Technology" ed. L.I. MAISSEL and R. GLANG (Mac Graw-Hill, New-York 1970).
18. B. LEWIS and J.C. ANDERSON in "Nucleation and Growth of Thin Films" (Ac. Press., New-York 1978) 442.
19. N.C. HALDER and C.N.J. WAGNER in "Advances in X-Ray Analysis" volume 9 ed. G.R. Mallett, M. Fay and W.M. Mueller (Plenum Press, New-York, 1966) 91.
20. A. GANGULEE, J. Appl. Cryst. $\underline{7}$ (1974) 434.

SURFACE ELECTRONIC FUNCTION PROPERTIES CHARACTERIZATION FROM DEFECT HETEROGENEITY DOMINATED SPECULAR/GLANCING/GRAZING VERSUS BULK TRANSMISSION SMALL-ANGLE-SCATTERING(SAS) DIFFRACTION-PATTERN VIA THE STATIC SYNERGETICS ALGORITHM/EXPERIMENTAL MODEL

EDWARD SIEGEL
Static Synergetics Research Ltd.,183-14thAvenue,San Francisco,CA.94118

ABSTRACT

The ability of Static Synergetics,the universal,reversible,scalable mathematical algorithm/(used as an)experimental model,connecting external radiation small-angle-scattering diffraction-pattern/static structure factor S_{SAS}(k),dominated by defect heterogeneities/clumps/clusters,to electronic (magnetic,mechanical) complex,frequency-dependent electrical,dielectric, noise,optical...Function properties,can be enhanced by utilizing a comparison/difference of bulk transmission versus surface specular/glancing/grazing incidence small-angle-scattering diffraction-patterns.

INTRODUCTION

Static Synergetics is a new subject.[1] It connects Pattern to Function in a universal,reversible,scalable mathematical algorithm utilized as an experimental model,where **experimentally** measured Patterns are input yielding Function properties output,experimentally measurable for verification,or the reverse. At least four universal phenomena are understood by this unification/synthesis of the universal classic Brillouin-like generalized-disorder collective-boson mode-softening universality-principle[2] (G...P)with the universal infra-red divergence principle[3] of Wigner and Dyson,the static,time-independent limit of Haken's dynamic synergetics:
- Universal Functional Properties(Experimentally Measured):
  (1) . 1/f Flicker Noise Power Spectrum[3]
  (2) . Dielectric(Magnetic,Mechanical/Viscoelastic) Susceptibilities[4]
- Universal Derived Functional Properties(Experimentally Measured):
    - A.C.Electrical Conductivity
    - D.C.Electrical Conductivity
    - Dielectric Constant(Function) (Relative Dielectric Permeability)
    - Dielectric Dispersion
    - Dielectric Dissipation(Loss)
    - Dielectric Loss Tangent
    - Dielectric Response
    - Capacitance
    - Impedance
    - Admittance
    - Conductance
    - Current Flicker Noise Power Spectrum
    - Voltage Flicker Noise Power Spectrum
    - Optical Refractive Index, Extinction Coefficient, Refelctivity
  (3) . Universal Anderson Localization(criteria)
  (4) . Universal Existence of Multi-Level System & Associated Low Temperature/Frequency Thermal,Acoustic,...Property Anomalies (in Glasses, Powders,Composites,...)

These are <u>COMPLEX</u> and <u>FREQUENCY-DEPENDENT</u>($w \leq 10^{11}$Hz.) Function properties that are <u>corrections</u> to the bulk material Function properties due to processing introduced defect heterogeneities/clumps/clusters.
How can we use these universal Function properties,explicit functions of process introduced defect disorder distribution/Pattern(to be explicitly calculated here)to analyze real-time,interactively,providing feedback on material,device or system processing(**introduced defect distribution**) opt-

imizing yield? The key is the symmetry-breaking the defect disorder distribution performs the material,device or system on whatever scale they occur,at some set of heterogeneous-disorder heirarchy wavevectors in the external radiation scattering-diffraction pattern-static structure factor $S(\underline{k})$ ie.at some configuration-space distribution of scales.

STATIC SYNERGETICS

Static Synergetics allows derivation of explicit equations relating defect generalized-(heterogeneous)-disorder scattering-diffraction pattern morphology ie. static structure factor $S(k;\theta)$, implicitly a function of(through the particular process model)processing parameter set θ:deposition temperature T,deposition pressure P,concentration of jth solute/dopant species X_j,grain size R,deposition voltage V,deposition current J...in some usually unknown functional form. Static Synergetics embodies three properties:

Universality:means that Static Synergetics is applicable quite generally to any heterogeneous-disorder symmetry-breaking defect distribution introduced during processing,whatever its type and however produced.

Reversibility:means that the Static Synergetics algorithm can be used as an experimental model either forward or backward,ie.to calculate (experimentally measured)output Function properties from (experimentally measured)external radiation diffraction-pattern/static structure factor $S(k;\theta)$ (or from Fourier transform configuration-space defect Pattern morphology photograph $g(r;\theta)$) input;or visa versa. In either direction Static Synergetics algorithm works!

Scalability:means that the heterogeneous-disorder(defect distribution heirarchy)so critical for the infra-red divergence principle part of Static Synergetics (as seen in the small-angle-scattering(SAS)dominating part of $S(k;\theta)$) can be considered as low wavevector(long wavelength)scaled with respect to total system,device or material sample size/diameter.Thus microscopic defect distributions within a material may be treated equivalently to material defect distributions within a larger device,which in turn is equivalent to treating device defect(ive) distributions within a circuit/system,...

We list the Function/functional properties-Pattern/morphology diffraction-pattern/static structure relations derived previously[5],where $S(k;\theta)$ is related to defect configuration-space photomicrograph by

$$S(\underline{k};\theta) = \int \bar{g}(\underline{r};\theta) \exp(-i\underline{k}.\underline{r}) \, d\underline{r} \qquad (1)$$

and we abbreviate inverse density of states/group velocity,totally containing diffraction-pattern/static structure factor contribution to properties factor

$$\left[2k \, S_{SAS}(k;\theta) - k^2 \, \partial S(k;\theta)/\partial k \right] / S_{SAS}^2(k;\theta) = [\cdots] \qquad (2)$$

the critical exponent in universal $1/f$ flicker noise power spectrum and relaxation response susceptibilities $n=g(w)/w=1/v_g(w).w=1/v_g(\underline{k}).(\underline{k}^2/S(\underline{k})$

COMPLEX , FREQUENCY DEPENDENT , DIFFRACTION-PATTERN DEPENDENT , IMPLICITLY DEPOSITION PARAMETER (PROCESS) DEPENDENT :

A.C.Electrical Conductivity:
$$\sigma_{A.C.}(w;\theta) \cong \sigma_{D.C.}(0;\theta) + \epsilon_0 / w^{(2m|V|^2/wh^2)}[\cdots] \qquad (3)$$

Dielectric Constant (Function) (Relative Dielectric Permeability):
$$\epsilon(w;\theta) \cong \epsilon_0 \left[1 + (1+i)/ w^{(2m|V|^2/wh^2)}[\cdots] \right] \qquad (4)$$

Dielectric Dispersion:
$$\epsilon'(w;\theta) \cong \epsilon_0 \left[1 + 1/ w^{(2m|V|^2/wh^2)}[\cdots] \right] \qquad (5)$$

Dielectric Dissipation (Loss):
$$\epsilon''(w;\theta) \cong \epsilon_0 / w^{(2m|V|^2/wh^2)}[\cdots] \qquad (6)$$

Dielectric Loss Tangent:
$$\tan\delta(w;\theta) \cong \left\{ 1/w^{(2m|V|^2/wh^2)}[\cdots] \right\} / \left\{ 1 + 1/ w^{(2m|V|^2/wh^2)}[\cdots] \right\} \qquad (7)$$

Dielectric Response:
$$\chi''(w;\theta) / \chi'(w;\theta) \cong \cot(n\pi/2) \neq (w;0) \text{ (independent of w and } \theta) \qquad (8)$$

Dielectric Relaxation Response Susceptibilities:
$$\chi'(w;\theta) \cong 1/w^{n(w;\theta)} \; ; \chi''(w;\theta) \cong 1/w^{n(w;\theta)} \; ; \; n = g(w)/w \qquad (9)$$

Capacitance: (over area A and width W)
$$C(w;\theta) \cong (A/W)\,\epsilon_o \left[1 + (1+i)/w^{(2m|V|^2/w\hbar^2)}[\cdots]\right] \quad (10)$$
Impedance:
$$Z(w;\theta) \cong R_o + 1/iw(A\epsilon_o/W)\left[i/w^{(2m|V|^2/w\hbar^2)}[\cdots] + \{1+1/w^{(2m|V|^2/w\hbar^2)}[\cdots]\}\right] \quad (11)$$
Conductance:
$$G(w;\theta) \cong (A/W)\,\epsilon_o / w^{(2m|V|^2/w\hbar^2)}[\cdots] \quad (12)$$
Admittance:
$$Y(w;\theta) \cong 1/Z(w;\theta) = 1/(11) \quad (13)$$
Optical Refractive Index:
$$N(w;\theta) \cong \epsilon_o^{1/2}\left\{1 + (1+i)/w^{(2m|V|^2/w\hbar^2)}[\cdots]\right\}^{1/2} \quad (14)$$
For nonzero extinction coefficient one has
$$N(w;\theta) + i\,K(w;\theta) = \epsilon^{1/2}(w;\theta) = [\epsilon'(w;\theta) + i\epsilon''(w;\theta)]^{1/2} \quad (15)$$
causing difficulty in deriving a simple functional law such as (14) without further equation relating $N(w;\theta)$ optical refractive index to $K(w;\theta)$ optical extinction coefficient

1/f Flicker Noise Universal Power Spectrum:
$$P(w;\theta) \cong 1/w^{n(w;\theta)}; n = g(w)/w$$
$$\cong 1/w^{(2m|V|^2/w\hbar^2)}[\cdots] \quad (16)$$

1/f Voltage Flicker Noise Power Spectrum:
$$P_V(w;\theta) \cong (4k_B T/g)/w^{(2m\,V^2/w\hbar^2)}[\cdots] \quad (17)$$
in terms of geometrical factor g, where V is the voltage applied by electric field $\underline{E} = -\underline{\nabla}V$, where m=ħ=1 in atomic units. Very important is the realization that the approximate "$\cong$" rather than actual equalities indicates that the Static Synergetics algorithm is for FM properties only; the amplitude, the numerator of each expression, is not universal and is not known explicitly as a function of defect symmetry-breaking diffraction-pattern/static structure factor $S(k;\theta)$ though Dutta and Horn, and Hooge point out that the numerator is a constant, Hooge's constant $\gamma = 2 \times 10^{-3}$, dimensionless, and seemingly universal AM for semiconductors and metals.

Generically (3)-(17) can be summarized as
$$Q(k,w;\theta) = Q(w;\theta) = Q(g(w;\theta)/w = Q([\cdots]) = Q(S(k;\theta)) = Q(\theta) = Q(n(\theta)) \quad (18)$$
for the generalized Function in terms of generalized Pattern Fourier transform diffraction-pattern $S(k;\theta)$ and hence in terms of plasma deposition parameter set $\theta = \{T, P, X_i, R, \ldots\}$ implicitly. These provide the coupling to process model specifics for particular plasma deposition systems equipment.

Static synergetics covers the following frequency bands with FM Function calculation as explicit function of Pattern(diffraction-pattern) and parameter set for deposition: **BAND**

very low frequency	(V.L.F.)	: 10^{-5}Hz. – 10^0Hz.
low frequency	(L.F.)	: 10^0Hz. – 10^{+3}Hz.
audio frequency	(A.F.)	: 10^{+3}Hz. – 10^{+5}Hz.
radio frequency	(R.F.)	: 10^{+5}Hz. – 10^{+8}Hz.
short wave	(S.W.)	: 10^{+8}Hz. – 10^{+10}Hz.
microwave-infra-red	(M.W.)-(IR)	: 10^{+10}Hz. – 10^{+12}Hz.

With a conservative cut-off in validity and applicability at the quantum/inter-band transition limit of approximately 10^{11}Hz. in the infra-red (IR) band, Static Synergetics spans some 16 decades/orders of magnitude in applicability to calculation of relative FM electronic and optical Function properties explicitly as functions of small-angle-scattering(SAS) defect diffraction-pattern. Above approximately 10^{11}Hz. absorption edges of quantum interband transitions occur and are untreatable by Static Synergetics, though it has recently been proposed that local coordination number EXAFS data be tried as an input to calculate higher frequency properties. But we caution that with no logical theory of inclusion of quantum processes in Static Syhergetics, an essentially long wavelength limit approach, such attempts to extension to still higher frequency bands must be open to rigorous scrutiny; the linear density of states of the infra-red divergence principle becomes "universally" Gaussian with $n \neq 1$ for higher energies. Yet the possibility exists for extension since ^{2}the linear density-of-states is the low frequency limit of a Gaussian density-of-states in general, first having been utilized for nuclear energy level structure in the Mev energy range, far above those applicable for solids and other condensed matter.

Static Synergetics summarizing equation set for generic Function properties as explicit function of Pattern morphology diffraction-pattern/static structure factor $S(\underline{k};\theta)$, itself some implicit function of plasma deposition processing parameter set $\theta = \theta(T,P,X_i,R,\ldots)$ known by independent plasma process model on a particular deposition system or by trial and error :

$$Q(\theta) = Q(w;\theta) = Q(\mathbf{X}'(w;\theta), \mathbf{X}''(w;\theta)) \cong Q(1/w^{n(S(k;\theta))}) \quad (3)-(17) \quad (18)$$

$$n(\theta) = n(S(k;\theta)) = |\mathbf{v}|^2 g(w)/w \cong |\mathbf{v}|^2 / v_g(w) \cdot w \cong |\mathbf{v}|^2 / w \cdot \partial w(k;\theta) / \partial k$$
$$\cong |\mathbf{v}|^2 / w \cdot \partial(k^2/S(k;\theta))/\partial k \quad \text{(Infra-Red Divergence Principle)} \quad (19)$$

$$g(\theta) = g(w;\theta) \cong 1/\partial w(k;\theta)/\partial k \cong 1/\partial(k^2/S(k;\theta))/\partial k \quad (20)$$

$$w(\theta) = w(k;\theta) = k^2/S(k;\theta) \quad \text{(First Principle of Static Synergetics)} \quad (21)$$

$$S(\theta) = S(k;\theta) = \iiint \bar{g}(r;\theta) e^{-ikr} d^3r = S^{SAS}(\theta) \quad \text{(Diffraction-Pattern)} \quad (22)$$

This will be the basic equation set of the Static Synergetics mathematical algorithm, usable as an experimental model with proper Pattern morphology diffraction-pattern $S(k;\theta)$ input resulting in Function property $Q(\theta)$ output, or reversibly with proper Function property $Q(\theta)$ input resulting in Pattern morphology diffraction-pattern $S(k;\theta)$ output, the latter always dominated by its small angle scattering (SAS) part, corresponding to heterogeneous-disorder clumps or clusters of defects.

We summarize equation set (18)-(22) in Figure 1 and in the following nested parentheses parametric equation set:

$$\text{OUT} \leftarrow Q(\theta) = Q(n(\theta)) = Q(n(g(\theta))) = Q(n(g(w(\theta)))) = Q(n(g(w(S(\theta))))) \quad (23)$$
$$n(\theta) = n(g(\theta)) = n(g(w(\theta))) = n(g(w(S(\theta)))) \quad (24)$$
$$g(\theta) = g(w(\theta)) = g(w(S(\theta))) \quad (25)$$
$$w(\theta) = w(S(\theta)) \quad (26)$$
$$S(\theta) = S(\bar{g}(\theta)) \leftarrow \text{IN} \quad (27)$$

These form the algebraic heirarchy of equations in the forward use of the Static Synergetics mathematical algorithm. Figure 2 summarizes their method of use, as well as inverse function nested parentheses equations of the reverse mathematical algorithm (the reversibility property), obtained by reading nested set (23)-(27) with inverse functions from bottom to top (inverse = backwards):

$$\text{OUT} \leftarrow g_{-1}(\theta) = \bar{g}_{-1}(S(\theta)) \quad (28)$$
$$S_{-1}(\theta) = S_{-1}(w(\theta)) \quad (29)$$
$$w_{-1}(\theta) = w_{-1}(g(\theta)) \quad (30)$$
$$g_{-1}(\theta) = g_{-1}(n(\theta)) \quad (31)$$
$$n_{-1}(\theta) = n_{-1}(Q(\theta)) \leftarrow \text{IN} \quad (32)$$

We must strongly emphasize that this algorithm, culminating explicitly in one equation into which these nested sequences collapse, Function properties equation (3)-(17), depending upon Function property desired to monitor continuously as function of processing variable set θ, always contains term (2), so that only two multiplications and one derivitive need be done with input data reducing data manipulation to a simple minimum. The input diffraction-pattern $S(k;\theta)$ may utilize whatever external-radiation wavelengths appropriately resonate/match geometrically the average inter-defect clump size scale; X-rays, electrons, neutrons, laser photons, microwaves and even ultrasound(conducted in the substrate) are all candidated. However, since the SAS part of the diffraction-pattern will dominate, it must be measured, entailing intense beams of external radiation(synchrotron light/X-ray sources, lasers,...) since unfortunately $S^{SAS}(k;\theta)$ is hardest to resolve. But with development of small X-ray cameras for in-system use, currently being done for a host of deposition system types, such sources may soon be readily available "off-the-shelf" accessories. Very positively, the time scales for measurement can be extremely short; with current state-of-the-art synchrotron light sources, $S_{SAS}(k;\theta)$ is measurable in a few milliseconds, with possible extension into the microsecond regime with the advent of new wiggler magnets. Since the mathematical algorithm (23)-(27) or its reverse (28)-(32) collapses to a single expression (3)-(17) for each desired Function property to be computed, and since all of these involve function (2) with only simple algebra and one derivitive, data processing times of a few millisecinds are estimated, totally sufficient for repeated sampling of $S_{SAS}(k;\theta)$ during material, device...processes. Development of in-system X-ray sources and detectors should open this technique to use in a wide spectrum of in-plant material, device,...fabrication processing.

SURFACE VERSUS BULK DEFECT HETEROGENEITY PATTERN CORRECTIONS TO PROPERTIES

Surface versus bulk defect heterogeneity distribution/Pattern corrections to material intrinsic Function properties can be differentiated by sampling/measuring $S_{SAS}(k;\theta)$ experimentally by specular/glancing/grazing incidence small-angle-scattering off the surface(or near surface layer) versus bulk transmission through the material. This requires two nearly orthogonal incident external radiation beams with detectors for each arranged in the forward scattering geometry. This crossed-forward scattering geometry will permit simultaneous separate measurement of surface defect distribution/Pattern(dominated by heterogeneity/clumping/clustering of whatever defects have been process introduced, however they were produced) $S_{SAS}^{surf}(k;\theta)$ and of bulk defect distribution(also dominated by heterogeneity/clumping/clustering of whatever defects have been process introduced, however they were produced) $S_{SAS}^{bulk}(k;\theta)$. This measurement could be sequential as opposed to simultaneous, but the latter mode of measurement is to be preferred lest the defect distributions/Patterns alter between measurements; an "instantaneous" sampling measurement set is the desired goal.

The difference between these two small-angle-scattering diffraction-patterns, surface versus bulk, the differential SAS diffraction-pattern,

$$\Delta S_{SAS}(k;\theta) = \left[S_{SAS}^{bulk}(k;\theta) - S_{SAS}^{surf}(k;\theta)\right] = \left[(d\sigma/d\Omega)_{SAS}^{bulk} - (d\sigma/d\Omega)_{SAS}^{surf}\right] \quad (33)$$

the difference between forward scattered external radiation current per solid angle in the bulk transmission path versus that in the surface specular/glancing/grazing incidence path, $\Delta(d\sigma/d\Omega)_{SAS} = \Delta\left(\frac{d\sigma}{d\Omega}(\phi \approx 0)\right)$, can be utilized to differentiate between surface layer versus bulk Function properties for any of the Function properties computable from the Static Synergetics Algorithm (3)-(17)

$$\Delta Q(k,w;\theta) = \left[Q^{bulk}(k,w;\theta) - Q^{surf}(k,w;\theta)\right] = \left[Q(S_{SAS}^{bulk}(k;\theta)) - Q(S_{SAS}^{surf}(k;\theta))\right]$$

$$= \left[Q(d\sigma/d\Omega)_{SAS}^{bulk} - Q(d\sigma/d\Omega)_{SAS}^{surf}\right] = \left[Q\left(\frac{d\sigma}{d\Omega}^{bulk}(\phi=0)\right) - Q\left(\frac{d\sigma}{d\Omega}^{surf}(\phi=0)\right)\right] \quad (34)$$

$$= Q\begin{pmatrix}\text{transmitted forward scattered}\\ \text{external radiation current}\\ \text{per unit solid angle}\end{pmatrix} - Q\begin{pmatrix}\text{specular/glancing/grazing}\\ \text{forward scattered external}\\ \text{radiation current per}\\ \text{unit solid angle}\end{pmatrix}$$

which can be rewritten by factoring as

$$\Delta Q(k,w;\theta) = Q\left[S_{SAS}^{bulk}(k;\theta) - S_{SAS}^{surf}(k;\theta)\right] = Q\left[(d\sigma/d\Omega)_{SAS}^{bulk} - (d\sigma/d\Omega)_{SAS}^{surf}\right]$$

$$= Q\left[\left(\frac{d\sigma}{d\Omega}^{bulk}(\phi=0)\right) - \left(\frac{d\sigma}{d\Omega}^{surf}(\phi=0)\right)\right] \quad (35)$$

Depth profiling of Static Synergetics Algorithm computed Function propperties (3)-(17) is possible using this technique. The surface specular/glancing/grazing incidence external radiation forward/small-angle-scattering diffraction-pattern $S_{SAS}^{surf}(k;\theta) = S_{SAS}^{surf}(k;\theta,z)$ is a function of depth of external radiation penetration during the surface specular/glancing/grazing small-angle-scattering diffraction-pattern. By adjusting external radiation angle incidence with respect to surface plane ϕ from 0, while keeping its compliment $\pi-\phi$ as near π as possible by moving the location of the diffracted radiation detector to insure small-angle-scattering(in as forward a direction as possible), the relation between angle of incidence and depth z z = 1 sin ϕ in terms of path length for external radiation 1, allows rewriting of relative Function property of surface layer versus bulk computed via the Static Synergetics Algorithm in terms of external radiation penetration depth/surface layer thickness z as

$$Q(k;w;\theta;z)^{surf} = Q\left[\frac{d\sigma}{d\Omega}\left(Arcsin \frac{z}{1}\right)\right]_{SAS}^{surf} \qquad (36)$$

so that the depth profiled differentiated surface versus bulk Function properties (3)-(17) can be reexpressed from (34) and (35) as

$$\Delta Q(k;w;\theta;z) = Q\left[\frac{d\sigma}{d\Omega}^{bulk}(\phi=0) - \frac{d\sigma}{d\Omega}^{surf}(Arcsin \frac{z}{1})\right] \qquad (37)$$

This relation will hold until bulk absorption of external radiation intensity along path length l precludes further depth profiling by totally absorbing the external radiation; this maximum depth z_{max} of profiling will vary with type of external radiation chosen and type of material being probed (atomic number, atomic weight, density,...). So, with clever manipulation of external radiation source-detector geometry with respect to material surface while insuring as much as possible forward/small-angle-scattering required as input to the Algorithm, differentiated surface versus bulk depth profiling of the Function properties (3)-(17) computable using the Algorithm, their defect distribution/Pattern deviations from intrinsic values, is measurable.

PROCESSING FLOWCHART AND MATHEMATICAL ALGORITHM

In Figure 3 we flowchart the use of the Static Synergetics Algorithm as the Pattern-Function direct connection it is. In the forward mode, input Pattern yields output Function(s). In the reverse mode, input Function(s) yields output Pattern. Of course, the Pattern used is actually the actual Pattern Fourier transform small-angle-scattering diffraction-pattern $S_{SAS}(k;\theta)$.

Given the universal, scalable, reversible properties of this Static Synergetics Algorithm, used as an experimental model/mathematical algorithm, how can we utilize it in practical materials, device,...processing? In addition, given the ability to differentiate surface versus bulk Function properties, how can this be used to augment application of the Algorithm to materials or device processing?

The Algorithm can be utilized in two ways. Firstly, with rapid small-angle-scattering diffraction-pattern measurements, it can function as a <u>real-time feedback quality-analysis/assurance</u>(Q.A.) <u>during</u> processing. As the processing technique, whatever it is, introduces defect distributions/Patterns accidentally, whatever their specific processes/mechanisms of formation, migration and agglomeration are, the Algorithm permits evaluation of Function properties (3)-(17) modification from the defect-free intrinsic Function properties of the material or device due to the formation of the defect Pattern. Secondly, when utilized in parallel with a specific process model(which includes the material or device specific mechanisms/processes specifically excluded from consideration in the Algorithm), it can function as a <u>real-time feedback interactive quality-control</u>(Q.C.)<u>during</u> processing.

Q.A. is achieved by performing a rapid small-angle-scattering diffraction-pattern measurement with a suitable external radiation source and type, chosen to resonate geometrically with expected defect and inter-defect dis-

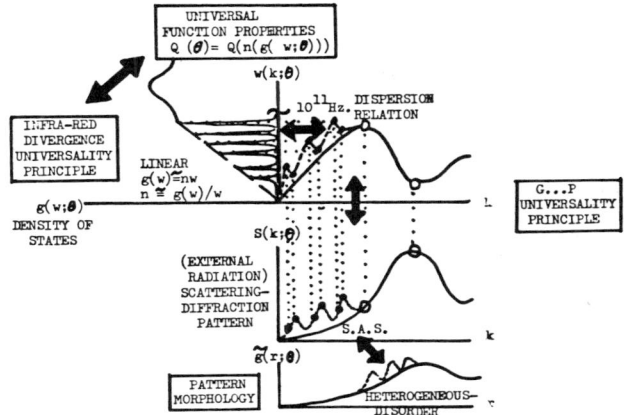

Figure 1. Pictorial Summary of Universal, Reversible, Scalable, Static Synergetics as a Union of G...P with Infrared Divergence Universality Principles to Relate Pattern Morphology to Universal Function (Dielectric, Electronic, Flicker Noise... (Magnetic, Mechancial...)) Properties

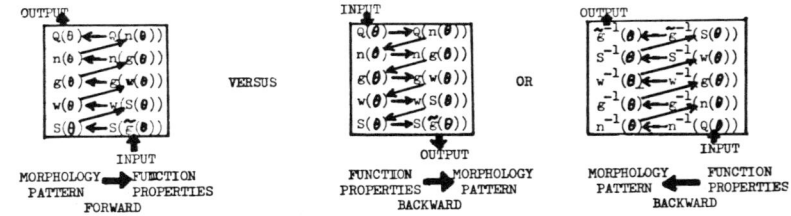

Figure 2. Static Synergetics Universal, Reversible, Scalable, Mathematical Algorithm. Reversibility Allows Use of Algorithm and Flowcharts in Either Direction as Experimental Model

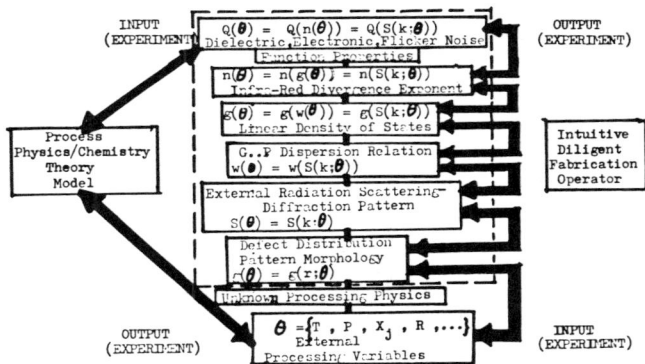

Figure 3. Processing Flowchart of Static Synergetics Mathematical Algorithm as Experimental Model

tances specific to that material, or taking the Fourier transform configuration-space photomicrograph and then Fourier transforming it, and then inputting it into the Algorithm. The output, one or more of the designer specified Function properties (3)-(17), can then be compared with designer specifications on the limits of tolerance of the relevant Function properties(and can as well be independently measured to verify the Algorithm computed values) to ascertain how successful the processing has been in producing the desired material or device properties. That this can be done real-time during the processing provides an exciting new tool in processing of materials and devices, heretofore unavailable in materials characterization.

Q.C. is achieved by performing the previously outlined application of the Algorithm backwards(using the reversibility property)also, measuring the Function properties of interest experimentally directly, and using them as input to backcompute the defect distribution/Pattern as output. Then if these Function properties are not within designer specifications/tolerances desired, one understands which defect distribution/Pattern heterogeneities are the cause. Steps can then be taken to remedy this situation by modification of the processing through alteration of the process variable/parameter set θ values chosen during the previous processing duration. To perform Q.C. using the Algorithm, the Algorithm must be used in parallel with a mechanism/process specific, material/device specific, processing technique specific process model embodying the microscopic physics/chemistry of that processing technique. But most importantly, it must be stressed that none of these specifics are part of the Algorithm, nor are they required to impliment the Algorithm for Q.A. applications. Q.C. applications however, do require a detailed process model embodying these microscopic specifics.

This is the major advantage of the Algorithm. It forms a universally applicable(to any and all processing) flexible, versatile tool for Q.A. whose real-time success is dependent upon the ingenuity of the processor. In its Q.C. mode of use, it however must require the specific process model of necessity, since only that process model can tell how to vary the process variables/parameters to modify the processing introduced defect distribution/Pattern. What then does the Algorithm do by itself? It allows non-contact sampling of the Function properties a material or device is being processed to perform/possess during the processing by diffraction-pattern measurement of defect distribution/Pattern of heterogeneities/clumps/clusters which dominate the small-angle-scattering regime.

Additionally, the ability to differentiate surface versus bulk Function properties (34)-(35) from surface versus bulk small-angle-scattering diffraction-patterns of suitable chosen(type and spectrum) external radiation(33) can allow real-time feedback Q.A. evaluation of the success and differentiation of surface versus bulk materials or device processing techniques. And, with suitably differentiated specific process models, real-time feedback interactive Q.C. differentiation between success of surface versus bulk processing techniques can be performed. Lastly, the ability to depth profile(36)-(37) material or device Function properties by depth profiling the small-angle-scattering diffraction-pattern of suitably chosen(type and spectrum) external radiation could allow depth profiled real-time feedback Q.A. and interactive Q.C. during processing.

1. H.Haken, Series in Synergetics, Springer-Verlag(1970's-1980's)
2. E.Siegel, J.Noncryst.Sol.40,453(1980); Intl.Conf.Lattice Dynamics, Paris(1977)
3. K.L.Ngai, Comm.S.S.Phys.9,4,127(1979);9,5,141(1980);N.R.L.Rept.#3917(1979)
4. A.K.Jonscher, Contemp.Phys.24,1,75(1983); in Physics of Dielectric Solids, I.O.P.#58,I.O.P.(1979); Chelsea College Dielectrics Group Handbook(1979)
5. E.Siegel, Proc.Electrochem.Soc.83,8,497(1983); Intl.Conf.Lasers & E.O.(1983)
6. P.Dutta and P.Horn, Rev.Mod.Phys.53,3,497(1981)
7. F.N.Hooge, Phys.Lett.A,29,139(1969); P.Handel, Phys.Rev.A22,2,745(1980)
8. B.O.Kolbesen and M.Strunk, Solid State Devices(1980), I.O.P.#57,I.O.P.(1981)

CORE-LEVEL ELECTRON BINDING ENERGY CHANGE OF EVAPORATED Pd

SHIGEMI KOHIKI
Matsushita Technoresearch Inc., Moriguchi, Osaka 570, Japan

ABSTRACT

Positive core-electron binding energy shifts in small palladium clusters supported on cadmium telluride substrate are shown to arise from the initial-state effects those are more sensitive to cluster size than are the final-state properties and the mean valence band electron binding energy is primarily responsible for the Pd $3d_{5/2}$ electron binding energy in lower coverage region (Pd $\leq$ 1×10^{15} atoms·cm^{-2}).

INTRODUCTION

Small metal clusters on substrates are presently a subject of great interest in the transition of electronic state from the isolated atoms to the bulk metal. It is reported that the electron binding energies (BE) for small metal clusters supported on poorly conducting substrates generally diminish with the increase of cluster atoms [1-7]. Two kinds of possibility could be predicted as origins for the BE shift. One is based on a size dependence of the initial-state electronic structure. An alternative is that the shift is due to variations in the final-state relaxation processes.

Egelhoff and Tibbetts [1] reported that the core-level electron BE of Pd changed larger for amorphous carbon substrate than for crystalline carbon substrate. Amorphous carbon is the most widely used substrate, and the noble metals and group VIII metals are the most thoroughly studied metals.

In this experiment Pd clusters on amorphous CdTe substrate were investigated. CdTe is a wide band gap semiconductor and poorly conducting. The bond ionicity (f_i) of CdTe is 0.675 [8]. It is almost ionic contrary to the bond of carbon (f_i=0) [8]. The final-state extra-atomic relaxation energy observed in an ionic solid is smaller than that observed in covalent solid. It is expected that the extra-atomic relaxation energy in photoemission final-state do not dominate the electron BE shift for small Pd clusters on amorphous CdTe substrate in contrast with the case such as amorphous carbon substrate. This paper presents experimental results which suggest that photoemission final-state extra-atomic relaxation is not effective at smaller coverage and that the shifts in core-level electron BE are result of changes in initial-state properties. Mean valence band electron binding energy changes are primarily responsible for the Pd $3d_{5/2}$ BE shifts in lower coverage.

EXPERIMENTAL

The photoemission spectra measurements were made on a VG ESCALAB-5 electron spectrometer using unmonochromatized AlKα radiation. The linewidth (FWHM) for the Ag $3d_{5/2}$ photopeak was 1.15 eV. No attempt has been made to remove the instrumental broadening. The spectrometer was calibrated by utilizing the quantum energy difference (233.0 eV) between AlKα and MgKα radiation. Then, the core-level BEs of Pd, Ag and Au foils were measured. The Pd $3d_{5/2}$, Ag $3d_{5/2}$ and Au $4f_{7/2}$ BEs were, respectively, 355.4, 368.3 and 84.0 eV relative to the Fermi level. The probable electron energy uncertainty amounts to 0.1 eV. The normal operating vacuum was less than 3×10^{-8} Pa.

Firstly, the single crystal CdTe (110) surface was sputtered with 7 keV Ar^+ ions in the sample preparation chamber of the spectrometer at room temperature. The sputtered substrate was not annealed to maintain the amorphous surface. Ar^+ ion sputtering produced a clean surface of the substrate and removed the native oxide layer of the CdTe. Spectra of the valence-band (VB), Cd 3d, Te 3d, Pd 3d, Pd M_5VV, Ar 2p, O 1s and C 1s regions were recorded to monitor the condition of the substrate. No carbon and oxygen contamination could be detected. The atomic concentration of implanted Ar was 2.6%.

The composition of sputtered CdTe surface before Pd deposition was measured by varying the photoelectron take-off angle (θ=10, 25, 35, 50, 90°). The effective sampling depth (λ^θ_{eff}) at θ equals to $\lambda^{90°}\cdot\sin\theta$. The electron escape depth ($\lambda^{90°}$) of Cd $3d_{5/2}$ and Te $3d_{5/2}$ is 14.9 and 16.6 Å, respectively [9]. It is possible to determine atomic concentration near the surface [10]. The value of Cd-to-Te intensity ratio was 0.71 for various θ. The composition near the surface is constant and almost stoichiometric [11]. The effect of preferential sputtering reported is serious for relative low energy (≤ 1.5 keV) and small atomic number (He, Ne) primary ion. In this experiment relatively high energy (7 keV) and heavier (Ar) primary ion was used for sputtering. The effect of preferential sputtering was negligible for the CdTe substrate in this experiment.

The Pd was deposited by resistive evaporation in the sample preparation chamber at room temperature. The sample was transferred between the analyzer chamber and the preparation chamber under the vacuum below 3×10^{-8} Pa. The coverage of the Pd was determined from the Pd $3d_{5/2}$ peak intensity [12].

VGS1000 data system was used for data acquision and data processing. Determination of core-level peak position and spectrum intensity (peak area) was accomplished after smoothing and subtracting a smooth background. The Pd clusters contribution to VB spectra was obtained by subtracting the spectrum obtained on the clean CdTe surface, with amplitude determined by the attenuation of the Cd 4d signal by the overlayer. The subtraction of background in the VB region induced by the satellite X-ray (AlK$\alpha_{3,4}$) irradiation has already performed. The VB spectra of the Pd clusters were then precisely determined.

RESULTS AND DISCUSSION

All spectral features of the substrate are unchanged by the adsorption of Pd.

Pd is one of the most thoroughly studied metals. Pd M_5VV Auger electron could be excited by the AlKα X-ray. The Auger electron spectra were also recorded on the instrument. The Auger energies were taken as the peak in the second-derivative spectra.

The BE of a level j, BE (j), is the difference in the total energy of the system in its ground state and in the state with one electron missing in orbital j. For most situations encountered in photoemission, the approximation

$$BE(j) = -\varepsilon(j) - R(j)$$

is close enough to discuss the chemical shift. Here $-\varepsilon(j)$ is the term for orbital energy calculated by solving the Hatree-Fock equations by Koopmans' theorem. $R(j)$ is the term for relaxation energy, which is the result of a flow of negative charge towards the hole created in the photoemission process in order to screen the suddenly appearing positive charge. The screening lowers the energy of the hole state left behind and therefore lowers the measured BE as well.

The relaxation energy (R) can be partitioned into two terms: intra-atomic relaxation energy (R_{in}) and extra-atomic relaxation energy (R_{ex}). The intra-atomic relaxation energy is constant for the core-level electrons of a given atom.

It is generally stated that the actual photon absorption process occurs nearly instantaneously ($\leq 10^{-17}$ s) and the hole switching on time is very much less than 10^{-16} s. The localized screening response ($10^{-16} \sim 10^{-15}$ s) is very fast in contrast to the delocalized screening responce ($10^{-13} \sim 10^{-12}$ s). Delocalized screening is accompanied with Core-Valence-Valence (CVV) Auger transition [13]. This is the time scale which determines how long the hole remains localized on the source atom.

In the photoemission final-state, hole state of the Pd core-level in the cluster should be screened with the valence electrons of Pd cluster and

the conduction electron of the support. This relaxation shift depends on
the relative magnitude of the polarizability of the substrate and the Pd
metal. A metal has a density of states at the Fermi level and itinerant
electron states. In an ionic solid extra-atomic relaxation cannot easily
take place via electronic relaxation [14]. Fadley et al. [15] pointed out
that electrons on neighboring ions will respond to sudden creation of a
positive charge during photoemission by moving away from their equilibrium
positions so as to change the electrostatic potential at the site of the
ionized atom. The sudden relaxation will proceed within the cluster.

In the simplest approximation, the change in extra-atomic relaxation
energy can be derived from the combination of core-level electron BE and CVV
Auger electron KE referenced to the Fermi level. These quantities define
the modified Auger parameter [16] α=BE+KE. The difference in the modified
Auger parameters for a given element in two different environments is twice
the difference in extra-atomic relaxation energies (R_{ex}), $\Delta\alpha = 2\Delta R_{ex}$ [17].

Wigner and Bardeen [18] obtained good values for the work function by
assuming that the valence-band hole is completely delocalized. The $\overline{BE}$(VB)
of a filled valence state is reduced from the sum of the atomic electron
BE, BE(atom), and the cohesive energy (E_c) by an amount which represents
the Coulomb and exchange interaction of a missing electron with all the
remaining valence electrons [14] to give

$$BE(atom) + E_c - \overline{BE}(VB) = \{(3 \times e^2)/(5 \times r_s)\} - \{(0.458 \times e^2)/r_s\},$$

where r_s is the Wigner-Seitz radius which means the radius of a sphere containing one electron per atom and increases with decreasing cluster atoms.
The cohesive energy per atom is expressed as the negative value of the energy
required to the out going of one valence electron from a metal by breaking
down effectively electron bonding to either one of the atoms, and leaving
an ion in the cluster, which is correspond to the Coulomb and exchange
energies accompanying a valence-band hole. Therefore, the $\overline{BE}$(VB) shift is
inversely depending on the change of r_s. The $\overline{BE}$(VB) decreases with increasing cluster atoms where the electron configuration in the cluster is
varying from atomic like to bulk metal like one.

From these derivation, extra-atomic relaxation accompanied to the
ionization of an localized core-orbital electron is to be distinguished to
that due to the ionization of an electron from the valence-band in a metal.
Extra-atomic relaxation energy is constructed with two components: the
localized extra-atomic relaxation energy (R_{ex}^{loc}) and the delocalized extra-atomic relaxation energy (R_{ex}^{deloc}).

It is possible to derive following equation about the relation between CVV Auger KE shifts ΔKE and core-level photoelectron BE shifts ΔBE [19].

$$\Delta KE = \Delta BE - 2\Delta\overline{BE}(VB) + \Delta R_{ex}$$

Here ΔR_{ex} is equivalent to the change in the localized extra-atomic relaxation energy, ΔR_{ex}^{loc}, and $\Delta\overline{BE}(VB)$ is the change in the mean valence band electron binding energy. $\Delta\overline{BE}(VB)$ is equivalent to ΔR_{ex}^{deloc}.

Figure 1 shows the Pd $3d_{5/2}$ BE versus the Pd coverage. The Pd $3d_{5/2}$ BE increases by 0.6 eV positively with decreasing coverage in the region less than $\approx 1\times 10^{15}$ atoms·cm^{-2}. The Pd $3d_{5/2}$ BE for Pd clusters in the coverage region more than $\approx 1\times 10^{15}$ atoms·cm^{-2} is constant and identical to that obtained bulk Pd metal.

Figure 2 shows the Pd M_5VV Auger KE versus the Pd coverage. The Pd M_5VV Auger KE shifts almost linearly with the coverage increase.

Figure 3 shows the modified Auger parameter versus the Pd coverage. α is almost constant at the coverage below than $\approx 1\times 10^{15}$ atoms·cm^{-2}. Therefore, the extra-atomic relaxation energy did not change. At the larger coverage region (Pd $\geq 1\times 10^{15}$ atoms·cm^{-2}), α increases by 0.7 eV with the increase of coverage. In this coverage the change of extra-atomic relaxation energy amounts to 0.35 eV.

If the number of occupied d orbitals is dependent on cluster size, BE shifts expected from such increased d orbital occupation qualitatively account for the core-level and mean valence band electron BE changes. In the region of coverage less than $\approx 1\times 10^{15}$ atoms·cm^{-2}, ΔKE (Pd M_5VV) is +0.6 eV, ΔBE (Pd $3d_{5/2}$) is -0.6 eV and ΔR_{ex} is zero with increasing coverage. Therefore, $\Delta\overline{BE}(VB)$ amounts to -0.6 eV. In the region of coverage more than $\approx 1\times 10^{15}$ atoms·cm^{-2}, ΔKE (Pd M_5VV) is +0.7 eV, ΔBE (Pd $3d_{5/2}$) is zero and ΔR_{ex} is +0.35 eV with increasing coverage. $\Delta\overline{BE}(VB)$ amounts to -0.18 eV.

The change of Pd $3d_{5/2}$ BE is connected to the sum of changes of mean valence band electron binding energy and extra-atomic relaxation energy in this experiment. In table I the changes of $\Delta\overline{BE}(VB)$ and ΔR_{ex} and Pd $3d_{5/2}$ BE are listed. In low coverage region a -0.6 eV shift for Pd $3d_{5/2}$ BE is ascribed to the change (-0.6 eV) of $\Delta\overline{BE}(VB)$. In high coverage region the difference between ΔBE (Pd $3d_{5/2}$) and the sum of ΔR_{ex} and $\Delta\overline{BE}(VB)$ amounts +0.17 eV, which is slightly larger than the experimental error (0.1 eV). The correlation energy term and the term of barrier potential due to surface dipole layer are not included in this simple approach. This discrepancy may suggest that the energy change of these two terms should be considered in high coverage region.

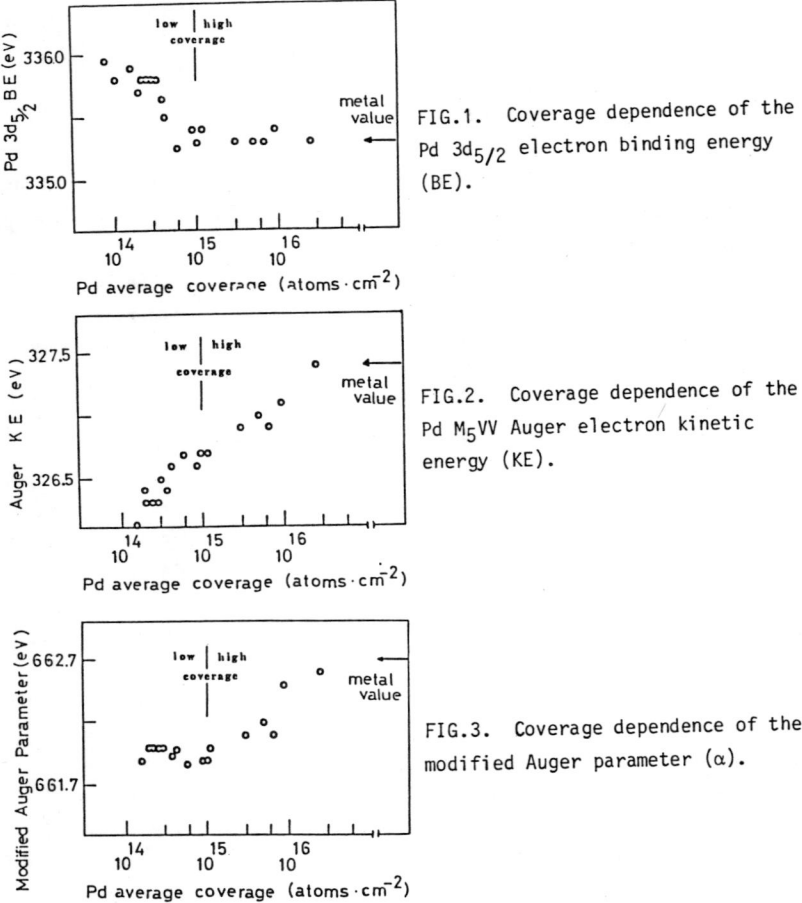

FIG.1. Coverage dependence of the Pd $3d_{5/2}$ electron binding energy (BE).

FIG.2. Coverage dependence of the Pd M_5VV Auger electron kinetic energy (KE).

FIG.3. Coverage dependence of the modified Auger parameter (α).

Table I

Changes of extra-atomic relaxation energy and mean valence band electron and $3d_{5/2}$ electron BEs for Pd with increasing coverage. Energies are in eV.

coverage (atoms/cm^2)	ΔR_{ex}	$\Delta \overline{BE}$(VB)	ΔBE (Pd $3d_{5/2}$)
low ($\lesssim 1\times 10^{15}$)	0	−0.6	−0.6
high ($\gtrsim 1\times 10^{15}$)	+0.35	−0.18	0

The extra-atomic relaxation energy change, $\Delta R_{ex}=\Delta\alpha/2$, is not sensitive to the initial-state valence electron density change, whereas the mean valence band electron binding energy change, $\Delta\overline{BE}(VB)$, is more sensitive to the initial-state valence electron count change because the $\overline{BE}(VB)$ is connected to the renormalization energy through the r_s. Hybridization of the d-band with the empty s-p conduction states will cause the true d-electron count to be lower. The increased electron density in the Wigner-Seitz cell volume with increasing coverage decreases core-electron BE.

CONCLUDING REMARKS

Present observation of BE shift with cluster size may be interpreted as a result of photoemission initial-state properties for the smaller size ($Pd \leq 1\times10^{15}$ atoms·cm^{-2}) metal clusters on poorly conducting CdTe substrate. It also may be concluded that mean valence band electron binding energy is primarily responsible for Pd $3d_{5/2}$ BE in the lower coverage.

ACKNOWLEDGMENTS

The author thanks Professor S. Ikeda, Osaka University, for his helpful discussions and suggestions and Miss K. Oki for her assistance and Dr. F. Konishi for his support in this work.

REFERENCES

1. W.F. Egelhoff, Jr. and G.G. Tibbetts, Phys. Rev. B19, 5028 (1979).
2. G.K. Wertheim, S.B. Dicenzo and S.E. Youngquist, Phys. Rev. Lett. 51, 2310 (1983). Wertheim et al. pointed out that the electron BE shift for small Au clusters on amorphous carbon substrate was due to final-state effect.
3. M.G. Mason, Phys. Rev. B27, 748 (1983).
4. Y. Takasu, R. Unwin, B. Tesche, A.M. Bradshaw and M. Grunze, Surf. Sci. 77, 219 (1978).
5. H. Roulet, J.-M. Mariot, G. Dufour and C.F. Hague, J. Phys. F 10, 1025 (1980).
6. L. Oberli, R. Monot, H.J. Mathieu, D. Landolt and J. Buttet, Surf. Sci. 106, 301 (1981).
7. K.S. Liang, W.R. Salaneck and I.A. Aksay, Solid State Commun. 19, 329 (1976).
8. J.C. Phillips, Bonds and Bands in Semiconductors, (Academic, New York, 1973).
9. D.R. Penn, J. Electron Spectrosc. Relat. Phenom. 9, 29 (1976).

10. S. Kohiki, K. Oki, T. Ohmura, H. Tsuji and T. Onuma, Jpn. J. Appl. Phys. 23, L15 (1984).
11. The number of Cd and Te atoms near the CdTe surface was calculated from the peak intensities of Cd $3d_{5/2}$ and Te $3d_{5/2}$ using the method described in Ref. 12.
12. S. Kohiki, Appl. Surf. Sci. 17, 497 (1984).
13. J.W. Gadzuk, in Photoemission and The Electronic Properties of Surfaces, edited by B. Feuerbacher, B. Fitton, and R.F. Willis (John Wily, Chichester, 1978) p.111.
14. D.A. Shirley, in Photoemission in Solids, edited by M. Cardona, and L. Ley (Springer, Berlin, 1978) Vol.1, p.177.
15. C.S. Fadley, S.B.M. Hagstrom, M.P. Klein, and D.A. Shirley, J. Chem. Phys. 48, 3779 (1968).
16. C.D. Wagner, Faraday Discuss. Chem. Soc. 60, 291 (1975), and C.D. Wagner, L.H. Gale, and R.H. Raymond, Anal. Chem. 51, 466 (1979).
17. T.D. Thomas, J. Electron Spectrosc. Relat. Phenom. 20, 117 (1980).
18. E. Wigner and J. Bardeen, Phys. Rev. 48, 84 (1935).
19. S.P. Kowalczyk, L. Ley, F.R. McFeely, R.A. Pollak, and D.A. Shirley, Phys. Rev. B9, 381 (1974).

DEPTH OF PENETRATION OF THE PLASMA
FLUORINATION REACTION INTO VARIOUS POLYMERS

EVE A. WILDI, GERALD J. SCILLA and ALAN DeLUCA
Xerox Corporation, 800 Phillips Road, Webster, NY 14580

ABSTRACT

The surfaces of a variety of polymers presenting either primary, secondary, tertiary, vinyl or aromatic protons to the reactive gas environment were fluorinated by glow discharge treatment. Depth profiles of the plasma surface reaction were obtained by SIMS with oxygen ion sputtering. Further information was obtained from the appearance of ESCA traces of these substrates before and after the reaction.

INTRODUCTION

Modification of materials surfaces in a plasma atmosphere is attractive for a number of reasons. Due to the high concentrations of reactive species generated in such an atmosphere, only short times at low process temperatures are required to drastically alter surface properties. For example, the surface of polyethylene may be made to approximate that of Teflon® after one minute in a plasma of 5% F_2 in He [1] at ambient temperature whereas the pure molecular gas does not react to this extent in tens of hours.[2] Furthermore, a given surface may be made more adhesive or more releasing with equal ease with just a change of plasma reactant gas. A thin skin may be grafted on, or removed, depending on the nature of the gas, the discharge power, and other controllable variables.[3]

In applications where wear of treated parts will occur the depth, therefore durability, of the new surface layer will be of interest. Angle resolved ESCA (probing the 50 A nearest a surface) was utilized in such a study of non-discharge molecular fluorination [2] which procedes to a depth of about 40 A in five minutes. Plasma fluorination is too fast for this technique. We have, therefore, shown the feasibility of a SIMS/-oxygen ion sputtered depth profile into organic polymeric media such as polystyrene, polyisobutylene, co-poly(hexafluoropropylene/vinylidene fluoride) and their plasma surface modified cogeners. The depth of the reaction of two gases into two polymers is reported.

Also, the selectivity of various reacting plasma gases toward a variety of C-H bonds was examined by surface ESCA. Polymers were systematically chosen to present mainly -CH_3 groups (polyisobutylene), -CH_2 and -CH groups (polynorbornene), vinyl groups, e.g., -CH=CH-(polycisbutadiene), or phenyl groups (polystyrene) to the reactive gas environment. These substrates were then exposed to a variety of fluorine atom source gases in a matrix experiment.

EXPERIMENTAL

Polymers were purified by dissolution in toluene followed by reprecipitation from methanol. The precipitates were dried in vacuo over $CaCl_2$ before preparation of 1-5% solutions in toluene. Thin films were spun from these solutions onto clean glass plates bearing an aluminum mirror. Poly(norbornene) was used as received and spun from N,N-dimethylformamide. Thicknesses and refractive indexes were characterized by ellipsometry using a Rudolf Research Auto El III. These results were corroborated for some samples by profilometry on a Dektak surface profilometer or visual examination by SEM.

A two layer sample for the SIMS characterization of an abrupt interface was prepared. Poly(paraxylylene) was deposited on an aluminized glass slide from the vapor. The thickness and refractive index of this material was repeatedly measured (TL=736±28 A, nL=1.73). A fluoropolymer similar to Viton®, Fluorel FT 2481, kindly supplied by Paul Tuckner of 3M Commercial Chemical Division, was spin coated (6000 rpm) from a 5% solution in toluene onto the insoluble Paralene® layer. Since the real part of the refractive index of the Fluorel is 1.37, sufficiently different from that of the underlayer, the thickness of this top layer could be measured by ellipsometry. It proved to be 383±23 A, about the same as the thickness produced at 6000 rpm on a clean slide.

Plasma modification was performed in an 2 l (gas volume) reactor which could be pre-pumped to 10^{-7} torr, then switched to an automated gas handling system maintaining, for these experiments, 2 torr at 15 sccm 5% F_2 in He, and 10 sccm of CF_4 or SF_6. The 3" diameter capacitively coupled electrodes were held 2 cm apart, and were powered such that the glow from the normal glow discharge just filled the space between the electrodes and was not evident between any other surfaces in the reactor. This required (RF power at 1.56 MHz) 5 watts for 5% F_2/He and CF_4, and 25 watts for SF_6.

ESCA data were collected on a McPherson 36 instrument employing polychromatic MgK x-rays. Sputtering in the ESCA was carried out using a normal incidence Ar^+ ion beam at 5 KeV. Window regions for C_{1S}, F_{1S} and O_{1S} were collected for all polymers studied.

The SIMS data were obtained utilizing an Applied Research Laboratories Ion Microprobe Mass Analyzer (IMMA). The primary ion species employed was $^{16}O^-$ at 8.5 KeV and an intensity of 1.5 mA reacted over an area of 140 x 140 μm. The $^{12}C^-$, $^{19}F^-$ and $^{31}CF^-$ secondary ions were monitored for each polymer.

RESULTS AND DISCUSSION

In order to assess the depth of the plasma fluorination reaction, SIMS spectra with oxygen ion beam sputtering of several types of controls were first produced. The SIMS' sputtering rate was assessed by observing the time of appearance of an $^{43}AlO^-$ signal from 3 samples of Fluorel FT 2481 coated at different thicknesses over an aluminum layer on glass. The disappearance of the $^{12}C^-$ signal was simultaneously monitored. The time at half-maximum of each of these two signals from the three samples was taken as the duration of sputtering required to reach the interface under the known thickness of polymer. This procedure yielded a sputtering rate of 0.21 ±.01 A/sec.

A thicker sample of the model fluorinated polymer FT 2481 was probed to ascertain that a constant composition of fluorinated polymer yields a constant analytical signal. The constant fluorine to carbon ratio for this 1800 A film matched the bulk elemental analysis.

Figure 1 shows the depth profile into a poly(styrene) sample fluorinated for two minutes in a plasma of five percent fluorine in helium. Figure 2 shows the same material after exposure instead to a plasma of sulfur hexafluoride. Both traces show fluorine content falling rapidly within the first few molecular layers of the sample surface.

A third experiment shows the effect of a known abrupt polymer-polymer interface. Sample preparation is described in the experimental section. Figure 3 shows a change of one and one half orders of magnitude in the intensity of the F/C signal centered at 316 A into the depth of the two layer structure. The measured thickness of the fluorinated top layer was 383 ± 23 A. The error of approximately 50 A might be accounted for in noting the long tail of a few percent fluorine found in the bulk of the initially unfluorinated polymer underlayer. This tail is attributed to fluorine atoms sputtered back into the bulk during etching.

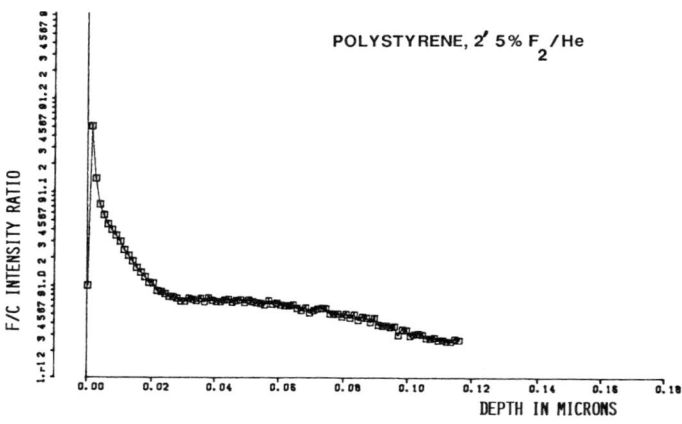

Figure 1

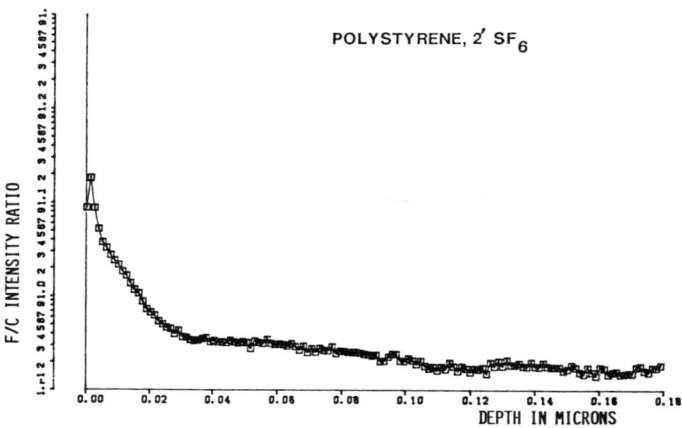

Figure 2

Comparing Figure 1, then, to Figures 2 and 3, it is evident that the plasma fluorination, though rapid, does not penetrate deeper into the bulk than about 300 A in two minutes for either of the two polymers examined. Neither does the identity of the gas make a significant difference, although at the surface 5% F_2/He is much a faster fluorinating agent than SF_6. The 5% F_2/He plasma was found to etch FT 2481 at 200 A/min. Thus, the original surface disappears at nearly the same rate as the carbon-fluorine bond front is penetrating the bulk. Other reactor configurations and plasma conditions will be tried in order to achieve greater depths of fluorination.

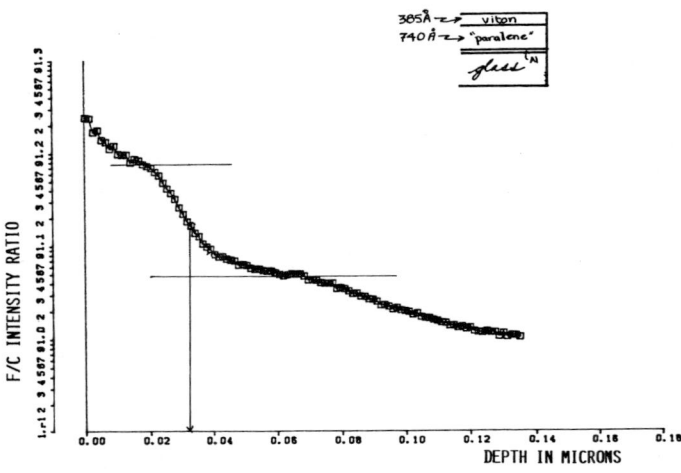

Figure 3

The relative rate of fluorination by three fluorine atom source gases, 5% F_2/He, CF_4 and SF_6, was determined by ESCA spectroscopy. Figure 4 shows that in two minutes at a poly(isobutylene) surface (plasma conditions are listed in the experimental section) 5% F_2/He produces the greatest extent of fluorination. In fact, the appearance of this spectrum changes little at longer times indicating that the reaction is complete. The atomic composition of perfluoro-poly(isobutylene) is 33% C, 66% F. Crosslinking reactions, well known to occur under these conditions, can account for the missing 10% F atoms at this plasma modified "perfluoro-poly(isobutylene)" surface. The ellipsometric index of refraction of these polymer thin films steadily increases during the course of the reaction, from 1.3-1.5 before exposure to the plasma, to 1.7-2.0 after such exposure. This is also consistent with a polymer crosslinking reaction. Fluorination is somewhat less than complete in two minutes with CF_4 as the plasma F atom source gas, and just beginning when SF_6 is used. Very similar trends were observed for each of the other polymers.

Figure 5 shows the effect of one gas (CF_4) on four different polymers. Polyisobutylene presents a largely permethylated surface to the reactive gas, while polynorbornene consists of roughly equal numbers of methylene and methine type protons. The behavior of these two saturated alkyl polymers contrasts with that of polybutadiene, where half the carbons are π-bonded (but isolated) and that of polystyrene, where 3/4 of the carbons are aromatically π-bonded. The π bonded materials fluorinate faster and to a greater extent than do the σ-bonded polymers. Note, however, that if the starting polymer structures were saturated with fluorine atoms, the atomic percent C would be lower and the atom percent F higher still at the end of the reaction. Oxidative passivation of dangling bonds accounts for some of the missing bonds, but the crosslinking reaction must account for the remainder. Very similar trends were observed for the other two reactive gases.

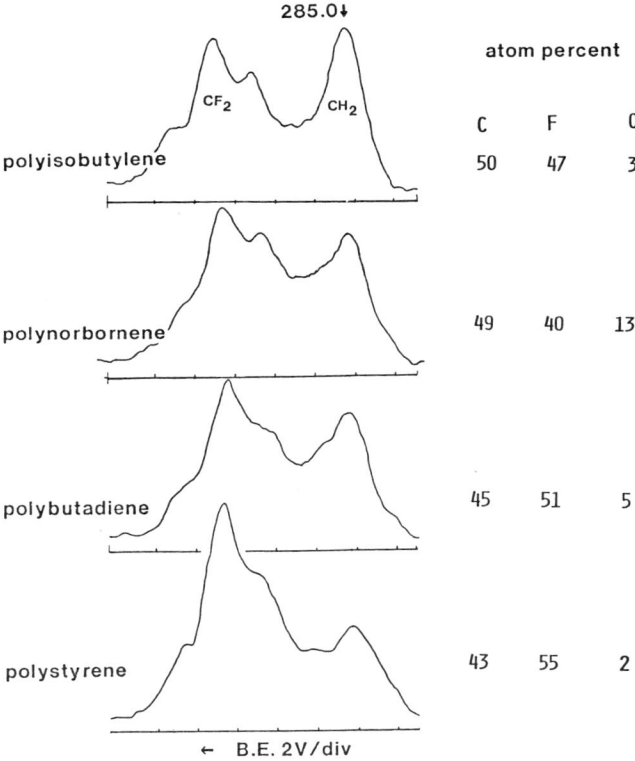

Figure 4. Response of Various Polymers

CONCLUSION

The plasma fluorination reaction has also been shown to discriminate between bond types, π bonds being more reactive than σ bonds. Plasma oxidation of liquid alkenes has recently been shown strongly to distinguish between π and σ type C-C (or C-H) bonds.[4] The fluorination can be made slow enough to study its rate by judicious choice of reactant gas. The mechanistic details of these reactant selectivities are being studied.

The plasma surface fluorination reaction has been shown to extend only a few hundred angstroms beyond the etching front of polymer materials in two minutes. Greater fluorination depths may be achived by supression of the etch rate. Work is in progress on this problem.

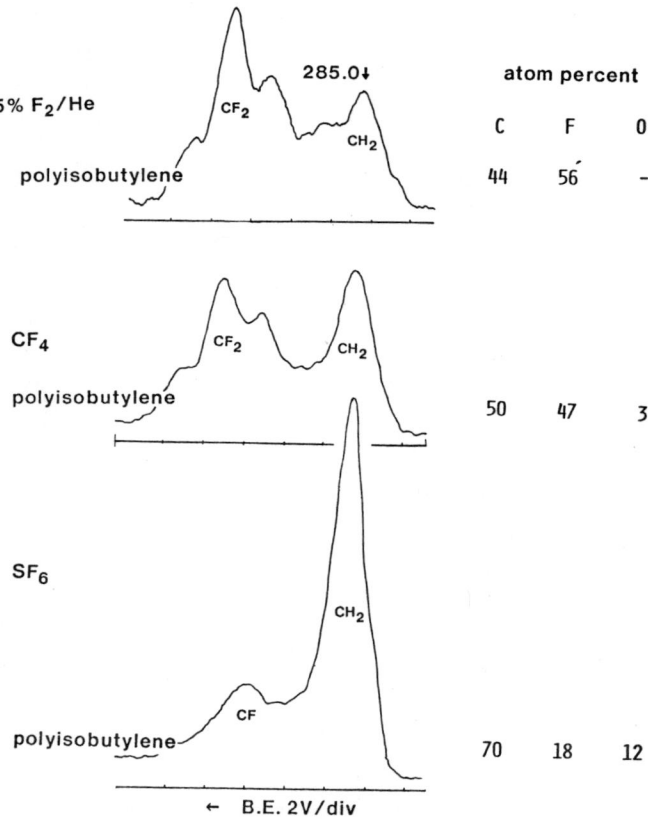

Figure 5. Response to Various Gases

REFERENCES

1. G.A. Corbin, et al., Polymer, **23**, 1546-48 (1982).

2. D.T. Clark, et al., Journal of Polymer Science, Polymer Chem. Ed., **13**, 857-90 (1975).

3. H. Yasuda, J. Macromol. Sci.-Chem., A10(3), pp. 383-420 (1976).

4. H. Suhr, et al., Plasma Chemistry and Plasma Processing, **4**, 285 (1985).

SURFACE COMPOSITION OF CARBURIZED TUNGSTEN TRIOXIDE AND ITS CATALYTIC ACTIVITY

MASATOSHI NAKAZAWA AND H. OKAMOTO
Central Research Laboratory, Hitachi Ltd., Tokyo 185, Japan

ABSTRACT

The surface composition and electronic structure of carburized tungsten trioxide are investigated using X-ray photoelectron spectroscopy (XPS). The relationship between the surface composition and the catalytic activity for methanol electro-oxidation is clarified. The tungsten carbide concentration in the surface layer increases with the carburization time. The formation of tungsten carbide enhances the catalytic activity. On the other hand, the presence of free carbon or tungsten trioxide in the surface layer reduces the activity remarkably. It is also shown that, the higher the electronic density of states near the Fermi level, the higher the catalytic activity.

INTRODUCTION

Transition metal carbides are interesting materials for wide applications since they have great hardness and very high melting points[1]. In particular, tungsten carbide (WC) is known as an effective catalyst for certain chemical reactions that are readily catalyzed by such noble metals as Pt and Pd [2]. The catalytic behavior of WC and its surface electronic structure have been investigated [3-8]. In addition to those works, Miles investigated the suitability of using highly dispersed tungsten carbide as inexpensive alternative catalyst to Pt for use in methanol fuel cells [9]. The catalytic activity of such a heterogenous catalyst as WC generally depends on the surface composition, and the geometric and electronic properties at the surface [10]. In addition, the activity is influenced by the properties of gas-adsorption during catalytic reactions. In order to clarify the mechanism of catalytic reaction on a solid surface, it is of great importance to classify those factors and to examine the influence of the factors on the catalytic activity. Therefore, from a practical point of view, it is necessary to investigate the relationship between the surface composition of WC and its catalytic activity. In the present study, the surface composition of tungsten trioxide containing various amounts of WC is systematically examined using X-ray photoelectron spectroscopy. The relationship between the electronic structure and the catalytic activity is also clarified.

EXPERIMENTAL

The samples studied in this work were prepared by heating tungsten trioxide in carbon monoxide atmosphere at 700 °C. Since a change in the carburization time makes a great difference in the surface composition of the samples, the carburization time was increased in a stepwise fashion from 2 to 11 h. The catalytic activity of the samples for the methanol electro-

oxidation was measured. The measurement method was the same as that reported previously [11].
The XPS spectra of the W 4f, C 1s and O 1s core levels were obtained using a VG ESCA 3 spectrometer with Mg K$_\alpha$ X-ray source. The valence band spectra were also measured.

RESULTS AND DISCUSSION

Photoelectron spectra. The XPS spectra of the W 4f, C 1s and O 1s core levels are shown in Figs. 1-3. The crystal structure and the catalytic activity of each sample are listed in Table I in order of the carburization time. Changes in the surface composition are generally consistent with those in the crystal structure.
Since the intensity of the W $4f_{7/2}$ peak labeled as 1 (the binding energy of 31.3 eV) increases with the carburization time, peak 1 is considered to be associated with W-C bonding. This can be expected from the chemical shift in the W $4f_{7/2}$ level. On the other hand, the intensity of peak 2, whose $4f_{7/2}$ binding energy is 35.2 eV, decreases as the carburization time is increased. Peak 2 is concluded to be due to the W 4f photoelectrons ejected from WO$_3$, since the W $4f_{7/2}$ binding energy measured for WO$_3$ as reference was 35.2 eV. The chemical shifts in the binding energies for WC and WO$_3$ are in general agreement with those reported by other workers [12,13].
The C 1s spectra in Fig. 2 consist of three overlapping components. The intensity of the lowest binding energy peak labeled as 1 (the binding energy of 282.4 eV), which is considered to be due to the photoelectrons ejected from the carbon in WC, increases as the carburization time is increased. Peak 3 is considered to correspond to free carbon (non-carbide) because its binding energy is almost equal to the value measured for graphite as reference. The peak labeled as 4 near the binding energy of 285 eV is associated with adsorbates on the surface such as CO, CO$_2$ and hydrocarbons.
The O 1s spectra have a main peak (labeled as 2), as shown in Fig. 3. This peak corresponds to the photoelctrons from the oxygen in WO$_3$. The peak labeled as 4 is associated with adsorbates such as CO and CO$_2$.
Fig. 4 shows the relationship between the intensities of the photoelectron peaks and the carburization time. The peak intensities were obtained through the peak separation [14].

Table I Crystal Structure and Catalytic Activity of Each Sample

Carburization time (h)	Crystal structure [a]	Catalytic Activity [b] (mA/g)
2	WO$_2$ + WO$_3$	0
4	WO$_2$ + WC	0.02
5	WC + WO$_2$	0.25
6	WC	0.30
8	WC	0.24
11	WC	0.03

a) Examined by X-ray diffraction analysis
b) Current density for methanol electro-oxidation (CH$_3$OH + H$_2$O → CO$_2$ + 6H$^+$ + 6e$^-$)

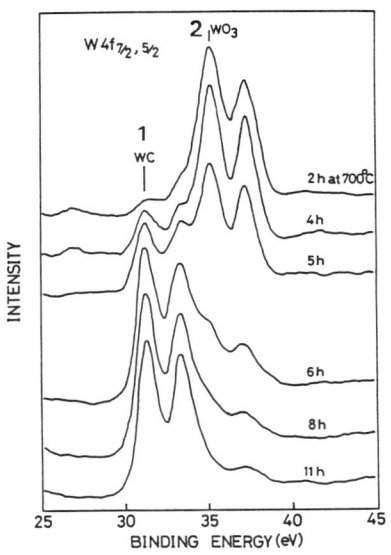

Fig. 1. XPS spectra of W 4f core level from samples carburized for various carburization times.

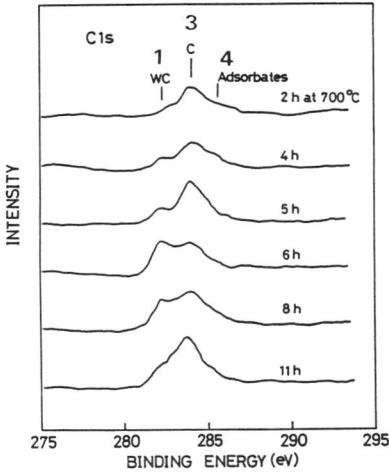

Fig. 2. XPS spectra of C 1s core level from samples carburized for various carburization times.

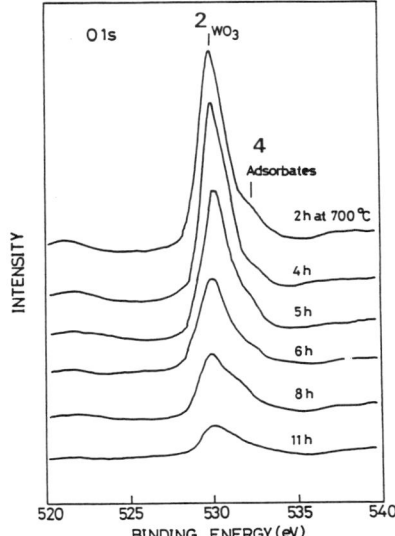

Fig. 3. XPS spectra of O 1s core level from samples carburized for various carburization times.

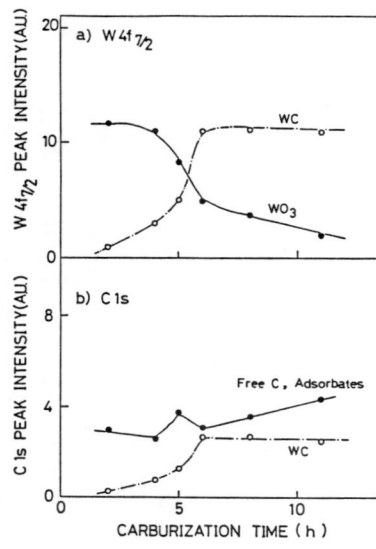

Fig. 4. Relationship between carburization time and intensities of photoelectron peaks

Relationship between surface composition and catalytic activity. Fig. 5 shows the relationship between the surface composition and the catalytic activity. The abscissas of Figs. 5a and 5b indicate the surface composition (i.e., the ratio of the W concentration in WC to the total W concentration and the ratio of the C concentration in WC to the total C concentration).

The catalytic activity begins to increase when the WC concentration approximates around 0.25, as shown in Fig. 5a. However, it begins to decrease rapidly at the concentration of 0.75. The carburization time that corresponds to the rapid decrease is 6-8 h. Furthermore, Fig. 5b indicates that the ratio of the C concentration in WC to the total C concentration decreases as the carburization time exceeds 6 h. This means that the deposition of free carbon on the surface, which is formed more beyond the carburization time of 6-8 h, reduces the activity remarkably.

Valence band spectra. Fig. 6 shows the changes in the photoelectron valence band spectra of carburized tungsten trioxides. The peak labeled as 1 is interpreted as the O 2p band where electrons are transferred from W, because the results for the core levels indicate that the samples carburized for 2 and 4 h are similar to WO_3. The broad peak labeled as 2 represents the mixing of the C 2p with the W 5d and 6s bands.

A close relationship between catalytic activity and the density of d-like states near the Fermi level has been reported [15-17]. The valence electrons in WC seem to contribute to the methanol electro-oxidation reaction. Fig. 6 indicates that the electronic density of states near the Fermi level becomes higher with the increased carburization time. The comparison between the valence band and the catalytic activity listed in

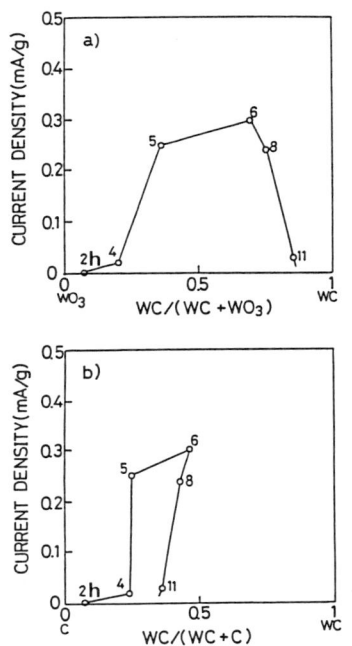

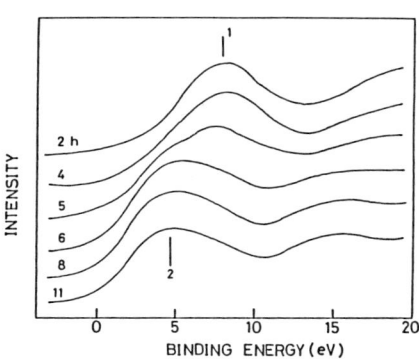

Fig. 6. Valence band spectra of samples prepared for various carburization times.

Fig. 5. Relationship between surface composition and catalytic activity during methanol electro-oxidation.

Table I shows that, the higher the density of states near the Fermi level, the higher the catalytic activity. However, both samples carburized for 2 and 11 h possess no activity in spite of the great difference between the density of states. Concerning the latter sample, the deposition of free carbon reduces the activity. On the other hand, no tungsten carbide is formed with the carburization time of 2 h. For this reason, neither of these two samples acts as a catalyst for methanol electro-oxidation.

CONCLUSION

The relationship between the surface composition of carburized tungsten trioxide and the catalytic activity for methanol electro-oxidation has been clarified. The catalytic activity has also been correlated with the valence band structure. The results are summarized as follows.
(1) The concentration of WC in the surface layer increases with increased carburization time. The formation of WC enhances the catalytic activity. However, the concentration of free carbon at the surface increases when the carburization time exceeds 8 h. As a result, the catalytic activity begins to decrease because of the deposition of free carbon on the surface at the carburization time above 8 h.
(2) The electronic density of states near the Fermi level for the carburized tungsten trioxide becomes higher with increased

carburization time. The comparison between the valence band and the catalytic activity indicates that, the higher the density of states, the higher the activity.

ACKNOWLEDGEMENTS

We wish to thank Drs. S. Yamamoto, G. Kawamura and F. Nagata for useful discussions. We also express our gratitude to Drs. S. Harada and T. Kudo for their encouragement and suggestions during this work.

REFERENCES

1. L.E. Toth, Transition Metal Carbides and Nitrides (Academic Press, New York, 1971).
2. R.B. Levy and M. Boudart, Science 181(1973)547.
3. J.E. Houston, G.H. Laramore and R.L. Park, Science 185 (1974)258.
4. L.H. Bennett, J.R. Cuthil, A.T. McAlister, N.E. Erickson and R.E. Watoson, Science 184(1974)563.
5. R.J. Colton, J.J. Huang and J.W. Rabalais, Chem.Phys. Letters 34(1975)337.
6. P.N. Ross and P. Stonehart, J.Catalysis 48(1977)42.
7. V.SH. Palanker, R.A.Gajyev and D.V. Sokolsky, Electrochim. Acta 22(1977)133.
8. E.I. Ko, J.B. Benziger and R.J. Madix, J.Catalysis 62 (1980)264.
9. R. Miles, J.Chem.Tech.Biotechnol. 30(1980)35.
10. G.C. Bond, Catalysis by Metals (Academic Press, New York, 1964).
11. T. Kudo, G. Kawamura and H. Okamoto, J.Electrochem.Soc. 130(1983)1491.
12. R.J. Colton and J.W. Rabalais, Inorg.Chem. 15(1976)237.
13. K.T. Ng and D.M. Hercules, J.Phys.Chem. 80(1976)2094.
14. M. Nakazawa and H. Okamoto, to be published.
15. G.M. Schwab, Disc.Faraday Soc. 8(1950)166.
16. D.A. Dowden and P.W. Reynolds, Disc.Faraday Soc. 8(1950) 184.
17. J.H. Sinfelt, J.L.Carter and D.J.C. Yates, J.Catalysis 24(1972)283.

INSTANTANEOUS IMPEDANCE OF ALUMINIUM IN ANODIC POLARIZATION STATUS

ZHU YINGYANG   WANG KUANG   ZHU RIZHANG, and ZHANG WENQI
Beijing University of Iron and Steel Technology, Beijing, China

ABSTRACT

The instantaneous impedance of Al in 1N NaCl at various polarization times has been measured using LapLace transformation method. In an instant of beginning polarization, the dependance of interfacial resistance, R_f, on polarization current, i_p, follows:

$$\log R_f = K - \log i_p$$

Interfacial capacitance, C_f, however, independs of i_p, only depends on film's thickness. In the course of polarization, R_f changed hardly with time, C_f changed greatly. This is due to little difference between the resistance on corroded region and the resistance on the region with film, and great difference capacitance on both regions. A reasonable model has been proposed to explain the experimental results.

INTRODUCTION

The impedance of aluminium in aqueous solution has been extensively studied[1-6]. The kinetic prosess of impednace change with time during anodic polarization, however, was hardly studied for lack of rapid measurement method. The measurement method usually used such as frequency response analysis, phase sensitive detection method and so on, is not suitable to measure the impedance change behaviour in this prosess, because of their relatively slow measurement rate. The study on the instantaneous impedance behaviour of Al in anodic polarization prosess, in which the film on surface was broken and corroded region increase succesively,is important and interest for us to understand the action of film for electrochemical dissolution and its status in the course of corrosion and the mechanism of passive. Thus it is necessary to use a rapid impedance measurement method which is able to measure instantaneous impedance of the system.

The aim of this paper is to report the results of study on the instantaneous impedance of Al/NaCl system during anodic polarization, relationship between interfacial impedance and polarized current as well as impedance of the electrode with different corroded region. According to our experimental results, the equevalent circiut of the Al/NaCl interface, the mechanism of influence of passive film on reaction rate and the relationship between interfacial impedance and status of film broken are discussed.

EXPERIMENTAL

The impedance of system was measured by a quick impedance measurement method based on the LapLace transformation[7]. In the present work, each measurement, from which impedance diagram can be obtained by later calculation treatment, was accomplished on the order of millisecond magnitude. The block diagram of measurement system was shown in Fig.1.

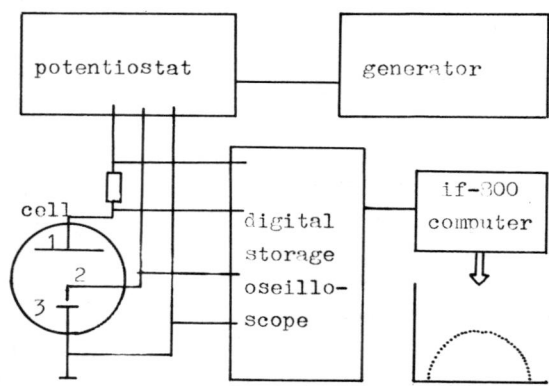

Fig.1 Block diagram of measruement system (1:auxiliary electrode, 2:reference electrode, 3:specimen)

The material used in experiment is pure aluminium (99.999%

purity). The specimens were prepared in following way, first annealed at 400°C for 4 hours, then deformed by 5% and again annealed at 590°C for about 50 hours to get a coarse grain structure. The area of the electrode used is 1cm^2. The auxiliary electrode used is made of platinum flake that has a large surface area. The solution is 1N NaCl. The experiment was performed at room temperature.

RESULTS

Relationship between interfacial impedance and polarization time

The interfacial resistance and interfacial capacitance of aluminium electrode in 1N NaCl solution at room temperature have been investigated as a function of anodic polarization time extending from 1 sec. to 80 min. at various anodic polarized currents by measuring instantaneous impedance at various times. The results were shown in Fig.2 and Fig.3. The Fig.2 is

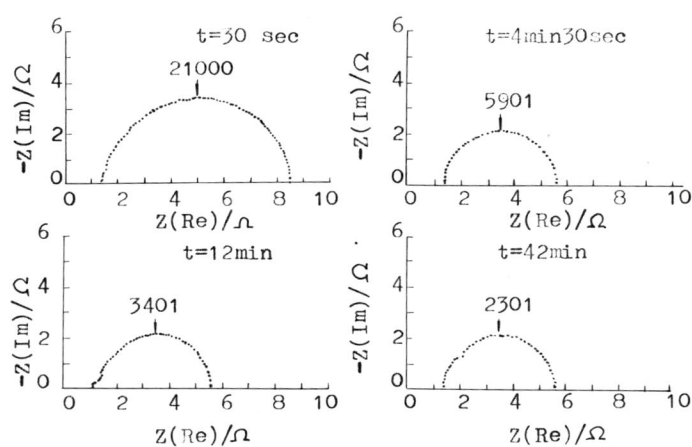

Fig.2 Impedance diagrams of Al in NaCl solution on 32mA/cm^2 polarization current at various times

the instantaneous impedance diagrams of aluminium electrode

during polarization with $32mA/cm^2$ anodic current density at the times of 30 seconds, 4.5 minutes, 12 minutes, and 42 minutes. These plots exhibit the hemicircle and its diameter changed little with time but the frequency at the top of the hemicircle or the frequency of the real part in the center of hemicircle changed greatly with time. So it is clear that during polarization, the interfacial resistance changed small with time but interfacial capacity changed greatly. At different polarized current density, the change of the interfacial resistance, R_f, and the interfacial capacity, C_f, with time are shown in Fig.3. The change of R_f is small but the change of C_f is obvious. The plot of C_f vs. polarized time, tp, was found to be a curve consisting of two straight line with different slope and the second slope is the same for various polarized current density. This shows that there are two different stages in anodic polarizing prosess. When aluminium was anodicly polarized in NaCl solution, the film on surface was destroyed and the corroded region increased continually. This phenomena can be clearly observed with eyes. During this course, the small changes of the interfacial resistance suggests that it be independance of the status of the film and the great changes of the interfacial capacity present that it is dependance on the status of film.

Relationship between interfacial impedance and polarization current density

The impedance in an instant of beginning polarization was measured at various polarized current densities. Two kinds of the aluminium electrode were used in this measurement. The surface of one electrode has a thin film formed in the air. The surface of another was polished with abrasive paper just before the measurement. The results of measurement are shown in Fig.4.

The interfacial resistance of either surface with film or the surface polished decrease fastly with increasing of anodic polarization current density and the relationship of log R_f vs. log i_p is linear with slope of -1. (see Fig.4)

By contrast to the obvious changes of R_f, C_f hardly changes with polarized current density. The curves of C_f vs. i_p are horizontal lines, whether the electrode is with film or polished.

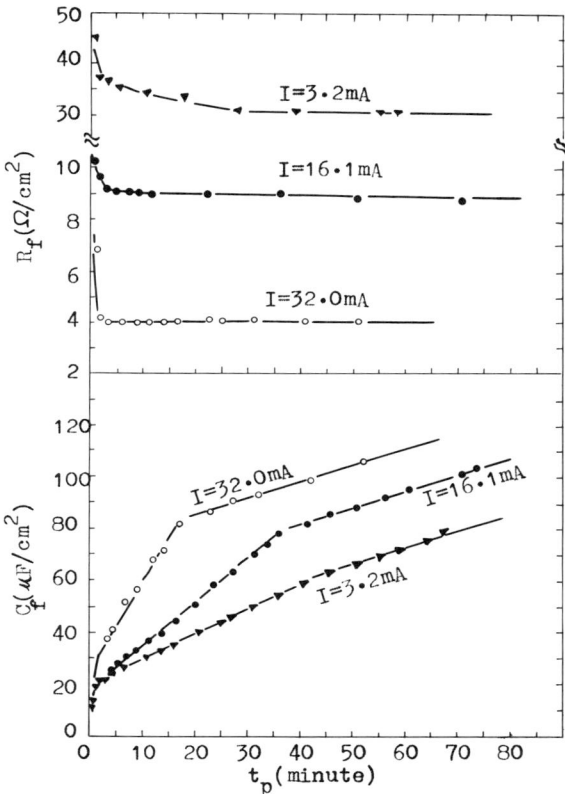

Fig. 3 The Change in interfacial resistance, interfacial capacitance of Al in 1N NaCl with polarized time at various polarized current

But the C_f of polished surface is far larger than that of surface with film, this shows that the film's thick of the former is far thiner than that of the latter.

From the fact of C_f independing of i_p, it is evident that in an instant of beginning polarization, the film on the surface is not broken and its status are the same as that before polarized. Nevertheless the R_f changed fast with i_p. This suggest that this resistance have the property of faradaic resistance.

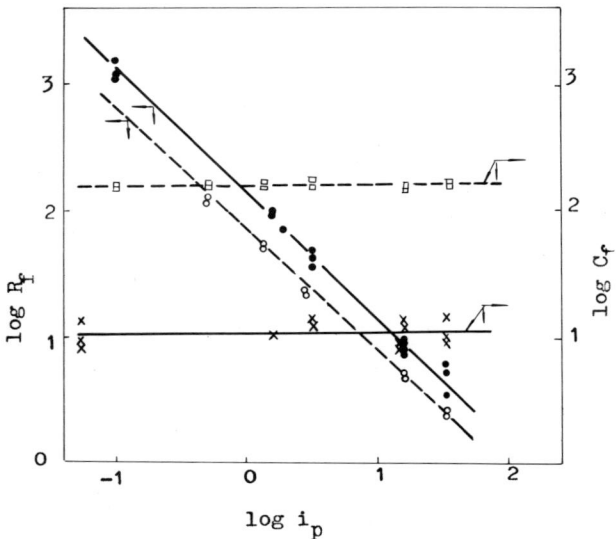

Fig.4 The change in interfacial resistance and interfacial capacitance of Al in 1N NaCl with polarized current

Relationship between interfacial impedance and status of film

  The interfacial impedance in the instant of beginning polization was measured immediately after polishing a part of surface by abrasive paper. Fraction of surface area, θ, is equal to the new polished area divided by all surface of specimen. The results are shown in Fig.5.

  Fig.5 shows that interfacial capacity, C_f, increases fast with increasing of θ, the curves of C_f vs. θ are linear, but the changes of interfacial resistance are very small. This reveals the status of film has a obvious effect on C_f but little effect on R_f.

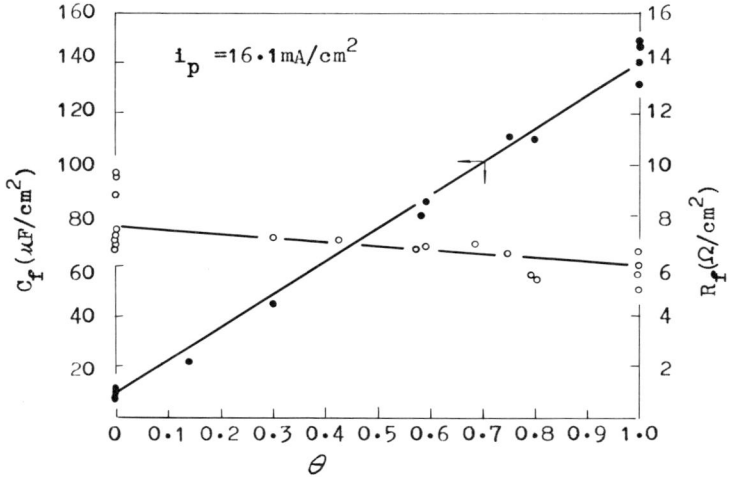

Fig.5 The Change in interfacial resistance and interfacial capacitance of Al in 1N NaCl with fraction of surface area, θ

DISCUSSION

According to the results above mentioned, the interfacial resistance of Al/NaCl system, R_f, depends on i_p but nearly is independance of status of film (thick, degree of dectroy), it has a property of faradaic impedance. On the contrary, interfacial capacity, C_f, depends greatly on film's status. So it is necessary to further discuss the property of R_f and the contribution of film to R_f and to C_f.

For a long time, the passive film on the surface of Al (it formed in the air) was known to be a insulator, the passivity of Al in aqueous solution is attributed to locking action of this insulator for movment of electron and ion through the film. [8]

However from the results of our experiment before mentioned, the interfacial resistance, R_f, of Al with this kind of passive film is related to the polarized current by the equation

$$\log R_f = K - \log i_p$$

where K is a constant relative to film's thick. it is difficult to interpret that with the view of the instulator film.

In order to explain and analysis the results from the experiment, we consider that the passive film on Al is the semiconductor film and make the reckoning based on the electrochemical principle as follow:

When the surface have not any film, the equation for speed of electrochemical reaction in polarized status is

$$i_p = i_o \exp(\beta nF\eta/RT)$$

where i_p is a polarized current and the other symbols have their usual meanings. This is Butter-Volmer equation neglecting the contrary reaction on strongly polarizing status.

If the i_p increase by Δi, the speed equation on new polarizing status will become

$$i_p' = i_o \exp(\beta nF\eta'/RT) \qquad (1)$$

since Δi can be written as

$$\Delta i = i_p' - i_p$$
$$= i_o \exp(\beta nF\eta'/RT) - i_o \exp(\beta nF\eta/RT)$$

and $\eta' = \eta + \Delta\eta$, equation (1) becomes

$$\Delta i = i_p(\exp(\beta nF\Delta\eta/RT) - 1) \qquad (2)$$

When $\Delta\eta$ is very small, Eq.(2) can be simplified as

$$\Delta i = i_p \beta nF\Delta\eta/RT$$

Thus in this case, R_f is given by

$$R_f = \Delta\eta/\Delta i = RT/\beta nFi_p \qquad (3)$$

Eq.(3) can be written as

$$\log R_f = \log(RT/\beta nF) - \log i_p \qquad (4)$$

Eq.(4) shows that the curve of log R_f vs. log i_p is linear and its slope is -1.

When a semiconductor film exists in the interface of metal and solution, Helmholtz layer potential ϕ_h will not be equle to metal surface potential ϕ_m (Fig.6)

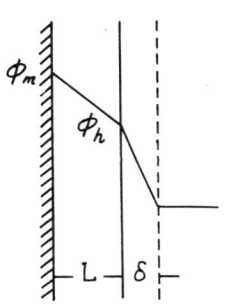

ϕ_m-metal potential

ϕ_h-Helmholtz layer potential

L-passive film thickness

δ-Helmholtz electric double layer's thickness

Fig.6 Potentials in interfaces of metal/film/solution (ϕ_m and ϕ_h are relative to potential of solution)

In the simple case of a thin intrinsic semiconductor film having no electronic surface state, there is a relationship between ϕ_m and ϕ_h [9]

$$\phi_m = \phi_h(1+L\varepsilon_H/\delta\varepsilon_F) \quad (5)$$

where ε_H is film's dielectric constant, and ε_F double layer's. According to equation (5), the following equation may be written

$$\Delta\eta_M = \Delta\eta_H(1+L\varepsilon_H/\delta\varepsilon_F) \quad (6)$$

where $\Delta\eta_M$ is the overpotential applied to metal, $\Delta\eta_H$ is real overpotential in Helmholtz layer. The substitution of Eq.(3) written for $\Delta\eta_H$ in Eq.(6) gives

$$\begin{aligned}R_f &= \Delta\eta_H/\Delta i = RT/\beta nF\ i_p = \Delta\eta_M/\Delta i(1+L\varepsilon_H/\delta\varepsilon_F) \\ &= R_f'/(1+L\varepsilon_H/\delta\varepsilon_F)\end{aligned} \quad (7)$$

where R_f' is apparent reaction resistance measured.

From Eq.(7)

$$R_f' = \frac{RT}{\beta nFi_p}(1+L\varepsilon_H/\delta\varepsilon_F) = R_f(1+L\varepsilon_H/\delta\varepsilon_F) \qquad (8)$$

It can be written as

$$\log R_f' = \log(\frac{RT}{\beta nF}(1+L\varepsilon_H/\delta\varepsilon_F)) - \log i_p \qquad (9)$$

Two conclution can be derived from the Eq.(9). The first, a plot of log R_f vs. log i_p should be a straight line and its slope is -1. The second, as the increase of film's thick the slope of the line will don't change, but its intercept increase. These conclutions agree with the results of experiment (Fig.4). This represent that it is correct to assume the thin oxide film on Al formed in the air to be a semicontuctor.

According to the foregoing disscussion, it can be constructed that the action of passive film on Al is to change the potential of Helmholtz layer and make the real overpotential decrease, so that reaction rate is affected by the action. Thus, it is reasonable to consider passive film to be a reaction resistance, R_m, in series with the electrochemical reaction resistance, R_{fra}, and the interfacial impedance consist of R_{fra} and R_m.

Interfacial impedance of the film failed locally

When passive film on Al was destroyed partly due to dissolution, the interfacial impedance consists of the impedance on dissolued area and the impedance on area with film by parallel connection. Both interfacial resisdance R_f and interfacial capacity C_f are expressed by

$$R_f = \frac{R_{dis} \cdot R_{film}}{(R_{film}-R_{dis}) \cdot + R_{dis}} \qquad (10)$$

$$C_f = C_{film} + (C_{dis} - C_{film}) \cdot \qquad (11)$$

Where R_{film} and C_{film} are unit resistance and capacity on area with film, R_{dis} and C_{dis} are unit resistance and capacity on dissolved area, respectively. θ is the fraction of surface area.

It is apparent from Eq.(10) that if R_{film} is approximately equal to R_{dis}, R_f will be approximately equal to R_{film} for any value of θ. This can be use to explain the results shown in Fig.3 and Fig.5. Small change of R_f with θ means that resistance on corroded area is approximately equle to the resistance on area with film. In fact, on condition of large current polarization, the difference of R_{film} and R_{dis} (from the result obtained for the polished samples) is not large. For example, on condition of $16.1 mA/cm^2$ polarized current, $R_{film}-R_{dis} = 1.75 \Omega$. From Fig.4, it is known that difference between R_{film} and R_{dis} increase with the decrease of polarized current. Thus, it can be understanded why interfacial resistance hardly changed with polarized time on condition of $32 mA/cm^2$ polarized current, but the changes is relatively obvious on $3.2 mA/cm^2$.

In the referrence [6], R_{film} was assumed to be far greater than R_{dis} and from this, the linearity of log R_f vs. log i_p with slope which is -1 was also derived. But the foregoing results shows that relation of $R_{film} \gg R_{dis}$ don't always exists but dependes on polarized current i_p.

From Eq.(11), it is known that interfacial capacity, C_f, increases following the increase of dissolved region Linear relationship can be obtained by ploting the C_f against θ. The experimental results consist with this conclusion (Fig.5).

It can be observed from Fig.3 that there are two stages in prosess of anodic polarization. In first stage the greater the polarized current is, the faster the C_f changes with time. In second stage, the changes of C_f with time are the same for any value of polarized current. Because of connecting of C_f with θ, Fig.3 represent that the corrosion prosess of Al in NaCl solution can be divided into two stages : the first, the corrosion rate increase with increasing of polarized current, i_p, the second, the rate independs on i_p.

CONCLUSIONS

The instantaneous impedance measurement at various polarization times extending from 1sec. to 80 min. was made on Al/NaCl system by using LapLace transformation method. The conclusions can be introduced as follow:

1. The corroded region on the surface of Al electrode increase constantly with the time during the anodic polarization. The interfacial resistance changes small with time, but the interfacial capacity changes greatly. It is clear that the difference between the resistance on the soluble region and the resistance on the region covered by the film is small, but the difference in capacity of these regions is great.

2. In an instant of beginning polarization, the film on the surface don't be broken and the dependance of interfacial resistance, R_f, on polarized current density, i_p, follows

$$\log R_f = K - \log i_p$$

where K is a constant proportional to the film's thick. The interfacial capacitance, however, is independent of polarized curren density, but only dependent on the film's thickness.

3. Interfacial capacity represents the status of film broken by dissolution. From the measurement results of C_f as a function of polarization time, it is established that the corrosion prosess of the Al/NaCl system during anodic polarization can be divided into two stages: the first, the corrosion rate is polarized current dependent, the second, is independent.

4. Assumming the passive film of aluminium formed in the air is a semiconductor, the results above mentioned can be explained very well. The electrochemical dissolution happend on aluminium surface covered by the film, still, follows Tafal's law. The presence of the film in metal/solusion interface makes the real overpotential of reaction decrease, so it behave as a resistance of reaction, R_m, proportional to its thickness. Even if so that, the electrochemical resistance, R_{fra}, known as faradaic resistance, is predominant over R_m and other impedance.

REFERENCES

1. M.J.Pryor: in Localized corrosion, R.W.staehle et al: eds., NACE Houston TX 1974, p.2.

2. G.C.Wood, W.H.Sutton, J.A.Richardson, T.N.K.Riley, A.G.Malhevbe: ibid, p.526.
3. Xu Naixin, G.E.Tompson, J.L.Dawson and G.C.Wood: Journal of chinese society of corrosion and protection, 1982, vol.2, ho. 4 p.27.
4. M.A.Heine, M.J.Pryor: J.Electrochem. Soc., 1963, vol.110, p.1205.
5. A.J.Brook, G.C.Wood: Electrochim,Acta, 1967, vol.12,p.395.
6. Shen Xingsu, Chang Yugin and Gao Xiauyue: Jounal of Chinese Society of Corrosion and Protection, 1983, vol.3, no.1, p.16.
7. Zhu Yingyang, Zhu Rizhang Wang Kuang and Zhang Wenqi:"Quick Measurement Method For AC Impedance of a Corrosion System", to be published
8. H.Kaesche: in Passivity in Metal, The Electrochemical Soc. Princeton. N.J. 1978, p.714.
9. Norio Sato: ibid. P.29.

PART II

Microstructures of Materials and Chemical Analysis

PROGRESS AND PROSPECTS OF MATERIALS CHARACTERIZATION AT SUBNANOMETER SPATIAL RESOLUTION USING FINELY FOCUSSED ELECTRON BEAMS

MICHAEL ISAACSON
School of Applied and Engineering Physics and the National Research and Resource Facility for Submicron Structures, Cornell University, Ithaca, New York 14853

ABSTRACT

The prospects of microanalysis using 0.5 nm diameter electron beams are reviewed. This paper will discuss the various characterization possibilities with emphasis on incoherent imaging, energy loss spectroscopy and convergent probe diffraction. Some examples indicating characterization at a 0.5 nm lateral spatial resolution scale will be presented and we will conclude with a discussion of the limits of nanometer characterization.

INTRODUCTION

The study of materials is increasingly becoming concerned with the relationships between nanometer scale structure and macroscopic properties since properties of materials are very dependent on local inhomogeneties on an atomic scale. It is the main premise of this meeting to look into the old and new methods for characterizing materials on an even smaller scale.

This paper is aimed at reviewing the principles behind an emerging technique capable of performing analysis on a scale approaching atomic dimensions. It is not a "trace" method in the usual sense of the word since it cannot detect even 10 PPM analysis over the volume it probes, but rather it is capable of probing such small volumes that individual atom species may be identified and chemical analysis can be achieved on aggregates of dozens of atoms or less. This technique is called Scanning Transmission Electron Microscopy (STEM) and utilizes the fact that fast electron beams can be focussed to diameters less than 0.3 nm and appropriate analysis can be carried out about the structural, chemical and electronic character of the material through which the electrons have just passed.

In a sense, STEM should be thought of as a hybrid between microscopic and spectroscopic methods, and my discussion will hinge along the lines that STEM is a spectroscopic/diffraction tool in which the scattering comes from nanometer scale volumes. An "image" is just an extension of incorporating scattering information from NxN adjacent volumes. (for a general discussion see ref. 1). In fact, essentially almost any spectroscopic or diffraction experiment that can be performed with a wide beam can be performed with a subnanometer diameter beam, the main difference being the number of atoms probed and therefore the ultimate signal available.

If we detect scattered electrons (or secondary products), it can be shown that the signal is given by:
$$S = NJ\sigma YF$$
where N is the number of atoms probed under a beam of current density J electrons/unit area, σ is the cross-section for the primary excitation, Y is the yield for the secondary product (for example, Y=1 for energy loss electrons; Y= the fluorescent yield for Xrays) and F is the efficiency with which we can detect the end product. In this paper we will consider only elastically scattered electrons and inelastically scattered electrons as our signal, so Y=1. However, detailed discussions of many other modes can be found in reference 2.

SCATTERING MECHANISMS FOR CHARACTERIZATION

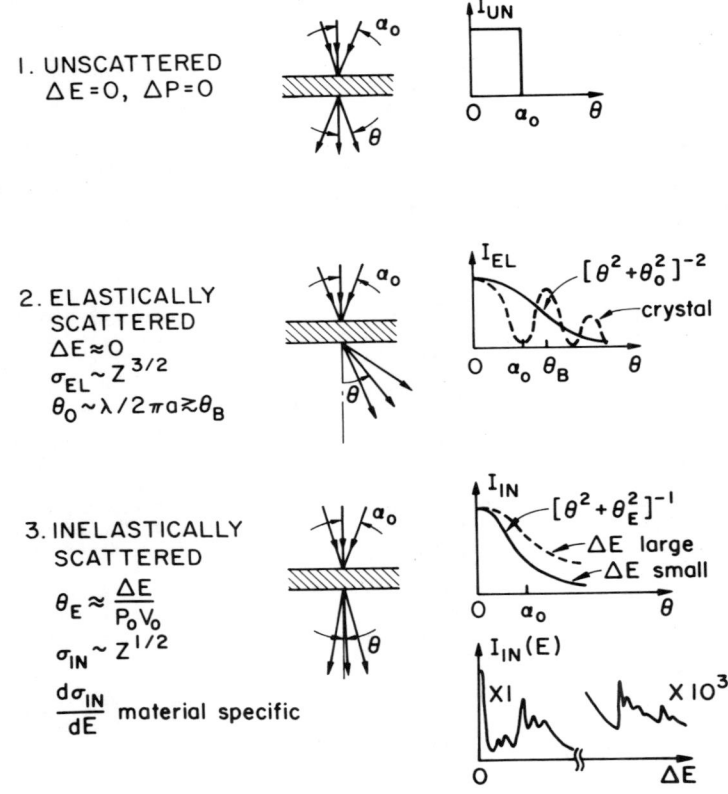

Figure 1. Schematic illustration of the various scattering mechanisms available for characteriztion in the STEM (apologies to the purists). The incident beam illumination half angle is α_0. 1. The angular distribution of the unscattered electrons is just $\theta = \alpha_0$, assuming uniform illumination. Typically this is 5-10 mradians at 100keV. 2. The angular distribution of electrons elastically scattered from isolated atoms falls off as $[\theta^2 + \theta_0^2]^{-2}$ in the Wentzel atomic model where the characteristic scattering angle $\theta_0 \approx \lambda/2\pi a$ (λ is the electron wavelength, a is the atom "radius"). Of course, in crystals, the characteristic angle is the Bragg angle θ_B. Both are typically of the order of degrees and greater than α_0. The total cross-section for elastic scattering is approximately proportional to $Z^{3/2}$. 3. The angular distribution for inelastic scattering is generally less than α_0, with the characteristic angle θ_E being dependent on the energy loss. The total inelastic cross-section is slightly material dependent, whereas the differential cross-section $d\sigma/dE$ is extremely material specific.

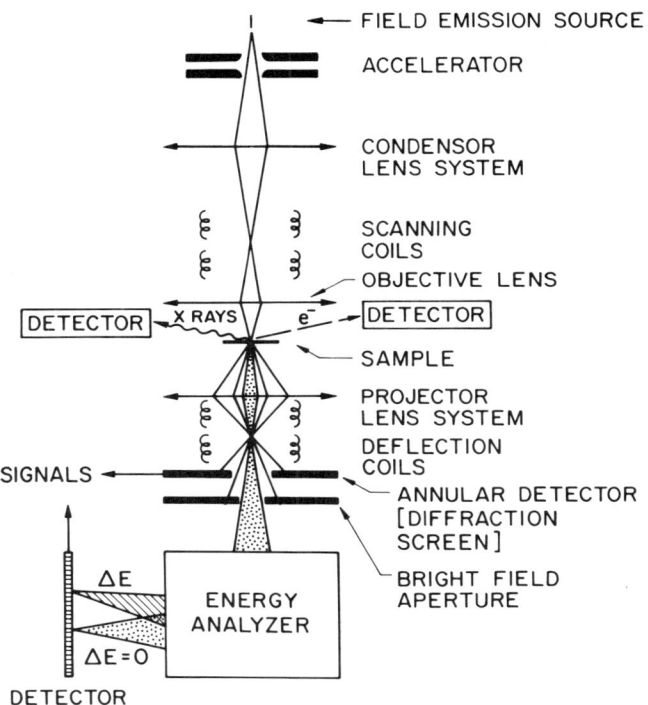

Figure 2. Schematic diagram of a scanning transmission electron microscope.

We can subdivide the scattered electrons into three groups (as schematically illustrated in figure 1). Although this is a gross simplification, it allows to give a simple, general description of the method. We define the "elastically scattered" electrons to be those that have undergone Coulomb collisions with the nuclei of the sample atoms with little energy loss, while the "inelastically scattered electrons" are those that have undergone collision with the sample electrons with finite probability of losing a considerable amount of energy. These are the electrons responsible for Xray and Auger Spectroscopy and both of them directly can give complementary information. The main point to be noted here, is that for thin enough samples (thickness $\leq$ one mean free path for scattering), there is somewhat of a spatial separation of these groups of electrons after they have left the sample. Thus, we can detect the predominant energy loss electrons through relatively smaller scattering angles (say several degrees or less), while the wider angle elastic scattering can be collected on an annular detector (see figure 2). Or with suitably positioned scan coils, we can record the entire scattering distribution (for instance a diffraction pattern).

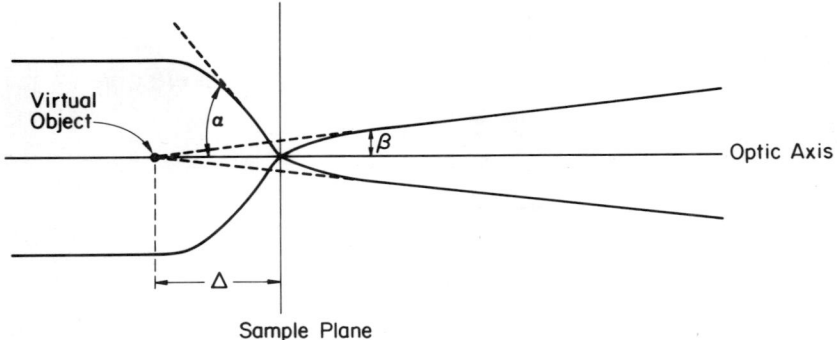

Figure 3. Illustration of beam convergence illumination angle α. The ratio α/β is determined by the exit properties of the objective lens.

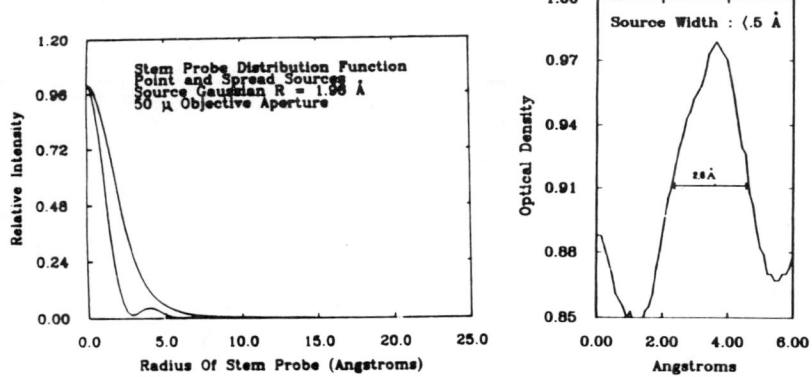

Figure 4. Calculated and measured electron probe distributions in the VG HB5 STEM. (From Reference 4).

It should be remembered however, that even if we could form a probe of zero diameter, we would not necessarily reduce our analysis volume below that with a finite beam. The reason is that, small diameter probes are formed using converging illumination, the smaller the diameter, the larger the convergence angle. (see figure 3). Thus, for finite thickness samples, the "depth of field" of the probe may result in the limit to the probed volume. For example, in the HB5 STEM (located in NRRFSS) at Cornell, one can achieve (measured) probe diameters of 2.6-4.8Å with a convergence half angle at the sample of 7.5 mrad [3,4]. With such a convergence angle, if the beam were focussed in the center of a 1000Å thick sample, it would be 7.5Å at the entrance face. Reducing the beam diameter to 1Å would not reduce that. In addition, to the extent that the inelastic scattering is non-local, the probed volume will be slightly larger. In spite of this, it has been shown that one can record energy loss spectra that vary on a subnanometer scale (eg, 5,6,7).

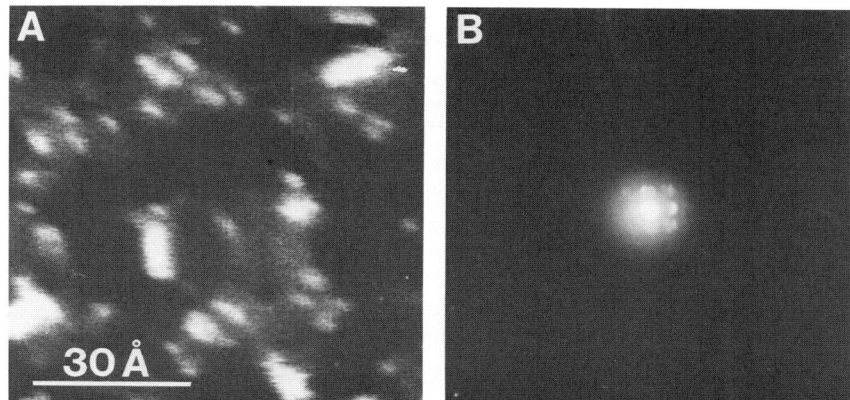

Figure 5. A. Annular dark field micrograph of a field of osmium atoms and atom clusters on a thin carbon support. B. Microdiffraction pattern from a cluster shown in figure A. The beam diameter is about 3-4 Å.

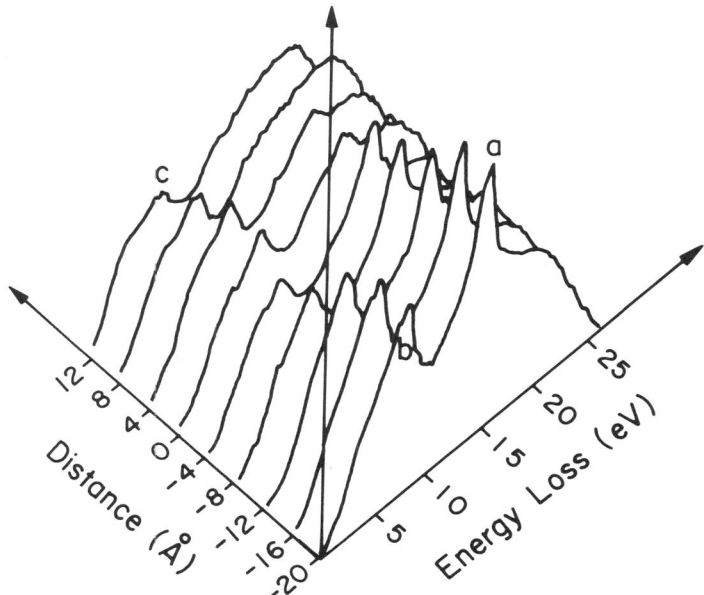

Figure 6. A series of electron energy loss spectra taken in 4Å increments as a 4.6Å diameter beam is moved perpendicular to an Al-AlF$_3$ interface. "a" is the bulk plasmon of Al (15eV), "b" is the surface plasmon of Al relaxed by the carbon substrate (8.0eV) and "c" is the surface plasmon of Al in AlF$_3$ (7.5eV) (from ref. 6).

Experimental Method

All the data presented in this paper have been obtained using a modified dedicated Scanning Transmission Electron Microscope (VG Scientific HB-5) located in the National Research and Resource Facility for Submicron Structures at Cornell. The microscope is equipped with a specially designed aberration corrected wide gap sector magnet spectrometer, digital beam blanking and positioning facilities and a Faraday cup to measure the absolute beam current. The cold field emission source gun is capable of providing one namp of beam current in a 0.46nm diameter spot. Details have been described before (eg. ref. 3).

Experimental Results

The STEM at Cornell has a final probe forming lens with a 3.5 mm spherical aberration constant, resulting in a measured minimum probe size of 2.6Å (see figure 4). For energy loss spectroscopy and microdiffraction, we generally use a small source demagnification resulting in more probe current but a slightly larger probe diameter (4.6Å). All the spectra and diffraction patterns shown were taken using such a 4.6Å diameter probe.

In figure 5A we show an incoherent dark field annular detector image obtained using electrons scattered from 2.5° to 10°. The background modulations are the thickness fluctuations in the carbon substrate, and the small brighter spots are indicative of single and multiple osmium atoms. Positioning the probe on a larger cluster (of the order of 1.5nm in size) results in the diffraction pattern in figure 5B. From the cluster size, volume considerations and scattering intensity, we estimate less than two dozen atoms in this cluster. Energy loss spectra from this cluster indicates clear peaks corresponding to the $Os_{L_{23}}$ levels.

In figure 6 we show energy loss spectra from an $Al-AlF_3$ interface [8]. The spectra were taken as the probe moved across the interface in 4Å steps [6] and it is clear that within +4Å of the "interface" the spectra changes considerably. Similar results have been demonstrated for other interfaces using both valence shell excitations and deeper core excitations. (see for example ref. 4,10 and figure 7). In fact, even though we are signal limited, M. Scheinfein has demonstrated that spectra obtained from 0.5nm diameter areas can be correlated reasonably well with optical data (eg. 3,4). In figure 8 we show the real and imaginary parts of the dielectric function ϵ derived from energy loss data taken from a 0.5nm diameter by 30nm thick volume of CaF_2. In figure 9 we show that L_{23} optical absorption edge of MgF_2 and AlF_3 derived from energy loss spectra obtained from a volume approximately $7nm^3$ [4] compared with xray absorption data taken from macroscopic samples [9].

CONCLUSIONS

We have given some selected examples of the use of 1/2 nm diameter probes for characterizing materials via dark field electron microscopy, energy loss spectroscopy and diffraction. Although, with a probe of 100 keV electrons of full width at half maximum of 0.46nm, 90% of all the electrons are within a 0.9nm diameter cylinder, it is still possible to detect significant changes in the electron energy loss spectra over lateral distances as small as 0.4nm. Such examples have been shown here and elsewhere [4,5,6,10]. The point to be emphasized is that because of the large converging angle of the illumination in the subnanometer diameter probe

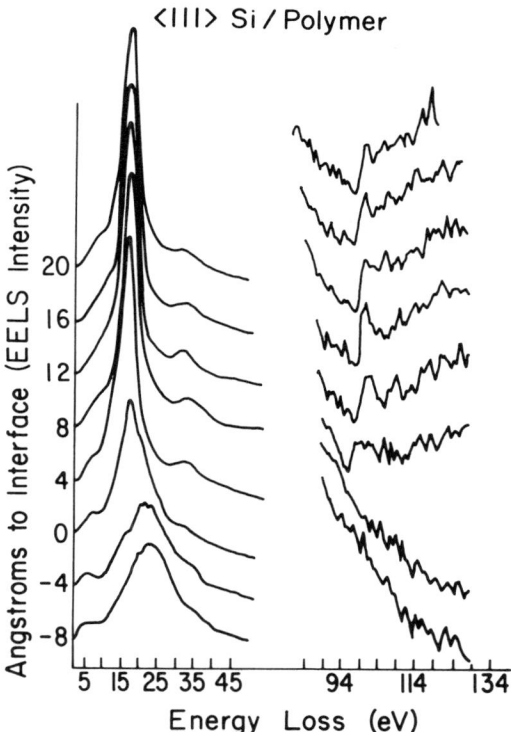

Figure 7. Low loss electron energy loss spectra obtained as a 5Å diameter probe is moved across a Si-polymer interface in 4Å steps. Left spectra are due to valence shell excitations; right spectra are in the vicinity of the Si L_{23} edge. Each spectrum corresponds to one 4Å step (from reference 10).

STEM, the beam shape in the sample is more of an hourglass figure. Moreover, this shape is convoluted with the scattering profile as well. Thus, for quantitative evaluation of concentration profiles across an interface, great care must be taken in the interpretation at the nanometer scale. Similar considerations of the probe shape on the determination of concentration profiles using xray spectroscopy have been made for larger probe diameters (>1nm) by Fries et al [11]. However, we have not yet gone quantitatively through the entire analysis for the subnanometer diameter beam case using energy loss spectroscopy.

In addition to the beam shape and scattering profile effects in small beam analysis, consideration also needs to be taken of the non-localization of the inelastic scattering event itself. Arguments based entirely upon impact parameter considerations (eg. 12) have to be evaluated carefully since low intensity contributions to scattering at large distances (>1nm) from the probe center may just be buried in the background noise signal. The point to

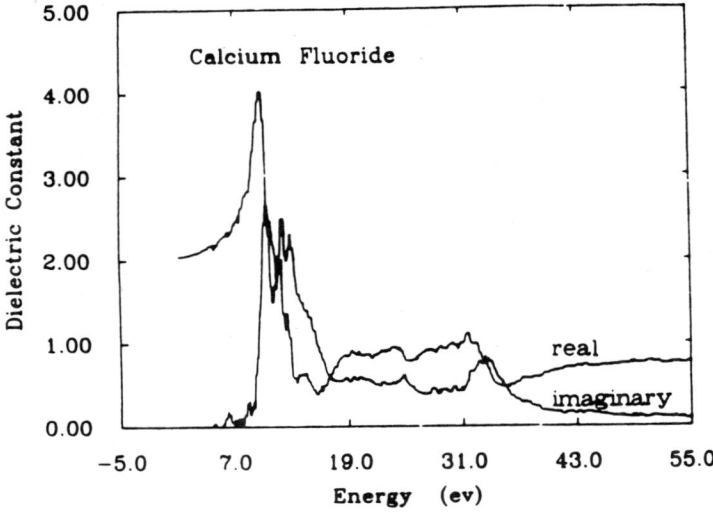

Figure 8. The real and imaginary parts of the complex electric constant of CaF$_2$ determined from electron energy loss spectra using a 5Å diameter probe of 100keV electrons. The measurement corresponds to the analysis of less than 400 molecules (from ref. 3).

be made here, is that selective low energy <10eV) surface excitations may be quite useful as a sensitive interface probe, yet such excitations may correspond to impact parameters >1nm.

Reducing the probe diameter below 0.3nm, will be useful from the microscopy point of view, but may not be any advantage for electron energy loss analysis. Since the convergence illumination angle will increase, the entrance beam size may still be as large as before. Of course, one may reduce the probe diameter, by increasing the electron accelerating voltage. In that case since the optimum probe illumination half angle is given by $\alpha=(4\lambda/Cs)^{1/4}$, where λ is the electron wavelength, then one can conceivably get a reduction in probe diameter without increasing the illumination angle. The main disadvantage in using higher energy electrons is the greater probability of knock-on collision damage (eg. 13), which would make it extremely more difficult to obtain useful analytical information from subnanometer areas. Perhaps an alternative approach might be to go to lower accelerating voltages (say <1keV) for surface type analysis. There are efforts to construct such low voltage scanning microscopes with 1-2nm diameter probes (eg. 14). However, the prospects of reducing such probes below the 0.5nm scale will most certainly require considerations of aberration correction and new novel spectrometer design. Neither one of these items appear to be insurmountable.

In summary then, it appears that we are entering a new era in which subnanometer area spectroscopic characterization is feasible. The task at hand is to refine the techniques of analysis to make such characterization practical.

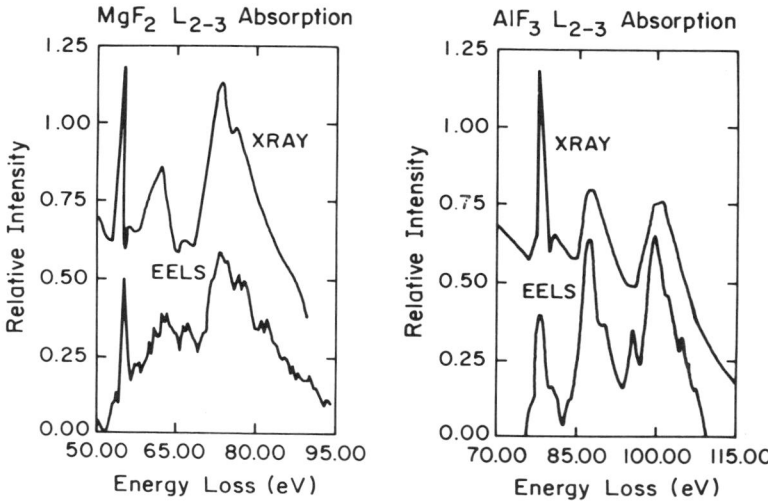

Figure 9. Calculated optical absorption in the vicinity of the metallic L_{23} edge for AlF_3 and MgF_2 from electron energy loss spectra (using a 5Å diameter probe) (ref. 4) and Xrays (ref. 9).

ACKNOWLEDGEMENTS

This work was supported by the National Research and Resource Facility for Submicron Structures under NSF grant #EC58200312. The author wishes to thank M. Scheinfein for his tireless efforts and A. Muray and H. F. Johnson for their always useful discussions.

REFERENCES

1. M. Isaacson, Analytical Chemistry (1985) submitted for publication.

2. M. Isaacson, "Microcharacterization: The Physics and Methodology of Materials Characterization from Volumes<$1\mu^3$." Plenum Publishing Company (to be published).

3. M. Scheinfein and M. Isaacson, SEM 1984/IV. ed. O. Johari. (SEM, Inc., AMS Ohare, Illinois, 1984) p. 1681.

4. M. Scheinfein, Ph.D. Dissertation, Cornell University, May 1985.

5. M. Scheinfein, A. Muray and M. Isaacson, Ultramicroscopy, 16, 223 (1985)

6. M. Scheinfein and M. Isaacson, Advanced Photon and Particle Techniques for the Characterization of Defects in Solids. (J. B. Roberto, R. W. Carpenter and M. C. Wittels, eds.) (Materials Research Society, Pittsburgh, 1984), p. 343.

7. J. C. H. Spence and J. Lynch, Ultramicroscopy 9, 267 (1982).

8. A. Muray, M. Isaacson, I. Adesida and M. Isaacson, J. Vac. Sci. Tech. B.3. 367. (1985).

9. R. J. Bartlett, O. G. Olson and D. W. Lynch, Phys. Stat. Sol. (b) 107.93 (1981).

10. M. Isaacson and M. Scheinfein, Proc. 43rd Ann. EMSA Meeting, Louisville (1985 in press).

11. E. Fries, A. J. Garratt-Reed and J. B. Vandersande, Ultramicroscopy. 9. 295 (1982).

12. J. P. Langmore, M. Isaacson and H. Rose, Optik 41, 92 (1974).

13. M. Isaacson EMAG 83 (Inst. Phys. Conf. Ser. 68) ed. P. Doig (1983) p. 1.

14. E. D. Boyes EMAG 83 (Inst. Phys. Conf. Ser. 68) ed. P. Doig (1983) p. 485-488.

TRANSMISSION ELECTRON MICROSCOPE INVESTIGATION OF
SPUTTERED Co-Pt THIN FILMS

P. ALEXOPOULOS*, R. H. GEISS* AND M. SCHLENKER**
*IBM Research Division, 5600 Cottle Road, San Jose, CA 95193
**Laboratoire Louis Neel, CNRS/USMG, Grenoble, France

ABSTRACT

Thin films of Co-10 at% Pt, ranging from 15 to 90 nm in thickness, have been DC-sputtered at various temperatures on to carbon-coated mica, carbon substrates on copper grids, or (001) silicon single crystals under 3 μm pressure of Ar, using targets of the alloy in the hexagonal phase, at growth rates of 9 nm/min. The samples were investigated by TEM, using bright- and dark-field imaging, lattice imaging, selected area diffraction and both Fresnel and focussed Lorentz modes. The primary structure of the films was found to be hexagonal, with a = 0.255 nm and c = 0.414 nm. For the samples sputtered at room temperature, the grain sizes were on the order of 0.1 μm on carbon-coated mica and carbon-substrate grids, and approximately an order of magnitude smaller on silicon substrates. Heavy streaking along the [001] of the hexagonal matrix was observed on diffraction patterns for grains having the [001] parallel to the surface; this streaking was found to be associated with the presence of a high density of faults parallel to the (001). In films sputtered on to carbon-coated mica at 225 °C, where a substantial reduction of the coercivity is observed, the overwhelming majority of the grains had the (001) basal plane parallel to the surface. Lorentz microscopy showed the magnetic domain structure in films grown on silicon to be markedly different from those grown on the carbon substrates, and further changes occurred for the films grown at elevated temperatures.

INTRODUCTION

The magnetic properties of Co-Pt alloys has been studied extensively recently [1-5]. RF sputtered films having very high coercivities with Pt in the range of 5-50 at% have been reported.
In this paper we report on an investigation of the microstructure of DC sputtered Co-10 at% Pt thin films using transmission electron microscopy, with both conventional and Lorentz techniques, for films prepared on various substrate materials, and varying the film thickness and deposition temperature. The observation of heavy streaking in the [001] of some of the diffraction patterns was determined to be associated with the presence of a high density of faults parallel to the (001) basal plane. Bulk magnetic measurements for these films will be discussed in view of the microstructural observations.

EXPERIMENTAL

Film Deposition

Thin films of Co- 10 at% Pt were deposited using conventional DC sputtering techniques. The target was an alloy of the desired composition in the hexagonal phase. Film deposition rates on the order of 9 nm/min were used to deposit films onto carbon-coated mica (CCM), carbon substrates on copper grids (CSG) and (001) silicon single crystal coupons. The film thickness was varied over the range of 15 to 90 nm and deposition temperatures ranged from room temperature to 250 °C. A rotating substrate holder (60 rpm) insured the uniformity of the circumferential composition of the deposited films. The sputtering chamber was typically evacuated to a base pressure between 5×10^{-7} and 1×10^{-6} torr prior to the introduction of high purity argon at a controlled pressure of $\approx$ 3 mtorr.

Characterization of the Films

The magnetic properties of all the sputtered films were measured using a vibrating sample magnetometer in both in-plane and transverse directions. Microstructural characterization was done for films on silicon substrates via x-ray diffraction and for films removed from the substrate via TEM with a Philips 301 S(TEM).

Observations of the micromagnetic structure were made using Lorentz electron microscopy in both the phase contrast (Fresnel) mode and in the diffraction contrast (focussed) mode. In doing experiments on magnetic materials inside the TEM it is necessary to turn off the objective lens of the microscope in order to preserve the remanent state of the material. In so doing the maximum magnification available for imaging is on the order of 8000X.

Films deposited on CSG could be inserted directly into the TEM; films deposited onto CCM were floated off on water and picked up on 200 or 300 mesh grids; films sputtered on to silicon crystals were prepared using an elaborate technique incorporating mechanical polishing, chemical thinning and ion-milling.

RESULTS AND DISCUSSION

Magnetic Properties

The room temperature coercivity for Co-10 at% Pt thin films sputtered on to CCM or (001) Si single crystal substrates is plotted in figure 1. The strong influence of the substrate is easily seen. In the case of films grown on Si, the room temperature coercivity shows a large decrease as the film thickness increases, in agreement with previous reports [1,5].

Figure 2 is a plot of the room temperature coercivity for 15 nm Co-10 at% Pt films sputtered on to the CCM or (001) Si substrates as a function of deposition temperature. For both substrates a large decrease in the coercivity is observed with increasing substrate temperature.

The squareness of Co-Pt films on silicon varied between 0.8 and 0.9, but it was only in the range of 0.7 to 0.85 for CCM.

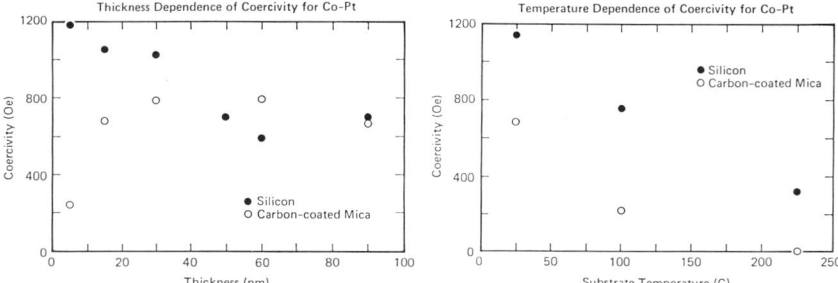

Fig.1 Room temp. coercivity vs film thickness for Si and CMM substrates.

Fig.2 Room temp. coercivity vs deposition temp. for Si and CCM substrates.

The squareness was found to decrease with increasing film thickness.
Magnetic data for films deposited directly onto the CSG is not available due to problems associated with the size of the specimen.

Structural Studies

A typical x-ray diffraction pattern of 15 nm Co-Pt film deposited at room temperature is shown in figure 3. Analysis of the peaks showed that the structure was hexagonal with a = 0.2542 nm, c = 0.415 nm. The strong 002 reflection suggests that there may be some preferred orientation of the (002) parallel to the film surface. Most notable in the x-ray pattern, however, is the broad asymmetric 101 peak extending through the nearby 101 and 002 peaks. This asymmetrical broadening of diffraction peaks toward lower angles, or higher d-values, suggests that compositional disorder accompanies the displacement disorder of the stacking faults; in agreement with observations from electron diffraction presented next.

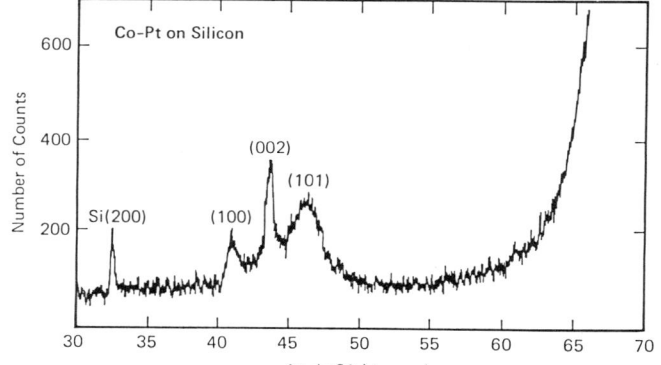

Fig. 3 X-ray diffraction chart from Co-10 at% Pt film deposited on a silicon (001) substrate.

Extensive microstructural characterization was performed with transmission electron microscopy, including bright and dark-field imaging, lattice imaging, SAD and magnetic imaging using the various Lorentz modes.

The film structure determined by electron diffraction shows the matrix to be hexagonal with a = 0.255 nm and c = 0.414 nm, in complete agreement with the x-ray data.

Figures 4a and b show electron micrographs of 25 nm films sputtered at R.T. on Si and CCM, respectively. The differences in these micrographs clearly show the influence of the substrate on the microstructure. Whereas films grown on Si crystals, Fig. 4a, show a uniform distribution of equiaxed grains with an average grain size of ≈ 30 nm, films grown on CCM, Fig. 4b, show two distinctly different grain structures. One consisting of an agglomerate of small grains having diameters in the range of 10 to 30 nms, the other is composed of larger (50 to 200 nm) diameter grains, some with extended areas displaying parallel B/W striations.

Electron diffraction patterns from these films are inserted on the figures. The diffraction pattern from the film grown on the Si substrate indicates a random in-plane orientation of the grains. The 101 ring is weakly diffuse, in support of the x-ray observations. Weak diffuse diffraction rings can also be seen for all reflections of the type h0l on the original pattern. Superposed on the diffraction pattern is the [001] pattern from the Si substrate.

The two different grain structures of the film on CCM give two distinctly different diffraction patterns as can be seen in the inserts on Fig. 4b. The areas of the film with the agglomerated small grain structure give rise to diffraction patterns showing a basal plane (001) orientation parallel to the film surface. A typical diffraction pattern from an area approximately 1 µm in diameter is given and indicates a very high degree of alignment of neighboring micro-grains. Areas of the film showing the heavily striated microstructure give diffraction patterns in which the [001] is in the plane of the film and show heavy streaking in the [001] direction at particular hkl reflections. (All statements made for films deposited onto CCM apply equally to films deposited on CSG under the same conditions with the exception that, in the case of films on CSG, the striated grains are somewhat larger.)

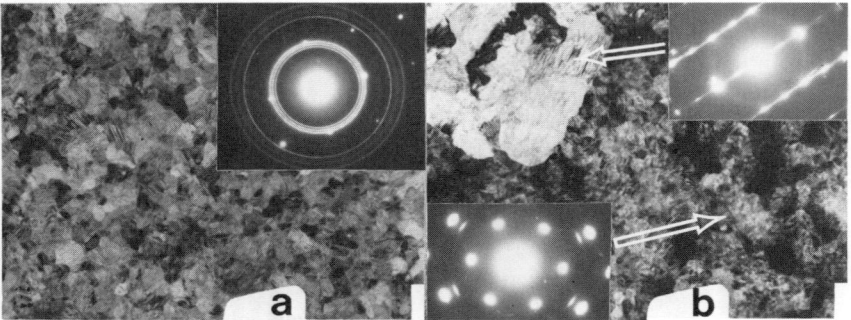

Fig. 4 Micrographs and SAD of 25 nm thick Co-10 at% Pt films deposited at R.T. onto (a) silicon (001) and (b) CCM substrates. The marker indicates 100 nm. on all the micrographs in this paper.

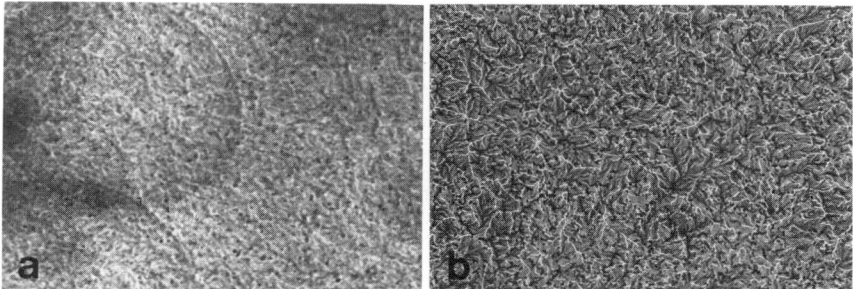

Fig. 5 Defocussed Lorentz images showing the magnetic domain structure in films deposited onto (a) silicon (001) and (b) CMM substrates at room temperature.

Figures 5a and b show defocussed Lorentz images from films deposited on the Si and CCM substrates, corresponding to the films discussed in Figs. 4a and b, respectively. Films grown on Si substrates exhibit a magnetic structure consisting of white and black dots with short segments of domain walls connecting the dots. The white and black dots are thought to be Bloch lines having the magnetization direction normal to the film plane, with the Lorentz contrast arising due to the magnetic curling around the Bloch line converging (or diverging) the electron beam to a point (hole). The domain wall segments are thought to be Neel walls arising from the in-plane curling of the magnetization around the Bloch lines. Typical separation of the Bloch lines is several 10's of nms. High coercivity is usually found to be associated with this type of domain structure [8]. The film grown on CCM, Fig. 5b, shows a similar domain pattern, but on a much larger scale with average domain diameters on the order of 0.5 to 1.0 µm.

Investigation of films deposited at room temperature over the thickness range of 5 to 90 nm showed that the average grain size remained essentially unchanged over the whole thickness range. There was an increase in the volume of the non-striated phase, that is, the phase which has the (001) basal plane parallel to the film surface, as the film thickness increased. This increase in the volume of the [001] normal phase was accompanied by a decrease in the coercivity of the same films. An additional feature observed in the case of thicker films (≥ 60 nm thick) was the presence of a regular pattern of voids, see Fig. 6, in the [001] normal phase.

Fig. 6 Image showing voids in a 90 nm film made at room temp. The voids outline prior grain boundaries in the nucleation and growth stages. The marker indicates 100 nm.

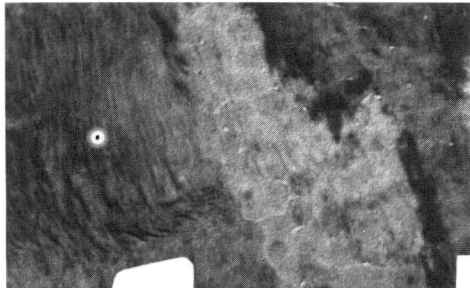

Since such regions display single crystal-like diffraction patterns with only a few degrees of arc in the diffraction spots,(see the diffraction pattern in Fig. 4b) this suggests that in the early stages of nucleation and growth of the island grains, there was a very good agreement between orientations of neighboring grains. From the figure one can estimate that the islands grew to approximately 15 to 20 nm in diameter before touching.

The effect of deposition temperature on the microstructure and micromagnetic structure will be discussed using as an example a 15 nm film deposited on CCM at 225 °C. The microstructure, shown in Fig. 7a, in contrast to films deposited at R.T. shows only the non-striated phase. This observation is also supported by electron diffraction where no evidence for streaking, associated with the striated, non-normal [001] phase, was found. The defocussed Lorentz image, Fig. 7b, showed a cross- tie domain wall structure typical of strong in-plane magnetization and indicative of a low coercivity film.

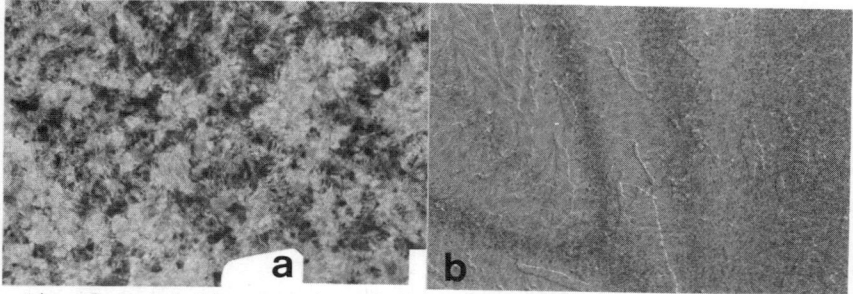

Fig. 7 (a) Micrograph and (b) Lorentz image from a 15 nm film deposited at 225 °C onto a CCM substrate. The marker indicates 100 nm.

Defect Analysis

From the previous discussion it is apparent that the features associated with high coercivity are grains that show a heavily striated microstructure and are oriented with [001] in the plane of the film. Typical examples of this structure, shown here at higher magnification, are given in Figs. 8a and b for films deposited onto Si and CCM, respectively. The difference in grain size is immediately seen and the striations in the silicon substrate film are quite apparent in almost all of the grains. An even higher resolution micrograph taken from a film on CCM is shown in Fig. 9. Inspection of this image, which shows the (001) lattice of the Co-Pt solid solution at a spacing of 0.415 nm, indicates how the periodic arrangement of the Co-Pt (001) planes, region A in the figure, is interrupted by a fault, area B, and then returns to the regular period of the Co-Pt lattice. As previously discussed, SAD from such regions shows heavy streaking in the [001] for certain of the diffraction spots. Detailed analysis of the streaking and its extinction under different conditions of specimen tilt showed that streaking does not occur for planes of the type $\{001\}$ and for planes of the type $\{hkl\}$ streaking occurs when $h-k \neq 3n$.

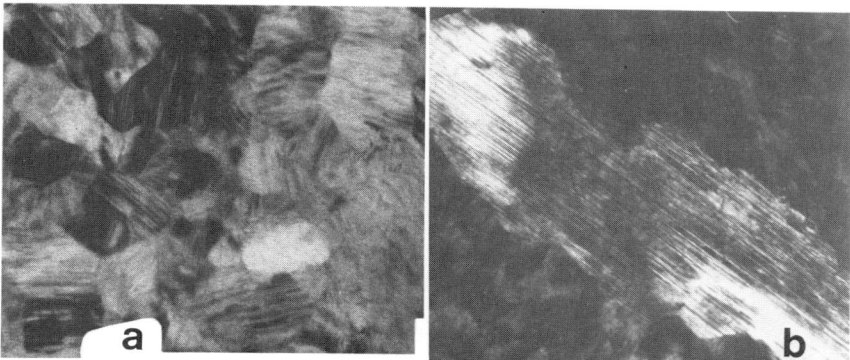

Fig. 8 High magnification micrographs from 25 nm thick films deposited at R. T. on (a) silicon and (b)CCM substrates. Note the heavily striated structures, especially in (a). The marker indicates 100 nm.

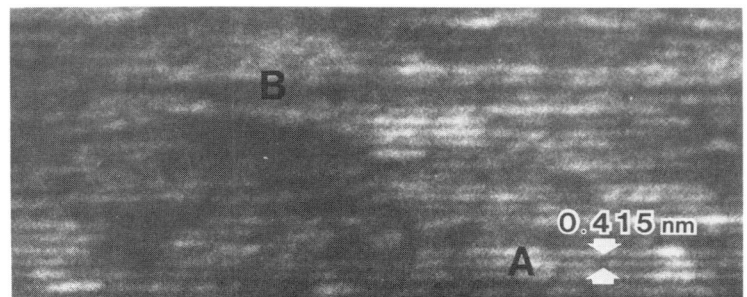

Fig. 9 Lattice image of faulted structure from film deposited on CMM substrate. Region A shows hexagonal stacking and B indicates a faulted area.

The extinction conditions found here are in accord with those determined by Warren and Guinier [9,10] when considering the effects of distortion and twin faulting in HCP materials.

One model that would explain the observations is shown schematically in Fig. 10. In this model, the hexagonal ABAB.. stacking is interrupted by a FCC stacking fault with ABCABC.. stacking for a short distance. While the experimentally observed streaking of certain diffraction spots is in agreement with the Warren-Guinier extinction conditions, analysis of streaking due to displacement disorder alone does not allow streaking of the 000 spot. Inspection of the diffraction pattern given in Fig. 4b from the faulted region shows strong streaking through the 000 spot in the [001]. Careful tilting experiments were done to eliminate the possibility of double diffraction, and as shown in Fig. 11, streaking does indeed exist at the 000 spot. This observation suggests that substitutional disorder is correlated with the displacement disorder of the stacking fault. This is also in agreement with the previously discussed asymmetry in the x-ray diffraction pattern.

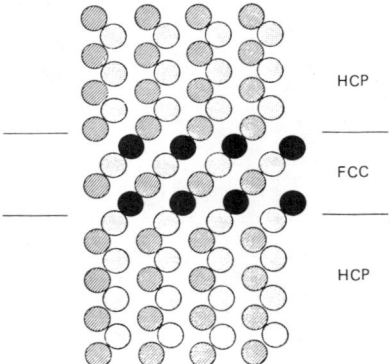

Fig. 10   Schematic depicting proposed FCC stacking fault.

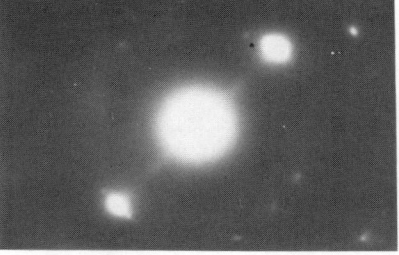

Fig. 11   Diffraction pattern taken under single diffraction conds. The only streaking observed passed through the 000 central spot, suggesting substitutional disorder.

The substitutional disorder is most likely manifest by a phase separation of the Co and Pt, especially in the formation of the stacking fault. It is easy to imagine that the Pt separates to form a Pt-rich FCC phase which is accommodated by the faulting of the Co-rich hexagonal phase. Although we have no unambiguous electron diffraction data to support this postulate, the occurrence of a FCC phase has been observed in the study of alloy films with higher Pt concentrations [1]. We have also used optical diffraction to simulate both cases. For pure displacement disorder, which would be the case in a homogeneous alloy, no streaking is observed through the central spot; while for a combination of displacement and substitutional disorder, such as is being suggested here, there is definite streaking through the 000 spot. The streaking always occurs normal to the habit plane of the stacking fault.

Coercivity and Microstructure

Faulting, and the accompanying dislocation structure, along with a modulation in composition should provide effective barriers to either domain wall movement or spin reorientation and thus result in higher coercivities. The data presented here suggest a direct relation exists between the volume of defects, via the volume of the striated grains, and coercivity.

CONCLUSIONS

Transmission electron microscopy has been used to investigate the crystallographic microstructure and micromagnetic domain wall configuration of an assortment of sputtered thin films of Co-10 at% Pt alloy. Bulk magnetic measurements show that the films have a very high coercivity, which is dependent on both the deposition temperature and film thickness. Two distinct orientations of the hexagonal matrix phase were seen, one with [001] normal to the film plane and one with [001] lying in the film. Associated with the latter is a heavily striated defect structure thought to be due stacking faults on the (001) planes of the hexagonal structure. Both electron and x-ray diffraction show that substitutional disorder is associated with the displacement disorder of the stacking faults.

ACKNOWLEDGEMENTS

The authors would like to acknowledge the help and support of Mohammad Moshref, William LaFontaine, Kathy Barron, Grace Lim and Drs. William Parrish and David Bullock.

REFERENCES

1. J. A. Aboaf, S. R. Herd and E. Klockholm, IEEE Trans. Magn. MAG-19, 1514 (1983).
2. T. R. McGuire, J. A. Aboaf and E. Klockholm, J. Appl. Phys. 55, 1951 (1984).
3. M. Kitada and N. Shimizu, J. Appl. Phys. 54, 7089 (1983).
4. M. Yanagisawa, N. Shiola, H. Yamaguchi and Y. Suganuma, IEEE Trans. Magn. MAG-19, 1638 (1983).
5. V. Tutovanet and V. Georgescu, Thin Solid Films 61, 133 (1979).
6. J. P. Jakubovics, Electron Micr. Matls. Sci., Part IV (Pub. by the Commission of the European Communities, Luxembourg, 1976), p.1303.
7. A. Guinier, X-Ray Diffraction, (W. H. Freeman Co.,San Francisco, 1963), p. 279-292.
8. H. Riedel, phys. stat. sol. (a) 24, 449 (1974).
9. B. E. Warren, X-Ray Diffraction, (Addison-Wesley, Menlo Park, CA, 1969), p. 298-303.
10. A.Guinier, X-Ray Diffraction, (W. H. Freeman Co., San Francisco, 1963), p. 226-233.

CHARACTERIZATION OF THE Ni/NiO, INTERFACE REGION IN OXIDIZED HIGH PURITY NICKEL BY TRANSMISSION ELECTRON MICROSCOPY

HOWARD T. SAWHILL AND LINN W. HOBBS
Department of Materials Science and Engineering
Massachusetts Institute of Technology
Cambridge, MA 02139, USA

ABSTRACT

Ni/NiO interface structures were investigated using TEM, and the observed structures were compared with current heterophase interface models. Relative magnitudes of Ni/NiO interfacial energies were obtained from measurements of dihedral angles at triple grain junctions between Ni and NiO grains. Extra reflections in diffraction patterns from oxide grains adjacent to the Ni/NiO interface were compared with kinematical structure factor calculations for several proposed structures.

INTRODUCTION

Metal-metal oxide interfaces constitute a current area of interest in interfacial science, and their properties are of considerable technological importance to the oxidation and corrosion communities. The structure and properties of such heterophase interfaces have received little prior attention principally due to the difficulties in producing suitable specimens for such studies. Nickel oxidation provides a convenient method for studying such a heterophase interface because the crystallography is inherently simple (both NiO and Ni have FCC Bravais lattices), the oxide scales that grows on pure nickel are protective and consists of a single phase of oxide (NiO), and the overall properties of both Ni and NiO are reasonably well characterized. Recent developments in specimen preparation techniques [1] have provided methods for viewing transverse sections through oxidation scales, allowing a direct view of the interfacial region. In this paper, experimental results of Ni/NiO interfacial structures, measurements of interfacial energies, and structural modifications of NiO grains adjacent to the Ni/NiO interface are presented.

EXPERIMENTAL DETAILS

Strip (20x5x1mm) and disk (3mm) specimen were cut from 99.995% pure polycrystalline nickel sheet (30-μm average grain size), mechanically hand polished through 1-μm diamond paste, vacuum annealed at 1100K, electrochemically polished, and oxidized in pure flowing oxygen at 3kPa pressure at 1273K for 15 minutes. Details of the thinning procedures for transverse and parallel sections are described in detail in reference [2]. Boundary and dihedral angle measurements were determined by projection analysis following tilting about two independent axes, checking for consistency using vector algebra representations for tilted thin foils [3]. TEM was performed using 200 keV electron energies in a JEOL JEM200CX instrument.

Ni/NiO INTERFACE STRUCTURES

Oxide grains formed during the oxidation of nickel have been observed to favor specific orientation relationships with the underlying metal substrate known as topotaxies [4]. The scalloped nature of the Ni/NiO interface illustrated in Fig. 1 shows that the orentations of the Ni/NiO interface planes are often inclined to the nominal specimen surface orientation. The geometric matching across these inclined interfaces differs from the matching across interfaces parallel to the substrate surface, yet it was predominantly in the inclined interfaces that periodic structural features were observed. For example, the crystallographic orientation of the Ni and NiO grains in Fig 2 lies close to the $(1\bar{1}1)[110]$ NiO//(001)[110] Ni topotaxy, but the interface plane is quite inclined, resulting in high-index planes in both the structures lying parallel to the interface plane. It is unlikely that coincidence or near coincidence of atomic positions is an important aspect of the matching between the Ni and NiO in this case, but near parallelism of low-index planes having nearly parallel traces in the interface plane (in a plane matching geometry) appears to be a more important criterion. Plane matching models [5], which are special cases of CSL-DSC models [6] in homophase interfaces, consist of parallel traces of low index (equivalent in homophase cases but may differ for heterophase cases) planes lying in the boundary plane. Twist misorientations of these traces form a moire pattern, and the introduction of interface dislocations in periodic arrays acts to locally shear these traces back into alignment, thereby accommodating the misorientation away from lower energy configurations by localization of mismatch into line defects. The analysis of features in Fig.2 suggests that the twist misorientation between the traces of the (001)NiO and (111)Ni planes (which differ in interplanar spacing by less that 1.4%), is accommodated through the presence of a set of parallel interface screw dislocations. Maximal contrast of these dislocations was observed with g parallel to the line sense of the dislocations, while no visible contrast was observed with g perpendicular to the line sense; however the Burgers vector for these line defects was not unambiguously determined.

In a second interface shown in Fig. 3, the line features are observed to lie nearly parallel to the direction perpendicular to the trace of the (111) planes in both Ni and NiO grains. For this particular case, the orientation of the grains is not one of the more commonly observed topotaxies. The interface orientation is changing along its length, but it appears that plane matching between (111) planes may account for the presence of the periodic dislocations which again possess predominantly screw-type behavior. In this case, the two interplanar spacings differ by 18% . It is possible that some of the features present are associated with steps in the interface, but transverse sections (higher magnifications of Fig.1) show interfaces to consist of relatively planar sections and step heights on the order of 1 nm would result in the interface stepping out of the thin foil. Also, images such as shown in Fig. 4 show uniform thickness fringes along the length of the interface. The line features in Fig. 4 lie parallel to the projection of [110] directions in the Ni, suggesting that these interface defects may have Burgers vectors equal to lattice vectors of the Ni lattice.

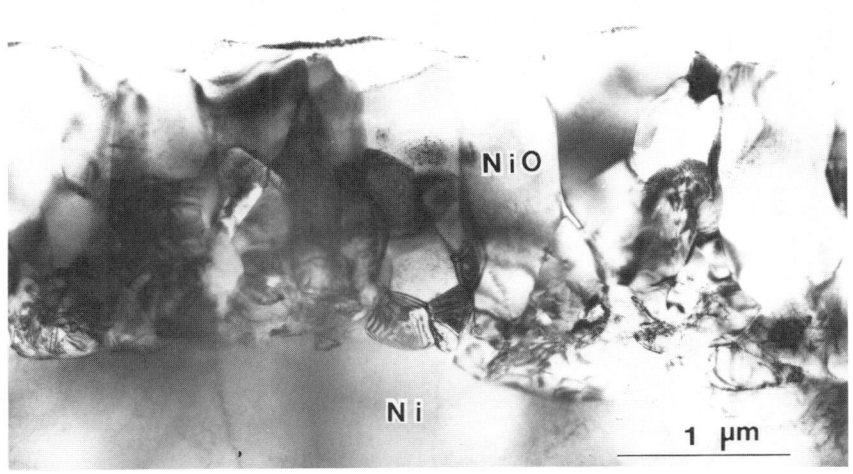

Fig. 1 Transverse section of oxidation scale formed after 15 minute oxidation of Ni at 1273K.

Fig. 2 Dark field image of a Ni/NiO interface exhibiting periodic interface dislocations.

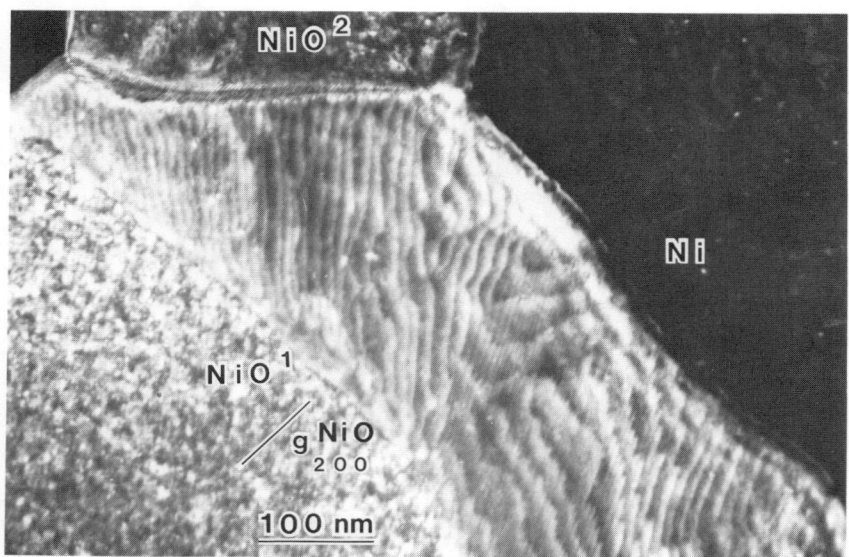

Fig. 3 Dark field image of a Ni/NiO interface exhibiting changes in the interface structure with changes in the orientation of the interface.

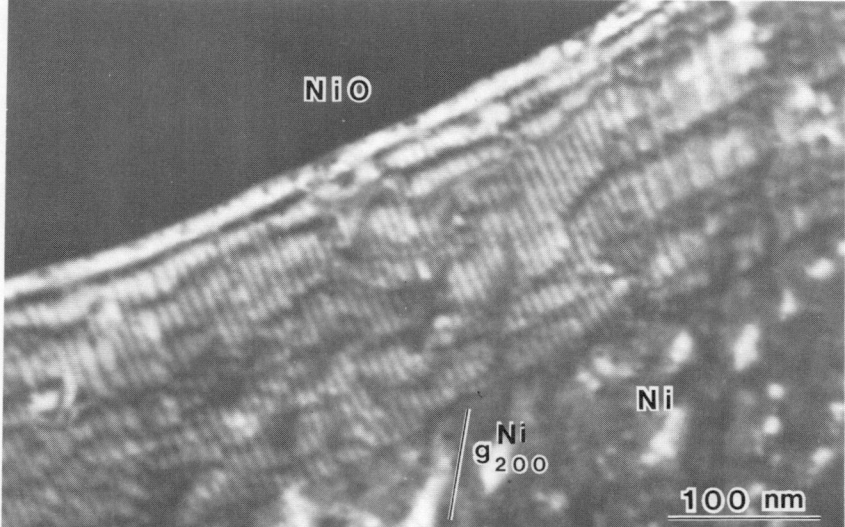

Fig. 4 Dark field image of a Ni/NiO interface illustrating uniform thickness fringes along the length of the interface.

Ni/NiO INTERFACIAL ENERGIES

The relative energies of Ni/NiO interfaces in oxidized nickel samples were determined from dihedral angle measurements at triple grain junctions between Ni and NiO grains. The energies are determined as ratios with Ni or NiO grain boundary energies by use of the Herring equilibrium equation [7] which includes interfacial torque terms. Torque terms arise from the variation of interfacial energy with orientation of the interface plane and are representative of the angular derivatives in the Wulff plot constructions. Little is currently known about the geometry of the Wulff plots for Ni/NiO interfaces due to the lack of interfacial energy measurements. The influence of symmetry constraints on interfacial energies for Ni/NiO interfaces has been considered [8], but in most cases the symmetry (including most of the topotactic orientations except epitaxial) is quite low and does not impose noticeable constraints. The triple grain junction where two Ni/NiO interfaces meet a Ni grain boundary is shown in Fig.5. The curvature of the two Ni/NiO interfaces is gradually changing and no sharp changes in the morphology are observed, so it is believed that the energy measured is representative of an average value over a reasonably wide angular range, and the torque terms are considered small. With torque terms neglected, the relative ratios of the two Ni/NiO interfacial energies with the Ni grain boundary energy are as follows:

$$\gamma^{(1)} \text{ (Ni/NiO)} = 1.16 \; \gamma\text{(Ni-gb)}, \; \gamma^{(2)} \text{ (Ni/NiO)} = 1.12 \; \text{(Ni-gb)}$$

A second triple grain juction where two Ni/NiO interfaces meet a NiO grain boundary is shown in Fig. 6. The NiO grain boundary is an asymmetric 1.8 tilt boundary [(110) plane with a [001] tilt axis]. The edge dislocations with 1/2<110> type Burgers vectors and average spacing of 9.2 nm are shown at a higher magnification in Fig. 7. Interface structure is also observed in the Ni/NiO interface shown at a higher magnification in Fig. 8. In this case, it appears that the largest torque term is associated with the Ni/NiO(1) interface because it remains reasonably planar up to the triple grain grain junction, while the other two interfaces exhibit substantial curvature. Also, there is an increase in the dislocation spacing in the NiO tilt boundary near the triple junction, whereas the torque term should otherwise act to rotate the boundary into a more symmetric, lower energy position. The magnitude of the torque terms for Ni/NiO interfaces is not known, but related experiments [9] have revealed a substantial range of interface orientations deviating from exact topotactic positions, which suggests that the torque terms are small (<10%). On the other hand, torque terms near deep cusps in the Wulff plot can reach values as high as 20% for surface energies in metals [10] or for highly ordered boundaries such as twin boundaries. With the torque terms omitted, the relative interfacial energies for the two Ni/NiO interfaces in Fig. 6 are as follows:

$$\gamma^{(1)} \text{ (Ni/NiO)} = 0.85 \; \gamma\text{(NiO-gb)}, \; \gamma^{(2)} \text{ (Ni/NiO)} = 1.34 \; \gamma\text{(NiO-gb)}$$

Judging from these ratios of energies and values of Ni and

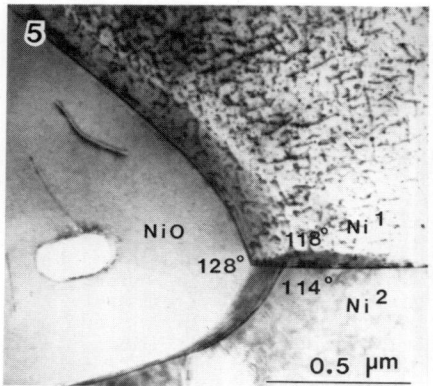

Fig. 5 Triple grain junction showing two Ni/NiO interfaces meeting a Ni grain boundary. The dihedral angles are used to calculate relative ratios of interfacial energies.

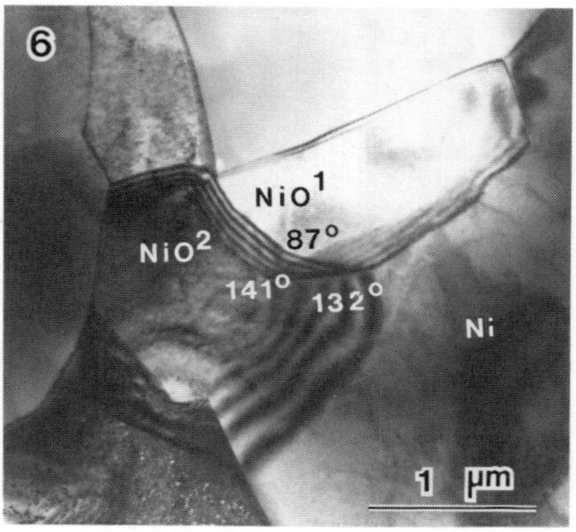

Fig. 6 Triple grain junction showing two Ni/NiO interfaces meeting a NiO grain boundary.

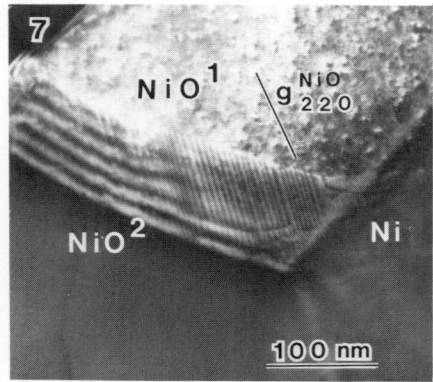

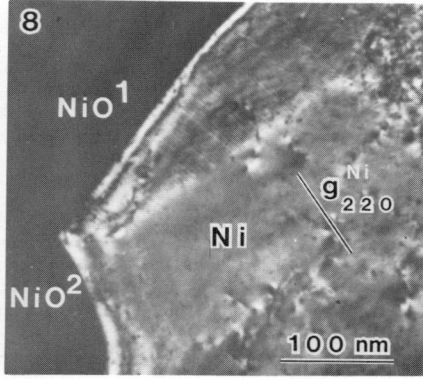

Fig. 7 Higher magnification of the NiO tilt boundary in Fig. 6

Fig. 8 Higher magnification of the Ni/NiO (1) interface in Fig. 6

NiO grain boundary energies from previous studies [10,11], the Ni/NiO interfacial energies are in the neighborhood of 1000 mJ/m^2, with slightly lower values for more structured interfaces such as shown in Fig. 8. Oxidation stresses and growth anisotropy may influence the oxide grain orientations and hence the type of boundaries impinging upon triple grain junctions, but their influence on energy measurements per se is considered minor since boundary curvature to establish equilibrium dihedral angles is quite localized near the triple grain intersections.

STRUCTURAL MODIFICATION OF OXIDE GRAINS AT THE NI/NIO INTERFACE

Electron diffraction patterns from a moderate fraction of nickel oxide grains adjacent to the Ni/NiO interface, in oxide scales formed during 1273K oxidation of nickel, were observed to possess extra reflections forbidden to the rocksalt structure of NiO. Electron diffraction patterns of several low-index poles containing the weaker 'extra' reflections are shown in Fig. 9. The extra reflections lie systematically at half the distance (from the origin) of allowed NaCl reflections. Similar forbidden reflections have been reported in oxidation studies of nickel [12] and annealing studies of pure NiO [13,14]. In these studies, the extra reflections were attributed to the presence of a spinel phase of nickel oxide. Some inconsistencies were found with the spinel indexing scheme in the current study, so it was decided to calculate kinematical structure factors for several structures including spinel and defective rocksalt. The results are presented in table 1 along with microdensitometry measurements taken from the experimental diffraction pattern shown in Fig. 10. The table contains three sections: the top section contains hkl reflections which are allowed for the NaCl structure, and the bottom two sections correspond to the extra reflections- these are divided into two groups, one with all odd values of hkl indices and one with all even values of hkl indicies, with the former set exhibiting somewhat higher intensities. The entries correspond to the squares of the structure factors ($\propto$ intensity) which have been normalized to the strongest extra reflection which is designated 100. The observed intensities for the allowed reflections flooded the densitometer readings, so these are simply reported as xxs. Dynamical diffraction effects were not included in the calculation but multi-slice n-beam programs are available for calculations of structures containing a distribution of defects [15]. The purpose of these calculations is to identify possible structures which could account for the presence of the extra reflections and for such a purpose the kinematical approximation is a good first step; dynamical factors for different types of unit defects can vary widely and multi-slice programs can be quite costly for such probing chores. From table 1, it is apparent that the spinel structure shows very poor agreement with the experimentally observed diffracted intensities. The presence of 10% vacancies on 16c sites (spinel notation) in the NaCl structure predicts the presence of the extra reflections with all odd hkl indices but not those with all even hkl indices. Nominally, the best fit was found using a NaCl structure containing both octahedral 16c vacancies and occupying tetrahedral sites in a four to one ratio. This result suggests that the presence of 4:1 tetrahedral defects which have been proposed as the building block defects in the rock salt insulating oxides may be partially responsible

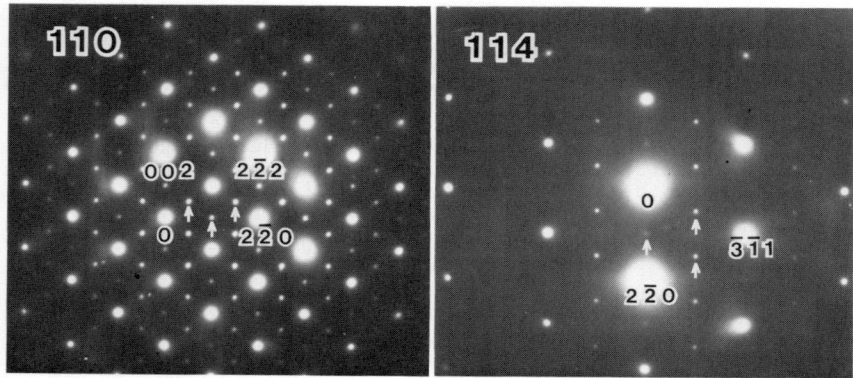

Fig. 9 Diffraction patterns exhibiting forbidden NaCl reflections noted by arrows.

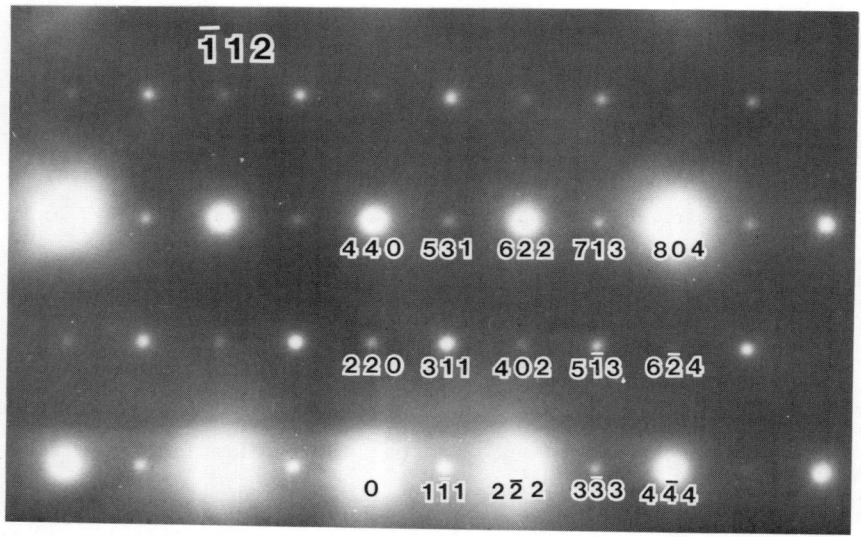

Fig. 10 Diffraction pattern illustrating variations in the intensities of diffracted spots along a row of extra reflections. The index scheme is based on a spinel unit cell which has cell edge twice the length of the NaCl structure.

Table 1 Squared values of calculated kinematical structure factors for several proposed structures. The entries represent values relative to the highest value for an extra reflection which are denoted 100 for purposes of comparing relative intensities. The relative intensities observed experimentally and measured using microdensitometry are listed in the far right column.

Relative Magnitude of the
Squares of Structure Factors for hkl
Reflections from Different Structures

ALLOWED REFLECTIONS (h,k,l)	SPINEL	NaCl DEFECTIVE 10% VACANCIES ON 16c SITES	10% VACANCIES ON 16c SITES WITH 1/4 NO. TET. SITES FILLED	OBSERVED INTENSITIES
222	10	4840	4950	xxs
400	152	20977	20523	xxs
440	241	10239	10827	xxs
444	46	6122	5989	xxs
622	2	1055	1073	xxs
662	1	560	573	xxs
666	1	305	309	xxs
EXTRA REFLECTIONS ODD INDICIES				
111	6	100	67	75
311	100	48	100	100
331	2	29	19	93
333	40	29	56	100
511	40	19	36	93
533	24	17	26	87
EVEN INDICES				
200	0	0	21	68
220	44	0	14	68
422	16	1	9	80
442	0	1	9	62

for the observed variations in intensities seen in Fig. 10. Clearly a complete structural determination will require considerably more research with allowance for dynamic effects and structural modifications. However, the results from this preliminary study have shed some light on possible defect assemblages and discounted earlier suppositions that the presence of the extra reflections was due simply to the presence of a spinel phase of nickel oxide.

ACKNOWLEDGEMENTS

This work has been supported by the National Science Foundation through the research grant DMR-8309461.

REFERENCES

1. M.T. Tinker and P.A. Labun, Oxid. Met., 18, (1982), 27-40

2. H.T. Sawhill and L.W. Hobbs, Proc. 9th Int. Cong. Met. Cor., Toronto, Canada, June 3-7 (1984) Vol.1, 21-26

3. S.M. Allen, Proc. 10th Int. Cong. Elec. Micrs., Hamburg, FRD, August 17-24, (1982), 353-4

4. H.T. Sawhill, L.W. Hobbs and M.T. Tinker, Adv. in Ceramics, Vol.6, Amer. Ceram. Soc., Columbus, OH, (1983), 128-38

5. R. Schindler, J.E. Clemans and R.W. Balluffi, Phys. Status Solidi A, 56, (1979), 749-61

6. R.W. Balluffi, A. Brokman and A.H. King, Acta Metall. 30 (1982), 1453-70

7. C. Herring, The physics of Powder Metallurgy, (1951), 143

8. H.T. Sawhill and L.W. Hobbs, J. de Physique (in press, 1985)

9. H.T. Sawhill, Ph.D. thesis, Mass. Inst. Tech. (1985)

10. K. Hodgson and H. Mykura, J. Mater. Sci., 8, (1973), 565-70

11. D.M. Duffy and P.W. Tasker, Philos. Mag. A, 48, (1983), 155-62

12. L.B. Garmon, Ph.D. thesis, Univ. of Virginia (1966)

13. K. Katada, Nakahigashi and Y. Shimomura, J. Appl. Phys. Jpn. 9, (1970), 1019-28

14. K. Tagaya and M. Fukada, J. Appl. Phys. Jpn. 15, (1976), 561-62

INTERFACE STUDY OF Mo/GaAs.

PEICHING LING*, JYH-KAO CHANG**, MIN-SHYONG LIN** AND JEN-CHUNG LOU**
* Department of Materials Science and Engineering
** Department of Electrical Engineering, National Tsing Hua University,
Hsinchu, Taiwan 300, R.O.C.

ABSTRACT

The electrical characteristics and the microstructure of Mo/GaAs Schottky diodes fabricated by electron-beam evaporation have been studied. The barrier height, ideality factor, deep trapping levels and intermetallic compounds of these annealed or unannealed Mo/GaAs Schottky diodes are obtained by using the I-V, C-V, Rutherford backscattering spectroscopy (RBS), Auger electron spectroscopy (AES), deep level transient spectroscopy (DLTS) and transmission electron microscopy (TEM) analyses. An obvious interdiffusion at Mo/GaAs interface is observed in Mo/GaAs Schottky diodes annealed above 500°C for 10 min. DLTS results show that there are two electron traps [Ec-(0.52 ± 0.02)·eV and Ec-(0.86 ± 0.02) eV] and one hole trap [Ev+(0.92 ± 0.02) eV] are demonstrated for 300°C, 400°C post-annealed Mo/GaAs diodes. TEM results also indicate that the disappearance of these deep trapping levels may correlated to the formation of intermetallic compounds $GaMo_3$ and $MoAs_2$ existed in Mo/GaAs diodes post-annealed above 500°C. It is believed that the metal-semiconductor interdiffusion and the intermetallic compounds play the major roles for the thermal degradation of Mo/GaAs Schottky diodes.

INTRODUCTION

Recently, the Mo/GaAs structure has been studied as an ideal Schottky diodes [1,2] and shown its potential in the microwave device field. However, this Mo/GaAs device need to be operated at high temperature environment and to maintain stable and reliable electrical characteristics, the study of the thermal stability and the thermal induced degradation of Mo/GaAs Schottky diodes, therefore, becomes very important. In this study Mo/GaAs Schottky diodes, fabricated by electron-beam evaporation, are used to investigate the mechanism of thermal degradation. It is found that the Mo/GaAs Schottky diodes begin to degrade at 500°C, and can correlate the thermal degradation to the metal-semiconductor interdiffusion and the formation of intermetallic compounds. By using the I-V characteristics measurements, C-V measurements, Rutherford backscattering spectroscopy (RBS) techniques, Auger electron spectroscopy (AES) evaluations deep level transient spectroscopy (DLTS) analysis [3] and transmission electron microscopy (TEM), it is revealed that the electrical characteristics of Mo/GaAs Schottky diodes are strongly influenced by the metal-semiconductor interdiffusion and the formation of the intermetallic compounds.

EXPERIMENTAL PROCEDURES

The n-type <100> GaAs substrates with a carrier concentration of $1.2 \times 10^{17}/cm^3$ were used in this study. The substrates were subjected to a series of cleaning and degreasing processes, then etched in a $HCl:3H_2O$ solution with the temperature of 75-80°C for 5 minutes to remove the native oxide layer and finally rinsed in deionized water. Ohmic contacts were obtained by evaporating Au-Ge-Ni eutectics onto the substrates with a subsequent 450°C thermal annealing for 5 min in a nitrogen ambient. Schottky barriers with area 1.25×10^{-3} cm^2 were formed with a 1500 Å Mo layer, which is deposited onto the substrate at room temperature by electron-beam evaporation with a rate of 3Å/sec. Subsequently, a 700 Å SiO_2 film was encapsulated onto the Mo layer to prohibit the Mo film from oxidation during the annealing process. In order to study the thermal stability of Mo/GaAs Schottky barriers, the specimens were annealed at temperatures ranging between 300 and 700°C in flowing N_2 gas for 10 min. In addition, for specimens analyzed by TEM method, 300Å of Mo and then 700 Å SiO_2 were evaporated onto GaAs substrates with the same evaporation condition mentioned as above. After annealing, the SiO_2 layer of TEM specimens was removed by dilute HF solution (HF: H_2O=1:50) and the specimens were subsequently etched in Bromine-methanol (12:88) solution. Furthermore, the TEM specimens were then slowly etched in a HNO_3:HCl:HG:H_2O (6:1:2:5) solution until a suitable hole appeared.

In order to investigate the mechanism of the thermal degradation of Mo/GaAs Schottky barriers, the barrier height, ideality factor and deep trapping levels of these annealed or unannealed Mo/GaAs Schottky barriers are obtained by using the I-V, C-V and deep level transient spectroscopy (DLTS) analysis, respectively. The metal-semiconductor interdiffusion phenomenon and the intermetallic compound formation are investigated by Rutherford backscattering spectroscopy (RBS), Auger electron spectroscopy (AES) and transmission electron microscopy (TEM), respectively.

RESULTS AND DISCUSSIONS

1. I-V and C-V characteristics. As shown in Fig. 1, the forward I-V characteristics of Mo/GaAs diodes with and without annealing in the temperature range from 300°C to 700°C for 10 min. are measured at room temperature. The dependence of the ideality factor n, derived from the slope of I-V characteristics, on the annealing temperature is shown in Fig. 2.

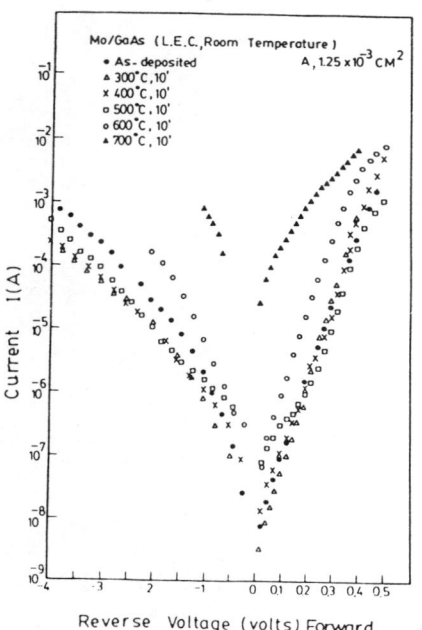

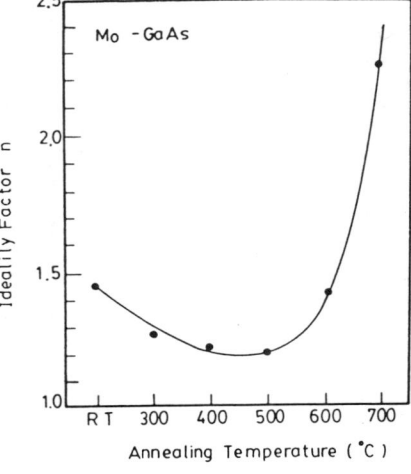

←Fig. 2 Variations of ideality factor vs. annealing temperature.

←Fig. 1 I-V curves for as-deposited Mo-GaAs Schottky diodes with and without annealing in the temperature range 300-700°C.

It is found that the ideality factor n initially decreases from 1.45 5o 1.20 and finally increases from 1.20 to 2.30 in the temperature range from room temperature to 500°C and from 500°C to 700°C, respectively. Furthermore, it is obvious that the 500°C annealed specimen shows a better ideality factor n. The reason is that the annealing temperature of 500°C is in agreement with the temperature at which both the gallium oxide and the arsenic oxide evaporate from the GaAs surface and migrate into the Mo metal layer [4].Similary, the same effect appears in both 300°C and 400°C annealed specimens. The large ideality factor n=2.35 for 700°C annealed Mo/GaAs diodes is probably due to the decomposition of GaAs substrate and the formation of the intermetallic compounds.

The variation of barrier height Φ_{bc} versus annealing temperature obtained by C-V measurements is shown in Fig. 3. It shows that the barrier height Φ_{bc} is almost kept a value around 0.95 eV for as-deposited specimens and is very close to the two-thirds of the bandgap-1.42 eV for GaAs. This result is mainly due to the existence of the surface states pinning the Fermi level to the surface of the covalent semiconductors [5]. In addition, the barrier height of 700°C annealed Mo/GaAs diodes drops to 0.84 eV and the corresponding ideality factor n rises to 2.35. Such increased value of n, clearly shown in the forward bias region of Fig. 1, can be explained as being caused by enhanced

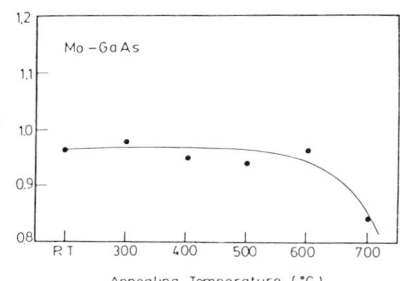

Fig. 3 Variations of barrier height values at 300°K as a function of annealing temperature.

thermionic field emission which is favored by a small barrier height and/or the existence of intermetallic compounds induced by interdiffusion at the interface. The reverse current of 700°C annealed specimen, mainly due to the field emission, is comparable with the forward current at low fields, then there is a clear tendency toward Ohmic behavior. This phenomenon is probably due to the formation of intermetallic compound and/or the existence of generation-recombination centers in the depletion region. Here, RBS measurements, AES evaluations, TEM analysis and DLTS technique are used to confirm this consideration.

2. RBS, AES and TEM analysis. In order to investigate the metal-semiconductor interfiffusion, the Rutherford backscattering technique is performed on Mo/GaAs diodes by using a 2-MeV $4He^+$ ion beams with 160° backscattering angle. It is found that a very small indiffusion of Mo into GaAs is observed only in 700°C annelaed Mo/GaAs diodes. The RBS spectra of as-deposited Mo/GaAs and 700°C annealed Mo/GaAs are shown in Fig. 4. It indicates that there are no detectable intermetallic compounds and Ga outdiffusion under the resolution of RBS technique. However, the Ga out-diffusion effect in metal-gallium arsenide has been known to cause degradation as it operates in high temperature [2,6]. In order to further study whether there is a phenomenon of Ga outdiffusion,Auger electron spectroscopy in conjunction with Ar^+ ion sputtering has been employed to confirm Ga outdiffusion in Mo/GaAs diodes. A depth profile is shown in Fig. 5 for as-deposited and 700°C 10-min. annealed specimens. Both significant Mo indiffusion and consequential As, Ga outdiffusion are observed in annealed specimens. This is because the annealing process effects the thermal dissociation of GaAs at the interface of Mo. Consequently, the electropositive Ga atoms migrate to electronegative Mo layer and leave a large amount of electrically active Ga vacancies (V_{Ga}). Meanwhile, Mo atoms indiffuse into GaAs and occupy these V_{Ga} sites to form Mo_{Ga} which may act as an acceptor type of impurity in the interface of GaAs. This effect seems to correlate with rather profound variations in the electrical behavior of these specimens.

140

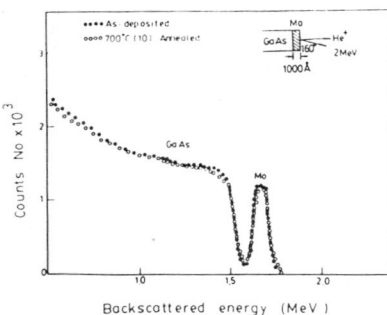 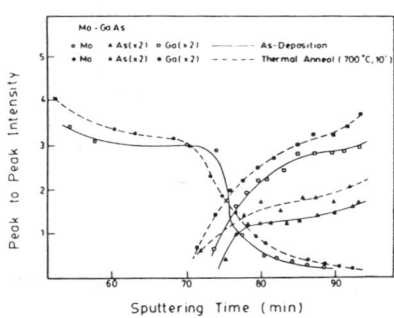

Fig. 4 Rutherford backscattering spectra of 2-MeV He⁺ ions from as-deposited Mo/GaAs with and without 700°C annealing.

Fig. 5 Auger spectra of as-deposited Mo/GaAs with and without 10-min 700°C annealing.

However, there also may still be a possibility for the formation of intermetallic compounds after the annealing treatment, the TEM analysis is therefore used to identify the existence of intermetallic compounds.

The TEM diffraction pattern and the bright field image of as-deposited Mo/GaAs is shown in Fig. 6, it indicates that the Mo film evaporated on (100) GaAs at room temperature is polycrystalline with the average grain

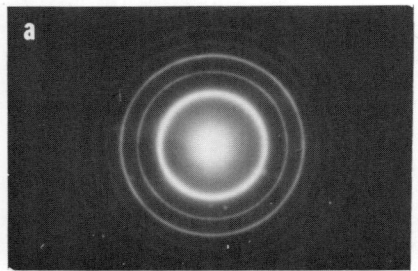 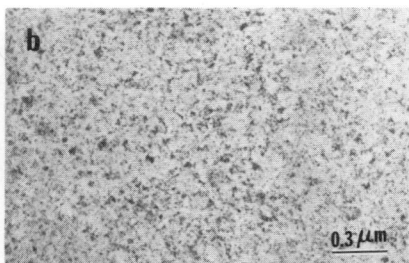

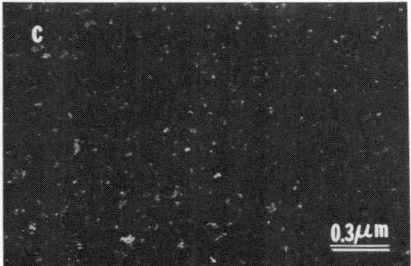

Fig. 6 (a) The TEM diffraction pattern , (b) the bright field image and (c) the dark as-deposited Mo/GaAs Schottky diode.

size of 160 Å. Furthermore, there is no reaction and no intermetallic compounds between Mo and GaAs interface because only Mo diffraction rings are observed. The TEM diffraction pattern of 400°C annealed Mo/GaAs diodes indicates no reaction and no intermetallic compounds between Mo and GaAs interface, too. The bright field image of 400°C annealed specimen indicates the Mo grain size has no obvious change. In Fig. 7, for 500°C or 600°C annealed Mo/GaAs diodes, an additional phase of polycrystalline $GaMo_3$, a very small amount, with average grain size of 110Å is formed. Even though

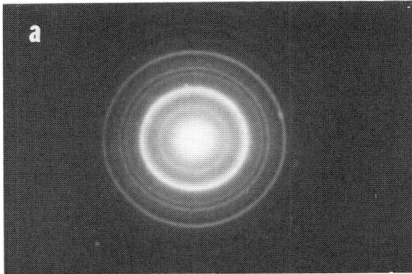 

Fig. 7 The TEM diffraction patterns for (a) 500°C and (b) 600°C annealed Mo/GaAs diodes.

the intermetallic compound $GaMo_3$ has occurred, but for such occurring the barrier height shown in Fig. 2 has no much influence. Again, the Mo grain size has no obvious change. The TEM results of 700°C annealed Mo/GaAs diodes, shown in Fig. 8, indicate a phase of intermetallic compound $MoAs_2$ interface. Comparing with above results, we therefore speculate that this $MoAs_2$ phase plays a major role in the decrease of barrier height. In addition, it is also found that the grain size of $GaMo_3$ increases to 160Å and the quantity also increases. However, the Mo grain still keeps the same size. Furthermore, the TEM dark field images of as-deposited Mo/GaAs with and without subsequently thermal annealing show no change in Mo grain size. This result correlated with that of RBS and AES indicates that the

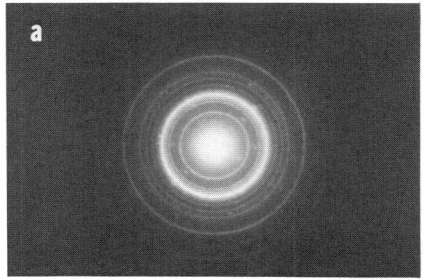 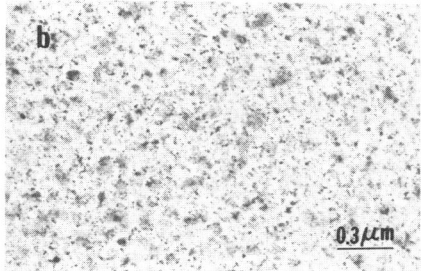

Fig. 8 (a) The TEM diffraction pattern and (b) the bright field image of 700°C annealed Mo/GaAs diode.

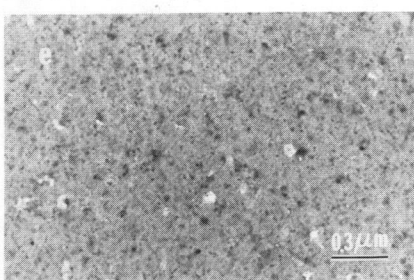

Fig. 8 (c) The dark field image of Mo, (d) the dark field image of $GaMo_3$ and (e) the dark field image of $MoAs_2$ of 700°C annealed Mo/GaAs diode.

Mo does not react completely.

3. DLTS analysis. On one hand, in order to find out whether there is a correlation between the degradation of Mo/GaAs diodes and the presence of deep levels, and on the other hand, to confirm the defect which is induced by the metal-semiconductor interdiffusion as well, DLTS is performed on each of these Mo/GaAs diodes that are fabricated by different processes and being separated into An and Bn series. A_1, A_2, A_3, A_4 and A_5 are the Mo/GaAs diodes fabricated by predepositing Mo on the GaAs substrates and then subsequently being annealed at 300, 400, 500, 600 and 700°C, respectively. B_1, B_2, B_3, B_4 and B_5 are the Mo/GaAs diodes fabricated by preannealing the GaAs substrates, which have been pre-deposited a SiO_2 encapsulated layer, at 300, 400, 500, 600 and 700°C, respectively, then the SiO_2 encapsulated layers are etched by dilute HF and the GaAs substrates are subsequently deposited by Mo film. The annealing process is proceeded in N_2 for 10 min. and the deposition of Mo film goes through room temperature at 10^{-7} torr for all specimens.

The characteristics of these traps existed in Mo/GaAs diodes are listed in Table I and II. It is found that only two electron traps [$E_c-(0.52\pm0.02)$ eV] and [$E_c-(0.86\pm0.02)$ eV] are observed in as-deposited Mo/GaAs diodes and Bn series specimens. Furthermore, the preannealing process only induces a slight increase in trap concentration of these two electron traps. Here, the electron trap $E_c-(0.52\pm0.02)$ eV is attributed to the $V_{Ga} \cdot V_{As}$ divacancy [7] and the electron trap $E_c-(0.86\pm0.02)$ eV, EL2, is identical with the anion antisite defect As_{Ga} [8,9]. It is believed that the increase in the concentration of these electron traps is mainly due to the increasing of the preannealing temperature, and so, the V_{Ga} is induced by the migration of Ga atoms from GaAs substrate into SiO_2 encapsulated layer during the preannealing process. However, for An series specimens, both the electron traps existing in Bn series specimens and an additional hole trap [$E_v+(0.92\pm0.02)$ eV] are observed in A_1 specimens (300°C post-annealed) and A_2 specimens

TABLE I. DLTS results of pre-annealed Mo/GaAs diodes.

Sample	Activation Energy(eV)	Concentration N_T (cm^{-3})	Thermal capture Cross section σ_∞ (cm^2)
As-deposition	$E_c-(0.52\pm0.02)$	1.06×10^{14}	3.63×10^{-14}
	$E_c-(0.86\pm0.02)$	1.20×10^{14}	1.19×10^{-14}
300°C preannealing	$E_c-(0.52\pm0.02)$	4.30×10^{14}	4.07×10^{-15}
+ R.T. deposition	$E_c-(0.86\pm0.02)$	2.30×10^{14}	5.04×10^{-14}
400°C preannealing	$E_c-(0.52\pm0.02)$	3.75×10^{14}	3.40×10^{-15}
+ R.T. deposition	$E_c-(0.86\pm0.02)$	3.21×10^{14}	1.31×10^{-14}
500°C preannealing	$E_c-(0.52\pm0.02)$	3.80×10^{14}	1.82×10^{-15}
+ R.T. deposition	$E_c-(0.86\pm0.02)$	4.0×10^{14}	1.28×10^{-14}
600°C preannealing	$E_c-(0.52\pm0.02)$	5.08×10^{14}	4.60×10^{-15}
+ R.T. deposition	$E_c-(0.86\pm0.02)$	3.95×10^{14}	1.56×10^{-14}
700°C preannealing	$E_c-(0.52\pm0.02)$	6.50×10^{14}	1.07×10^{-15}
+ R.T. deposition	$E_c-(0.86\pm0.02)$	4.2×10^{14}	3.55×10^{-14}

TABLE II. DLTS results of post-annealed Mo/GaAs diodes.

Sample	Activation energy (eV)	Concentration N_T (cm^{-3})	Thermal capture Cross section σ_∞ (cm^2)
R.T. deposition +300°C post-annealing	$E_c-(0.50\pm0.02)$	2.70×10^{14}	1.5×10^{-14}
	$E_c-0.86$	6.76×10^{13}	9.7×10^{-15}
	$E_v+0.94$	4.18×10^{16}	5.9×10^{-14}
R.T. deposition +400°C post-annealing	$E_c-(0.50\pm0.02)$	2.24×10^{14}	8.0×10^{-14}
	$E_c-0.86$	1.68×10^{16}	2.5×10^{-14}
	$E_v+0.94$	4.38×10^{16}	1.2×10^{-14}
R.T. deposition +500°C post-annealing	$E_v+(0.94)$	2.21×10^{16}	4.2×10^{-14}
R.T. deposition +600°C post-annealing	$E_v+0.90$	9.60×10^{15}	3.6×10^{-14}

(400°C post-annealed). In addition, only the Mo phase is found from the TEM diffraction pattern for specimens A_1 and A_2. Therefore, it is believed that this hole trap [$E_v+(0.92\pm0.02)$ eV] is associated with Mo interdiffusion [6]. During the thermal annealing, these Mo atoms diffuse into GaAs and become trapped by reacting with gallium vacancies or vacancy complexes. Furthermore, only the hole trap [$E_v+(0.92\pm0.02)$ eV] is observed in A_3 (500°C post-annealed) and A_4 (600°C post-annealed) specimens, and the trap concentration apparently decreases with increasing the annealing temperature. From the corresponding TEM diffraction patterns, it shows that an additional phase GaMo$_3$ with the grain size of 110Å is observed. We therefore speculate that the decrease in the trap concentration of [$E_v+(0.92\pm0.02)$ eV] may be due to the formation of GaMo$_3$ which performs as a diffusion barrier for the indiffusion of Mo atoms. Under these annealing conditions, the Mo atoms may prefer to combine with outdiffused Ga atoms to form GaMo$_3$ intermetallic compound than indiffuse into Ga vacancies to form Mo$_{Ga}$ hole traps. However, the disappearance of Ga vacancy-correlated electron traps [$E_c-(0.52\pm0.02)$ eV] and [$E_c-(0.86\pm0.02)$ eV] is still unclear. For 700°C annealed Mo/GaAs diodes, the DLTS spectrum shows no detectable signal, the TEM results, however, show the GaMo$_3$ and the MoAs$_2$ phases. Therefore, the disappearance of these traps may be correlated with the existence of the intermetallic compounds. But, the mechanism of those various changes described above is not clear. Further study of the Mo/GaAs diodes is still going on with follows.

CONCLUSIONS

In this study, Mo/GaAs diodes are thermally annealed from 300°C to 700°C. A systemic investigation has been made to study the degradation phenomenon of these annealed diodes. From the experimental results, the degradation of thermally annealed Mo/GaAs diodes can be explained on the basis of metal-semiconductor interdiffusion and the formation of intermetallic compounds. In addition, the following conclusions can be obtained from this study.

(1) The heat treatments at 300 and 400°C for 10 min. do not cause any obvious variation in the electrical properties of the Mo/GaAs diodes. RBS, AES and TEM analysis also indicate no detectable metal-semiconductor interdiffusion and the formation of intermetallic compounds. These results show that the Mo/GaAs diodes are thermally stable until 400°C. Furthermore, the Fermi level pinning effect induced by the existence of surface states between Mo and GaAs is observed, it is the reason that the barrier height is almost kept a value around 0.95 eV for as-deposited specimens and specimens annealed below 600°C.

(2) A thermally induced degradation is observed in these Mo/GaAs diodes annealed above 500°C. AES shows the significant metal-semiconductor interdiffusion and Ga outdiffusion, furthermore, the TEM shows the existence of $GaMo_3$ phase in these diodes and of an additional $MoAs_2$ phase observed only in 700°C annealed Mo/GaAs diodes. It is found that the electrical characteristics of Mo/GaAs are strongly influenced by the metal-semiconductor interdiffusion and the formation of intermetallic compounds.

(3) DLTS analysis indicates that there are two electron traps [$E_c-(0.52\pm0.02)$ eV and $E_c-(0.86\pm0.02)$ eV] and one hole trap [$E_v+(0.92\pm0.02)$ eV] are observed in degraded Mo/GaAs diodes annealed below 400°C. It is believed that the hole trap [$E_v+(0.92\pm0.02)$ eV] is induced by the indiffusion of Mo atoms into the Ga vacancies during the annealing process. However, these traps have the tendency to disappear as the intermetallic compounds $GaMo_3$ and $MoAs_2$ occur in those Mo/GaAs diodes annealed above 500°C. Finally, the metal-semiconductor interdiffusion and the formation of intermetallic compounds are believed to play the major roles in the degradation of annealed Mo/GaAs diodes.

REFERENCES

1. P.M. Batev, M.D. Ivanovitch, E.I. Kafedjiiska, and S.S. Simeonov, Int. J. Electron. 45, 511 (1980).
2. M.S. Lin, W.H. Su, J.C. Lou, and T.F. Lei, Jpn. J. Appl. Phys. Supplement 22-1, 397 (1982).
3. D.V. Lang, J.Appl. Phys. 45, 3023 (1974).
4. A. Munoz-Yague, J. Piqueras and Fabre, J. Electronchem, Soc., Vol. 128, No. 1, 149 (1981).
5. J. Bardeen, Phys. Rev., 71, 717 (1947).
6. J.C. Lou, M.S. Lin, and W.H. Su, J. Appl. Phys. 54, 4482 (1983).
7. L.L. Chang, L. Esaki, and R. Tsu, Appl. Phys. Lett., 19, 143 (1971).
8. E.J. Johnson, J. Kafalas, R.W. Davis, and W.A. Dyes, Appl. Phys. Lett., 140, 993 (1982).
9. E.R. Weber, H. Znnen, U. Kaufmann, J. Windschief, J. Schneider and T. Wosinski, J. Appl. Phys., 53, 6140 (1982).

DETERMINATION OF THE COMPOSITION AND THICKNESS OF THIN POTASSIUM POLYPHOSPHIDE FILMS

KLARA KISS* AND PAUL M. FIGURA
Stauffer Chemical Co., Dobbs Ferry, NY 10522

SUMMARY:

An energy dispersive X-ray microprobe (EDX) analysis was developed to determine simultaneously the lateral uniformity of the thickness and the composition of thin potassium polyphosphide (KP_x) films. The EDX analysis was based on theoretical calibration curves generated by the Monte Carlo simulation approach developed by Kyser and Murata and extended by Miller and Koffman. Simultaneous determination of both composition and thickness was possible for this binary-element thin film due to the concentration independence of the theoretical intensity ratio of phosphorus $\left(\dfrac{I_{p\ film}}{I_{p\ bulk}} \right)$

The EDX results were compared to macro techniques applied routinely in the characterization of thin films, i.e., piezoelectric thickness measurement and compositional analysis via X-ray fluorescence spectroscopy (XRF). A special XRF technique was developed to determine the weight ratios from fluorescent intensity measurements "directly" using "bulk" standards instead of thin film standards. The accuracy of this technique was demonstrated for an indium phosphide standard film since no certified KP_x standard films are available.

The comparison was carried out on films of widely different thicknesses (0.26-2.0 um) and compositions (KP_5-KP_{83}). A Student's t test demonstrated that the compared techniques were identical at the 95% confidence level for the determinations of both thickness and composition. Thus, EDX analysis can be used to complement the macro techniques when the lateral uniformity of the thin films is to be determined at the micron scale.

INTRODUCTION:

Alkali polyphosphides are a new class of compound semiconductors which have potential application in large area electronic devices or in conjunction with other semiconductors containing group V elements, e.g., InP. They can be grown as thin films on substrates by various deposition techniques such as physical vapor deposition, chemical vapor deposition, thermal evaporation, and sputtering (1,2).

The potassium polyphosphide (KP_x) films on various substrates are being developed for use in thin film transistors and electronic devices based on III-V's. The potassium/phosphorus (K/P) ratio has a significant effect on the stability and on the electronic properties of the films, and, therefore, their stoichiometry must be monitored. The objective of this work was to establish the suitability of energy dispersive X-ray microprobe analysis (EDX) to determine the composition and lateral uniformity of KP_x films on silicon and on glass substrates at the micron scale. The EDX analysis was based on the well established Monte Carlo simulation approach developed by Kyser and Murata and extended by Miller and Koffman (4-8).

EXPERIMENTAL:

Thin films of KP_x were deposited on glass microscope slides and silicon plates by physcial or chemical vapor deposition, thermal evaporation or sputtering. Details of the deposition processes will be published elsewhere (2). Film thickness was determined by piezoelectric measurements. X-ray fluorescence analysis was carried out in a Siemens SR-1 wavelength dispersive X-ray fluorescence spectrometer at 40 KV, and 40 mA, using a chromium X-ray tube. Areas of 0.4-0.5 cm^2 were exposed to the Cr X-rays in He atmosphere. Intensity measurements corrected for background were taken on the $P_{K\alpha}$ and $K_{K\alpha}$ X-ray emission lines.

Standards for the XRF analysis were prepared as follows: separate solutions containing 0.1g/ml of phosphorus and potassium were prepared by the dissolution of $NH_4H_2PO_4$ and KNO_3 (NBS standards) in dilute nitric acid, respectively. Aliquots of these primary standards were combined to produce a range of compositions from KP_5 to KP_{50}. Two microliters of each mixed standard were deposited on alkali-free glass slides. The slides were dried at 80°C for 4 hours and stored in a desiccator until analyzed.

An Amray 1000A scanning electron microscope interfaced with a Philips-Edax energy dispersive X-ray analyzer system was used for EDX measurements. Background corrected integrated intensities of the $P_{K\alpha}$ and $K_{K\alpha}$ lines were determined both on the samples and on a bulk standard at 20 KV accelerating voltage, 50 degrees takeoff angle and 100 x magnification. A 12 mm working distance was maintained to avoid unreliable data due to variations in the working distance, i.e., takeoff angle. The standard was a pellet pressed from KP_{15} single crystals of exact stoichiometry. The preparation of the KP_{15} single crystals was reported before (3). The standard was mounted on each sample to permit sample and standard to be analyzed under identical conditions.

Theoretical calibration curves were constructed based on a Monte Carlo simulation approach developed by Kyser and Murata (4,5,6) and extended by Miller and Koffman (7,8). A FORTRAN computer program was used to generate a database for KP_{15} on the glass/silicon substrate system. Theoretical intensity ratios were obtained for both potassium and phosphorus X-ray fluorescence ($K_{K\ theor.}$ and $K_{p\ theor.}$ respectively). These were calculated by dividing the intensity of theoretical X-ray scattering from the thin film (I_{film}) by scattering from the bulk KP_{15} reference material (I_{bulk}). The theoretical intensity ratios were calculated for a film thickness range of 0.01-2 um and for composition range of 0.1-20 wt% potassium. Input data of 20 KV accelerating potential, 50° takeoff angle, and 1000 trajectories were used.

RESULTS AND DISCUSSION:

The theoretical background of the Monte Carlo simulation of the electron-solid interaction and scattering process has been discussed elsewhere (4-9). A typical background corrected EDX scan (Figure 1) shows the well-resolved phosphorus and potassium peaks which are easy to quantify.

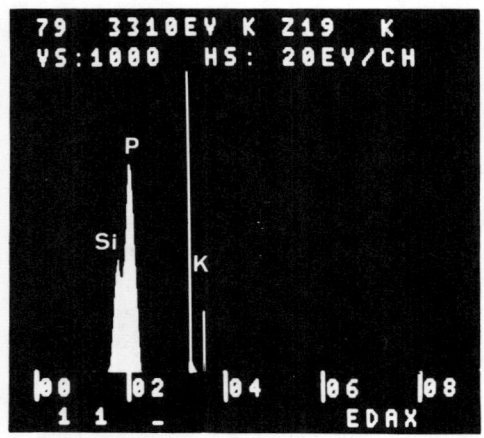

Figure 1

Typical EDX Scan of KP_x Film

The intense silicon peak originates from the substrate. The phosphorus and silicon peaks are deconvoluted by the computer of the Edax system. Thus, no interference from silicon takes place.

Theoretical calibration curves of $K_{theor.}$ vs. log composition are shown in Figure 2. Both the potassium and phosphorus calibration curves are series of straight lines; each line corresponds to a particular film thickness. The figure reveals that the $K_{K\ theor.}$ values vary with both composition and thickness, while the $K_{P\ theor.}$ varies only slightly with the concentration of phosphorus in the investigated 80-99.9 wt% concentration range. The $K_{P\ theor.}$ is very sensitive, however, to the film thickness. The $K_{P\ theor.}$ vs. concentration plots are straight lines practically parallel to the abscissa. Their vertical position along the Y axis depends on the film thickness. This fact permits the simultaneous determination of both thickness and composition for this two-component compound via single EDX

measurements on the sample and on the standard respectively
(i.e., two measurements with a total analysis time of about
200 seconds). The usual iterative techniques to establish
both thickness and composition of thin films from
t(t=thickness) vs composition curves (4,5,8) is not necessary
in this case.

The film thickness is established by the comparison of
the experimental K values of phosphorus ($K_{P\ exp.}$) with
$K_{P\ theor.}$. Once the film thickness has thus been determined,
the unknown potassium concentration (and thus the composition
of the film) can be obtained by comparing the experimental
potassium values ($K_{K\ exp.}$) with $K_{K\ theor.}$ at the appropriate
thickness. The Monte Carlo calculation was carried out only
for film thicknesses of 0.01, 0.05, 0.075, 0.1, 0.2, 0.5, 1.0,
and 2.0 um. Intermediate thicknesses were estimated by
interpolation.

An example of the above determination is shown in Figure
2 by the heavy lines. $K_{P\ exp.}$ and $K_{K\ exp.}$ were 0.235 and
and 0.361 respectively. $K_{P\ theor.}$ = 0.235 corresponds to a
thickness between 0.5-0.6 um (piezoelectric measurement gave
0.7 um). Thus the unknown potassium concentration must
lie on the 0.5 um line of $K_{K\ theor.}$ vs. concentration plot-
series. The intercept of this line at y=0.361 (the $K_{K\ exp.}$)
projected onto the abscissa yields a potassium concentration
of 8.2 wt%, which corresponds to $KP_{14.1}$.

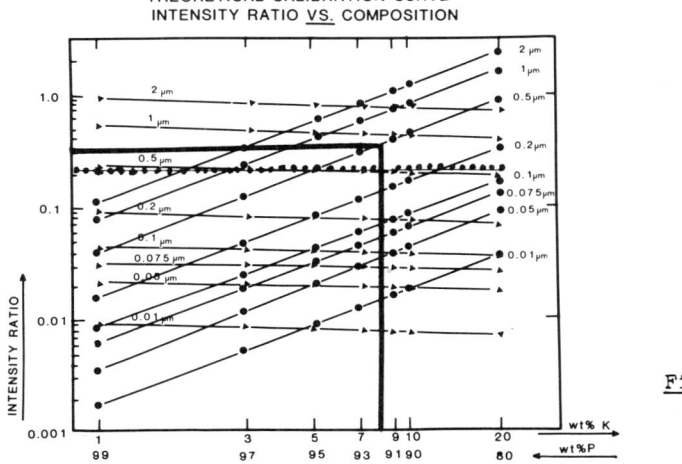

Figure 2

This composition was compared to the composition obtained by the Kyser-Murata iteration method. The latter derives a unique fit for both composition and thickness for binary systems from the experimental K values. $K_{theor.}$ vs. thickness curves are constructed for a range of compositions and the thickness values are determined for each component. A difference in the thickness values corresponds to an incorrect composition.

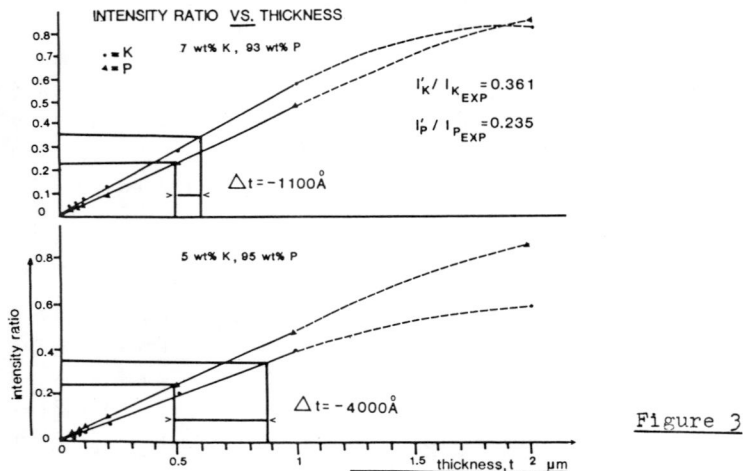

Figure 3

Figure 3 depicts examples of plots of intensity ratio vs. thickness and the determination of Δt. Obviously both compositions are incorrect since the Δt values are high (-1100A° and -4000A°, respectively). The value of Δt immediately reveals how far the plotted composition deviates from the true value: the smaller the Δt, the closer are the values of the plotted and true compositions. The unknown concentration of one of the components (potassium, in our example) is determined from the Δt vs. elemental concentration plot at $\Delta t=0$ (Figure 4). The iteration method yielded 8.4 wt% potassium which corresponds to $KP_{13.8}$. This value compares well with $KP_{14.1}$ and $KP_{14.0}$, obtained by the simplified EDX and XRF methods, respectively.

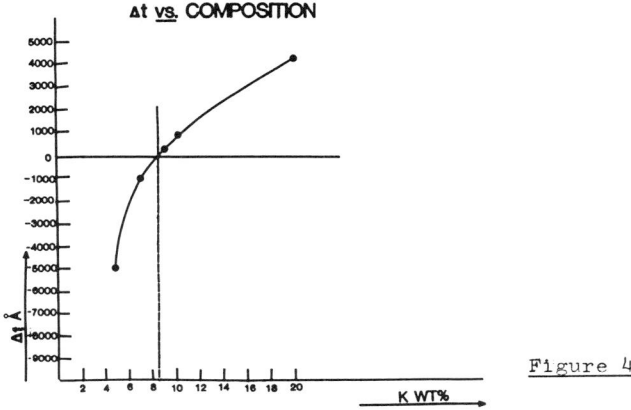

Figure 4

The films subjected to EDX analysis were also analyzed by XRF to establish the comparability of the two techniques. For the XRF analysis a special technique was developed to "directly" determine the weight ratios from fluorescent intensity measurement made on thin film standards prepared from appropriate salt solutions of known compositions.

XRF analyses of thin films of various compositions and thicknesses are described in the literature (10-15). These methods are based on determining the amount of each element present in the film. This is generally accomplished by a comparison of intensity measurement from standards of known thicknesses and compositions similar to the film being analyzed. Such standards are extremely difficult to prepare and standardize.

An XRF method has been developed which permits the "direct" determination of weight ratios of binary element thin films without recourse to determining the absolute amount of each element. With this method, calibration standards can be easily prepared and thickness measurements are not required.

The "direct" method is based on the relationship between fluorescence line intensity and concentration of an element in a film of t thickness given by Equation 1. (16).

$$I_i = \frac{C_i W_i}{\bar{\alpha}}\left(1-\exp[-\bar{\alpha}\varrho t]\right) \qquad (1)$$

Where:
I_i = intensity for λ_i excited by λ_o
λ_i = wavelength of the fluorescent X-radiation
λ_o = wavelength of the exciting X-radiation
C_i = a constant for the element in a given instrument
W_i = weight fraction of element i
$\bar{\alpha}$ = total absorption coefficient
ϱ = density of the sample
t = thickness of the sample

For sufficiently thin films $\exp[-\bar{\alpha}\varrho t] \approx 1-\bar{\alpha}\varrho t$
Substitution in Equation 1 yields
$$I_i = C_i W_i \varrho t \qquad (2)$$

This equation is independent of matrix absorption and enhancement terms. It can be conveniently used for thin film analysis since the relationship between I_i and w_i is dependent only on the mass thickness, t, of the specimen.

For a binary system the ratio of the fluorescent intensity of element 1 relative to element 2 is given by Equation 3.

$$\frac{I_1}{I_2} = \frac{C_1 W_1}{C_2 W_2} = C \frac{W_1}{W_2} \qquad (3)$$

C can be experimentally determined by measuring the fluorescent intensity ratio of elements 1 and 2 from a set of thin film standards containing different weight ratios of the elements. A plot of I_1/I_2 vs. W_1/W_2 yields a straight line with a slope equal to C.

Once C has been determined, the composition of an unknown thin film can be easily obtained from Equation 3 by measuring the intensity ratio of the two elements.

The results of XRF measurements on standards from chemical solutions are shown in Figure 5. A least square plot of I_K/I_P vs wt. K/P yields a straight line with a slope of 6.68. By measuring the intensity ratio of K/P, the K-P composition of the sample films can be easily obtained using Equation 3.

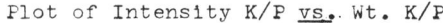

Plot of Intensity K/P <u>vs.</u> Wt. K/P

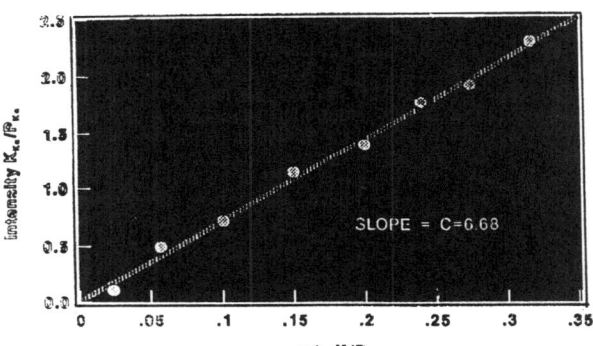

Figure 5

This equation is valid only for samples where the approximation, $\exp(-\bar{\alpha}\varrho t) \approx (1-\bar{\alpha}\varrho t)$ is valid. For the K-P films of the composition range of KP_4-KP_{50}, the above equation is satisfied for film thicknesses < 2 um. Above a thickness of 2 um the simple relationship between fluorescence intensity and composition is no longer linear due to absorption and enhancement effects. The method cannot be used below 0.1 um because the intensity of the K_{K_α} line is too low to obtain accurate results.

The accuracy of the "direct" XRF method was tested on the InP system because no certified KP_x film standards are available. The InP film has a definite theoretical In/P weight ratio of 3.7. Calibration standards were prepared as they were for the KP_x system from $In(NO_3)_3$ and $NH_4H_2PO_4$ solutions. To achieve a good comparison with the KP_x system, the In_{L_β} line was monitored. The energies for K_{K_α} and In_{L_β} are 3.31 and 3.49 keV, respectively. An In/P weight ratio of 3.5 was obtained with the "direct" method, demonstrating its accuracy.

TABLE 1

COMPARISON OF EDX AND XRF METHODS
FOR FILM - COMPOSITION DETERMINATION

SAMPLE #	XRF	EDX	XRF-EDX	(XRF-EDX)^2
1	16	14.3	+1.7	2.89
2	14	14.1	-0.1	0.01
3*	11.5	25	-	-
4	34	41	-7.0	49.00
5**	8	5	-	-
6	21	17	+4.0	16.00
7	24	24	0	0
8	26	29	-3.0	9.00
9	23	24	-1.0	1.00
10	20	18	+2.0	4.00
11	36	41	-5.0	25.0
12	63	62	+1.0	1.00
13	24	24	0	0
14	13	15	-2.0	4.00
15	37	40	-3.0	9.00
16	9	8	+1.0	1.00
17	14.5	12	+2.5	6.25
18	28	35	-7.0	49.00
19	23	22	+1.0	1.00
20	16.5	17	-0.5	0.25
21	24.5	30	-5.5	30.25
22	81.5	83	-1.5	2.25
23	34	40	-6.0	36.00

```
# of Samples              21
D.F. (degree of freedom)  20

σ                         3.2287
t                         1.9195  identical at the 95% C.L.

t$^{90}_{20}$ = 1.73

t$^{95}_{20}$ = 2.09

*   too thick for XRF (2 um)    not included in stat analysis
**  too thin for XRF (0.07 um)
```

TABLE 2

COMPARISON OF EDX AND DEKTEK METHODS
FOR FILM - THICKNESS DETERMINATION

SAMPLE #	THICKNESS um		D-E	$(D-E)^2$
	DEKTEK	EDX		
1	1.09	1.00	+0.04	0.0081
2	0.70	0.55	+0.15	0.0225
3	2.00	2.00	0	0
4	N.A.	0.90		
5	N.A.	0.07		
6	0.70	0.85	-0.15	0.0225
7	0.60	0.90	-0.3	0.0900
8	0.30	0.50	-0.2	0.0400
9	0.89	1.10	-0.21	0.0441
10	N.A.	1.20		
11	0.82	0.95	-0.13	0.0169
12	0.40	0.70	-0.30	0.0900
13	1.02	1.10	-0.08	0.0064
14	1.12	1.00	+0.12	0.0144
15	0.32	0.40	-0.08	0.0064
16	0.50	0.50	0	0
17	1.10	1.00	+0.10	0.0100
18	1.30	1.30	0	0
19	1.50	1.40	+0.10	0.0100
20	N.A.	1.80		
21	0.55	0.70	-0.15	0.0225
22	0.21	0.26	-0.05	0.0025
23	0.40	0.70	-0.3	0.0900

```
# of samples    19
D.F.            18
σ               0.1481
t               2.1538  identical at the 95% C. L.

 95
t     =  2.10
 18

 98
t     =  2.55
 18
```

Tables I and II compare the EDX data with those obtained by XRF composition and piezoelectric thickness measurement (DEKTEK) determination respectively. The data of the Student's t test are also included in the Tables. The statistical analysis shows that the two different techniques give data that are identical at the 95% confidence level for both thickness and composition determinations. The results of the EDX microprobe determination agree well with those obtained with macro techniques and thus it complements the latter when the lateral uniformity of the thin film is to be established at the micron scale.

CONCLUSION

Composition and thickness values for potassium polyphosphide thin film, determined by an EDX method based on Monte Carlo simulation compare well with those obtained by macro methods (piezoelectric thickness and XRF composition determinations). A simplified EDX method was shown to yield simultaneous thickness and composition data for the KP_x binary system from single determinations on the thin film and on a standard. Statistical analysis shows that the EDX and the macro technique give results that are identical at the 95% confidence level. Thus, EDX can be used to complement the macro techniques when the lateral uniformity of the KP_x films is to be established at the micron scale.

REFERENCES:

1. Schachter, R. et al. Appl. Phys. Lett. 45, (3) August 1, 1984.

2. To be published in J. of Appl. Phys., June 1985.

3. von Schneering, H. G. and Schmidt, H., Angev. Chem. Int. Ed. 6 356 (1967).

4. Kyser, D. F. and Murata, K. "Quantitative Electron Microprobe Analysis on Thin Films on Substrates" IBM J. Research and Development, 78, 352 (1974).

5. Murata, K., Sato, T. and Nagamie, K. "A Simple Quantitative Electron Microprobe Analysis of Multielement Thin Films on Substrate" Japan J. Appl. Phys. 15 No. 11 (1976).

6. Kyser, D. F., and Murata, K. "Quantitative Electron Microprobe Analysis of Thin Films with Monte Carlo Calculations" Proceedings of the 8th MAS Conference, 116, (1977).

7. Miller, N.C., and Koffman, D.M. "Determination of Thin-Film Composition or Thickness from Electron Probe Data by Monte Carlo Calculations" Microbeam Analysis, Dale E. Newberry, Ed. 1979, p. 41.

8. Phil, C. and Cvikevich, S. "Monte Carlo Simulation Approach to Quantitative Electron Microprobe Analysis of Ternary Alloy Thin Films" Microbeam Analysis, David B. Wittry, Ed. 1980, p. 161.

9. Heinrich, K. F. J., Newberry, D. E. and Yakovitz, H. "Use of Monte Carlo Calculations in Electron Probe Analysis" in Microanalysis and Scanning Electron Microscopy" NBS Special Publication 460 (1976).

10. Honig, R. E. Thin Solid Films, 31, 89 (1976).

11. Stankiewicz, W. et al. X-Ray Spectrometry, 12, 92 (1983).

12. Verheijke, M. L., and Witmer, A. W. W. Spectrochimica Acta, 33B, 817 (1978).

13. Laguitton, D. and Parrish, W. Analytical Chemistry, 49, 1162 (1977).

14. Kalnicky, D. J. and Monsteles, T. D. Analytical Chemistry, 53, 1782 (1981).

15. Hwang, T. C. and Parrish, W. Advances in X-Ray Analysis, 22, 43 (1979).

16. Jenkins, K. er al. "Quantitative X-Ray Spectrometry" Marcel-Dekker, Inc., New York, (1981).

DEFECTS IN PLATINUM SILICIDE FORMATION

MICHAEL J. WARBURTON
MOTOROLA INC, Discrete & Special Technologies Group,
Analytical Laboratories, Surface Analysis Lab
5005 East McDowell Road, Phoenix, AZ 85008

ABSTRACT

Four types of cause related defects in the formation of platinum silicide were examined by scanning Auger Microscopy (SAM). These are: "Dark Platinum", a dark rough patch on the platinum silicide, "Aqua Regia Attack", the apparent etch of the platinum silicide by the aqua regia used to remove pure Pt metal, "Spot Defects", holes in the Pt-Si layer and Ti-W strip defects, the apparent defects left in the Pt-Si by the stripping of a Ti-W overlayer. Elemental Auger maps show that silicide formation was prevented, in the first three cases, by remaining oxide from a masking step. The fourth case was caused by a layer of titanium oxide overlying the Pt-Si.

INTRODUCTION

Platinum silicide is in common use to form Schottky barrier and Ohmic contacts to Si [1-3]. The most common method of formation is to deposit a thin film of pure platinum on silicon (doped or undoped) and anneal it at 600-700 degrees centigrade [4-6]. It has also been known for some time that platinum does not react with silicon dioxide [7]. This provides a ready method of producing patterns of silicide formation on a silicon wafer. An oxide is grown on the silicon wafer and then patterned using standard photoresist masking techniques. Pure platinum is then deposited and the wafer is heated in an inert atmosphere. Silicide formation takes place only in those areas where platinum is in contact with silicon [8]. The unreacted platinum may then be removed with aqua regia leaving behind the platinum silicide.

Several casually related types of defects are found at the end of these processing steps. Three of these defects occur often enough to have acquired names among process engineers. The first of these three was given the name "dark platinum" because it appears optically as if the platinum silicide turned dark or black in selected areas. The second "aqua regia attack", appears along the edges of small openings in the oxide and looks as if the platinum silicide was etched by the aqua regia used to etch unreacted platinum. The third; "spot defects", appears as small holes in the platinum silicide. An unnamed defect is similar in appearance to spot defects, but on a slightly larger scale.

EXPERIMENTAL

A scanning Auger spectrometer (Physical Electronics Model 590) was used for examination of samples acquired directly from production lots of integrated circuits. Scanning Auger spectrometry has the capability of looking at small geometries and also mapping elemental distribution on a surface [9-11]. All samples were sputtered briefly with 2Kev Ar+ ions to remove any ambient contamination. This pretreatment would be sufficient to remove approximately 50 angstroms of silicon dioxide. The primary electron beam energy varied from 3Kev to 10Kev depending upon the image magnification required. Primary beam diameter varied from a nominal 2 micron to about 2500 angstroms and beam current varied from 1 microamp to about 50 nanoamps. Data acquisition time varied with the size of the sample geometry. All data was taken in the DN(E)/DE mode since the instrument lacked the capability to use N(E) data.

RESULTS

A representative optical micrograph of "dark platinum" appears in Fig 1. Auger elemental maps for oxygen (Fig 2) and platinum (Fig 3) show platinum present on one side of the dark transition area and oxygen on the other. The oxygen probably indicates the presence of SiO_2 which would interfere with silicide formation. The rough surface in this area would scatter light creating "dark platinum". We have observed similar results when silicon nitride is used rather than silicon dioxide (i.e., one side shows the presence of nitrogen and the other platinum). The oxide in areas other than scribegrids (spaces between individual integrated circuits) is thick enough to produce local charging when bombarded by the primary beam. This charging prevents the oxide from showing oxygen.

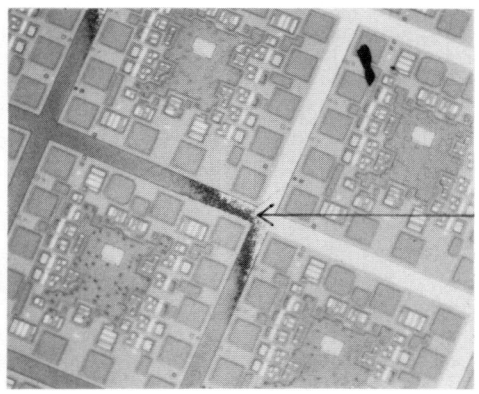

FIGURE 1. OPTICAL MICROGRAPH (400X)

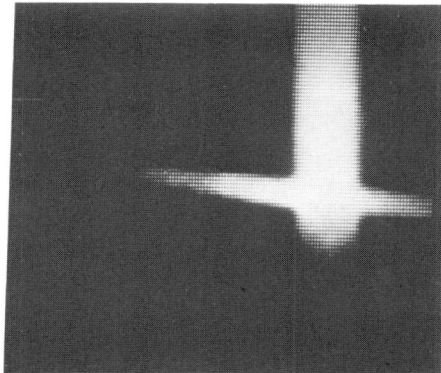
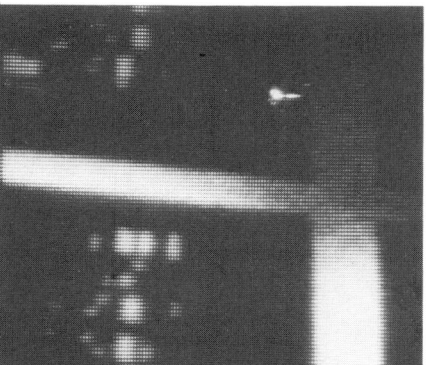

FIGURE 2
OXYGEN MAP (508ev)

FIGURE 3
PLATINUM MAP (62ev)

"Aqua regia attack" is observed at the edges of small pre ohmic openings (small openings in the mask oxide that later become contacts to underlying layers) in the oxide (nitride). A scanning electron micrograph (taken in the Auger system) is seen in Fig 4. The platinum silicide in the pre ohmic openings appears to have been etched along the edges of the openings by the aqua regia used to remove the unreacted platinum. However, elemental Auger maps for oxygen (Fig 5) and platinum (Fig 6) show oxygen in the etched areas, but little platinum. As above, the oxide (shown by the oxygen) would prevent silicide formation. The unreacted platinum would be removed by the aqua regia, causing the platinum silicide to appear etched. A defect that is similar to "aqua regia attack" in appearance, but quite different in origin appears on reworked wafers (Fig 7). Reworked wafers are those that have failed to meet specifications, generally because the metal layers did not cover steps in the oxide sufficiently to provide electrical continuity in the device. The metal overlying the platinum silicide (usually Ti-W/Al) is chemically stripped. When this is done, it appears as if the etchant used to strip the Ti-W (usually H_2O_2) has also etched the platinum silicide. Electron microprobe analysis shows platinum in these areas even though it does not appear optically or on an elemental Auger map (Fig 10). A titanium map (Fig 8) shows the surface to be covered with this element. Fig 9 (an oxygen map) shows the surface to be oxidized. A film of titanium oxide left behind from the Ti-W strip has covered the platinum silicide. Its thickness is greater than the escape depth of platinum Auger electrons. This film is also optically dense and obscures the platinum silicide underneath making it appear that the platinum silicide has been etched by the hydrogen peroxide used to strip the Ti-W.

Spot detects appear as small holes in the platinum silicide layer (Fig 11). Auger maps of the area for oxygen (Fig 12) and platinum (Fig 13) show oxygen in the hole, but no platinum. It appears that oxide has prevented silicidation in these areas resulting in holes in the silicide layer.

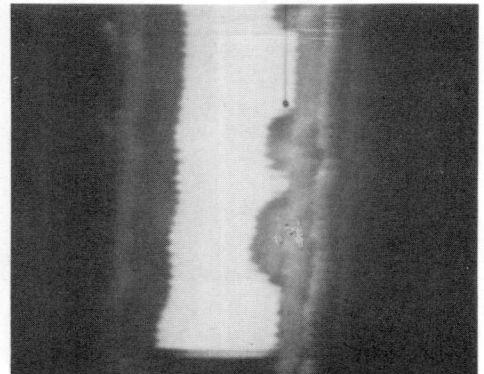

FIGURE 4. SEM (4000X)

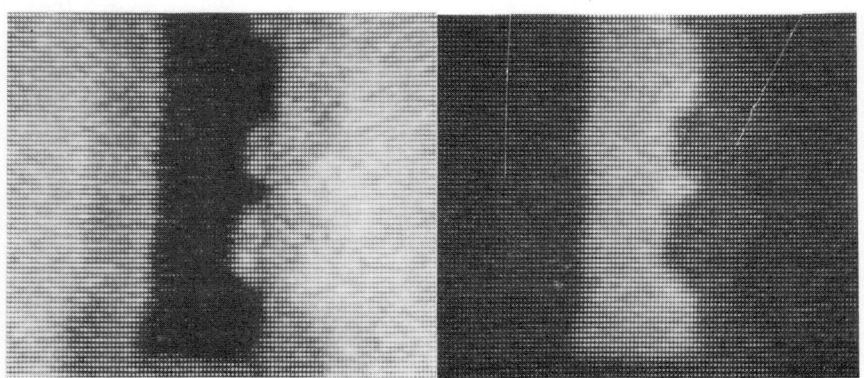

FIGURE 5
OXYGEN MAP (508ev)

FIGURE 6
PLATINUM MAP (62ev)

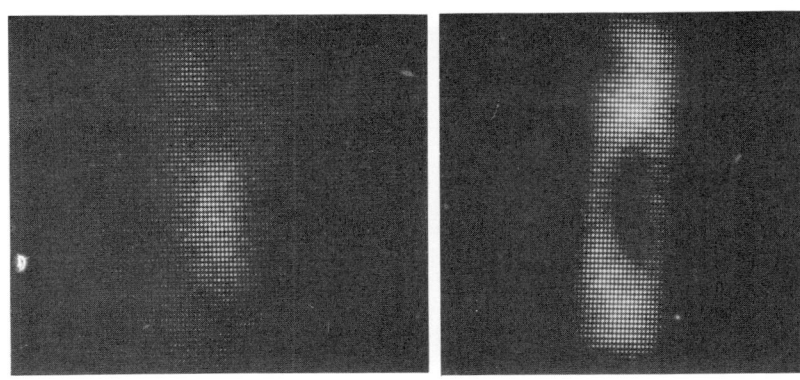

FIGURE 7. SEM (4000X) FIGURE 8. TITANIUM MAP (422ev)

FIGURE 9. OXYGEN MAP (508ev) FIGURE 10. PLATINUM MAP (62ev)

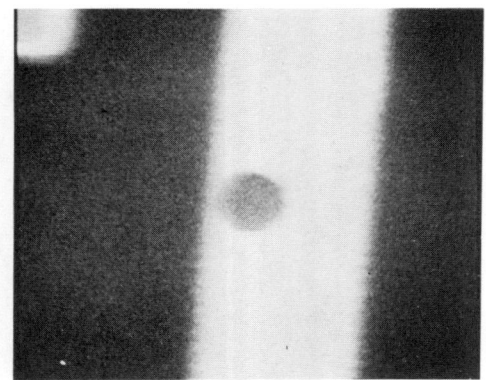

FIGURE 11. SEM (4000X)

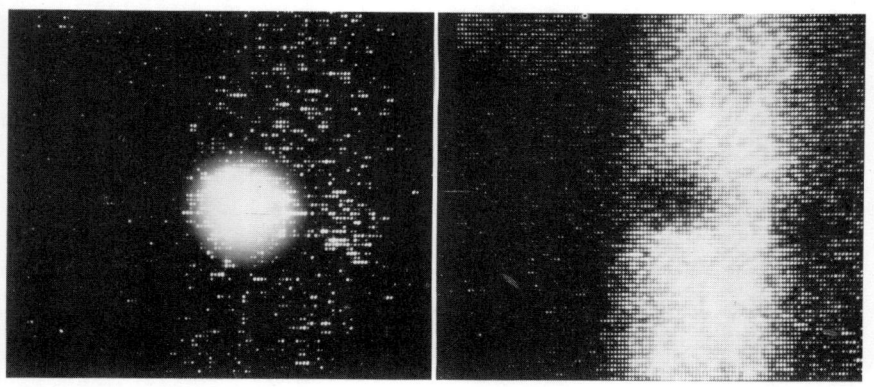

FIGURE 12. OXYGEN MAP (508ev)   FIGURE 13. PLATINUM MAP (62ev)

CONCLUSIONS AND DISCUSSION

"Dark Platinum" (Fig 14A) appears to result from unetched field oxide (nitride) on the silicon surface. Little or no silicide formation takes place in these areas. The partially etched boundaries between etched and unetched areas have small islands of oxide (nitride) remaining on the surface. Along these boundaries, partial silicide formation takes place, resulting in "dark platinum".

"Aqua regia attack" (Figure 14A) appears to occur when oxide remains only partially etched. No platinum silicide formation takes place in these areas. The unreacted platinum is removed by aqua regia etch resulting in the appearance that the platinum silicide has been etched.

"Spot defects" (Fig 14A) result from small spots of unetched oxide in the vias preventing silicide formation. Holes in the platinum silicide layer result.

The underlying cause of the failure of the oxide (nitride) to etch is a very complex problem so only a few possible causes are considered. One is that the photoresist used to pattern the oxide was not completely removed by the developer. the residue would inhibit the etch of the oxide until it was either dissolved by the etch or was undercut by it. In the latter case, the etch must remove a layer of material underneath the photoresist until it lifts off. Both cases would result in oxide remaining in areas where this occurred and this would inhibit silicide formation. Another cause could be that on the scale of the vias (<5u) the oxide is not of uniform composition especially if it is deposited by plasma or CVD, et cetera, rather than a thermally grown oxide. Etch times are chosen very carefully to minimize lateral etch (undercut). If the layer is not uniform, it may etch at a different rate resulting in a residue being left. This would inhibit silicide formation in this area. Either of these phenomenon would be exacerbated by small geometries.

The unnamed defect occuring after Ti-W removal (Fig 14B) may have equally complex origins. The defect is caused by the deposition and oxidation of Ti on the platinum silicide layer. This layer is optically dense so the platinum silicide is not visible thru it. Auger elemental maps (because of the shallow analysis depth of Auger microscopy) show only titanium oxide, but electron microprobe analysis clearly shows platinum silicide to be present.

One possible explanation for the presence of the Ti film is that W is more soluble in hydrogen peroxide than Ti. The W complexes with the peroxide displacing dissolved Ti and depositing it as an oxide. This would leave a film of titanium oxide covering much of the wafer as is found in some cases. The hydrogen peroxide bath may be used to etch many sets of wafers and could become saturated with Ti and W in just this fashion.

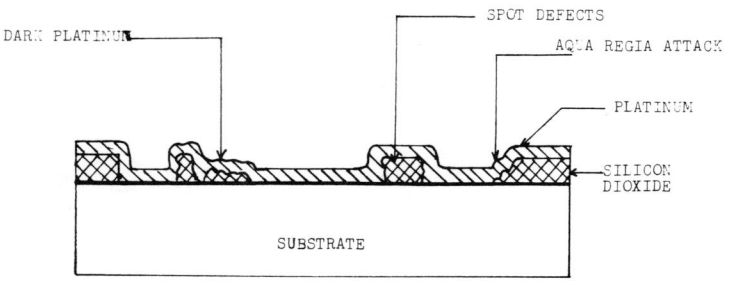

FIGURE 14A.

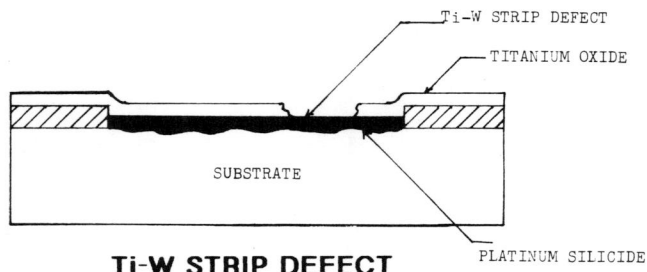

FIGURE 14B.

REFERENCES

1. M. P. Lepselter, Bell System Tech J. 45, 233 (1966).

2. D. Kahng and M. P. Lapselter, Bell System Tech J. 44, 1525 (1965).

3. M. P. Lapselter and J. M. Andrews in "Ohmic Contacts to Semiconductors," B. Schwartz, Editor.

4. C. Canali et al, App. Phys Lett. 31, No. 1 (July 1977).

5. T. J. Kingzett and C. A. Ladas, J. Electrochem Soc. 122, No. 12.

6. ibid (1)

7 ibid (1)

8 ibid (5)

9. N. C. MacDonald, G. E. Riach and R. L. Gerlach, Research/Development, 27, No. 8 (1976).

10. Kane and Larrabe "Characterization of Solid Surfaces", 1974, Plenum Press.

11. N. C. MacDonald, C. T. Horland and R. L. Gerlach, Scanning Elect. Microscopy, 1 (1977).

AUGER ELECTRON ANALYSIS OF OXIDES GROWN ON A DILUTE ZIRCONIUM/NICKEL ALLOY

R.A. PLOC, R.D. DAVIDSON AND J.A. ROY
Atomic Energy of Canada Limited, Chalk River Nuclear Laboratories,
Chalk River, Ontario, K0J 1J0   Canada

ABSTRACT

Auger electron analysis of oxides grown at 573 K in dried oxygen on the matrix and on precipitates in a 0.5 wt% Ni in zirconium alloy, is part of a larger program to study the effect of second phase particles on the oxidation and hydriding behaviour of Zircaloy-2*. The alloy consisted of a zirconium matrix containing randomly distributed Zr_2Ni precipitates of up to 10 μm in diameter. Oxidation simultaneously proceeded about the whole periphery of the precipitate and after approximately 50 days exposure was completely oxidized and of nearly uniform composition. The chemical structure of the oxides grown on the matrix and on the free and internal surfaces of the precipitate were determined as a function of depth in the oxide film.

INTRODUCTION

It has been known for many years that Zircaloy-2 can enter a regime of "breakaway" oxidation where the kinetic rate changes from parabolic/cubic to linear. In dry oxygen, a similar "breakaway" phenomenon has been seen in reactor grade, unalloyed zirconium [1,2]. Cox [3] has extensively reviewed the subject. Precipitates were suspected of possessing the ability to modify the oxidation process by providing easy oxygen access to the oxide/metal interface and also by enhancing electron transfer across the oxide film [4,5]. Nickel rich precipitates may also enhance hydriding, with serious consequences for reactors with Zircaloy-2 pressure tubes [6], since Zircaloy-2 is far more susceptible to this phenomenon than Zr-2½ wt% Nb (see [3] pages 235 and 336).

The morphology and composition of Zircaloy-2 precipitates have been reported by Chemelle et al. [7] where much of the earlier work is also referenced. Chemelle and workers found that two chemically different precipitates were present; one based on the Zr_2Ni structure and, the other, on $ZrCr_2$. In both instances, Fe partially replaced the non-zirconium element giving rise to compositions of $Zr_2Ni_{0.4}Fe_{0.6}$ and $ZrCr_{1.1}Fe_{0.9}$.

As a first step in the investigation of the role of precipitates, it was decided to study the oxidation, in dry oxygen at 573 K, of dilute binary alloys containing Zr_2Ni and $ZrCr_2$ as well as one further alloy containing Zr_3Fe (or Zr_4Fe) particles. Results from the Zr/Fe alloys have been reported [8]. The present study involved Auger electron analysis and Argon ion sputtering to obtain depth distributions of elements in both the matrix and in precipitate oxide films.

An explanation of the alloy preparation, the Scanning Auger Microprobe (SAM) and its use can all be found in [8].

RESULTS

Reference to Figure 1 of [8] will demonstrate that of the dilute alloys investigated, the Ni alloy oxidized most rapidly. The specimens were

---

*Zircaloy-2 is a dilute zirconium alloy containing approximately 0.05% Ni, 0.1% Fe, 0.1% Cr and 1.5% Sn (by weight).

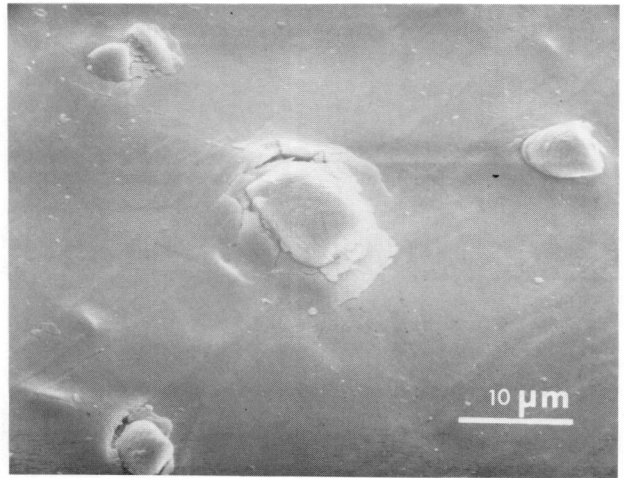

Figure 1 Surface of oxidized Zr-0.5 wt% Ni alloy showing cracked oxide on and about the oxidized Zr_2Ni precipitates.

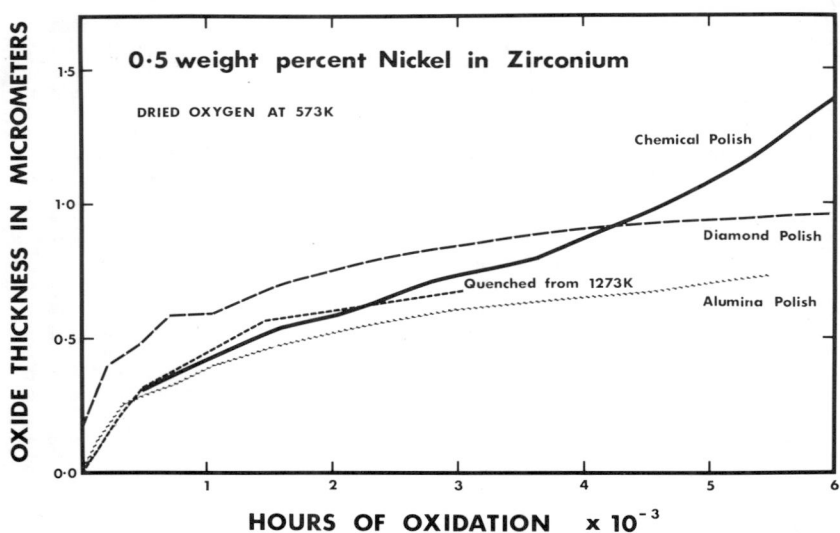

Figure 2 Oxidation kinetics for several surface preparations.

mechanically then chemically polished, leaving 5 to 10 μm diameter particles randomly distributed over the alloy surface. This observation is in contrast to earlier ones for Zircaloy-2 where the Zr/Ni/Fe precipitates were found to be selectively etched from the matrix surface. Figure 1 is typical of an oxidized surface (matrix oxide approximately ½ μm in thickness) showing oxide film cracking on and about the precipitates. To deduce the effect of this unusual geometry (particles standing proud of the surface), oxidation kinetics were determined for the same alloy but prepared by diamond and alumina polishing (without a chemical etch) to 1.0 and 0.05 μm, respectively. An additional test was performed on material water-quenched from 1273 K, mechanically polished and chemically etched, see Figure 2.

As part of our investigation, oxidized samples were cross-sectioned and slightly etched to remove a few microns of the metal core thereby revealing the inner surface of the oxide film. Figure 3 is a typical example where the foreground is the alloy matrix containing Zr_2Ni particles and the background shows the inside surface of a ½ μm thick oxide with oxidized inclusions protruding through. Note that the inclusion marked with an arrow in Figure 3b is not oxidized. In this manner, it was possible to perform Auger analysis of the inner and outer surfaces of the oxide on the precipitate.

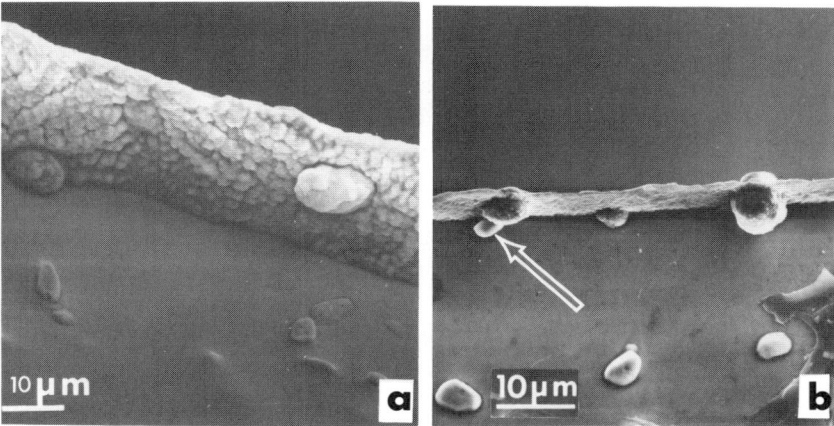

Figure 3 The inside surface (matrix/oxide interface) of an oxide grown on the Zr/Ni alloy. Foreground is the unoxidized metal containing Zr_2Ni precipitates. Arrow indicates an unoxidized precipitate lying near to the advancing oxide/metal interface.

Figure 4 is a summary of the depth profiling obtained by Argon ion etching and Auger electron analysis of oxides grown on the Zr_2Ni precipitates for the indicated times (dried oxygen at 573 K). These data were obtained beginning from and etching pependicular to the free surface. The results in Figure 4a for 64 days are unusual in that only one precipitate was found which was not completely oxidized and, in general, results such as Figure 4b were obtained. The matrix oxide film was usually thicker than that on the precipitate for at least the first 30 days of oxidation and thereafter, vice versa. This apparent reversal, however, needs to be verified by independent tests.

Results from profiling the inner precipitate oxide were fraught with interpretation problems and reproducibility. Figures 5a and 5b indicate how

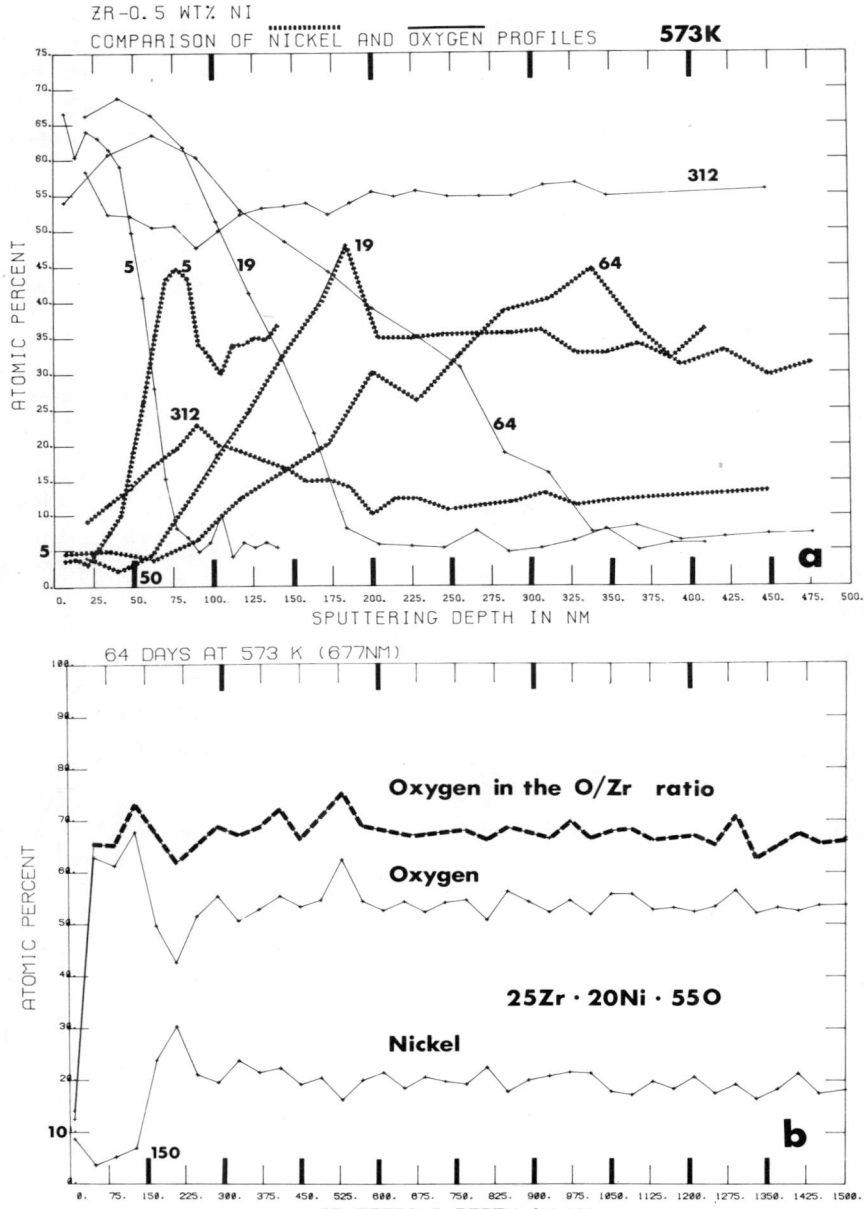

Figure 4 Element profiles for oxide formed on surface of Zr_2Ni precipitates oxidized for the indicated times. (a) shows only the nickel and oxygen profiles, (b) is typical for 64 days of oxidation where the O/Zr ratio suggests that the oxide is almost wholly ZrO_2.

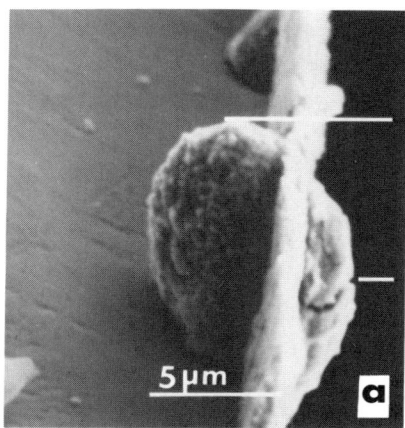 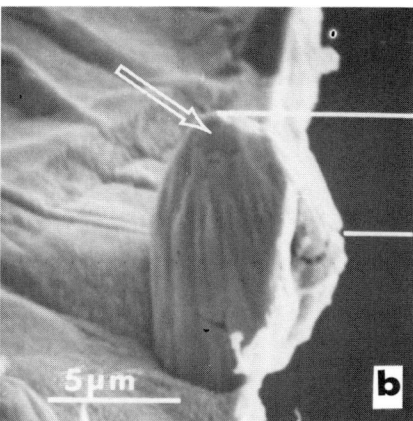

Figure 5 Cross-section of an oxidized sample with the metal core (on the left) slightly etched away revealing the outer and inner oxidized surfaces of the precipitates. (b) is the same precipitate after approximately 60 minutes of Ar ion etching. The bottom horizontal line marks the same location on the precipitate in both instances, the distance between the lines gives a general idea of the amount of erosion.

the profiling was performed. Figure 5a is without any Argon ion etching showing on the right the free surface and, on the left, the oxidized inner precipitate surface. The vertical bright band through the particle is the matrix oxide film in cross-section. The direction of the Argon ion beam can be deduced from Figure 5b. In all instances, the Auger analysis was taken from a point near the top, indicated by an arrow. If the precipitate is imagined in cross-section and viewed at 90° from Figure 5 (i.e. perpendicular to the free surface) then what might be seen is indicated in Figure 6. Clearly, as etching proceeds, the actual path the profile takes could yield misleading results (dotted lines) if the analysis point drifts between sampling times or if the Zr_2Ni sputters at a faster rate than the oxide, then the electrons would hit the oxide behind where the Zr_2Ni would have been expected. Our experiments have produced indirect evidence that would indicate this as a real possibility.

Auger profiles from the inner oxide of the precipitates varied. In some instances, the oxide appeared to be $5ZrO_2 \cdot ZrNi \cdot Zr_2Ni$ which would be similar to the oxidation process occurring near the free surface but for a very rough interface. Our most recent results, however, are shown in Figure 7.

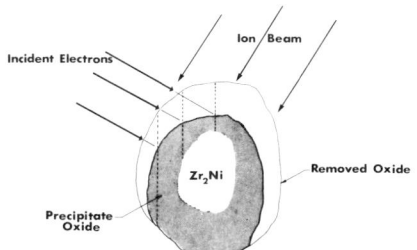

Figure 6 Relative geometry of the argon ion beam used for erosion, the incident electrons used for producing the Auger electrons and the possible profiling paths (dotted lines). The surface of the figure is considered parallel to the oxide/gas interface but profiling occurs when the same is perpendicular to the plane of the paper.

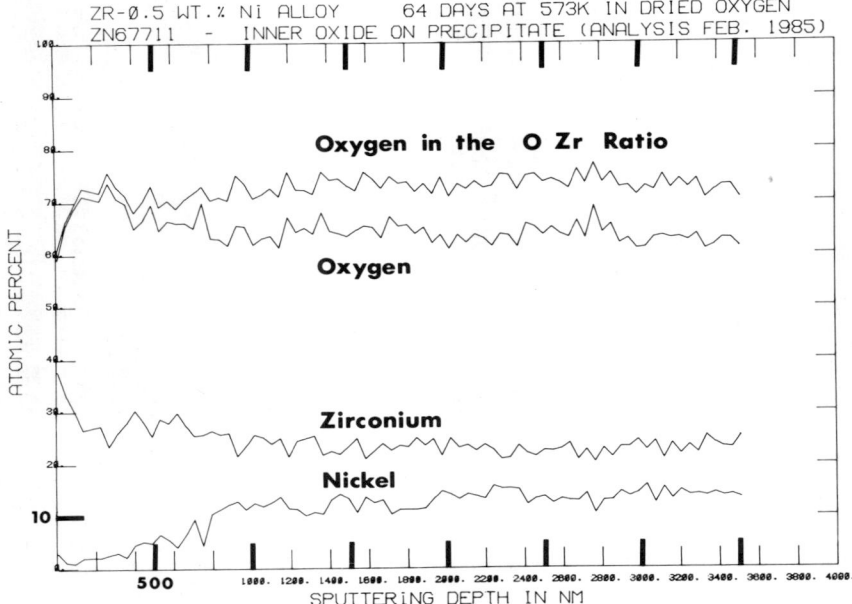

Figure 7 Profile of the elements in the oxide grown sub-surface on a Zr_2Ni precipitate.

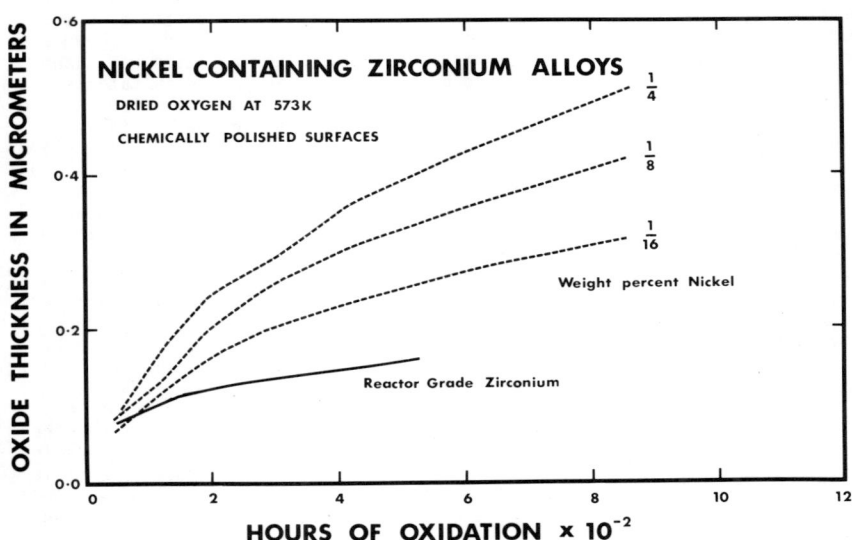

Figure 8 Oxidation kinetics for a series of dilute binary alloys of Ni in Zr.

Finally, Figure 8 dislays the oxidation kinetics for a series of low nickel content alloys.

DISCUSSION

As seen in Figure 4a, oxidation of the precipitate leads to a deficiency of nickel in the outer layers of the oxide. In this region of low to zero nickel content, the oxygen/zirconium concentration ratio is such as to suggest that the oxide is wholly ZrO_2. It is significant that as oxidation proceeds the nickel concentration peak remains at 50 at% indicating not rejection and pile-up of pure nickel at the oxide/particle interface but the formation of ZrNi. In other words, the Zr_2Ni is purged of half its Zr to form ZrO_2 leaving behind ZrNi which apparently oxidizes at a reduced rate.

If a simple mass balance is performed on the data of Figure 4a, it can be seen that the nickel missing from the outer oxide layers is not accounted for by the volume under the nickel peak (the unoxidized Zr_2Ni contains 33-1/3 at% Ni). For example, draw a line parallel to the 33 at% level, integrate and compare the area to the left of the mid-point of the nickel gradient to the area on the right (under the peak). A possible explanation for the large difference is that the nickel is precipitating as particles smaller than the electron beam diameter and being swept forward by the advancing oxide front; however, this is more unlikely than assuming that a layer of ZrNi is formed next to Zr_2Ni and Zr from the Zr_2Ni diffuses through the intermetallic to be oxidized at the nebulous ZrO_2/ZrNi boundary. The nickel profile slopes are indicative of a rough interface as well as an increasing thickness of the ZrNi layer. It is rather unusual that Zr should be diffusing; however, our observations seem to leave no other coherent explanation. This phenomenon is being investigated using single phase Zr_2Ni specimens.

After significant oxidation, in the present case 64 days, the nickel concentration stabilizes at about 20 at% (Figure 4b) and the oxygen to zirconium ratio suggests that the film is ZrO_2 containing a small amount of NiO. The displayed atomic concentrations suggest an oxide film of the composition $Zr_{25}Ni_{20}O_{55}$. This can be reduced to $25ZrO_2.5NiO.15Ni$ or $5ZrO_2.NiO.3Ni$, i.e., ¼ of the nickel is oxidized, the rest is in the zero oxidation state.

For even longer periods of oxidation (312 days), the results are more confusing and difficult to interpret. The origin of the slight nickel peak in Figure 4a is unknown unless it resulted from spalling of the surface oxide. It is difficult to imagine any mechanism such as diffusion, driven by chemical potentials, or partial oxidation that could have produced the peak at this position. The phenomenon is still under investigation.

Reference to Figure 3 will indicate that the precipitate, although extending more deeply into the metal than the penetration depth of the oxide, is yet oxidized. As long as the precipitate is in contact with the outer surface (gas/oxide) oxygen is channeled along the precipitate/matrix boundary to cause internal oxidation of both the matrix and precipitate. Precipitates lying near the surface but not in contact are not oxidized - see arrow in Figure 3b. Figure 7 represents our best data to date for profiling the inner oxide.

In the interpretation of data from the inner oxide, there are at least two major sources of error. First, as Figure 6 indicates, the actual point at which the Auger analysis is performed is important. If that point is not located on a line passing through the centre of the precipitate, or wanders from point to point, the profile may pass through a region of almost constant composition (for instance, the left-most dotted line in Figure 6). The second error involves selective etching. If the unoxidized alloy core is etched by the Argon ions at a faster rate than the oxide, then the analysis point will fall on the oxide, perhaps even at the oxide/precipitate interface. Using our present techniques, there is no way to guarantee these errors are not occurring.

The inner oxide profiles do not originate from the point furthest away from the free surface. Because of the geometry, the profile passes along a line through the side of the precipitate. The results shown in Figure 7 are consistent with the stoichiometric oxidation of Zr_2Ni. For instance:

$$2Zr_2Ni + 5O_2 \rightarrow 4ZrO_2 + 2NiO$$

where the atomic concentrations of zirconium, oxygen and nickel are 25.0, 62.5 and 12.5, respectively. This result is different from that found when etching the free surface, when it was noted that the Zr_2Ni was leached of some of its zirconium to leave ZrNi and ZrO_2. Continued oxidation of the ZrNi left about three-quarters of the nickel in the unoxidized state. The reason for this difference is unknown but, may be associated with the levels of stress in the two areas. In one instance, the precipitate oxide is 'locked' in place by compressive stresses caused by internal oxidation of itself and the surrounding matrix and, in the other instance, the zirconium diffuses to a region that could develop tensile stresses (toward the free surface) - note the cracks on the precipitate oxide surface (Figure 1).

An additional problem that creates interpretation difficulties is the size of the precipitate relative to the oxide film thickness. It would seem that some precipitates are completely oxidized and have been so for sufficient time to create an equilibration of the nickel concentration. Profile concentrations then would depend on the duration of oxidation relative to the precipitate size. The oxide front advances into the precipitate from all surfaces, but it is yet unclear what happens at the core prior to final oxidation.

With regard to the oxide film on the alloy matrix, nickel has not been detected although it must be present but at a very low concentration (solubility limit). One point is clear, however, the oxide/matrix interface is considerably rougher than for other alloys such as the Zr-0.5 wt% Fe.

Kinetics

One of the objectives of this study was to find a relationship between the physical structure of the oxide film and the oxidation behaviour of the alloy. It was evident from Figure 2 that specimen preparation had a dramatic effect on the oxidation rate. The smoother the free surface, the lower the oxidation rate. When the kinetics of the 0.05 μm alumina polished samples were compared to the alloys reported in Figure 1 of [8], the nickel alloy was no longer the fastest oxidizing but only slightly faster than the Zr-0.5 wt% Cr alloy, and slower than Zircaloy-2. This would be consistent with the observation that the oxide film thickness on the precipitates is thinner than on the matrix for the Ni alloy and vice versa for the Fe, but it still leaves the question as to what is limiting the oxidation rate. Clearly, oxygen seems to enjoy relatively free access to the oxide/metal interface as evidenced by the uniform film thickness right up to the edge of the precipitate. Electronic or charge transfer characteristics then must be limiting and of the two, probably the former for at least the initial stages of oxidation, since the outer oxide layers are apparently undoped ZrO_2.

Decreasing the nickel content of the alloy decreases the oxidation rate (Figure 8) and, since the same amount of nickel is present in the matrix in every instance (solubility limit) it is the precipitates themselves which are causing the increased oxidation rate. Since the underside of the precipitate is covered with a shroud of ZrO_2 from the oxidation of the matrix, the effectiveness of individual precipitates would be expected to decrease with time if simple electronic conduction was the only explanation. It is suspected that cracking of the surface oxide on the precipitate provides molecular oxygen access to the nickel which plays an important part in charge transfer. This theory remains to be proven. An additional observation which tends to support the charge transfer theory rather than simple electronic conduction is the fact that oxidized Zr_2Ni particles charge in an electron beam unless the first few nanometers of oxide are etched away (argon ion beam).

CONCLUSIONS

Preliminary Auger electron analysis of the oxidation of a Zr-0.5 wt% Ni alloy suggests that accelerated oxidation results from a charge transfer mechanism involving nickel located near the gas/oxide interface. For long term oxidation, nickel is uniformly distributed throughout the oxide formed on the precipitate except for a thin layer at the free surface; however, this region is heavily cracked due to oxidation stresses. A similar speculative charge transfer is involved for the oxidation of the Zr-0.5 wt% Fe alloys; however, in this instance most - if not all - of the iron from the precipitate ends up on the free surface.

[1] R.A. Ploc, J. of Nucl. Mater., 82, 411 (1979).
[2] R.A. Ploc, J. of Nucl. Mater., 91, 322 (1980).
[3] B. Cox, in Advances in Corrosion Science and Technology, edited by M.G. Fontana and R.W. Staehle (Plenum Press, New York and London, Vol.5, 1976) p.173.
[4] N. Ramasubramanian in Atomic Energy of Canada Limited report, AECL-3082, Chalk River Nuclear Laboratories, March 1968.
[5] N. Ramasubramanian, J. of Nucl. Mater., 55, 134 (1975).
[6] V.F. Urbanic and B. Cox, in press, Canadian Met. Quarterly, presented at 23rd Annual CIM conference, Quebec City, Aug.1984.
[7] P. Chemelle, D.B. Knorr, J.B. Van der Sande and R.M. Pelloux, J. Nucl. Mater., 113, 58 (1983).
[8] R.A. Ploc and R.D. Davidson, in press, presented at 17th Annual Technical Meeting of the International Metallographic Society, Philadelphia, July 1984.

AUGER SPUTTER DEPTH PROFILING APPLIED TO
ADVANCED SEMICONDUCTOR DEVICE STRUCTURES

D.K. SKINNER, C. HILL, and M.W. JONES
Plessey Research (Caswell) Limited, Allen Clark Research Centre,
Caswell, Towcester, Northants NN12 8EQ, England

ABSTRACT

The continuing miniaturisation of modern semiconductor devices is placing new demands on the spatial resolution of many existing assessment techniques. With device structures reduced to nm dimensions the need to optimise conditions for high resolution in-depth profiling is essential for accurate characterisation. In this paper we present Auger profiles of two micro-electronic structures and consider the choice of ion beam parameters necessary to minimise ion beam induced distortion. In VLSI (Very Large Scale Integration) technology where high dopant concentrations are ion implanted into shallow layer structures, Auger profiling can be used to establish dopant distributions accurately in the peak regions of the high dose implant. This information is shown to provide data for process modelling programmes.

Also presented are profiles of MOCVD (Metal Organic Chemical Vapour Deposition) grown GaAlAs/GaAs superlattices with 70 Å repeat structures. Interface width measurements are of particular importance in this case where sharp interfaces are a pre-requisite for good device performance. These structures have been found to be very vulnerable to enhanced atomic mixing, a consequence of the small mass of the Al atom. The Al content of the GaAlAs determining the obtainable in-depth resolution.

INTRODUCTION

The technique of Auger electron spectroscopy (AES) has become widely used in the semiconductor field since its inception in the late 1960s. During this period semiconductor technology has made enormous advances and to a large extent the development of AES has paralleled this progress. However, the circuit complexity and density of modern devices continues unabated as technology extends to smaller dimensions both laterally and in depth, with many device structures now reduced to nm dimensions. Assessment techniques are therefore going to find it increasingly difficult to cope with the demands of these new technologies. State-of-the-art AES with primary beam diameters of 50 nm and analysed lateral resolution of 100 nm is not likely to be improved upon. Extending analysis into the bulk of the material using AES in conjunction with sputter etching is, however, an area where advances can make a valuable contribution to device assessment. In this paper we are concerned primarily with the conditions necessary to reduce ion beam induced distortion in AES sputter depth profiling of two different semiconductor materials. In both cases specially fabricated, reduced cross-section test structures were used. The results presented from Si form part of a series of experiments aimed at providing input data for process modelling programmes! Process modelling uses computor programmes to model integrated circuit fabrication processes, the goal being to relate quantitatively the final device structure to the relevent variables of the fabrication process. Such programmes have a need for good basic experimental data in ion-implantation, diffusion and oxidation of small geometry structures both as input to the model and as the ultimate check on the correspondence between predicted and actual device structures.

EXPERIMENTAL

Polysilicon test structures grown by LPCVD with buried oxides produced by thermal oxidation for the 57 Å oxides and boiling H_2SO_4-H_2O_2 for the 12 Å oxide; thickness of the oxide being determined by lattice imaged TEM. The GaAlAs/GaAs superlattice structures were grown by MOCVD at atmospheric pressure using a fast switching, low dead space radial manifold. Layer thickness in this case determined by cross-sectional TEM.

Depth profile analysis was performed in a commercial AES system fitted with a sample entry vacuum lock, this ensures continuity of the hard vacuum conditions which are necesary when studying reactive materials. Gas purity for the ion gun was maintained by titanium sublimation pumping and ensuring gas flow by means of a separately valued diffusion pump.

In the profiling of the superlattice structures the 68 eV Al peak was monitored, chosen for its low elastic mean free path of ~6 Å. A primary electron source operated at 5 KeV, 0.1 µA, into a 40 µA spot was used and blanked during the sputter period. The ion beam which was rastered to remove any contribution from beam inhomogeneity was tured off during the recording of spectra. The ΔZ resolution term adopted throughout this work is taken as the depth for which the peak height changes from 84% to 16% of its maximum.

SPUTTER ETCH PROFILING

In view of the wide ranging complexities of the sputtering process[2] it is an obvious advantage to keep at a minimum the amount of sputtering involved in the characterisation of device structures. One way this can be realised is by the fabrication of special test structures with thinner cross sections acepting that there may be constraints in relating actual device behaviour to the test structure. A requirement in VLSI work is to be able to profile dopant distributions with a depth resolution of 1 nm. The realisation of such high resolution has only been achieved to date in the profiling of very thin amorphous layers[3] and represents the limit of the technique. Extending this high resolution to crystalline material and greater depths is only going to be accomplished by improving our knowledge of the various ion beam induced artifacts which occur during sputtering. In the absence of any ion beam induced roughing effects the major distortions arise from atomic mixing and the statistical nature of the sputtering process itself. The roughing factor is strongly dependent on the nature of the material being sputtered and nor4ally increases with depth. It has been often repeated in review articles[4-5] on sputter profiling that low energy ions are essential for optimum resolution in relation to the atomic mixing process. Our experience has shown that although this is generally an accurate statement, consideration must be given to the nature of the material being sputtered and that in order to optimise the sputtering conditions for high resolution it is necessary to characterise the ion beam induced broadening in terms of ion specie, energy and angle of incidence.

It has been shown[6] that an important parameter in the reduction of atomic mixing is to effect an increase in the sputter yield. Measurements of the sputter rate vs angle of incidence of the ion beam reveal that although the sputter rate goes through a maximum for most materials between 45^0 and 60^0 the sputter yield continues to increase up to very shallow angles[7].

This implies that improved resolution would be expected within the range affected by atomic mixing using shallow angle primary ions. Unpredictable material dependent factors, however, prevent uncritical application of this relationship and optimum sputtering conditions very often need to be determined empirically.

RESULTS AND DISCUSSION

The application of optimised sputtering conditions to polysilicson layers is shown in Fig 1a and b. These profiles are of special test structures and incorporate buried SiO_2 layers of 57 Å and 12 Å respectively at the polysilicon/silicon interface. The top layer of 100 nm of polysilicon has been heavily implanted with arsenic followed by a transient anneal cycle. The in-depth resolution obtained at the oxide interface is 3 nm, using 3 KeV Ar^+ ions inscident at 78^0 from the normal. The surface of polysilicon would normally, because of its polycrystalline nature, be expected to suffer micro roughing following the sputter removal of 100 nm. An explanation for the high interfacial resolution and absence of surface roughing, simply in terms of the increased sputter yield and reduced atomic mixing with 78^0 incident ions is clearly unsatisfactory, if only because the optimum sputtering conditions varies with the target material. The cluster removal process at small ion beam angles suggested by Heyes[8] may be a contributing factor but an adequate explanation must include reference to the nature of the target material. Of particular interest in these profiles is the redistribution of arsenic relative to the buried oxide following the anneal cycle. The pile up of arsenic at the interface of the 57 Å oxide in Fig. 1a provides evidence as to the effectiveness of the oxide as a barrier to the diffusion of arsenic into the silicon substrate. This can be compared with the 12 Å buried oxide in Fig. 1b with coincident arsenic and oxygen maxima suggesting penetration of arsenic into the silicon substrate, the accuracy of this profile, however, is almost certainly limited by the the available resolution.

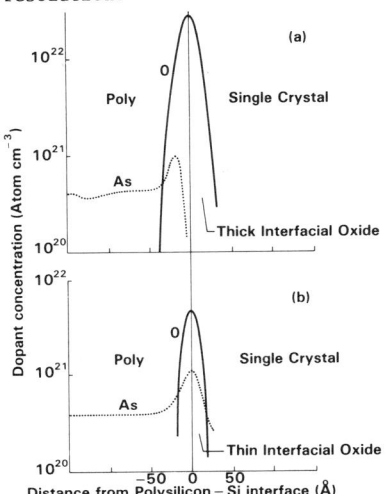

Fig. 1 AES dopant concentration profiles through the polysilicon/single crystal interface region of a poly emitter transistor structure for (a) 57Å and (b) 12Å interfacial oxides : polysilicon toplayer 1000Å thick

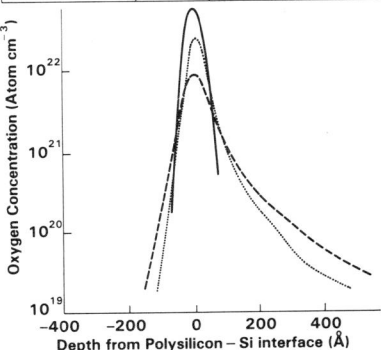

Fig. 2 Oxygen concentration profiles through a polysilicon–oxide–silicon structure : sputtering beam angle measured with respect to the surface normal

Analysis	Symbol	Width ÷ 2	Width ÷ 100
Auger 3KeV 78°	————	50A	135A
Sims 5KeV 5°	············	60A	290A
Sims 10,6KeV 5°	------	75A	450A

The very high implanted arsenic level in these structures is typical of many VLSI devices and effectively places dopant profiling within the domain of AES, an assessment normally exclusive to the SIMS technique.

In Fig. 2 a direct comparison with SIMS profiles is made using the 57 Å buried oxide structure. The sputtering conditions for the AES profile have been optimised as decribed previously. The comparison demonstrates the need for greater flexibility in the ion gun/analyser configuration of SIMS equipment, which can suffer a reduction in sensitivity with deviations from the standard configuration. Typical experimental concentration profiles for high arsenic implant levels in polysilicon are presented in Figs. 3 and 4, together with SUPREM[1] simulations (shown by the broken line) calculated from existing data. The SUPREM data in Fig. 3 is clearly seriously in error as to peak depth, profile width and profile shape, but establishes the importance of high in-depth resolution in locating accurately the peak region of the high dose implant in an inherently broad arsenic distribution. The transient anneal requirement for VLSI structures is concerned with the redistribution of dopant within the constraints of the greatly reduced device dimensions. The diffusion behaviour at such short annealing times cannot be extrapolated from existing data as the SUPREM simulations in Fig. 4 shows. The capability to model redistributions occuring at each process stage accurately is essential if the final device structure is to be adequately represented, but the complexities of high dopant concentrations, new materials and transient heat treatments of these structures means that the existing algorithms for the various process stages are not sufficiently sophisticated to allow exact modelling. It is clear, therefore, that the accuracy of the modelling programmes is critically dependent on the availability of high resolution, distortion free profile data, and that this in turn is dependent on a more detailed understanding of the complex mechanisms involved in the interaction of the ion beam with the sample.

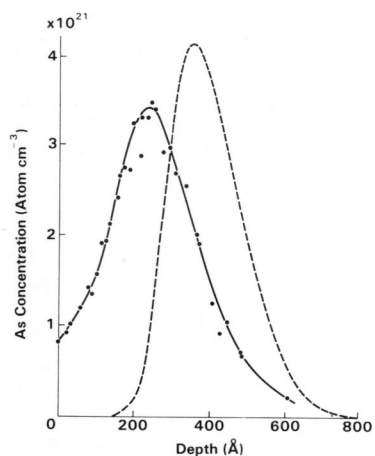

Fig. 3 AES arsenic implant profile (40 KeV, 10^{16} ions/cc)

Dotted curve = SUPREM simulation

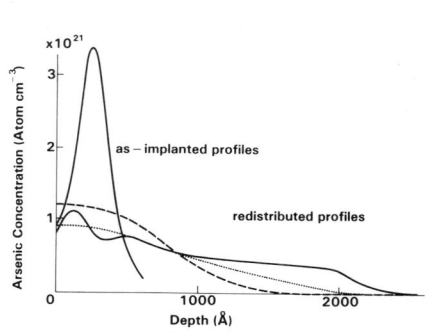

Fig. 4 AES arsenic implant profiles showing redistribution of arsenic by transient anneal : dotted lines are SUPREM simulations using different default values

GaAlAs/GaAs STRUCTURES

GaAlAs/GaAs superlattices with their regular repeat sequences of very thin layers are ideal structures for studying the ion beam broadening effects which dominate AES profiling. A compositional depth profile of the first six layers of a $Ga_{0.68}Al_{0.32}As$/GaAs structure obtained under optimised sputtering conditions is presented in Fig. 5. Ion beam parameters for this profile were 700 eV Xe^+ ions incident at $50°$ from the surface normal. Unlike the results obtained from polysilicon the superlattice structures were found to be very sensitive to the energy and mass of the incident ion. No improvement in resolutions was found with ion energies below 700 eV, due most likely to the rapidly reducing ion flux at low energies. A further disparity with polysilicon is that sputter angles greater than $50°$ are found to be detrimental, this is interpreted as being due to the material dependent variation in sputtering characteristics discussed earlier.

Plots of ΔZ vs Z from two superlattice structures, $Ga_{0.68}Al_{0.32}As$ and $Ga_{0.84}Al_{0.16}As$ are plotted in Fig. 6. Curve (a) with the higher Al content was plotted from the data in Fig. 5 and shows a $\Delta Z/Z$ relationship normally observed for polycrystalline metal layers[3]. Curve (b) with approximately half the Al content follows a Z dependence obtained from amorphous oxides[3-9] where the plateau region extends up to Z values of 400 nm before surface roughening effects take over. Interpretation of the polycrystalline metal $\Delta Z/Z$ relationship has been interpreted[10] as being a function of the developing surface topography. We have found no evidence for surface roughing and conclude that the degrading resolution with depth in Fig. 5 is caused by enhanced atomic mixing, brought about by the low mass of the Al atom. The relatively massive Xie^+ ion would be expected to reduce primary recoil implantation of Al into GaAs. This conclusion is supported by the improved resolution at lower Al levels and in changing from Ar^+ to Xe^+ primary ions.

In summary, we have shown that for the two semiconductor materials examined there is a considerable variation in the optimised sputtering conditions for high resolution profiling. We conclude from this that if very high in-depth resolution is to be realised, consideration must be given to the individual sputtering characteristics of the sample material.

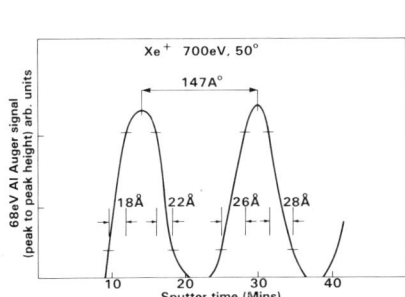

Fig. 5 AES profile of the top 4 layers of a $Ga_{0.68}Al_{0.32}As$/GaAs superlattice structure showing degrading interface resolution with increasing sputtered depth

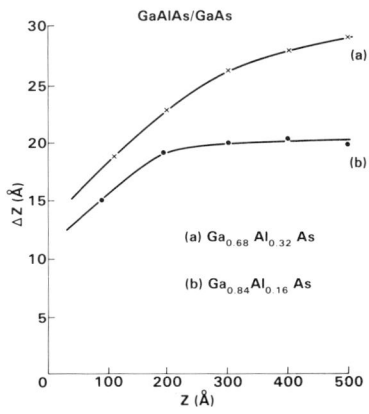

Fig. 6 The depth resolution as a function of interface depth using 700eV Xe+ ions incident at $50°$ from the surface normal for two different Al levels in GaAlAs/GaAs.

REFERENCES

1. C Hill and A L Butler. Published in Solid State Devices 1983 (Rhoderick E H ed) Inst of Physics, London (1984).

2. G Carter, A Gras-Marti and M J Nobes. Rad Efects, 62, 119 (1982).

3. M P Seah and C P Hunt. Surf-Interface Anal, 5, 33 (1983).

4. E Zinner. J Electrochem Soc, 130, No.5, 199C (1983).

5. S Hofmann. Surf Interface Anal, 2, 148 (1980).

6. K Wittmaack. J Appl Phys, 53 (7) 4817 (1982).

7. R E Chapman. J Mater Sci, 12, 1125 (1977).

8. D M Heyes, M Barber and J H R Clarke. Surf Sci, 105, 225 (1981).

9. J Fine, B Navinsek, F Davarya and T D Andreadis. J Vac Sci Technol, 20, 449 (1982).

10. H J Mathieu, D E McClure and D Landolt. Thin SOlid Films, 38, 281 (1976).

SURFACE SEGREGATION OF NI-CR ALLOY

N.Q. CHEN, Q.J. ZHANG, AND Z.Y. HUA
The Modern Physics Institute, Fudan University, Shanghai, China

ABSTRACT

The surface segregation of a Ni-Cr alloy used for commercial heating elements has been examined with a Scanning Auger Microprobe (SAM) and other electron spectroscopy. A variety of sample surfaces prepared by different ways, including fracturing, mechanical polishing, electrochemical polishing, annealing, oxidation and their combinations were investigated.

Results showed that on the surface treated with mechanical polishing followed by oxidation, a continuous, highly chromium-enriched layer would be formed in a few minutes. Three factors, i.e., mechanical polishing, oxygen environment and temperature, have been found all necessary to make this Cr enrichment happen rapidly.

Taking the advantage of high spatial resolution of SAM, other experimental facts have been obtained by comparing SEM image and Auger map. First, the Cr segregated at the area where abrading trace (about few micrometers in width) existed. Second, during the oxidation process at early stage Cr was enriched at the grain boundaries and spreaded laterally.

These results indicate that Cr was diffusing thru the grain boundaries to the surface and the process of mechanical polishing played the role for increasing a large number of grain boundaries. The chemical driving force of oxygen also promotes the diffusion rate at a relatively low temperature so as to complete the process in a short period.

INTRODUCTION

A variety of surface treatments, such as thermal, chemical and mechanical treatment, are conventionally used for improving the properties of alloys. Since the chromium is the chief addition to iron- and nickel-base alloys to improve corrosive resistance, the behaviour of oxidation of Cr and Ni in some Ni-Cr alloys have been investigated elsewhere /1-4/.

A new method of surface treatment has been developed in our laboratory to meet the requirement of vacuum-tight glass-to-metal seals for TV picture tubes /5/. It was found that suitable mechanical polishing followed by annealing in an oxygen environment would make highly enriched Cr on the surface of a Fe-Cr-Ni alloy to improve the surface adhesion of oxide films yet bulk expansion coefficient remain unchanged.

The effect of mechanical polishing on the surface segregation has not been investigated systematically so far. In order to understand the mechanism and possible applications this phenomenon has been **studied** in detail.

EXPERIMENTAL

The specimen was a typical Ni-Cr alloy used commercially as heating elements. Its nominal composition is 77 Ni-21 Cr with the rest of alloy, such as Si, Mn, Al, Ce, Fe, C, and S, less than 2 (wt%). The material was first cut in a suitable dimension and then abraded by carborundum papers simultaneously from coarse to fine grits, and finally it was polished by No.2 abrasive papers. After these procedures numerous traces would appear on the surface, as shown by the electron microscope (Fig. 1). The next step was electrochemical polishing with an etchant containing 10% $HClO_4$ + 90% CH_3COOH, 18 VDC, 30 sec. After that no visible surface defect from mechanical polishing was left. Finally, an additional electrolytic etching procedure (15% H_3PO_4 + 85% H_2O, 8 VDC, 10 min) should be done to make grain goundaries visible.

The thermal oxidation was performed at 600°C, 4 min in a quartz tube furnace with flowing oxygen at 1 atmosphere pressure. This furnace was also able to be evacuated to 10^{-6} Torr for vacuum annealing purpose.

The elemental composition and chemical state of sample after different treatments were analyzed by a PHI-590 Scanning Auger Microprobe and VG ESCALAB-5 Multifunction Electron Spectrometer. Before introducing into the test chamber, the specimen was agitated in acetone and then rinsed in deionized water thoroughly. Ion sputtering was also used if necessary. The data were taken by measuring the peak-to-peak amplitude H of Auger spectra in derivative form. The following Auger peaks were used for analysis: Cr(LMM)-528 eV, Ni(LMM)-848 eV, and O(KLL)-510 eV. To indicate how much Cr was precipitated, the ratio R was introduced,

$R = H_{Cr} / H_{Ni}$

and H_{Cr}, H_{Ni} are the amplitudes H for Cr and Ni respectively.

RESULTS

Five specimens have been used for a comparative study. The treatment used and the R measured are presented in Table I. Sputtering the surface for a long enough time, the R for bulk was measured and R=0.25.* After comparison the following results can be seen: After mechanical polishing at two different temperatures (Specimen A and B), the compositions are essentially the same. If afterwards this material is heated in vacuum (Specimen C), then the surface composition of Cr will be doubly increased. But the most dramatic result is: If thermal oxidation is followed after mechanical polishing (Specimen D), then R will be increased to a significant number, i.e., more than 1 order of magnitude as compared Specimen A and B. For the demonstration of the role played by the thermal oxidation, a clean fracture of the alloy (no mechanical polishing) was used for oxidation (Specimen E), and it was found that there was only a small increment of Cr content which was entirely different from the result of Specimen D.

Therefore the following conclusion will be easily obtained. There are effective Cr segregation over the Ni-Cr alloy surface after it was mechanical polished and then thermally oxidized. Nothing can be happened by simple mechanical polishing or thermal oxidation.

The depth profile of the Specimen D is shown in Fig. 2. From this figure one can clearly find that there is a Cr-enriched layer near the top surface where the Cr concentration is higher than that of Ni. When the sputter time increases the oxygen intensity decreases and gradually approaches to the noise level. In consistent with the oxygen signal the Cr decreases but the Ni rises up and surpasses the Cr intensity. Eventually all the signals of Cr, Ni and O remain unchanged and the composition reached is same as that of the bulk approximately. A XPS analysis showed that the layer on the top of specimen was in the form of Cr_2O_3.

In order to observe the process of Cr segregation, a Specimen F was made on which the grains with the size 30 - 100 micrometer were visible. Prior to thermal oxidation the Cr and Ni distributions were uniform but after thermal oxidation at temperature 670°C for 20 min, the distribution of Cr and Ni presented in Fig. 3(A) and Fig. 3(B) respectively were localized. The Cr was enriched at the area of grain boundaries with R=6.7 while the Ni occupied the rest of the area. During the initial stage of thermal oxidation the R increased with time duration and the width of Cr-enriched band was slightly widened.

A single abrading trace due to the mechanical polishing was also observed. If the time of electrochemical polishing was carefully controlled, the most of the scratches on the surface could be removed. Fig. 4(A) is the secondary electron image of Specimen G on which a single scratch with about 5 - 10 micrometer in

---

* According to E. Furman report /6/, the difference of composition between bulk and surface sputtered was less than 8% for Ni-Cr 25% alloy.

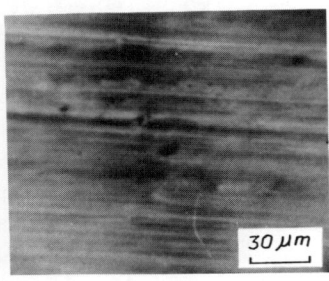

Fig. 1. Secondary electron image of the specimen abraded by No. 2 carborundum papers.

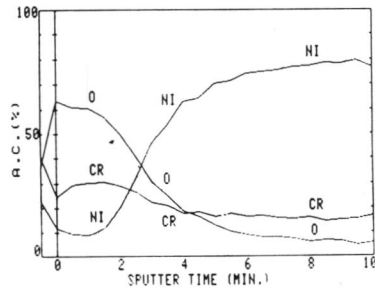

Fig. 2. The depth profile of Specimen D, mechanical polishing followed by thermal oxidation at 600°C, 4 min.

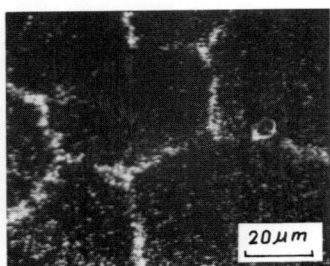
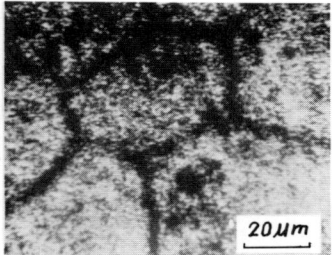

Fig. 3. The Auger map of CR(A) and NI(B) of specimen F, electrolylic etching followed by 20 min. thermal oxidation at 670° C.

TABLE I. Different Surface Treatments Used and R Measured

Specimen	Processes used	R
A	Mechanical polishing (under LN_2 temperature)	0.24
B	Mechanical polishing (under room temperature)	0.26
C	Vacuum annealing after process B	0.52
D	Thermal oxidation after process B	3.9
E	Thermal oxidation of a clean fracture surface	0.34

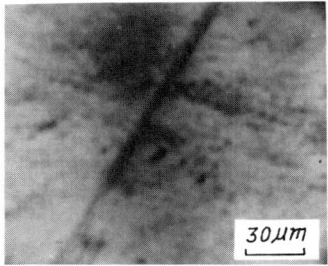 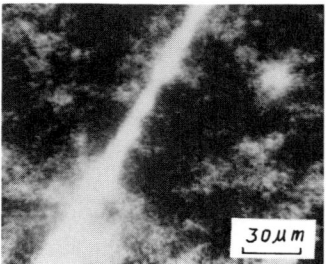

Fig. 4. The secondary electron image (A) and Cr Auger map (B) of specimen G, on its surface a scratch is visible.

width is shown. After thermal oxidation the Cr was enriched at the area where abrading trace existed, as shown in Fig. 4(B). The Cr enriched area is somewhat wider than the abrading one.

DISCUSSION

Results show that both mechanical polishing and thermal oxidation are the necessary conditions for fast Cr precipitation. Abrading, suitable temperature, and oxygen environment all played important roles to form a highly-enriched, continuous layer of Cr_2O_3 rather rapidly. The similarity of behaviour of Cr precipitation both at the grain boundaries and the scratched area indicates that the following model for Cr segregation is most probable: The mechanical polishing produces a numerous of small subgrains, then it makes the grain boundary increase hugely. As a result, Cr is able to diffuse outward along these grain boundaries under the effect of chemical drive force of oxygen to form a continuous oxide layer in a relatively short period. It should be mentioned that small grains not only make the outward diffusion much easily but also play an important role on lateral diffusion on the grain face to cover all the surface.

The effect of abrading on the surface has been investigated by Turley and Samuels /7/. Abrasion leaves a fragmented layer where a huge amount of small subgrains exist on the top surface with depth approximately same as to that of scratch. Giggins and Pettit /2/ found that both the grain size and oxidation rate of the grit-blasted specimen of Ni-Cr alloy were reduced. They then concluded that the low oxidation rate was due to the selective oxidation at grain boundaries. Bear and Merz /8/ also found that it was easier to form a protective Cr_2O_3 layer on the surface of stainless steel on which the grains were smaller.

However, the process of lateral diffusion is still somewhat not understandable. It was found that when one prolonged the oxidation time the Cr-enriched area did not spread all the time correspondingly. It seems that the lateral diffusion is a quite time-comsuming process. From Fig. 3(A) and Fig. 4(B) it can be seen that within limited area, just on the proximity of the grain boundaries or scratches, the lateral diffusion happens. These area where the Cr is easy to diffuse perhaps can be called as "diffusion-active zone". If the grains are so small or the scratches are so close that the diffusion-active zones touch each other then the continuous Cr_2O_3 layer can be formed on the top surface in a very short period. From this point of view two diffusion processes, upward and lateral, are both contributed to the formation of the Cr_2O_3 layer. However, the upward diffusion may be more important.

The oxygen would play two roles in this process, i.e., the selective oxidation and "Cr-holding effect". Since the free energies to form crystalline solid Cr_2O_3 and NiO at 25 °C are -252.9 kcal/mol and -50.6 kcal/mol respectively /9/, the formation of Cr_2O_3 is previaling in oxidation competition. As a result of Cr_2O_3 formation, these Cr which diffuse upward will be hold on the surface. It increases the gradient of concentration of mobile Cr and speed up the process of mass transportation.

CONCLUSION

Mechanical polishing and thermal oxidation afterwards are the indispensible conditions for the rapid formation of a continuous and highly Cr-enriched layer on the surface of Ni-Cr alloy. The Cr is in the form of Cr_2O_3. Direct observation of the Cr distribution on the surface by Auger map leads to a possible process for Cr segregation. Mechanical polishing produces numerous small grains in the fragmented layer in the selvage of the specimen. The huge amount of grain boundaries promote the upward diffusion of Cr. Selective oxidation and Cr-holding effect also speed up the formation of Cr_2O_3.

REFERENCE

/1/. C.S. Giggins and F.S. Pettit, TMS-AIME, 245, 2495(1969).
/2/. C.S. Giggins and F.S. Pettit, TMS-AIME, 245, 2509(1969).
/3/. B. Chattopadhyay and G.C. Wood, J. Electrochem. Soc., 117, 1163(1970).
/4/. K.L. Smith and L.D. Schmidt, J. Vac. Technol., 20, 364(1982).
/5/. X.L. Pan et al., To be published.
/6/. E. Furman, J. Mater. Sci., 17, 575(1982).
/7/. D.M. Turley and L.E. Samuels, Metallography, 14, 275(1981).
/8/. D.R. Bear and M.D. Merz, Met. Trans. A, 11A, 1973(1980).
/9/. CRC Handbook of Chemistry and Physics, R.C. Weast Ed., CRC Press, Inc., 1982-1983.

THE EFFECT OF OXYGEN ON DIFFUSION AND COMPOUNDING
AT Ni-GaAs(100) INTERFACES

J. S. SOLOMON, D. R. THOMAS, AND S. R. SMITH
University of Dayton Research Institute
300 College Park Avenue
Dayton, Ohio 45469

ABSTRACT

The effects of the presence of surface oxides and oxygen incorporated in deposited Ni films on intermixing and compounding at Ni/GaAs(100) interfaces was investigated with Auger sputter profile analysis. The Ni/GaAs structures were heated in vacuum with a high intensity incoherent lamp. Chemical changes and compound distributions were extracted from profile data by factor analysis of the Auger peaks. Diffusion coefficients for Ni at different temperatures were obtained from profile concentration gradients. The results showed that Ni diffusion was faster and activation energy lower for clean versus oxide covered GaAs substrates. Without the presence of oxygen, Ni_2 GaAs readily formed at 300°C while with oxygen, the formation of Ni-Ga and Ni-As compounds prevailed over the formation of a distinct Ni_2 GaAs layer.

INTRODUCTION

Segregation, interdiffusion, and chemical reaction of the constituents of a metal-semiconductor contact greatly affect the electrical properties of the metal-semiconductor structure. In the case of n-type GaAs, the Ni/Au-Ge system, first introduced by Braslau et al [1] and currently in wide use as an ohmic contact, has been shown to be spatially inhomogeneous because of phase separation and the formation of binary and ternary compounds during annealing [2-5]. The role and behavior of the individual elements, compounds, and phases in this contact system have been the subjects of many studies and controversies. For example, Robinson [6] reported that Ni increases the wetting properties of molten AuGe at the GaAs surface and enhances uniformity of the contact, while Wittmer et al [2] claimed that Ni acts as a sink for Ge and is important for producing uniform layers. Ogawa [7] concluded that Ni reacts with GaAs at the initial stage of annealing, whereupon an intervening layer on GaAs is eliminated, thus Ni assists the main contact constituent to react smoothly with GaAs. Heiblum et al [5] concluded that GeNi clusters formed during alloying are responsible for contact formation, and Kuan et al [8] showed conclusively that the element that causes contact degradation is Au.

The reactions between Ni and GaAs have been studied by a number of investigators [7-11]. Nickel reacts with GaAs at temperatures above 150°C to form the hexagonal ternary compound Ni_2 GaAs which grows epitaxially on (100) GaAs at temperatures of 150° to 550°C [7,8]. Above 450°C, Ni_2 GaAs decomposes into NiAs and β-NiGa [7].

In this work, the presence of oxides residing at Ni/GaAs interfaces and oxygen incorporated in the Ni films are shown to affect Ni diffusion and interactions with GaAs when subjected to rapid thermal annealing (RTA).

Three conditions were studied: (1) sputter deposited Ni on oxidized GaAs, (2) vapor deposited Ni on oxidized GaAs, and (3) vapor deposited Ni on clean GaAs. In the first case oxygen was present throughout the Ni film, while in the second case oxygen was only present at the Ni/GaAs interface. In the last case no oxygen was detected within the Ni film nor at the Ni/GaAs interface. The Ni/GaAs combinations were annealed in ultra-high vacuum. Diffusion coefficients for Ni at 300°C, 400°C, 500°C, and 700°C were extracted from Auger sputter profiles.

EXPERIMENTAL

Nickel was deposited to thicknesses of 100-250 nanometers (nm) by r.f. sputtering and electron beam evaporation in ultra-high vacuum (UHV) at pressures below 1×10^{-7} torr. The 1 cm^2 GaAs substrates were subjected to a wet chemical cleaning/etching treatment that included the following: methanol rinse-15 seconds, distilled H_2O-15 seconds, 10 minutes in $5H_2SO_4:1H_2O_2:1H_2O$ at 60°C, a second distilled H_2O rinse for 30 seconds, and a final rinse in $1HCl:5H_2O$ for 30 seconds. The substrates were blown dried with filtered dry nitrogen. Carbon and oxygen residuals were removed from electron beam deposited Ni films by UV ozone cleaning [12] and heat treatment [13] in UHV with a high intensity incoherent lamp. Substrates for r.f. sputtered nickel were backsputtered with 450 eV argon ions prior to deposition.

Annealing, Auger analysis, and sputter profiling were done in another UHV chamber. The experimental arrangement for this is shown in Figure 1. The Ni/GaAs structures were mounted on a rotatable manipulator that allowed the structures to be reproducibly positioned in front of the lamp assembly for heating and then rotated into position for analysis. The Ni/GaAs structure was mounted on a yttria stabilized cubic zirconia

(YSCZ) single crystal which was secured to a Ta holder. The insulating nature of the YSCZ ensured that little was lost to the Ta holder by thermal conduction. A time versus temperature calibration of the 600 watt lamp was done for a number of power levels by placing a thermocouple on a Ni/GaAs structure surface positioned 12.5 mm from the lamp assembly. Figure 2 shows the calibration curves for 50 and 100% power levels.

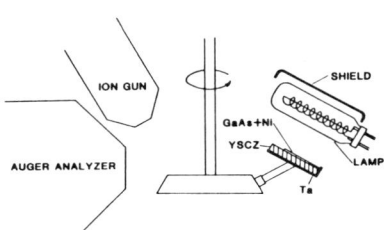

Fig. 1. Experimental configuration for in vacuo lamp annealing and Auger sputter profile analysis of Ni/GaAs structures.

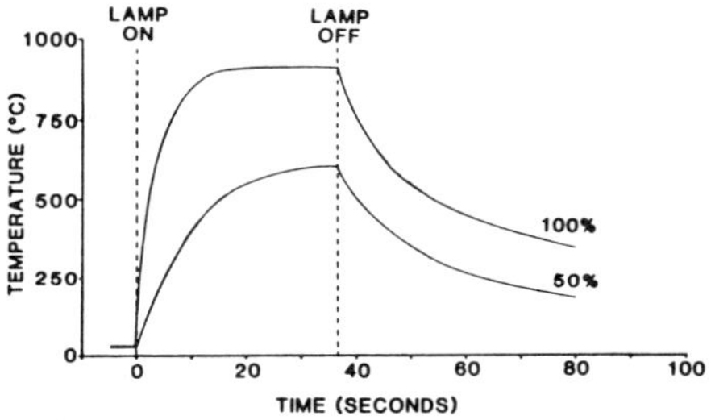

Fig. 2. Temperature versus time during heating of Ni on GaAs in vacuum with a 600 watt high intensity incoherent lamp at full and half power.

Auger electron measurements were made with a Physical Electronics Industries Model 10-150 analyzer using 4 keV electrons with a current density of 0.05 A/cm^2. Depth profiling was accomplished with argon ions generated with a Physical Electronics Industries Model 04-303 ion gun operated at 2 kV with a current density of 8.3×10^{-3} A/cm^2. The thickness of deposited Ni was measured with a quartz crystal oscillator during deposition and verified with a Sloan Dektak II surface profiler. The latter was also used to measure sputter crater depths to determine the sputter rates for pure Ni, GaAs and Ni$_2$GaAs. The estimated error in these measurements is ±5.0%.

RESULTS

Oxygen Free Ni/GaAs Interface

The Ni Auger sputter profiles from this structure are shown in Figure 3. The data shows a 40 nm broadening of the interface brought about by the diffusion of Ni, Ga, and As after a 500°C 30 second heat treatment. Using the method of Hall and Morabito [14] for determining diffusion coefficients (D) from sputter profiles, the values of D for Ni from Figure 3 is 2.9×10^{-13} cm^2/s. In order to facilitate the comparison of compositional variations, the elemental profiles in Figure 3, as well as those in following figures, were normalized by sensitivity factors (shown in each figure) which reflect the differences in ionization cross sections for the different Auger transitions used to construct the profiles.

The profiles in Figure 3, for both heated and unheated structures, show that the intermediate region between the interface and the bulk GaAs has a lower As:Ga ratio than the bulk where it should be 1:1. As shown in Figure 4, this apparent As depletion in the sub-interface region is due to preferential sputtering of As with higher ion beam potentials. In addition to As preferential sputtering, the higher ion beam potential significantly decreases the depth resolution by broadening all profiles.

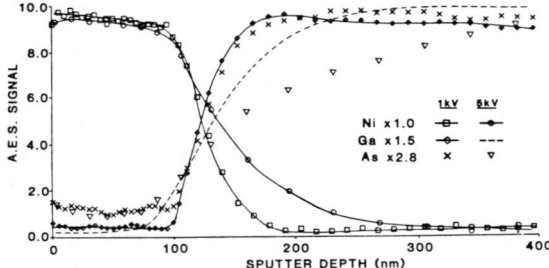

Fig. 3. Auger sputter depth profiles of Ni, Ga, As, and O from a 125 nm thick vapor deposited Ni film on clean (100) GaAs before and after a 500°C heat treatment for 30 seconds.

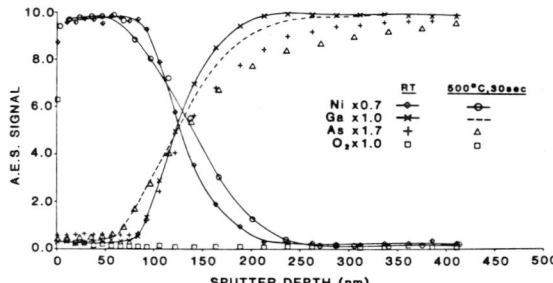

Fig. 4. Auger sputter depth profiles of Ni, Ga, and As from a 120 nm thick vapor deposited Ni film on clean (100) GaAs using 1 and 5 KV argon ion beams.

Heating at 500°C for 60 seconds, as shown in Figure 5, resulted in the rapid outdiffusion of Ga and As through the Ni film, with little or no further inward diffusion of Ni into the GaAs substrate. The composition of the resulting film structure was 2Ni:1Ga:1As. The above phenomena is not in agreement with work by Ogawa [7] who reported significant Ni diffusion into GaAs when the Ni/GaAs structure was heated above 300°C in a hydrogen atmosphere. It appears that with incoherent lamp heating, where most of the heat is initially concentrated in the outer surface of the Ni/GaAs structure, the condition for the outdiffusion of Ga and As is thermodynamically favorable over the inward diffusion of Ni. This may be due to the fact that the activation energy for Ga is less than Ni. (E_a for Ga = 0.1 [15] and E_a for Ni from this work 0.28 eV.) The rapid outdiffusion of Ga and As with incoherent lamp heating is further evident in Figure 6 which shows the Ni, Ga, As, and O profiles from a 240 nm Ni/GaAs structure before and after 300°C heat treatments for 2 and 4 minutes.

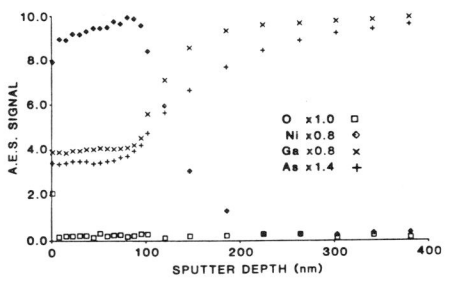
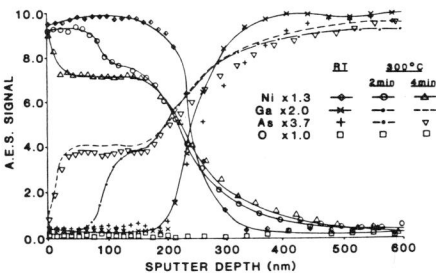

Fig. 5. Auger sputter depth profiles of Ni, Ga, As, and O from a 125 nm thick Ni film on clean (100) GaAs after a 500°C heat treatment for 60 seconds.

Fig. 6. Auger sputter profiles of Ni, Ga, As, and O from a 240 nm thick vapor deposited Ni film on clean (100) GaAs, before and after 300°C heat treatments of 2 and 4 minutes.

Figure 7 shows the profiles versus sputter time from the 300°C, 2 minute treated Ni/GaAs structure plotted versus sputter depth in Figure 6; also presented are the As_{LMM} peaks at corresponding sputter times that were used to construct the As profile. Because Auger peak shapes are affected by chemical bonding, the fact that the As Auger peaks exhibited slightly different shapes from the diffusion product region (compared to non-diffused bulk GaAs) indicates that the diffusion product is more than a physical mixture of three elements. Although not shown in Figure 7 the Ga and Ni peaks also showed subtle differences between diffusion product and unreacted bulk regions. Since the composition of the diffusion region was found to be 2 Ni:1 Ga:1 As, the most likely product was Ni_2GaAs, which is in agreement with work by Ogawa and Lahav et al. [9,10]. Figure 7 also shows the plots of the 2-component result of a factor analysis of the As_{LMM} peak. Factor analysis is a mathematical technique for studying matrices of data and has been applied successfully to chemical problems during the past 20 years.[16]. Basically it determines the number of different chemical components in sets of spectral data and calculates their concentrations. More recently it has been applied to Auger sputter profile analysis by Gaarenstroom [17]. The different peak shapes in Figure 7, and thus the components of different As chemical states, clearly distinguish the diffusion zone chemistry from unreacted Ni and bulk GaAs.

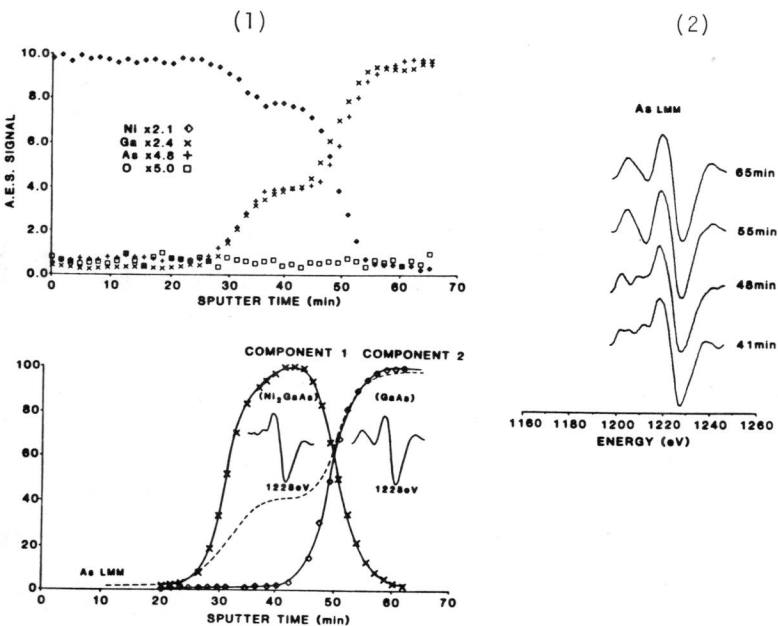

Fig. 7. (1) Auger sputter profiles of Ni, Ga, As, and O vs sputter time and the graphical two component solution of a factor analysis of the As_{LMM} Auger peak from a 240 nm thick vapor deposited Ni film on clean (100) GaAs heated to 300°C for 2 minutes and (2) The As_{LMM} peaks at indicated sputter times used to construct the As profile.

Oxygen Contaminated Ni/GaAs Interface

In this case the GaAs structure was not heated prior to Ni deposition to remove the surface oxide remaining after chemical and ozone treatments. The elemental profiles from this structure, before and after a 2 minute 500°C heat treatment, are shown in Figure 8. The oxygen concentration at the interface, based on published Auger sensitivity factors [18], was determined to be 10 atomic percent. The diffusion coefficient D for Ni in this case was determined to be 1.2×10^{-14} cm^2/s.

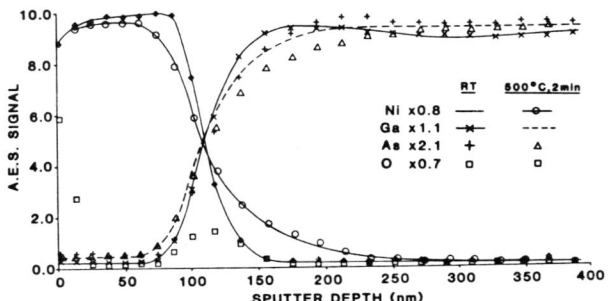

Fig. 8. Auger sputter profiles of Ni, Ga, As, and O from a 112 nm thick Ni film vapor deposited on oxide covered (100) GaAs, before and after a 500°C heat treatment for 2 minutes.

Unlike the previous case with an oxide free interface, 2-minutes heating resulted in neither Ga nor As diffusion throughout the entire Ni film and no evidence of any compounding was detected. After 5 minutes heating As and Ga were detected at the surface but not within the Ni film, and grain boundary diffusion is the probable mechanism. With a 10 minute heat treatment at 500°C the surface became enriched with Ga in the form of GaO. Just below this was a 1:1 Ni to Ga region, which was most likely NiGa. However, the vast majority of the remaining diffusion product appeared to be a mixture enriched with As whose atomic composition was 1 Ni:1.4 Ga:1.6 As. Oxygen was not detected within the diffusion layer, and no amount of heating generated a region containing just Ni_2GaAs. Prolonged heating (up to 1 hour) at 300°C also failed to promote significant outdiffusion of Ga on As or the formation of any compounds with Ni. However, As was detected at the Ni surface, probably the result of grain boundary diffusion since the heat treatment did enhance the formation of large Ni grains.

Sputter Deposited Ni on GaAs

As in the previous case, the residual oxide layer on the GaAs substrate was not removed prior to the sputter deposition of Ni, and therefore about 10 atomic % oxygen was present at the Ni/GaAs interface. About 5 atomic % oxygen was also incorporated into the Ni film as a result of the reaction of sputtered nickel with residual oxygen in the deposition chamber during deposition. Heat treatments of this structure showed similar effects on Ni, Ga, and As diffusion as with the structure with oxygen just at the interface. At 300°C, the Ni diffusion coefficient was 2×10^{-15} cm^2/s which compared to 8×10^{-16} cm^2/s for vapor deposited Ni on oxide contaminated GaAs. Figure 9 shows the profiles from this structure before and after a 300°C 15 minute heat treatment, and Figure 10 shows the profiles after 50 minutes at 300°C. The room temperature profile of Ni was renormalized in Figure 10 for ease of comparison.

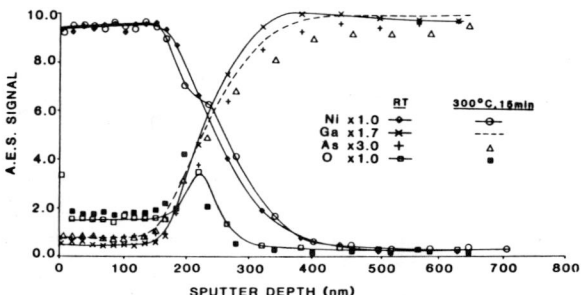

Figure 9. Auger sputter profiles of Ni, Ga, As, and O from a 250 nm thick sputter deposited Ni film on oxide covered (100) GaAs, before and after a 300°C heat treatment for 15 minutes.

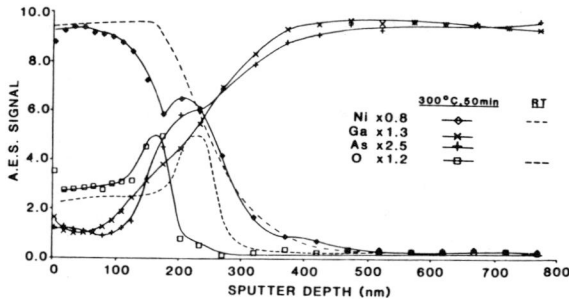

Fig. 10. Auger sputter profiles of Ni, Ga, As, and O from a 250 nm thick sputter deposited Ni film on oxide covered (100) GaAs following a 300°C heat treatment for 50 minutes. The superimposed Ni room temperature profile was normalized to the heat treated Ni profile for ease of comparison.

As shown in Figure 10, prolonged heating of this structure produces the outdiffusion of Ga, As, and the oxide contaminate layer. Ga appears to diffuse faster than As, while the latter is enriched with Ni at the interface, possibly as NiAs [7]. As with the previous case, no evidence of an isolated Ni_2 GaAs layer was found after various heating conditions. After heating to 500°C for 15 minutes, the surface became enriched with Ga and the oxide layer became more diffuse near the surface. The oxide layer was associated with Ga and Ni since As was not detected at or near the surface.

Ni Diffusivity

The method of Hall and Morabito [14] which was used to determine Ni diffusion coefficients, is based on the Ni concentration gradient at the Ni/GaAs interface before and after heat treatments. The advantage of this method is that it "subtracts out" broadening effects in a profile due to sputtering and surface roughness according to the following equation:

$$D = [(G_0/G_t)^2 - 1] / 4\pi G_0^2 t \qquad (1)$$

where D is the diffusion coefficient, t is the heating time, and G_0 and G_t are the concentration gradients before and after heating, respectively. Figure 11 shows the plot of the diffusion coefficients versus 1/T for Ni/GaAs structure with oxygen free (A) and oxygen contaminated (B) interfaces.

The activation energies E_a can be calculated from the diffusivity in Figure 11 according to the following:

$$E_a = -2.3Rm \qquad (2)$$

where R is the Rydberg constant and m is the slope of the log D vs 1/T line. This yields E_a values of 0.28 and 0.47 eV for the oxygen free and oxygen contaminated interface structures, respectively.

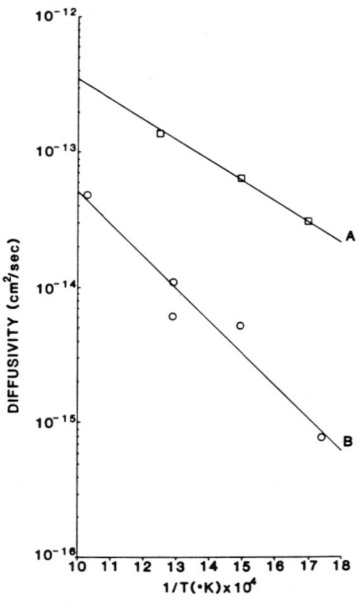

Fig. 11. Diffusivity versus 1/T (°K) plots for heat treated vapor deposited Ni on (A) clean and (B) oxide covered (100) GaAs.

CONCLUSIONS

Auger sputter profile analysis of heat treated Ni films on (100) GaAs has shown that diffusion and compounding is significantly affected by the presence of oxygen at the Ni/GaAs interface. Nickel diffusivities, extracted from the profile concentration gradients, were higher and the activation energy lower for clean versus oxide covered GaAs substrates. Without the presence of oxygen, Ni_2GaAs readily formed at 300°C, while with oxygen, the formation of Ni-Ga and Ni-As compounds prevailed over the formation of a distinct Ni_2GaAs layer.

ACKNOWLEDGEMENTS

This work was sponsored by the Air Force Wright Aeronautical Laboratories, Materials Laboratory, Air Force Systems Command, United States Air Force, Wright-Patterson Air Force Base, Ohio 45433. The authors also wish to thank P. T. Murray and J. T. Grant for their helpful discussions and D. A. Walsh for providing the sputter deposited Ni/GaAs structures.

REFERENCES

1. N. Braslau, J. B. Gunn, and J. L. Stables, Solid S. Electron. 10 (1967) 381.
2. M. Wittmer, R. Pretorius, J. W. Mayer, and M.-A. Nicolet, Solid State Electron. 20 (1977) 433.
3. F. Vidimari, Electron Lett. 15 (1979) 675.
4. M. Ogawa, J. Appl. Phys. 51 (1980) 406.
5. N. Heiblum, M. I. Nathan, and C. A. Chang, Solid State Electron. 25 (1982) 185.
6. G. Y. Robinson, Solid State Electron. 10 (1967) 381.
7. M. Ogawa, Thin Solid Films 70 (1980) 181.
8. T. S. Kaun, P. E. Batson, T. N. Jackson, H. Ruprecht, and E. L. Wilkie, J. Appl. Phys. 54 (1983) 695.
9. A. Lahav and M. Eizenberg, Appl. Phys. Lett. 45 (1984) 256.
10. A. Lahav, M. Eizenberg, and Y. Komen, to be published in Proceedings of the Materials Research Society, Symposium 10, 1984 Fall Meeting.
11. T. Kendelwicz, W. G. Petro, M. D. Williams, S. H. Pan, I. Lindau, and A. E. Spicer, Mat. Res. Soc. Symp. Proc. 25 (1984) 323.
12. Ultra Violet Ozone Cleaner Model 0306, UVOCS Inc., Montgomeryville, PA 18936.
13. A. Y. Cho, Thin Solid Films 100 (1983) 291.
14. P. H. Hall and J. M. Morabito, Surf. Sci. 54 (1976) 79.
15. G. Y. Robinson, Solid St. Electron. 18 (1975) 331.
16. E. R. Malinowski and D. G. Howery, Factor Analysis in Chemistry, John Wiley and Sons, NY (1980).
17. S. W. Gaarenstroom, Appl. Surf. Sci. 7 (1981) 7.
18. S. Mroczkowski and D. Lichtman, Surf. Sci. 131 (1983) 159.

THE CHARACTERIZATION OF ALLOYED NiGeAuAgAu OHMIC CONTACTS TO AlInAs/GaInAs
HETEROSTRUCTURE BY AUGER ELECTRON SPECTROSCOPY AND WAVELENGTH DISPERSIVE
X-RAY ANALYSIS

P.M. CAPANI[1], S.D. MUKERJEE[2], L. RATHBUN[2], H.T. GRIEM[3],
G.W. WICKS[2], L.F. EASTMAN[2], and J. HUNT[4]

[1] IBM Corporation, 1701 North Street, Endicott, NY 13760
[2] School of Electrical Enginneering and National Research and
Resource Facility for Submicron Structures, Cornell University,
Phillips Hall, Ithaca, NY 14853
[3] Highes Research Laboratories, Malibu, CA 90265
[4] School of Materials Science, Bard Hall, Cornell University,
Ithaca, NY 14853

ABSTRACT

In a lattice-matched <100> InP/$Ga_{0.47}In_{0.53}As/Al_{0.48}In_{0.52}As$ system
used for modulation-doped field effect transistors (MODFETs), low resistance
ohmic contacts to the two-dimensional electron gas have been fabricated
using alloyed NiGeAuAgAu metallization. In this work we examine the use
of Auger electron spectroscopy (AES) and wavelength dispersive x-ray
spectroscopy (WDX) analyses for studying the metal-semiconductor interactions and their correlation with measured ohmic contact resistance.

INTRODUCTION

Low resistance ohmic contacts are necessary in high performance
electron devices. Modulation-doped (MD) $Al_{0.48}In_{0.52}As/Ga_{0.47}In_{0.53}As$ is
superior to conventional GaAs, but the fabrication of ohmic contacts to the
heterostructure proved to be difficult. In MD structures, the ohmic contact
must be made to the two-dimensional electron gas (2DEG), which lies several
hundred angstroms beneath the surface of the epitaxial layer. In an earlier
study, we discovered that the AuGeNiAg metallization used for contacting
GaAs failed to contact the 2DEG in MD $Al_{0.48}In_{0.52}As/Ga_{0.47}In_{0.53}As$. This
was due to extreme alterations in semiconductor stoichiometry resulting
from the metals-semiconductor interactions. Adjustments to the
NiGeAuAgAu metallization and alloy schedule were made and used on the

---

[*] IBM Corporation, 1701 North Street, Endicott, NY 13760
[**] Hughes Research Laboratories, Malibu, CA 90265

$Al_{0.48}In_{0.52}As/n^+$-$Ga_{0.47}In_{0.53}As$ system to form very low resistance ohmic contacts [1]. Auger electron spectroscopy (AES)/sputter depth profiling was used to analyze the metal-semiconductor interaction processes.

Here, we report on the successful fabrication of low resistance ohmic contacts to MD $Al_{0.48}In_{0.52}As/Ga_{0.47}In_{0.53}As$ using alloyed NiGeAuAgAu metallization. The contacts were analyzed with AES and wavelength dispersive x-ray spectroscopy (WDX). Results of these analyses are presented in this paper.

EXPERIMENTAL

The following heterostructure was grown by molecular beam epitaxy (MBE) on <100> insulating InP: insulating $Al_{0.48}In_{0.52}As$ (3800Å)/ unintentionally doped $Ga_{0.47}In_{0.53}As$, $N_D \sim 1 \times 10^{16} cm^{-3}$ (800Å)/ insulating $Al_{0.48}In_{0.52}As$ (130Å)/n^+-$Al_{0.48}In_{0.52}As$:Si, $N_D \sim 2 \times 10^{18} cm^{-3}$ (65Å)/insulating $Al_{0.48}In_{0.52}As$ (320Å)/unintentionally doped $Ga_{0.47}In_{0.53}As$, $N_D \sim 1 \times 10^{16} cm^{-3}$ (160Å), as shown in Figure 1. Measured sheet carrier concentrations and Hall mobilities were $n_s = 1.64 \times 10^{12} cm^{-2}$, $\mu_H = 10,500 cm^2 \cdot V^{-1} \cdot s^{-1}$ at 300K and $n_s = 1.56 \times 10^{12} cm^{-2}$, $\mu_H = 54,100 cm^2 \cdot V^{-1} \cdot s^{-1}$ at 77K. Contact metallization was evaporated onto the structure sequentially: Ni (100Å)/Ge (450Å)/Au (800Å)/Ag (200Å)/Au (800Å) and alloyed in a thermal transient furnace. Temperature transients are shown in Figure 2. Electrical measurements were made by the transmission line method [2].

RESULTS

The surface morphology was acceptable for samples that were alloyed to maximum sample alloy temperatures below 500°C. Surface roughness increased with the maximum alloy temperature and globule formation was observed on the sample alloyed at 450°C. Scanning secondary electron microscope (SEM) photographs of sample surfaces are shown in Figure 3. For the sample alloyed to 520°C, however, we found that a pattern of rings had formed on the contact metallization, extending radially outward from the metal over the mesa region, as shown in Figure 4. The results of studies

Au	800 Å
Ag	200 Å
Au	800 Å
Ge	450 Å
Ni	100 Å
GaInAs	160 Å
AlInAs	320 Å
n-AlInAs	65 Å
AlInAs	130 Å
GaInAs	800 Å
AlInAs	3800 Å
InP:S.I.	< 100 >

Fig. 1. Heterostructure with deposited metals.

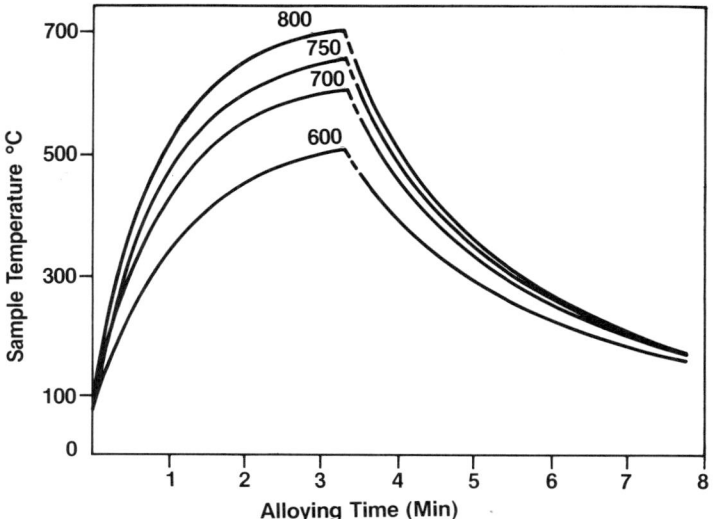

Fig. 2. Thermal transient alloy furnace response characteristics. (Curves for furnace temperatures T_f = 600°C, 700°C, 750°C and 800°C)

concerning these rings will be discussed later.

AES/sputter depth profiles were made of the as-deposited metallization and samples alloyed to 410°C and 450°C. The percent atomic concentrations as a function of sputter time are presented in Figures 5 and 6. There is little difference between the Auger profiles of the two alloyed samples and this is consistent with the electrical measurements. These profiles compared closely with those obtained for our contacts to $Al_{0.48}In_{0.52}As/n^+$-$Ga_{0.47}In_{0.53}As$ [3].

The measured specific transfer resistances of the ohmic contacts for the samples alloyed to temperatures from 400°C to 500°C was 0.20 ± 0.05 ohm·mm, providing low resistance contacts to the 2DEG.

The sample alloyed to a maximum temperature of 520°C was very interesting and worthy of further study. This sample exhibited gross structural and pronounced morphological changes, rendering it unsuitable for AES/sputter depth profile analysis (See Figure 4). Instead, lateral elemental distributions were obtained using AES and wavelength dispersive x-ray analysis (WDX). An example of a lateral map is shown in Figure 7b, which should be compared with the SEM photographs Figure 7a and Figure 4. With the help of Figure 4, the findings are tabulated in Table I.

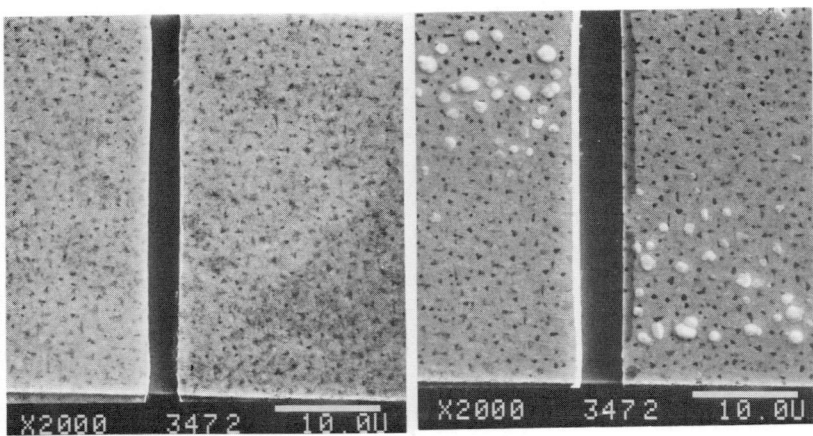

Morphology of sample alloyed to 410°C   Morphology of sample alloyed to 450°C

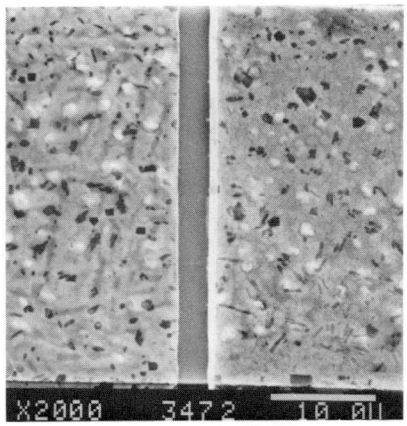

Morphology of sample alloyed to 520°C

Fig. 3. Scanning secondary electron microscope (SEM) photographs of samples alloyed to 410°C, 450°C and 520°C, showing morphologies.

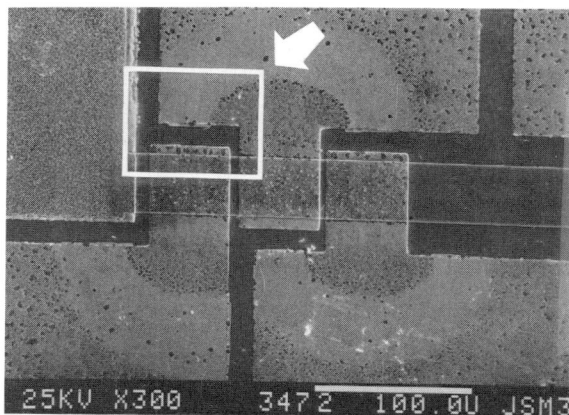

Ring pattern on contact metallization, extending radially outward from the metal over the mesa region.

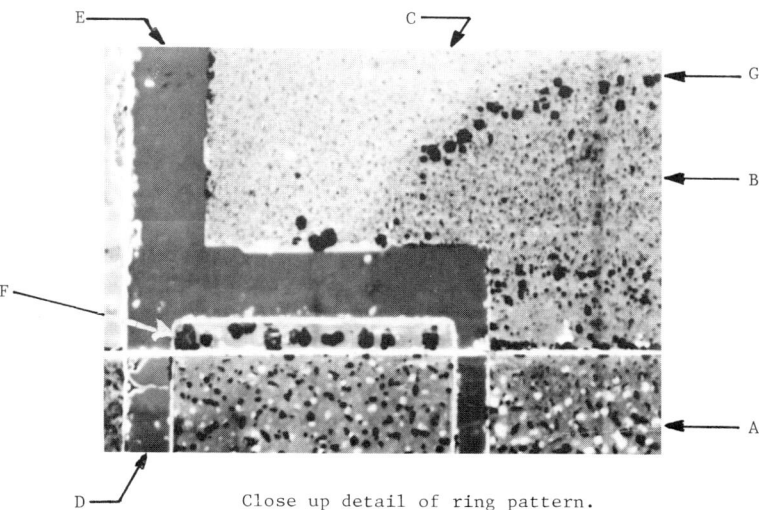

Close up detail of ring pattern.

Fig. 4. Scanning secondary electron microscope (SEM) photographs of the surface morphology of the sample alloyed to 520°C.

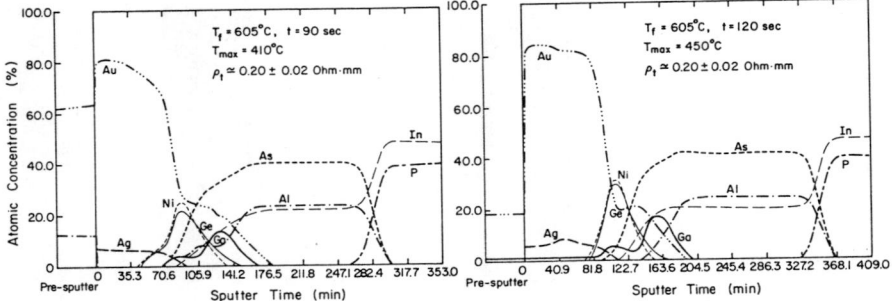

Fig. 5. AES/sputter profile, sample alloyed to 410°C.

Fig. 6. AES/sputter profile, sample alloyed to 450°C.

DISCUSSION

The analyses of the rings established that a sideways or lateral diffusion of the elements Ga and Al occurred for maximum alloy temperatures above 500°C with Ga moving farther than Al. By extending the alloy furnace curves in Figure 2, we found that the sample alloyed to a peak temperature of 520°C was held above 500°C for about t = 100 seconds. Let us, for the sake of a very simplistic treatment, consider this time of 100 seconds to be the diffusion time, however arbitrarily defined. During this period of time, the distance that Ga diffused laterally from the edge of the mesa to the contact pad was about 100μm (0.01cm). With this knowledge of diffusion time (t) and diffusion length (L_D), the lateral diffusion coefficient for Ga could be calculated:

$$D_{Ga} = \frac{(L_D)^2}{t} = \frac{(0.01)^2}{100} = 1 \times 10^{-6} cm^2 \cdot s^{-1}. \tag{1}$$

Although the diffusion coefficient ($1 \times 10^{-6} cm^2 \cdot s^{-1}$) is border line between liquid phase diffusion ($D_\ell \cong 10^{-5}$ to $10^{-4} cm^2 \cdot s^{-1}$) and grain boundary diffusion ($D_{gb} \cong 10^{-8}$ to $10^{-10} cm^2 \cdot s^{-1}$), in general we believe that the lateral movement of Ga was caused by liquid phase diffusion as the metal surface is found to be rugged. The surface morphology is less rugged for lower maximum alloy temperatures, suggesting that liquid formation, if any, is limited to interfaces. This experiment shows that a high chemical potential gradient exists between the Al and Ga metals on the mesa and the metal sitting on the InP substrate. Hence, lateral elemental migrations occurred.

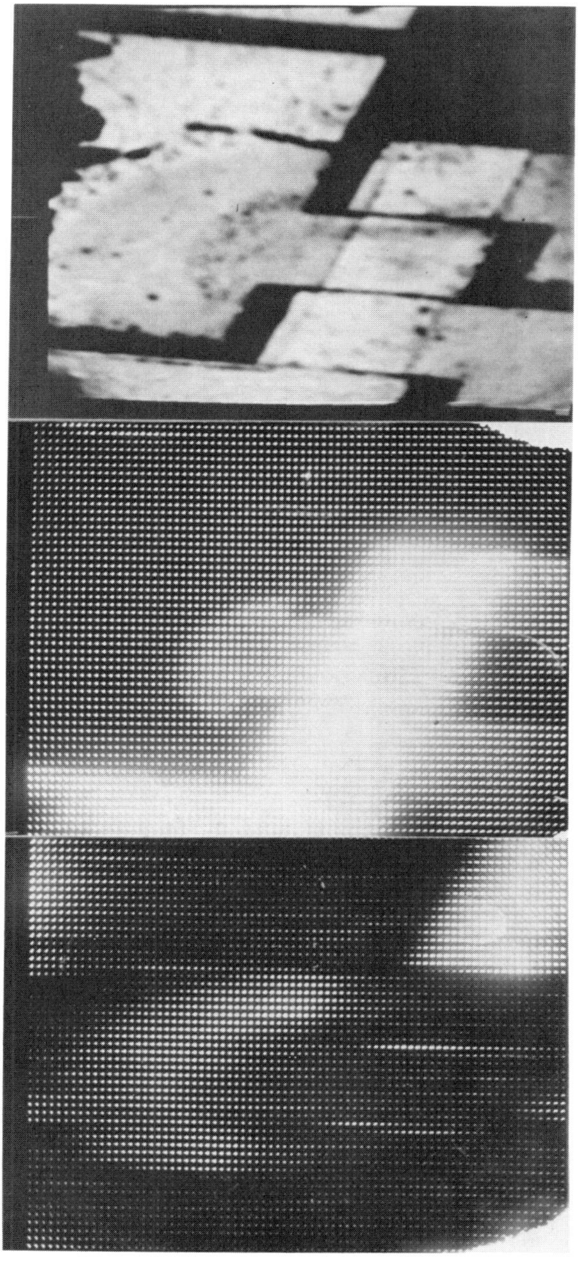

209

a. SEM photograph of sample surface, 520°C alloy sample.

b. Lateral elemental distribution of Al, 520°C alloy sample.

c. Lateral elemental distribution of Ga, 520°C alloy sample.

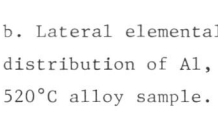

Fig. 7. Example of lateral elemental mapping, sample alloyed to 520°C.

Table I. Summary of AES$^{(a)}$ and WDX lateral mapping for sample alloyed to T_{max} = 520°C (Fig. 4)

Region $^{(b)}$	Structure and region	Observation $^{(c)}$	Discussion and plausible cause
A	Alloyed metal on mesa.	High Al, some Ga, P and O and smaller Ni, Ge, Au, In peaks. Ag is absent.	Changes in elemental composition from that of as-deposited layer are anticipated due to alloying process. Ga and Al appear to have outdiffused and surface segregated with surface oxidation acting as a surface sink process. The oxygen comes from the alloying furnace.
B	First ring on metallization on InP substrate.	More In and P with some Au,Ge,Ag and Ni. Also, Al,Ga,As and O peaks are observed. Al concentration is larger than Ga.	While In and P are believed to outdiffuse from the InP substrate to reach the surface, Al,Ga and As had to diffuse laterally from the mesa region to reach this region on the InP substrate. A higher affinity of Al for O probably explains the relatively higher Al concentration than Ga. Al being of lower atomic number, yields less secondary electrons, hence region B is darker than A in Figure 7a.
C	Second ring on metallization on InP substrate.	Same as above, exept lower P,Ge,Au and O signals, with higher Ag, In signals. Also Ga signal increased considerably with a lowering of Al.	Ga seems to have diffused laterally farther away from the mesa region than Al. The Ga is segregated over a larger area, thus reducing the masking effects, allowing more In and Ag to be detected in the AES spectra.
D	Mesa with no metallic overlayer.	Only Ga,In and As signals observed.	Auger electron escape depth being rather small (<30Å), only the top 160Å thick $Ga_{0.47}In_{0.53}As$ layer is sampled with no Al outdiffusion in the absence of metal/semiconductor interactions.

Table I. Summary of AES[a], and WDX lateral mapping for sample alloyed to T_{max} = 520°C (Fig. 4), continued.

Region [b]	Structure and region	Observation [c]	Discussion and plausible cause
E	InP substrate region.	In, with reduced P than stoichiometric InP, as determined by WDX does not indicate this.	The more volatile P has probably vaporized from the surface during the sputtering away of ~30Å of material prior to AES mapping. WDX does not show this due to its larger sampling depth.
F	Black spots on metal near mesa/substrate borderline.	Large concentrations of As and In, P and Al, and no Ag,Au,Ga or Ni (WDX results).	Probably some exchange of elements had occurred, e.g., Ag and As moved into the mesa region while As and Al diffused laterally from the mesa to fill these regions.
G	Black spots at the border between the 1st & 2nd rings B & C.	Large Ge signals from WDX suggesting Ge concentration extends deep into the metallization. Absence of Ag,Au and In. Some P present.	It is possible that the lateral diffusion front of Al has pushed the Ge present in the deposited metal in its wake, causing Ge buildup at its periphery.

(a) Sample sputtered for three minutes, approximately 30Å removed. (b) Areas defined in Figure 4.
(c) From lateral Auger electron or x-ray mapping and AE spectra taken on these areas.

One way to avoid this migration would be to stop etching the mesa partially through the bottom layer of insulating $Al_{0.48}In_{0.52}As$, instead of etching through to the substrate, as was done in this experiment. This would probably eliminate the Al lateral diffusion but perhaps not the Ga diffusion. Although planned, experiments to study the effects of etching partially into the bottom $Al_{0.48}In_{0.52}As$ layer have not been done yet. The lateral diffusion of Al and Ga apparently have not caused large differences in ohmic contact transfer resistances in our experiments.

CONCLUSION

We have successfully fabricated low transfer resistance (0.20 ± 0.05 ohm·mm) lateral ohmic contacts to the 2DEG in MD $Al_{0.48}In_{0.52}As/Ga_{0.47}In_{0.53}As$ heterostructures using alloyed NiGeAuAgAu metallizations. Surface morphology is excellent for the alloy temperature of 410°C and begins to deteriorate as higher alloy temperatures are used. For the 450°C alloy temperature, globule formations are seen on the metal alloyed into the mesa and are oriented toward the contact pads. This signifies some lateral diffusion. At 520°C, lateral diffusion probably occurs in the liquid phase, causing Al and Ga to diffuse out of the mesa to distances between 50 and 100μm, with Ga diffusing farther than Al.

One possible way of avoiding Al lateral diffusion is to limit the mesa isolation depth to the insulating $Al_{0.48}In_{0.52}As$ bottom layer, stopping short of the InP substrate. It would be impossible to halt the lateral Ga diffusion, because it is undesirable to have a lower bandgap layer of $Ga_{0.47}In_{0.53}As$ between the insulating $Al_{0.48}In_{0.52}As$ buffer layer and the InP substrate, as it might give rise to a second parallel conducting layer.

ACKNOWLEDGEMENTS

We express our sincerest thanks to P. Zwicknagl and J.D. Berry for assistance with our experiments and stimulating discussions. This work was supported by the Army Research Office through contract DAAG29-82-K-011 monitored by H. Wittmann. P.M. Capani acknowledges the IBM Corporation for fellowship support.

REFERENCES

1. P.M. Capani, S.D. Mukherjee, P. Zwicknagl, J.D. Berry, H.T. Griem, L. Rathbun and L.F. Eastman, "Low Resistance Alloyed Ohmic Contacts to $Al_{0.48}In_{0.52}As/n^{+}-Ga_{0.47}In_{0.53}As$", Electron. Lett. 20, 11, 446-447 (1984).

2. G.K. Reeves and H.B. Harrison, "Obtaining the Specific Contact Resistance from Transmission Line Model Measurements", IEEE Electron Dev. Lett. EDL-3, 111-113 (1982).

3. P.M. Capani, S.D. Mukherjee, P. Zwicknagl, J.D. Berry, H.T. Griem, G.W. Wicks, L. Rathbun and L.F. Eastman, "A Study of Alloyed AuGeNi/Ag/Au Based Ohmic Contacts on the $Al_{0.48}In_{0.52}As/Ga_{0.47}In_{0.53}As$ System", paper presented at Electronics Materials Conference, Santa Barbara, CA, June 1984, to be published in J. Electron. Maters.

EXPERIMENTAL INVESTIGATION OF GaAs SURFACE OXIDATION

S. Matteson and R.A. Bowling
Texas Instruments Incorporated, Materials Science Laboratory
P.O. Box 225936, MS 147, Dallas, Texas 75265

ABSTRACT

Experimental observations of the surface oxide chemistry of GaAs are reported for various commonly used chemical surface preparations. Auger electron spectroscopy (AES), X-ray photoelectron spectroscopy (XPS), and ellipsometry were employed to obtain information regarding the stoichiometry, depth distribution, and oxide growth kinetics of thin surface oxides. Previous observations of the segregation in depth of Ga and As oxides are corroborated. Arsenic oxides tend to be found near the surface while Ga_2O_3 is found near the GaAs-oxide interface. The presence of elemental As was frequently detected at this interface, as well. Surfaces essentially free from oxide are shown to be produced by certain chemical treatments, and the state of the surface in the solution is inferred. It is shown that the GaAs surface oxide stoichiometry can undergo several changes in a short time when exposed to water and air. In addition, the characterization of the oxidation of GaAs by ozone in the presence of intense ultraviolet illumination is reported. The oxide primarily consists of Ga_2O_3 and exhibits an interesting growth kinetics; the thickness of the oxide proceeds at first linearly, then logarithmically, then parabolically. This behavior is explained in terms of various mechanisms which are dominant at different thicknesses of oxide.

INTRODUCTION

Gallium arsenide (GaAs) is an important semiconducting material with applications for high frequency semiconducting devices as well as integrated electro-optical devices. The surface chemistry of this material has attracted much interest and activity in the last few years.[1-30] The present work is an effort to contribute to the understanding of the surface chemistry of GaAs in commonly used aqueous surface treatments. It is important to understand the extent to which these treatments modify the surface in order to design processes which fully account for the properties of the surface preparations. The nature of the oxide formed by exposure of a wafer to air, for example in storage, will be addressed, since unpassivated surfaces may be unstable and may require that the surface be prepared immediately before processing or even in situ. The character of the so-called native oxide will be compared to that produced by two methods of cleaning the surface by oxidation, viz. oxygen plasma "ashing" and UV-ozone photopyrolysis.[31,32] These latter techniques appear to have potential for the removal of hydrocarbon contamination from the surface of a wafer. However, the oxide that is produced may present difficulties in latter processing. The characterization of the oxide is therefore relevant to the application of these cleaning techniques.

EXPERIMENT

The materials used in this study were primarily undoped (100) semi-insulating GaAs wafers obtained from crystals grown by the liquid encapsulated Czochralski method (LEC). The wafers were degreased in boiling tetrachloroethylene and hot methanol for 10 minutes each, in order to remove gross organic contamination from the surface. Metal contamination was reduced by immersion of the wafer in concentrated HCl followed by a thorough rinse in flowing deionized (DI) water (>18 megaohm resistivity). Before surface treatment, in all cases, a chemical polish was performed in a solution of 8:1:1 by volume concentrated sulfuric acid, hydrogen peroxide and DI water to remove approximately eight to ten micrometers of material. At this point in the process, the various wafers underwent different surface treatments which will be discussed below.

After the surface treatment was completed, the wafer was introduced into a Phi Model 548 XPS system via an air lock. The wafer was in the vacuum system within one minute of its preparation. In several cases the wafer was withdrawn from the DI water rinse in a glove bag purged with dry nitrogen and transferred to the vacuum system without exposure to air. The surface was excited by Mg Kα radiation and the photoelectrons analyzed by a double pass cylindrical mirror analyzer. After the photoelectron spectrum of the original surface was obtained, the surface was sputtered by a 2 keV Ar ion beam incident at approximately 60° to the surface of the sample. After about 20 minutes the surface was again analyzed. The sputtered surface revealed approximately stoichiometric GaAs in all cases. The presence of a carbon photoelectron peak for the original surface of the wafers provided a convenient reference point for the monitoring and correction of small shifts in apparent binding energy due to specimen charging, for example. The sputtered GaAs surface was a cross check for the accuracy of the binding energy determination, as well. In all cases the binding energy of the Ga 3d and As 3d electrons were found to be 19.0 +/- 0.1 eV and 41.2 +/-0.1 eV, respectively.

The total photoelectron spectrum was obtained each time; in addition, expanded spectra of the 3d electrons for Ga and As were recorded, as well as spectra for 1s electrons from carbon and oxygen. The spectra energy was calibrated to center the C1s peak at 284.6 eV throughout the experiment. Only the spectra for As and Ga bands will be presented, since the oxygen band consisted of several convoluted peaks which were poorly resolved. The As and Ga spectra were analyzed by a fitting procedure in which Gaussian peak shapes of constant variance (obtained by a fit to data from sputtered surfaces) were used but with binding energies corresponding to the following compounds: Ga in GaAs (19.0 eV), Ga_2O_3 (20.0 eV); As in GaAs (41.2 eV), elemental or surface bound As (41.8 eV), As_2O_3 (44.0 eV), and As_2O_5 (45.4 eV). These values should be compared to previous values reported in the literature: Ga in GaAs (19.2[9], 19.5[20], 18.82[5], 19.0[23], 19.3[22], 18.7[21]) and As in GaAs (41.2[9,20], 41.1[23], 40.8[5,22]). The chemical shifts due to oxidation agree well those found previously by other groups. (See reference 6, for example). Thus, the peak assignment can be made with confidence. Figure 1 shows an example of the fitting procedure. The specimen shown there had been exposed to air that had been maintained at 40% relative humidity (RH) at 20°C for 5 days under constant ambient light. While the Ga signal consists primarily of Ga_2O_3 some evidence of unreacted Ga (in GaAs) is seen. Moreover, the As band is a composite of the arsenic oxides and some elemental or surface bound As as well as As in GaAs.

The surfaces of selected wafers were oxidized by exposure to air at different values of relative humidity (RH), by oxygen RF plasma oxidation at 50 W for 10 minutes, and by exposure to ultraviolet light in the presence of ozone. The UV-ozone oxidation was accomplished by illumination of the wafer with light from a Hg vapor discharge confined in a fused quartz tube, which transmits the lines at 254 nm and 186 nm. The UV flux was measured to be approximately 12 mW/cm^2 at 254 nm and about 1.0 mW/cm^2 at 186 nm. The shorter wavelength is responsible for the formation of ozone in the air, while the longer wavelength radiation causes the scission of hydrocarbon bonds. Unfortunately, the ozone absorbs strongly the 254 nm line. Therefore, to fully take advantage of the cleaning action of the system, the wafer must be very near the lamp surface. The level of ozone which is attained in the

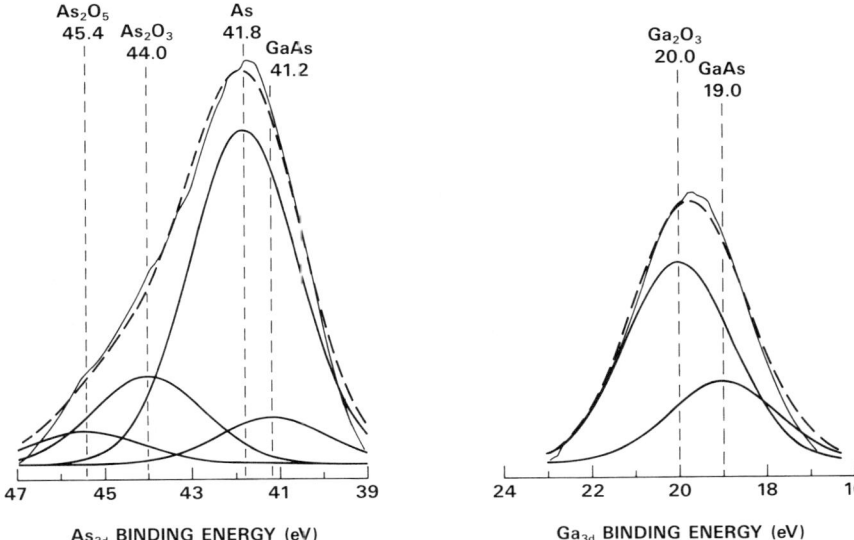

Figure 1. Expanded XPS spectra for As 3d and Ga 3d bands of oxidized GaAs (sample E). The experimental spectra were fitted by a least squares procedure using Gaussian functions.

exhasted lamp housing is 100 ppm as measured using air sample reaction tubes which are calibrated for the concentration of ozone present.[33]

On selected samples, computer automated Auger depth profiling analyses were performed. The composition of the surface was then determined as a function of depth. Data from such characterization of the relatively thick oxide formed by exposure of a slice to hard UV radiation and ozone for 19 hours will be discussed in the results section.

Ellipsometry was performed on the surfaces of wafers exposed to air and to ozone for various times. A dual mode automated Gaertner Ellipsometer Model L116A interfaced to an HP85B computer was used to obtain thickness and optical properties of the oxides using 623 nm radiation. The optical parameters were found by calculation using the vendor supplied computer routines for general absorbing layer measurement, which are based on standard algorithms.[34] The optical properties of the unoxidized but as-prepared wafers were found by repeated measurement of the ellipsometric parameters Ψ and Δ as a function of time for the same spot on a wafer etched in 8:1:1 then rinsed in DI water. The first measurement was made after three minutes and at 30 second intervals thereafter for 9 minutes, and at longer intervals up to an hour later. Extrapolation of the parameters Ψ and Δ by a logarithmic (in time) extrapolation back to 0.1 minute yielded optical parameters $n = 3.831$ and $k = 0.199$, for the as etched wafer. With exposure to air, n apparently declined while k increased in absolute magnitude. These values are to be compared with the values of $n = 3.857$ and $k = 0.198$ as reported by Aspnes and Studna for 1.96 eV (632.8 nm) radiation on cleaved GaAs.[35] The index of refraction of wafers prepared by other means however, varied from wafer to wafer. In light of the variation of the surface composition, however, this fact is understandable. The primary discrepancy appeared in the value of k for a particular specimen. The thickness of the oxide was determined by simultaneous solution for the value of thickness and index of refraction of the film.

RESULTS--COMPOSITIONAL ANALYSES

A total of twelve variations of surface treatment were made in the present work. The Ga 3d and As 3d bands for each treatment appear in figure 2. The letter annotating each spectrum corresponds to that appearing in Table I, where the treatment is identified. In the following we will examine each case and indicate the significant data and observations. Sample A was prepared in the standard way, as described above, but was withdrawn from the DI water in a nitrogen ambient. Subsequently, the wafer was transferred to the vacuum system without exposure to air. No significant oxide was detected. Only peaks corresponding to GaAs are observed. An apparent discrepancy appears in Table I at this point. While only peaks corresponding to GaAs are observed, the ratio Ga/As of the integrated signals is not unity. It would be erroneous to assume that this implied a major deviation from stoichiometry, however. The sensitivity factors which were applied to these data were determined by requiring that the specimen sputtered for 20 minutes (sample B), which also showed only peaks corresponding to Ga and As in GaAs, be stoichiometric. The apparent excess of As in sample A can be accounted for by a matrix effect, which is probably due to a thin surface contamination layer. A carbon signal in the unsputtered sample was observed. What is more, the absolute intensity of the two 3d band electrons was reduced in the present spectrum. The low energy electrons from the Ga 3d band would be attenuated more by a surface layer than would the higher energy electrons from the As 3d band. Thus, the sensitivity factor would vary slightly and the As would appear to be in excess. It is estimated that the accuracy of the compositional analysis is +/-0.05 in Table I.

TABLE I

Sample B is used for relative sensitivity correction. If the surface concentration is < .05 the value is not reported; if < 0.1 but > 0.5 the value is reported as < 0.1.

		Ga 3d Band		As 3d Band			
		GaAs 19.0	Ga_2O_3 20.0	GaAs 41.2	As 41.8	As_2O_3 44.0	As_2O_5 45.4
A	N_2 1 minute	0.4	--	0.6	--	--	--
B	Sputtered 20'	0.5	--	0.5	--	--	--
C	N_2-5 days	0.1	0.3	<0.1	0.5	<0.1	--
D	Air 1 minute	<0.1	0.3	--	0.4	0.1	<0.1
E	Air 5 days (40% RH)	0.1	0.3	<0.1	0.4	0.1	<0.1
F	Air 5 days (95% RH)	--	0.4	--	0.2	--	0.4
G	H_2O 5 days	--	0.7	--	<0.1	0.1	0.1
H	HCl (20 C)	0.2	0.2	0.2	0.4	--	--
I	O_2 Plasma	--	0.4	--	0.3	<0.1	0.2
J	I + NH_4OH (20° C)	0.3	0.2	0.3	0.2	--	--
K	Native Oxide	--	0.4	0.1	0.3	0.1	0.1
L	Ozone oxide	--	0.4	--	--	<0.1	0.5

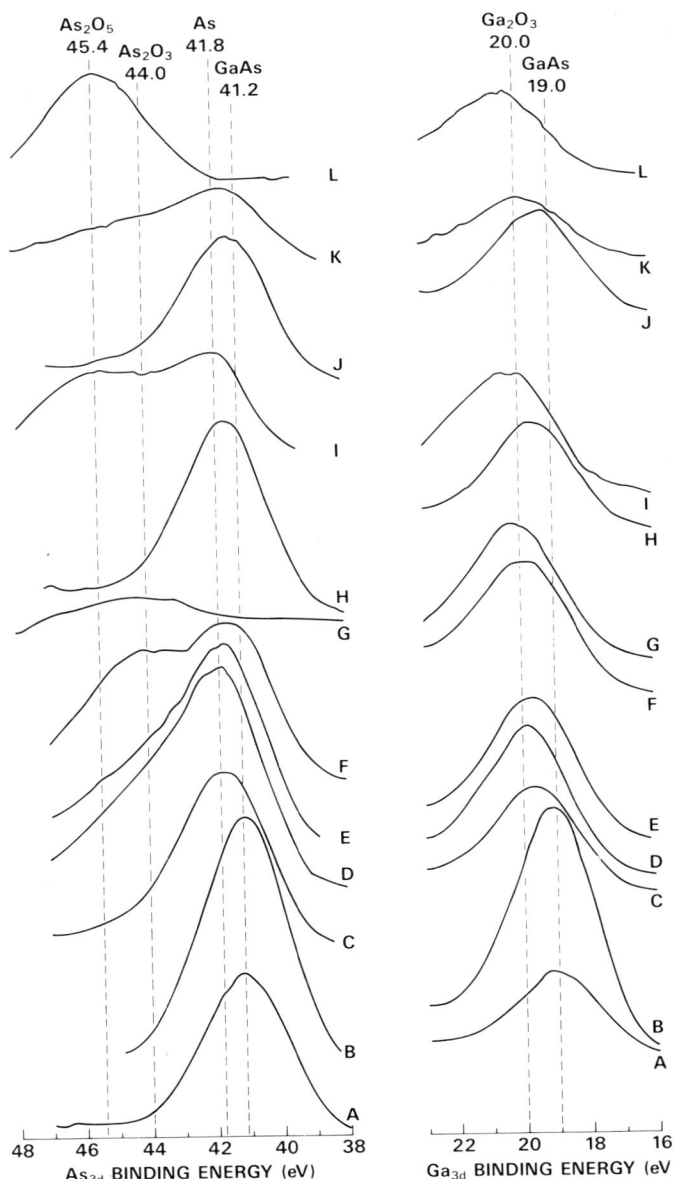

Figure 2. Composite XPS spectra for GaAs after various surface treatments, which are annotated in Table I. The spectra are discussed in the text.

As the surfaces were prepared by various treatments, the nature of the surface in regard to wetting by DI water was also noted. It has been common practice to judge the condition of the surface of GaAs after standard chemical treatments by the hydrophilic or hydrophobic character of the surface. We shall see that this practice can be misleading. When the wafer which has been etched in the 8:1:1 solution is first withdrawn from the water, the surface of the wafer is hydrophilic, i.e. water is observed to form a sheet over the surface. In a matter of 15 seconds with water on its surface but in nitrogen, the character of the surface changed to hydrophobic, with the water "beading up". It is possible that As oxides present on the surface in the etch (due to strongly oxidizing effect of the H_2O_2) were dissolved in the DI water and are not reformed by reaction with the ambient atmosphere. As_2O_3 is known to be quite soluable in water.[36]

In another related experiment, freshly etched wafers were chilled by contact on the back with dry ice, which caused small droplets of water to condense on the front surface of the wafer. When this was done in air, a "haze" appeared on the surface which was found to be rich in As by energy dispersive x-ray analysis in an electron microprobe. The haze was formed by crystalline particles, presumably As_2O_3 or As_2O_5. When this experiment was repeated on a companion wafer in nitrogen gas saturated with water, no haze was observed. It is inferred that the oxidation of the surface can take place very rapidly in air-saturated water. Furthermore, the arsenic oxides which form are easily dissolved in the water but precipitate when the droplet evaporates. This simple experiment was repeated several times using argon gas. The result was the same; only in air was a haze observed. Therfore, in air, uneven rinsing by DI water can leave the surface composition laterally inhomogeneous.

The data labelled sample B was obtained on the sample A after a subsequent 20 minute sputtering. The only compound detected was GaAs. The wafer was exposed to water after it was removed from the vacuum system. The surface was hydrophobic at first, then changed to hydrophilic in air in a matter of seconds, and then to hydrophobic again within minutes. This observation is indicative of the unstable nature of the GaAs surface in the presence of oxygen, water and light, especially when crystal damage is present.

Exposure of a wafer to utility grade nitrogen for 5 days resulted in the partial oxidation of the surface, as seen in the case of sample C. It is expected that trace quantities of oxygen will be present in the gas stream under the conditions of this experiment. Note that the primarily constituent of the As band is elemental or surface bonded As. The dominant surface oxide is Ga_2O_3 (with possibly $GaAsO_4$). This is consistent with earlier findings which reported that in weakly oxidizing environments, Ga_2O_3 was preferred.[8] Nevertheless, the surface is not completely oxidized, some GaAs remains.

Sample D is similar to sample A except that the slice was withdrawn from the DI rinse in the ambient air and was exposed to air for approximately one minute before the high vacuum was attained. The XPS data indicate that the surface is comprised of Ga_2O_3, of the arsenic oxides and elemental arsenic. This sample exhibited uneven wetting of the surface; parts were hydrophilic while other parts were hydrophobic. The existence of elemental or surface bound As (binding energy = 41.8 eV) could account for the hydrophobic character of the surface. All of the oxides of Ga and of As were found to be hydrophilic in character in wetting tests. Both GaAs and unoxidized As were found to be hydrophobic in nature.

Sample E was exposed to laboratory air at approximately 40% RH and 20°C under ambient light illuminated conditions for 5 days. The surface was observed to be partially hydrophobic. Some arsenic is still seen in the near surface as in the case of sample D. There is no essential difference in the nature of the surface oxide of the wafer whether it was exposed for one minute or five days to the standard laboratory ambient air. (Compare D and E in Table I). The oxidation of the surface is not yet complete after 5 days under these conditions.

Exposure of a slice to higher humidity conditions (95%) for 5 days in air (sample F), however, resulted in substantially more of the higher oxidation state As. The surface appears to be more thoroughly oxidized, e.g. no trace of GaAs is found. Still there remains approximately 20% unreacted As. This observation corroborates previous work which suggests that water vapor and oxygen have a synergistic effect on the oxidation of the GaAs surface.[9]

Immersion of a wafer in water at 20°C for 5 days resulted in the sample with spectrum G. Since the oxides of As are soluable in water, it is not unexpected to find that the surface stoichiometry has become definitely Ga rich in the oxide. The water was in equilibrium with air above it, and consequently contained a substantial concentration of dissolved oxygen. A thick oxide was observed to form of relatively uniform thickness as indicated by the uniform blue color of the surface. XPS analysis of the sample revealed total oxidation of the surface and a depletion of As in the surface. The surface was decidedly hydrophilic, probably due to the absence of both GaAs and elemental As on the surface.

Hydrochloric acid is commonly used to allegedly strip the oxide from GaAs. A wafer prepared by the standard method and exposed to air was then immersed in concentrated HCl at 20°C for 15 minutes. The wafer was rinsed in DI water and transported to the XPS system under nitrogen. As seen in figure 2 and Table 1 for sample H, no significant quantity of As oxide was found. However, Ga_2O_3 was observed in approximately equal abundance to GaAs. The surface of the wafer is rich in elemental or surface bound As. It has been proposed that in HCl without the presence of dissolved oxygen the surface is converted to an As-passivated surface by the leaching of Ga from the GaAs.[25] The present work supports the general trend to an As dominated surface after such treatment. The surface is observed to be hydrophobic in character, which does not, however, imply that the surface is oxide free. The presence of As is sufficient to render the surface hydrophobic.

The effectiveness of oxidation by an RF plasma was tested using sample I. The plasma was in pure O_2, at 50 MHz with a power of 50 W for 10 minutes. The Ga was completely oxidized, but much of the As remined unoxidized (slightly more than half the As). The oxidation of the Ga was enough to assure the complete destruction of the GaAs on the surface, however. Significantly, the higher oxidation state of As_2O_5 is preferred over As_2O_3 in this environment. The surface was slightly hydrophilic.

Part of wafer I was immersed in 1:40 $NH_4OH:H_2O$ at 20°C for 10 minutes; the resultant analysis of this sample is labelled J. Once again, the oxides of As have been dissolved by the treatment. The surface is closer to oxide free than any other sample except A and B. The stoichiometry is found to be composed of nearly equal parts Ga and As. The presence of Ga_2O_3 and elemental As however cannot be ignored, especially, when this treatment is used as the final step before a thin film deposition on the wafer surface, such as an anneal cap. The amount of Ga oxide present before exposure to NH_4OH may effect the surface composition after treatment. It was noted that the hydrophobic surface which resulted from the NH_4OH preparation was less contaminated with carbonaceous material than were surfaces exposed to acids. This surface was hydrophobic.

Wafer K was not etched; it was only degreased. Consequently, the surface of this wafer represents many months of exposure to air during storage. The analysis gives insight into the initial surface condition of a typical wafer. As can be seen in figure 2 and Table I, the oxide is very much like the oxide present on a slice freshly etched and exposed to air. The only noticeable difference is the more complete oxidation of the Ga on the surface in the present case than was observed in samples D and E. The significant presence of elemental or surface bound As is noted also. Typically, such surfaces are hydrophobic. One would anticipate moreover that the oxides formed would not be homogeneous in distribution, at least on a small scale. A relatively intense C signal is observed from a surface layer of absorbed hydrocarbon.

Upon exposure to hard ultraviolet radiation (254 nm and 186 nm) in an air ambient, the surface of sample K was oxidized. The analysis of the oxidized wafer is labelled L. The surface is free of hydrocarbon as indicated by the absence of an XPS signal for C. The surface oxide is in the highest oxidation state available for both Ga and As, and the surface oxidation is complete, i.e. no unreacted As remains.

RESULTS--GROWTH KINETICS

An ellipsometric determination of the oxide thickness as a function of time was undertaken in this experiment for wafers exposed to 40% RH air at 20°C and for wafers exposed to 100 ppm O_3 under 12 mW/cm^2 254 nm radiation. As described above, the ellipsometry was used to determine the film thickness. The thickness of the oxide which formed on the wafer in air was measured as a function of time for times varying from one minute to 40 minutes. In this relatively short time the oxide thickness obeys a logarithmic dependence on time. This observation is in agreement with that of Lukes, who measured the growth of the oxide on GaAs from a few minutes to several years.[37] The growth rate which we observe is approximately 0.6 nm/decade, in agreement with Lukes' value for cleaved (110) GaAs surfaces. Thus, in 10 minutes approximately .6-1 nm of oxide is grown (assuming an initial oxide film of 0-.5 nm), while in 24 hours about 2.5 nm results and in a week one can expect about 3 nm.

These values are to be contrasted with the observations of the growth kinetics of UV-ozone as previously described. Figure 3 illustrates the thickness as determined by ellipsometry as a function of the length of exposure to the radiation. Undoped, chromium-doped, silicon-doped, and tellurium-doped wafers from both horizontal Bridgman and LEC crystals were used to test the growth rate. No significant variation with type or crystal pulling method was found. Initially, the oxide may form a partial coverage which grows in equivalent thickness linearly in time. The growth rate for times greater than one minute and less than about 20 minutes can be modeled better, however, with a thickness which is dependent on the logarithm of the time with a rate of 2.5 nm/decade for an initial thickness of about 0.5 nm (at 1 minute). This model is qualitatively similar to that of air oxidation of GaAs. The present rate is larger by about a factor 4x for the same length of time larger, i.e. the time required for a like thickness of oxide to form using ozone is 10^{-4} that of ambient oxidation. The explanation of such behavior is found in a model due to Richie and Hunt [38] in which the oxide is modeled as a parallel plate capacitor. The rate controlling step is the formation of the oxygen negative ion. The electron concentration at the interface is controlled by a Boltzman distribution, whose mean energy is given by the potential drop across the dielectric, which is proportional to the thickness. Thus, a logarithmic dependence in time is derived after integration of the differential equation. This behavior holds until the oxide has attained a thickness of about 4 nm, which occurs after twenty minutes. At that time the kinetics change character and a growth curve is observed in which the thickness of the oxide is proportional to the square root of time. In 19 hours one finds that a thickness of approximately 28 nm has been grown. The temperature of the wafer was found to reach 60°C, the operating temperature of the lamp, during the exposure. This change in growth kinetics is also observed for air-reacted GaAs at about 4 nm, but this thickness is attained only after approximately 10 months of growth in air alone!

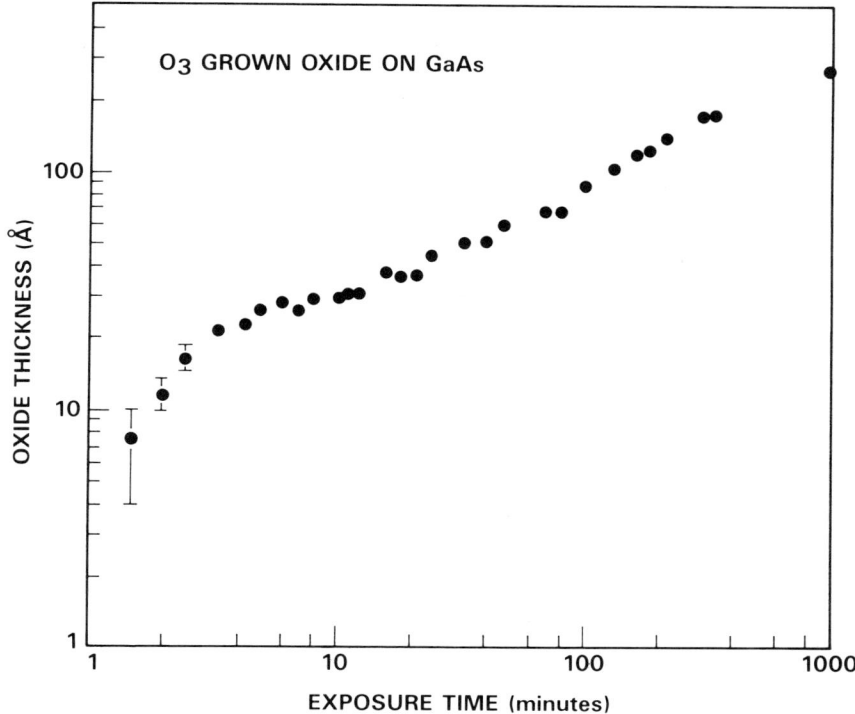

Figure 3. Oxide thickness on GaAs grown using UV-ozone treatment versus exposure time. The thickness was determined ellipsometrically.

Auger depth profile analysis was performed, the results of which are shown in figure 4. The film shown here was oxidized for 19 hours by UV-ozone exposure. The most striking feature of the figure is the loss of As from the oxide. This phenomenon is understandable when one considers the vapor pressures of As_2O_3, and As_2O_5, which are not insignificant at 60°C. Evaporation of the arsenic oxides could occur over a period of many hours and explain the excess of Ga. Note the characteristic excess of As at the interface. This has been observed in other oxidation processes, and is due to the preferential oxidation of Ga at the interface.[8,12,14]

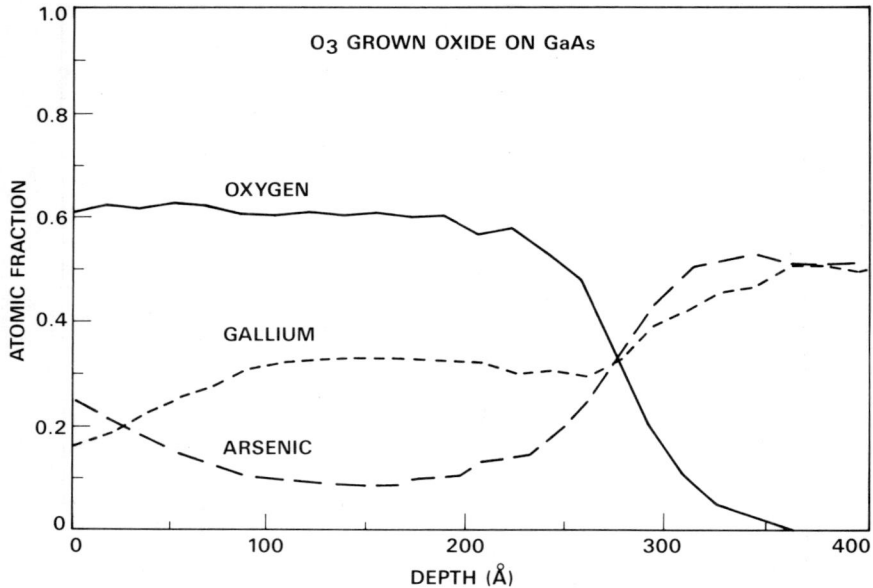

Figure 4. Auger depth profile of oxide grown on GaAs using UV-ozone treatment for 19 hours. Note the loss of As from the oxide and the accummulation of As at the oxide/GaAs interface.

CONCLUSIONS

We have reported observations of the composition of the GaAs surface after various aqueous surface treatments. It has been demonstrated that it is possible to produce a surface essentially free from oxide by etching in 8:1:1 solution sulfuric acid, hydrogen peroxide and water, if the surface is rinsed in DI water under nitrogen and the surface protected from exposure to air afterwards. It is conjectured that the mechanism of etching is by the oxidation and dissolution of As, while Ga is dissolved directly in acid or base solutions. Thus, the hydrophilic surface often seen immediately after etch is due to arsenic oxides which are water soluable and can be rinsed away, if not reformed by contact with the air.

Hydrophobic surfaces can be produced on GaAs by the presence of sufficient elemental or surface bound As, even in the presence of oxides. Hydrophilic surfaces, on the other hand, require a large measure of oxidation to have occurred. It was shown that while the oxides of As are relatively easy to remove, Ga_2O_3 is much more difficult to strip from the surface. Moreover, elemental As is frequently present on even oxidized surfaces, especially those treated with acidic solutions.

The oxide formed in one minute of air exposure differs little from that which forms after 5 days or many months at normal levels of humidity. Although the oxide may be the equivalent of 2.5 nm thick, after 5 days, the surface is not fully oxidized, in contrast to surfaces exposed to UV-ozone, which are fully oxidized and freed from hydrocarbons by photopyrolysis. The growth mechanism of oxides formed by UV-ozone treatment are not substantively different from those formed by exposure to ambient air, only greatly accelerated by the availability of atomic oxygen and photogenerated carriers in the semiconductor. The oxide formed by UV-ozone exposure may be deficient in As due to evaporation of its oxides, if the surface is heated for sufficiently long times. A As-rich layer remains at the interface between the oxide and the semiconductor, as has been observed with other oxidation techniques.

ACKNOWLEDGEMENTS

The authors gratefully acknowledge the technical assistance of J. K. Russell. The authors thank R. L. York, E.F. Schulte, and R. Strong for most helpful discussions.

REFERENCES

1. G. P. Schwartz, G. J. Gualtieri, G. W. Kammlott, and B. Schwartz, J. Electrochem. Soc. Solid State Sci. Technol. 126, 1737 (1979).

2. K. Watanabe, M. Hashiba, Y. Hirohata, M. Nishino and T. Yamashina, Thin Solid Films 56, 63 (1979).

3. G. Landgren, R. Ludeke, Y. Jugnet, J. F. Morar and F. J. Himpsel, J. Vac. Sci. Technol. B 2, 351 (1984).

4. C. Webb and M. Lichtensteiger, J. Vac. Sci. Technol. 22, 659 (1982).

5. L. Ley, R. A. Pollak, F. R. McFeely, S. P. Kowalczyk and D. A. Shirley, Phys. Rev. B9, 600 (1974).

6. P. Pianetta, I. Lindau, C. M. Garner and W. E. Spicer, Phys. Rev. B18, 2792 (1978).

7. J. R. Arthur, J. Appl. Phys. 38, 4023 (1967).

8. C. D. Thurmond, G. P. Schwartz, G. W. Kammlott, and B. Schwartz, J. Electrochem. Soc. 127,1366 (1967).

9. H. Iwasaki, Y. Mizokawa, R. Nishitani anf S. Nakamura, Jap. J. Appl. Phys. 17, 315 (1978).

10. I. Shiota, N. Miyamoto and J. Nishizawa, J. Electrochem. Soc. 124, 1405 (1977).

11. . W. G. Petro, I. Hino, S. Eglas, I. Lindau, C. Y. Su and W. E. Spicer, J. Vac. Sci. Technol. 21, 405 (1982).

12. X. Wang, A. Reyes-Mena and D. Lichtman, J. Electrochem. Soc. 129, 851 (1982).

13. S. P. Kowalczyk, J. R. Waldrop and R. W. Grant, J. Vac. Sci. Technol. 19, 611 (1981).

14. C. C. Chang, P. H. Citrin and B. Schwartz, J. Vac. Sci. Technol. 14, 943 (1977).

15. P. A. Betrand, J. Vac. Sci. Technol. 18, 28 (1981).
16. D. Flamm and E.-H. Weber, Phys. Stat. Solidi A 76, K163 (1983).
17. T. Oda and T. Sugano, Jap. J. Appl. Phys. 15, 1317 (1976).
18. H. Takagi, G. Kano and I. Teramoto, J. Electrochem. Soc. 125, 579 (1978).
19. M. Matsumura, M. Ishida, A. Suzuki and K. Hara, Jap. J. Appl. Phys. 20, L726 (1981).
20. G. Leonhardt, A. Berndtsson, J. Hedman, M. Klasson, R. Nilsson and C. Nordling, Phys. Stat. Solidi B 60, 244 (1973).
21. M. Cardona, C. M. Penchina, N. Shevchik and J. Tejeda, Solid Stat Comm. 11, 1655 (1972).
22. W. Gudat, E. E. Koch, P. Y. Yu, M Cardona and C. M. Penchina, Phys. Stat. Solidi. B 52, 505 (1972).
23. T. Lane, C. J. Veseley and D. W. Langer, Phys. Rev. B 6, 3770 (1972).
24. T. Suzuki and M. Ogawa, Appl. Phys. Lett. 33, 312 (1978).
25. J. M. Woodall, P. Oelhafen, T. N. Jackson, J. L. Freeouf and G. D. Petit, J. Vac. Sci. Technol. B 1, 795 (1983).
26. R. Oren and S. K. Ghandhi, J. Appl. Phys. 42, 752 (1971).
27. S. A. Schafer and S. A. Lyon, J. Vac. Sci. Technol. 19, 494 (1981).
28. M. Cardona and D. L. Greenway, Phys. Rev. 131, 98 (1963).
29. V. M. Bermudez, J. Appl. Phys. 54, 6795 (1983).
30. R. P. Vasquez, B. F. Lewis and F. J. Grunthaner, J. Vac. Sci. Technol. B 1, 791 (1983).
31. J. A. McClintock, R. A. Wilson and N. E. Byer, J. Vac. Sci. Technol. 20, 241 (1982).
32. R. Vig and J. W. Lebus, IEEE Trans. Parts Hybrids Packg. PHP-12, 365 (1976).
33. M. Horvath, L. Bilitzky and J. Huttner, Ozone (Elsevier, Amsterdam,1985) 104.
34. R. M. A. Azzam and N. M. Bashara, Ellipsometry and Polarized Light (North-Holland, Amsterdam,1977).
35. D. E. Aspnes and A. A. Studna, Phys. Rev. B27 , 985 (1983).
36. R. C. Weast,ed. CRC Handbook of Chemistry and Physics (CRC Press, Boca Raton, Fla.,1984) B-71.
37. F. Lukes, Surface Sci. 30, 91 (1972).
38. I. M. Ritchie and G. L. Hunt, Surface Sci. 15, 524 (1969).

PART III

Materials Characterization with Ion Beams

ON THE USE OF SECONDARY ION MASS SPECTROMETRY IN SEMICONDUCTOR DEVICE
MATERIALS AND PROCESS DEVELOPMENT

CHARLES W. MAGEE AND EPHRAIM M. BOTNICK
RCA Laboratories, Materials Characterization Research, Princeton, NJ 08540

ABSTRACT

This paper describes how secondary ion mass spectrometry can be used effectively in semiconductor device materials and process development. A short overview of the experimental technique is given followed by applications in the following areas:
1) Basic research
2) New Process Development
3) Trouble-shooting current processing
4) Analysis of competitor's devices.

INTRODUCTION

In semiconductor device materials and process development, characterization of surface layers and thin films is of vital importance. Areas of interest include: metallizations, passivation layers, contact regions, and of course, the electrically active regions of the actual semiconducting material. Secondary ion mass spectrometry (SIMS) is well suited for use in all of these areas. Due to its high sensitivity, it is the surface analytical technique best suited for investigation of dopant-level impurities in semiconductors. For this reason RCA has applied this technique to the study of semiconductor materials and process development.
There are four areas where SIMS plays a major role:
1) basic materials research
2) new device processing development
3) troubleshooting of existing processing
4) analysis of competitor's devices

This paper will endeavor to provide the reader a basic understanding of the technique and to show how SIMS can be applied successfully to the areas listed above.

EXPERIMENTAL

Secondary ion mass spectrometry [1] utilizes a beam of energetic (1-20 keV) primary ions to sputter away the sample surface producing ionized sputtered particles which can be detected with a mass spectrometer. This technique provides in-depth information on atomic constituents by recording one or more peaks of the mass spectrum as the sputtering process erodes the sample, thus producing the detected signal from increasingly greater depths beneath the original sample surface. This kind of analysis is called "depth profiling". Several critical reviews on depth profiling exist in the literature [2-5] which broadly cover most aspects of this mode of SIMS analysis.

Depth profiling using SIMS has many advantages over other methods of surface and thin film analysis. Its major strong points include:

1) Trace element sensitivity - parts-per-million concentrations are detectable for most elements.
2) Total elemental coverage - all elements, hydrogen to uranium can be detected.
3) Isotopically selective - can distinguish between different isotopes of the same element.
4) Good depth resolution - elemental information can be locallized in depth to 20 to 50 Å.
5) Can be made quantitative - through the use of standards.

These attributes make it a powerful technique for thin-film analysis, but not without its difficulties.

SIMS has two major problems which arise from: 1) the great variation in sensitivity from element-to-element within a given matrix; and 2) the variation in sensitivity of a single element contained in different matrices. The first problem makes it practically impossible to estimate, a priori, the concentration of elements within a sample only from their peak heights in the mass spectrum. This makes the SIMS technique a poor one for survey analysis ("What's in it?"). The second problem simply means that once the sensitivity for a certain element has been established within a given matrix, that element may (and indeed probably will) have a significantly different sensitivity in another matrix. These sensitivity variations are caused by orders-of-magnitude differences in the fraction of sputtered atoms which are ejected from the sample surface as either positive or negative ions [6], since only ionized particles can be separated and detected by a mass spectrometer. The ionization probability depends upon such things as the leaving atom's ionization potential (for positive ion formation) or electron affinity (for negative ion formation) as well as the electron population of the conduction band of the surface (for positive ion emission) or the work function of the surface (for negative ion emission) [7]. This means that in order to be absolutely sure of the sensitivity of a certain element contained in the matrix of interest, one must have a standard for that element contained in the same matrix. This subject is covered in the next section, "quantitation".

QUANTITATION

The raw data of a SIMS depth profile consist of intensities of the monitored mass peaks plotted versus the time during which the primary ion beam has been sputtering away the sample surface. For example, from the raw data of a SIMS depth profile of the thermal diffusion of phosphorus and boron into silicon, a device engineer might wish to learn the surface concentration of each dopant, how deeply each penetrates into the silicon substrate for these processing conditions, and above all, at what depth are the concentrations of the p-type acceptor atoms (B) and the n-type donor atoms (P) equal? That point, called a "p-n junction", has a critical effect on the electrical performance characteristics of a semiconducting device. Clearly, these raw data do not answer these questions. To transform these data into useful information we need 1) a method for determining the rate at which the sample material was eroded by the action of the primary ion beam thereby enabling one to transform sputter time into depth, and 2) a method relating signal intensity to volume concentration, thereby accounting for the sensitivity differences between boron and phosphorus.

The conversion of sputter time into depth can be accomplished simply by measuring, with a profilometer, the depth of the flat-bottomed crater produced by the primary ion beam. This method assumes that the sputtering rate during analysis remains constant. This is generally not true for samples composed of layers of signficantly different composition (such as InP on GaAs) or when the flux of primary ions to the surface is not constant during the analysis. For homogeneous samples and a well controlled primary beam, depths can generally be determined to an accuracy of 5%.

In order to convert secondary ion intensities into volume concentrations, standard samples of known composition similar to that of the unknown samples must be analyzed along with the unknown samples. (A matrix element is always monitored to adjust for small variations in overall instrumental sensitivity between sample and unknown.) Ion-implanted samples have been found to make ideal standards [8] and they are depth-profiled in exactly the same manner as the unknown samples. In these standards, the elements of interest are present in amounts expressed in atoms/cm^2 (called the "dose") and known to an accuracy of ± 2%. If the depth of analysis is sufficient to sputter away all of the implanted impurity atoms within the analyzed area of the samples, then the area under the intensity vs. sputter time profile for that impurity element is exactly equal to the dose (in at/cm^2). If one independently establishes the depth scale (in cm) by profilometry as mentioned above, the intensity axis of the profile can readily be converted to volume concentration (in atoms/cm^3), which is the desired quantity.

Once this relationship between intensity and concentration has been established for <u>each</u> element of interest in the unknown and the sputtering rate has been determined, the raw data can be replotted as concentration vs. depth as shown in Fig. 1. Presented in this form, the SIMS depth profile can answer all of the questions originally posed in this example. One can readily see that the junction occurs at a depth of 0.92 μm beneath the sample surface, where the concentration of each dopant type is 1.3 x 10^{18} atom/cm^3.

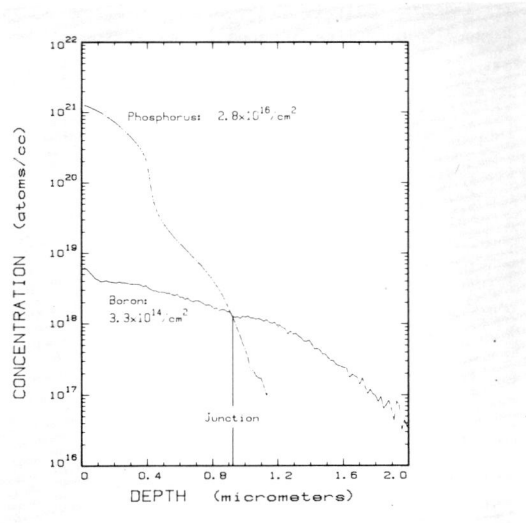

Fig. 1. *Quantitative depth profile of boron and phosphorus diffused into silicon. Quantitative concentrations for B and P were obtained after calibrating the instrument sensitivity for these elements by analyzing an ion-implanted standard. Areal densities of dopants diffused into the Si were determined by integrating the areas under the SIMS profiles. The depth scale was established by measuring the depth of the sputtered crater after the analysis.*

APPLICATIONS

Basic Materials Research

The first area of application of SIMS to semiconductor device problems has furthered our basic understanding of the materials and processes currently being used, or those being considered for use in the future. The following examples illustrate such uses.

1. Hydrogen migration under avalanche injection of electrons in silicon MOS capacitors [9].

Hydrogen impurities in the oxide layer of metal-oxide-semiconductor structures have been studied by a number of different workers and have been proposed as active agents in many phenomena. In particular, the role of hydrogen in generation of oxide charge has been the subject of long standing and recently active investigation [10]. The central issue is whether hydrogen species are redistributed in depth as a result of electron transport across the oxide film. A related question is whether such redistribution affects oxide charge generation, particularly at the SiO_2/Si interface. Because SIMS is the only surface analytical technique capable of detecting H, and because our SIMS instrument has been optimized for H analysis [11], we have studied H accumulation at the SiO_2/Si interface as correlated with electron transport and fixed oxide charge after avalanche electron injection.

The details of the experiment can be found in ref. 9, but the results can be summarized as follows:

1) The areal density of hydrogen accumulated at the SiO_2/Si interface increased linearly with the total injected electron fluence.
2) The resulting interfacial hydrogen could control the generation of interface trapped charge although the correlation is not one-to-one.

Of critical importance to this particular study was the ability to quantitatively measure the low levels of H at the SiO_2/Si interface. Such data are shown in Fig. 2. Note that hydrogen builds up at the interface with injected fluence, but the total amount of H detected is actually quite small. Experiments of this type have been tried before. They failed, however, to detect the low levels of hydrogen at this interface because of high instrumental background for this ubiquitous element [11], or in those cases where nuclear reaction analysis was used [12], the analysis technique simply failed to have sufficient sensitivity. Our SIMS instrument with its high sensitivity and low H background has made this study possible.

The analyses shown here were particularly difficult due to the insulating nature of the sample surface. Because the positive primary ion beam (in this example Cs^+) deposits electrical charge onto the insulator as the depth profile proceeds, the surface potential tends to rise to the breakdown potential of the oxide (several hundred volts). Such surface charge not only makes extraction of secondary ions from the charged-up surface impossible, but can cause migration of mobile species within the analyzed area. To neutralize such charging, a beam of electrons is simultaneously directed toward the sample [13]. The electron current is sufficient to just over-compensate for the positive charge of the primary ion beam thus causing the sample to charge slightly negative. At this point, the secondary electrons are no longer held to the surface thus keeping the surface at this potential regardless of any increase in the flux of incoming electrons. The success of this novel method of charge neutralization can be inferred from the constancy of the oxygen matrix-ion signal throughout the oxide layer. This constancy of sensitivity is required if quantitative integration of the detected H signal (areal density in at/cm^2) is desired.

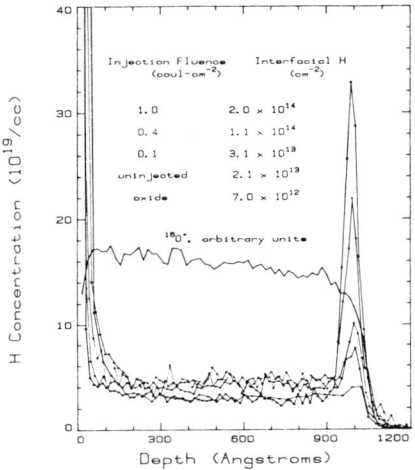

Fig. 2. Hydrogen concentration vs. depth for four Al/SiO$_2$/Si capacitors avalanche-injected with electrons to various fluences. Also shown is the ^{1}H' profile for an unmetallized region of the oxide, and (in arbitrary units) the ^{16}O' depth profile. The O' depth profile illustrates the consistency of instrumental sensitivity throughout the depth profile. The large peak appearing in all H profiles at the sample surface is due to adsorbed water.

2. Oxygen decoration upon annealing of ion-implantation-induced lattice damage.

When ion implantation is used to fabricate the sources and drains of FET's a screen oxide is often implanted through. These source-drain implants are usually of very high dose ($\geq 10^{15}$ at/cm^2) and can produce recoil implantation of oxygen from the oxide into the underlying Si, adversely affecting the sheet resistance of this heavily doped layer. SIMS analysis of silicon samples implanted with arsenic at 170 keV, 5 x 10^{15}/cm^2 through a 350 Å oxide show that no appreciable amount of oxygen is recoiled into the Si from the SiO$_2$ by the energetic As atoms. This result was obtained directly after implantation prior to any heat treatment. Annealing is, of course, required after such an implant because above arsenic doses of the order of 1-3 x 10^{14}/cm^2 the sample is rendered completely amorphous. Upon annealing this sample at 850°C for 30 min, a very unusual SIMS profile was obtained as shown in Fig. 3(a). The arsenic has been somewhat redistributed, but now large amounts of oxygen are detected in the silicon to depths of 0.2 micrometers. Since the oxygen was not detected in the "as-implanted" state, the annealing step must provide the driving force to rapidly diffuse the oxygen into the damaged layer of the implanted substrate. These results are in agreement with previous studies [14]. However, the peculiar peaked depth distribution of the oxygen led us to suspect that the oxygen was being gettered to residual damage sites remaining in the silicon after annealing. This hypothesis was checked by examining the sample in cross-section [15] by transmission electron microscopy (TEM) [16]. Fig. 3(b) shows the cross-section TEM micrograph of the sample for which the SIMS depth profile is shown in Fig. 3(a). One can see clearly from these data that oxygen has indeed been gettered to two distinct layers of dislocations, the deeper one of which is located at a depth corresponding to the deepest extent of the amorphous zone in the as-implanted sample. A sharp discontinuity in the arsenic depth profile is apparent at this deep dislocation layer. Previous studies [17] have theorized that the upper layer of dislocations is due to impurity segregation during annealing. The SIMS depth profiles seem to bear this out.

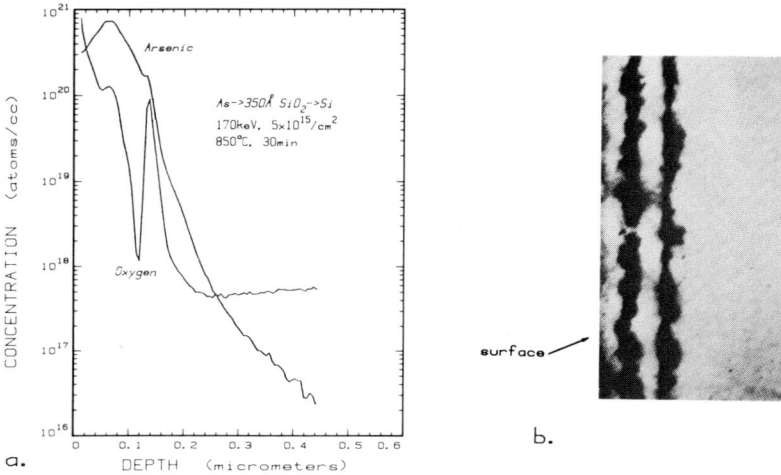

Fig. 3. (a) Arsenic and oxygen depth profiles for As implanted through a 350-angstrom SiO_2 into Si. The sample was then annealed at 850°C for 30 minutes in N_2. The oxide was stripped off prior to SIMS analysis. (b) Cross-section transmission electron micrograph of the sample shown in (a). Note the two layers of dislocations, and compare to the peaks in the oxygen profile of (a). The depth scales for (a) and (b) are the same.

New Process Development

New materials and processes are constantly being developed for producing semiconductor devices. Secondary ion mass spectrometry can be used in a constructive way to remove some of the guess-work or "black magic" from this often arduous task. Several illustrative examples follow:

1. IMPATT diode development

In the last two years, we have investigated and developed the technology base for silicon IMPATT diodes for use at frequencies of up to 220 GHz [18]. An outgrowth of this effort has been the development of a novel device technology which facilitates simple, well-controlled processing procedures for the fabrication of ultra-thin IMPATT devices with good heat sinking properties. The devices produced are capable of CW operation with 25-mW output power, at above 100 GHz, with a conversion efficiency of 2 percent. The work most responsible to this success includes accurate impurity concentration profiles and thickness control developed for multilayer vapor-phase expitaxial silicon devices. Throughout this process development, SIMS was used to depth profile p and n-type dopants simultaneously and thus determine electrical junction depths <u>directly</u> from the SIMS data. It is this unique capability which has been of most benefit to millimeter-wave device research. Such a depth profile is shown in Fig. 4. Here, using Cs primary ion bombardment of the sample at 10^{-10} torr, we have determined the depth profile for both B and As in an

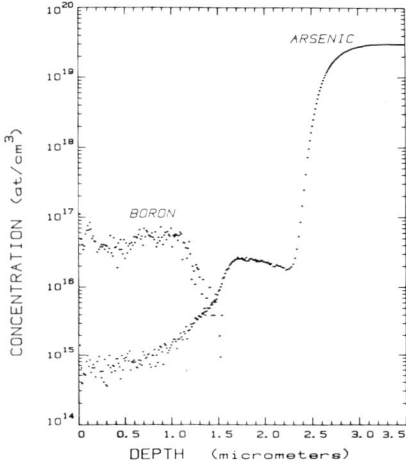

Fig. 4. SIMS depth profiles of B and As taken simultaneously on a sample to be used in fabrication of double-drift IMPATT diodes. The sample consists of one epitaxial Si layer continuously grown on an As-doped substrate. At the beginning of epi growth, the layer was doped with As. Midway through the growth the dopant was changed from As to B. The final top P⁺ layer needed to complete the double-drift structure is made by ion implantation.

epitaxial layer doped at different stages of the growth. For successful high frequency operation of these devices, the criticality of the epitaxial layer thickness and doping levels is such that without SIMS analysis, optimization of the epi-growth process would be impossible. Just to develop the growth technique to minimize arsenic out diffusion from the heavily doped substrate into the epi-layer required many SIMS analyses of samples grown under various temperatures, pressures, and process gases. An example of this use of SIMS can be found in ref. 19. Furthermore, when fabricating double-drift diodes by ion implantation, the As-implanted dopant distribution can only be predicted. SIMS is used to measure accurately the p and n-type dopant distribution after thermal annealing, which is important because diffusion can often alter the profiles in an unpredictable manner, smearing together the very thin layers needed for high gigahertz operation. This approach has been successful in determining whether or not the implantation and diffusion steps have produced the desired doping profiles prior to the difficult and tedious steps needed to process the wafers into diodes. SIMS will be of even greater benefit to high-energy ($\geq$1 MeV) ion implantation coupled with laser annealing, because of the less precisely known ion ranges at these energy levels and, as yet, relatively unpredictable diffusion behavior with laser annealing.

2. Emitter-Base Profile Optimization

Often new semiconductor processing schedules are developed at the research laboratory and then transferred to the factory. Frequently the process needs some "fine-tuning" in order to fit in with the processing for the device. The following example shows how SIMS can be used in such process development.

Our Solid State Division is introducing a new Laboratories-developed emitter diffusion process which has been shown to result in significantly lower substrate defect densities than the standard emitter diffusion process.

Measurements of surface phosphorus concentrations resulting from both processes were identical to within experimental error. In addition, angle-lap-and-stain measurements indicated that the junction depths produced by both processes were also the same. However, the breakdown voltage of the emitter-base junction produced by the new process was significantly higher than that produced by the old standard process. SIMS depth profiles of boron and phosphorus were taken simultaneously to determine the base and emitter profiles for both processes in hopes of discovering why different junction breakdown voltages were produced.

The SIMS measurements showed that the base diffusions of boron were identical, so the difference in electrical characteristics must be due to differences in the phosphorus emitter profiles. These emitter profiles are shown in Fig. 5 along with the base profile common to both processes. The SIMS depth profiles (made quantitative for concentration and depth by the methods described earlier) agreed with prior measurements in that they also measure nearly equal phosphorus concentrations on the surface of each sample, and that each has a junction depth of 1.65 μm. However, the SIMS profiles clearly show that, in the vicinity of the junction, the phosphorus concentration gradient produced by the old process is considerably greater than that produced by the new process. This greater concentration gradient produced the lower breakdown voltage for the old process. Without secondary ion mass spectrometry to show why the two processes produced different device properties, one could only speculate about the cause, and change the processing conditions in hope of randomly selecting the right parameters. It is obvious that SIMS can help provide a more direct route to the desired end.

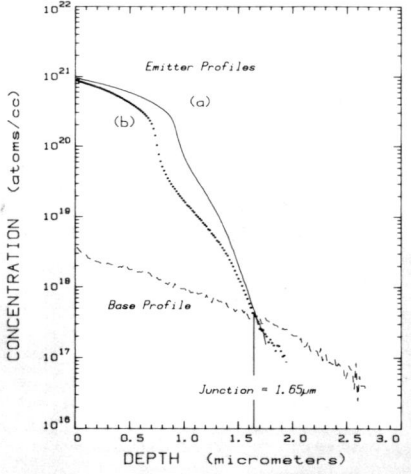

Fig. 5. Simultaneously obtained base (B) and emitter (P) profiles for two different emitter diffusion schedules. The "old" process (a) produced a sharper P concentration gradient in the vicinity of the junction than did the "new" process (b). This resulted in a lower breakdown voltage for the old process. The areal densities of P and B diffused into the silicon are obtained by integrating the area under the respective profiles. Since the base profile was found to be unaffected by the different emitter diffusion schedules it is shown only once.

Trouble-Shooting Existing Processing

Current processing schedules and the equipment used are not perfect and can inexplicably produce devices with poor electrical or mechanical properties. While it is not an ideal analytical tool for addressing such problems due to the large differences in elemental sensitivities, SIMS can be used in a trouble shooting mode provided the scope of the information desired is not too broad. If you ask "What's in this stuff that's messing up my device?" SIMS is not likely to produce a useful answer. Asking, "Is there any of element X in my device and if so, where?" is much more likely to yield a satisfactory answer. Shown below is an example of one such use of SIMS to elucidate a problem which occurred in device processing.

1. Disturbances in liquid-phase epitaxial growth of InGaAsP/InP 1.3 µm laser material [20]

It is well known that there can be irregularities in the growth of multiple layers by liquid-phase epitaxy (LPE). For example, there are occasional variations in the emission wavelength of AlGaAs devices caused by fluctuations in the aluminum concentration in the melt. These fluctuations may result from the sensitivity of aluminum dissolution to the residual oxygen content in the furnace, or from other causes such as pullover of material from an earlier bin in a multiple-bin boat. The objective of this example is to show how SIMS can be used to examine a type of growth disturbance seen in InGaAsP/InP 1.3-µm-laser material grown at this laboratory, and to elucidate its origin. Arguments will be presented to show that the disturbance is mostly attributable to pullover. The main effect observed was the occasional production of a wafer whose surface appearance was noticeably different from that of a good wafer, and which yielded lasers with unusually high threshold currents. The layered structure of these wafers is shown schematically in Fig. 6a. SIMS was used to look for evidence of pullover of Ga and As from the active layer into the p-InP cladding layer. The SIMS profiles are shown in Fig. 6b and 6c. Comparing the profiles for the good and anomalous wafers, one finds that the active layer in the disturbed sample is wider, with a tail reaching into the p cladding layer; the As and Ga concentrations first decrease and then both rise toward the cap layer. This behavior can be understood if one assumes that some melt had been pulled over from the previous bin. If this melt enters as a film adhering to the wafer surface, gradually decreasing Ga and As concentrations could result. Their subsequent rise indicates a further increase in these elements such as would result from the presence of more pulled-over melt causing the gradual diffusion of Ga and As to the growth interface. One might suppose that this could occur becuase of some property of the substrate could control the adhesion of the melt to the wafer surface. Measurement of the surface orientation performed on the two wafers shows that both are oriented to the (100) plane to within 0.3-0.4°, the difference in their orientations being 0.08°. Thus, it is not likely that there are any significant effects associated with wafer misorientation. Factors such as differences in dislocation density or other structural parameters cannot be ruled out at present. In any event, SIMS showed that interlayer contamination was the cause of the problem.

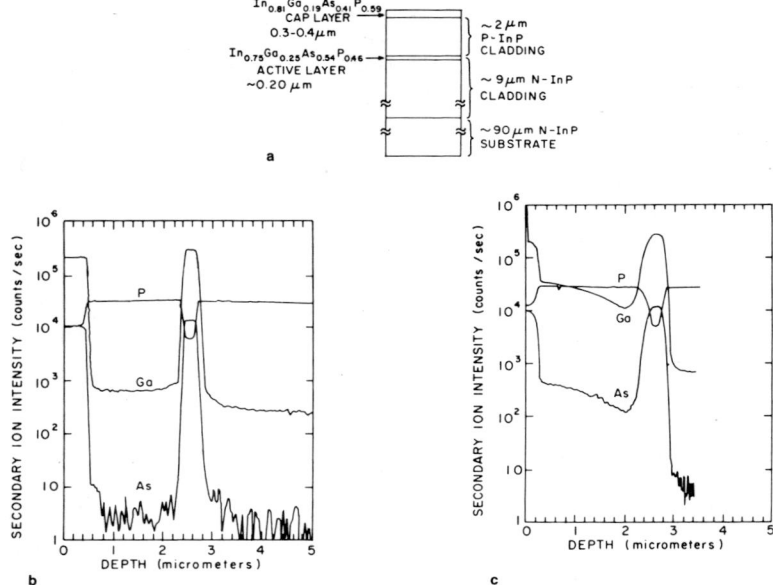

Fig. 6. Liquid phase epitaxial InGaAsP/InP 1.3-μm laser structures. (a) Schematic indicating layer thicknesses and composition of the active and cap quaternary layers. (b) SIMS depth profiles through a normal wafer. (c) SIMS depth profiles through an anomalous wafer showing evidence of pull-over.

Analyses of Competitor's Device

The examination of a competitior's device is important for many reasons. One may wish to know how sophisticated the competition's devices are becoming in order to keep a product in a leading position. The following is an example of how SIMS can be used for competitor's device analysis and how two of the technique's unique capabilities provided information about a competitor's device that would otherwise have gone undetected.

1. Passivation layer sequencing

The a competitor's device in question was known to have a series of passivation layers different from our own. The layers were thought to be silicon nitride and silicon dioxide but confirmation was needed of their composition as well as the arrangement of the layers. Since this is a major-element analysis, Auger electron spectrometry (AES) would have been the technique of choice, except for the fact that the sample was an insulating film several thousand Angstroms thick. Obtaining a depth profile by AES would have been possible, but quite tedious. On the other hand, the charge-compensation technique developed for the SIMS instrument [13] makes this analysis routine.

Upon first inspection of the device, it became apparent that the Si_3N_4 layer was outermost with SiO_2 underneath. However, there seemed to be a large amount of hydrogen in the sample as well as some phosphorus. It was decided to include these elements in the depth profile of the sample along with nitrogen, silicon and oxygen. The results are shown in Fig. 7. The concentration scale is accurate only for P because N and O are major elements, and we had no standard for H in silicon nitride. Nevertheless, SIMS with its

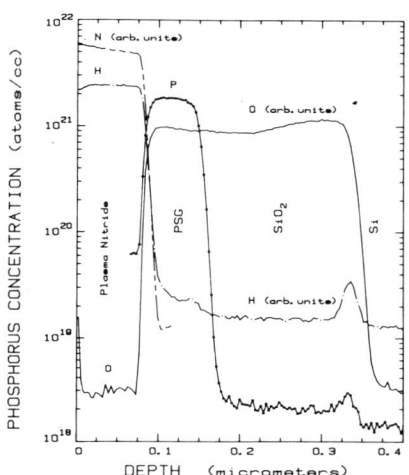

Fig. 7. SIMS depth profiles through a series of passivation layers on a silicon device.

high sensitivity and ability to detect hydrogen reveals two things that an AES depth profile would not have shown:

1) the silicon nitride top layer, because of its high hydrogen content, must be a low temperature plasma-deposited nitride which has not undergone any significant subsequent heat treatment (T $\leq$ 400°C). AES cannot detect H.

2) the "SiO_2 layer" was not actually all a thermal oxide, but composed of two separate layers; an SiO_2 layer directly on the silicon substrate, followed by a deposited phosphosilicate glass (PSG) layer in which the P concentration is of the order of three atomic percent. This low level of P would be missed easily in an AES depth profile.

SUMMARY

From the above examples, it is hoped that the reader has obtained an insight into secondary ion mass spectrometry, including its principles, its strengths, and its weaknesses. The section on quantitation should give the reader some assurance that SIMS can be made quantitative and under what analytical circumstances this is possible. These, together with the applications in basic research, new process development, trouble-shooting and competitor's device analysis, are intended to give the reader enough knowledge of, and exposure to, SIMS to allow him to make an intelligent decision whether or not secondary ion mass spectrometry can be of use in his particular problem.

ACKNOWLEDGEMENTS

Many helpful and instructive discussions were held with the following: J. F. Corboy, L. A. Goodman, W. L. Harrington, R. E. Honig, F. Kolondra, I. Ladany, J. T. McGinn, A. Rosen, P. Webb and C. P. Wu from RCA, and F. J. Feigl and R. O. Gale of Lehigh University. Thanks are also due to these last persons for release of data (ref. 9) prior to publication.

REFERENCES

1. For a useful review article on SIMS see: C. A. Evans, Jr., Anal. Chem. 44 No. 13, 67A(1972).
2. E. Zinner, Scanning 3, 57(1980).
3. S. Hofmann, Surf. Interface Anal. 2, 148(1980).
4. J. A. McHugh in "Methods of Surface Analysis", A. W. Czanderna, Ed., Elsevier, Amsterdam, 1975, pp. 223-278.
5. E. Zinner, J. Electrochem. Soc. 130, 199C (1983).
6. H. A. Storms, K. F. Brown and J. D. Stein, Anal. Chem. 49, 2023(1977).
7. C. A. Andersen and J. R. Hinthorne, Anal. Chem. 45, 1421(1973).
8. D. P. Leta and G. H. Morrison, Anal. Chem. 52, 514(1980).
9. This work was performed by Mr. R. O. Gale as part of a Ph.D. thesis project under Dr. F. J. Feigl, Dept. of Physics, Lehigh University. C. W. Magee served on Mr. Gale's thesis committee. The final summation of this work can be found in: R. Gale, F. J. Feigl, C. W. Magee and D. R. Young, J. Appl. Phys., 54, 6938 (1983).
10. F. J. Feigl, D. R. Young, D. J. Dimaria and S. Lai in "Insulation Films on Semiconductors 1981", M. Schulz and P. Pensl. Eds., (Spring-Verlag, New York, 1981).
11. C. W. Magee and E. M. Botnick, J. Vac. Sci. Technol. 19, 47(1983).
12. R. E. Benenson, L. C. Feldman and B. G. Bagley, Nucl. Instrum. Methods 168, 547(1980).
13. C. W. Magee and W. L. Harrington, Appl. Phys. Lett. 33, 193(1978).
14. J. C. Mikkelsen, Appl. Phys. Lett. 42, 695(1983).
15. M. S. Abrahams and C. J. Buiocchi, J. Appl. Phys. 45, 3315 (1974).
16. TEM work was performed by J. T. McGinn, Materials Characterization Research, RCA Laboratories.
17. D. K. Sadana, J. Washburn and C. W. Magee, J. App. Phys. 54, 3479(1983).
18. A. Rosen, M. Caulton, P. Stabile, A. Gombar, W. Janton, C. Wu, J. Corboy and C. W. Magee, IEEE Trans Microwave Theory Tech. MTT-30, 47(1982).
19. G. W. Cullen, J. F. Corboy and R. Metzl, RCA Review 44, 187(1983).
20. I. Ladany, R. T. Smith and C. W. Magee, J. Appl. Phys. 52, 6064(1981).

SIMS CHARACTERIZATION OF THIN THERMAL OXIDE LAYERS
ON POLYCRYSTALLINE ALUMINIUM

F. DEGREVE AND J.M. LANG
Cégédur Péchiney, Centre de Recherches, B.P. 27, 38340 Voreppe, France.

ABSTRACT

Thermal oxide layers of different thickness (T = 3.5-18 nm) covering polycrystalline aluminium sheets were investigated by dynamic SIMS in an ion microscope. Depth profiles of major elements (Al, O), segregated impurities in the oxide layer (Li, Na, Be, Mg) and implanted ^{18}O were recorded for different operating conditions : energy, incident angle, chemical (Ar^+ or O_2^+) and isotopic ($^{16}O_2^+$ or $^{18}O_2^+$) nature of the primary beam and oxygen jet to saturate the surface.

The main results are :

- the shape of the depth profiles is very sensitive to operating conditions : the best depth resolution is obtained at low energy (2.5 keV) and the minimal distorsion with an O_2 jet, even under O_2^+ bombardment. Deconvolution would be necessary to deduce the true depth profiles.

- a tremendous enrichment ($10^3 - 10^4$) of the oxide with respect to the Al bulk is observed for Li, Na, Be and Mg impurities via an efficient diffusion mechanism during heat treatment at 425°C. Their depth distribution within the oxide layer is heterogenous with surface enrichment for Li and Na and sub-layer (towards the oxide/metal interface) for Be and Mg. The lateral distribution below the interface is also heterogeneous e.g. at the grain boundaries for Be and Na.

- the depth profile of implanted $^{18}O^+$ versus operating conditions and oxide thickness gives direct information on the dynamic oxidation of Al induced by O_2^+ sputtering in the first 20 nm.

INTRODUCTION

The composition and the structure of the several nanometers thick oxide layer which covers aluminium and its alloys play a primordial role on the technological properties of these materials : protection, adhesion, corrosion, wear, vacuum brazing, secondary electron emission, ...

Surface analytical techniques such as AES, ESCA, ISS and SIMS have provided interesting results on the oxide thickness, the presence of some impurities, their chemical state and their spatial distribution [1-6].

In this work, we have investigated with an ion microscope the chemical characterization of thin thermal oxide layers covering polycrystalline aluminium ; i.e. the identification, the depth profile and the lateral distribution of major elements (Al, O) and impurities susceptible to segregation in the oxide layer.

1. EXPERIMENTAL PROCEDURE

Samples were prepared from commercial grade polycrystalline aluminium sheets of 500 µm thickness ; the main impurities were Si and Fe (= 0.1 %). Oxide layers of increasing thickness (3.5 - 18 nm) were obtained by heat treatment in air at 425°C between 2 and 64 hours. A 7 nm thick layer enriched in ^{18}O was also prepared at the same temperature in a vessel filled with $^{18}O_2$ at a pressure of one atmosphere. The thickness was determined by ellipsometry on an area of 0.1 cm^2 with a relative precision of ± 5 %.

SIMS analyses were carried out in a Cameca IMS3f ion microanalyser [7].

The samples were sputtered with a mass filtered primary beam of $^{40}Ar^+$, $^{16}O_2^+$ or $^{18}O_2^+$; the intensity was in the range of 20 - 50 nA and the ion density around 1x10^{-5} Axcm^{-2}. The surface of the sample could be saturated with an oxygen jet, i.e. with 99.52 % $^{16}O_2$; the base pressure (1x10^{-8} torr) then increases to an overpressure of 2x10^{-5} torr [8].

The sputter rates were determined in the range $\dot{Z}$ = 0.01-0.03 nmxs^{-1} from the final crater depths by a stylus method.

The oxide/metal interface was located in the depth (time) scale from the value of the oxide thickness T measured by ellipsometry. The estimated uncertainty (± 0.8 nm) is outlined in the depth profiles by a shaded area which takes into account the experimental errors in both $\dot{Z}$ and T.

The distinction between oxygen originally present in the oxide layer and oxygen arriving from an external source (e.g. primary ion beam) has been achieved in two ways :

- bombarding an ^{18}O enriched layer with $^{40}Ar^+$ or $^{16}O_2^+$.
- bombarding natural oxide layers (^{16}O) with $^{18}O_2^+$.

For each position in the depth profile, the intensities $I(^{16}O^+)$, $I(^{18}O^+)$ and $I(^{16}O^+) + I(^{18}O^+)$ provide values representative of the partial and total amount of oxygen at the surface.

2. RESULTS

It was observed that the shape of the depth profiles depends on both :
- the 3 operating parameters : chemical nature (Ar^+, O_2^+) and primary energy Ep (incident angle θ in an ion microscope [9]) of the primary beam and the use of an O_2 jet.
- The thickness T of the oxide layer.

A systematic investigation of the influence of these parameters revealed that the minimum distortion (matrix effects) in the depth profiles is obtained when the surface is saturated with an O_2 jet , even under O_2^+ bombardment. The best depth resolution is obtained at low energy (Ep = 2.5 keV) and high incident angle (θ = 57°). These effects are illustrated in the depth profiles of major elements (Al, O), segregated impurities(Li, Na, Be, Mg) and implanted oxygen 18.

2.1. Depth profiles of major elements (Al, O)

Fig. 1 corresponds to the analysis under Ar^+ bombardment at medium energy Ep(8keV) of a 7 nm thick layer enriched in ^{18}O. When the O_2 jet is not used (Fig. 1a), the drop in the intensity of every species is very sharp (10^2) and may be attributed to the progressive elimination of oxygen with etching. Under $^{18}O_2^+$ bombardment (Fig 2a), a more subtle variation occurs in the interface region for the intensity of Al^+ and total oxygen ($^{16}O^+ + ^{18}O^+$).

When the oxygen concentration at the surface is imposed by an O_2 jet, the intensity of Al^+ varies by up to only 30 % between Al_2O_3 and Al. This is true whatever the chemical nature or energy of the primary beam and the oxide thickness (Fig. 1b, 2b). Physically, this means that the surface oxide formed in-situ in the presence of sputtering is very similar to the natural oxide layer. This was confirmed by ESCA analysis of a natural oxide layer and of an Ar^+ sputtered cleaned Al surface oxidized in an oxygen ambient [10]. It may thus be concluded to a first approximation that the sputtering yield S and the ionization probability P of any species will vary little when going from Al_2O_3 to $Al + O_2$. From an analytical point of view, this is important since ion intensities on both sides of the interface can be directly compared. Distorsion in the depth profiles can be considered as minimal in these conditions.

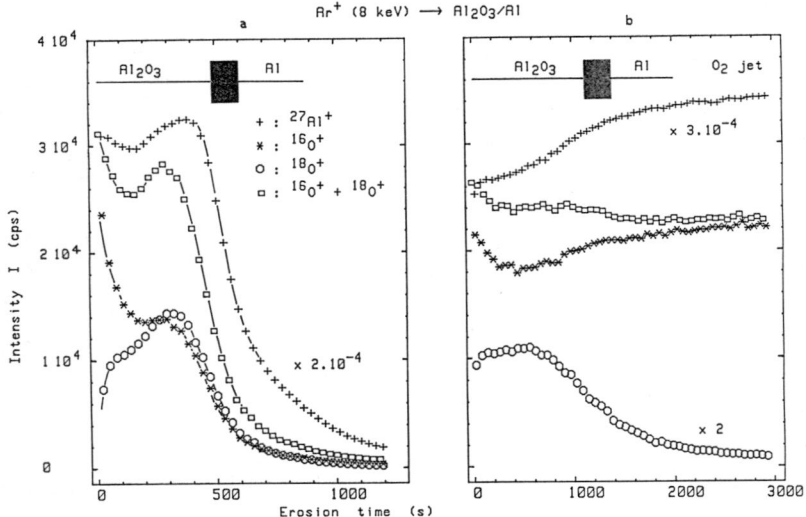

Fig. 1   Depth profiles of the major elements for an aluminium sheet covered by a 7 ± 0.5 nm oxide layer enriched in oxygen 18. a : Ar^+ bombardment at medium energy (8 keV), b : influence of an O_2 jet.

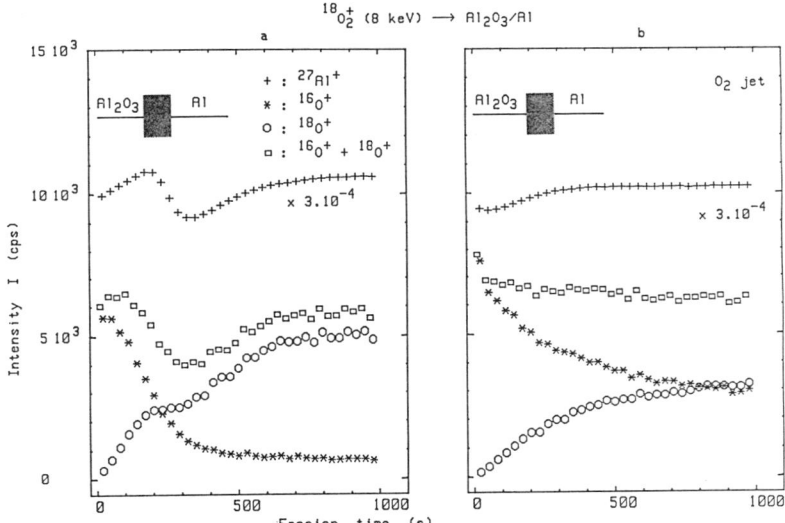

Fig. 2 Depth profiles of the major elements for an aluminium sheet covered by a 4.2 ± 0.5 nm oxide layer.
a : $^{18}O_2^+$ bombardment at medium energy (8 keV),
b : influence of an O_2 jet.

Note that the total oxygen intensity does not depend on the isotope used in O_2^+ bombardment : $I(^{16}O^+) + I(^{18}O^+)$ in $^{18}O_2^+$ bombardment and $^{16}O^+$ in $^{16}O_2^+$ bombardment. Fig. 3 shows that the normalized profiles of total oxygen are very similar for different oxide thickness.

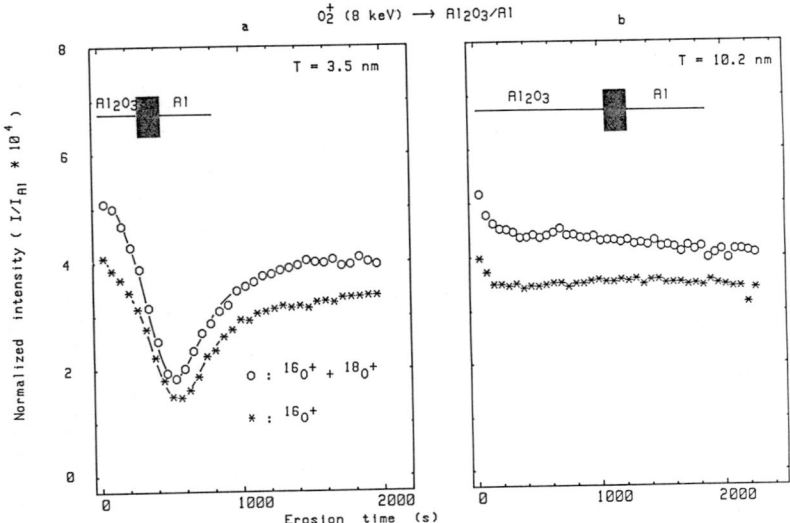

Fig. 3 Depth profile of total oxygen under O_2^+ bombardment at medium energy (8 keV) : $^{16}O^+$ for $^{16}O_2^+$ and ($^{16}O^+ + ^{18}O^+$) for $^{18}O_2^+$ bombardment. Intensities are normalized to the stationary $I(Al^+)$ in the Al bulk. a : oxide thickness $T = 3.5 \pm 0.5$ nm, b : $T = 10.2 \pm 0.5$ nm.

2.2. Depth profiles of thermally segregated impurities

Low mass resolution ($M/\Delta M = 300$) mass spectra revealed an oxide layer rich in impurities with respect to the Al bulk. This is particularly obvious for alkaline (Li, Na, K) and alkaline-earth (Be, Mg, Ca) impurities.

Fig. 4 presents the depth profiles of Li, Na, Be and Mg observed for an aluminium sheet covered by a 4.2 nm thick oxide layer bombarded under Ar^+ (2.5 keV) and with an O_2 jet. Very similar results are obtained under O_2^+ bombardment. The concentration scale was calibrated via a working curve established with standard aluminium samples [11]. The ionic intensities were high enough to record ion micrographs on both sides of the interface. The micrographs of Fig. 5 were taken at approximatively 5 - 10 nm below the interface, as indicated by an arrow in Fig.4. In order to reveal the cristallographic contrast between grains, the micrograph of $^{27}Al^+$ was taken in the metal bulk (Z > 50 nm) without an O_2 jet.

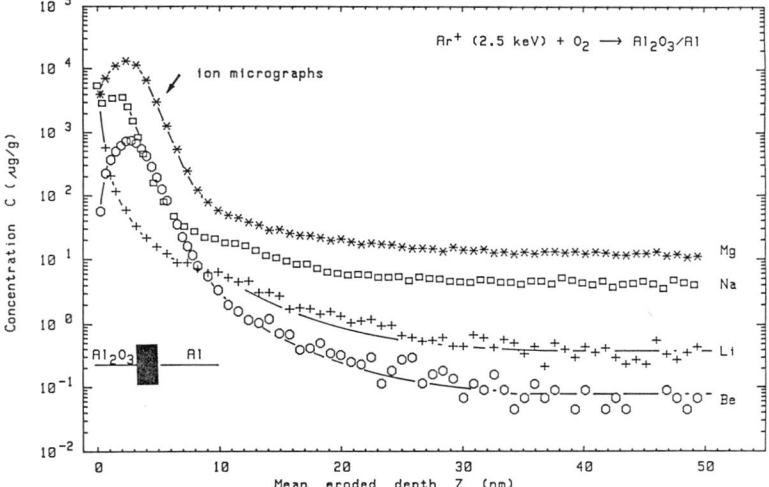

Fig. 4 Depth profile of segregated impurities Li, Na, Be and Mg for an aluminium sheet covered by a 4.2 ± 0.5 nm oxide layer. Ar^+ bombardment at low energy (2.5 keV) and O_2 jet. The arrow indicates the region where the ion micrographs of Fig. 5 were recorded.

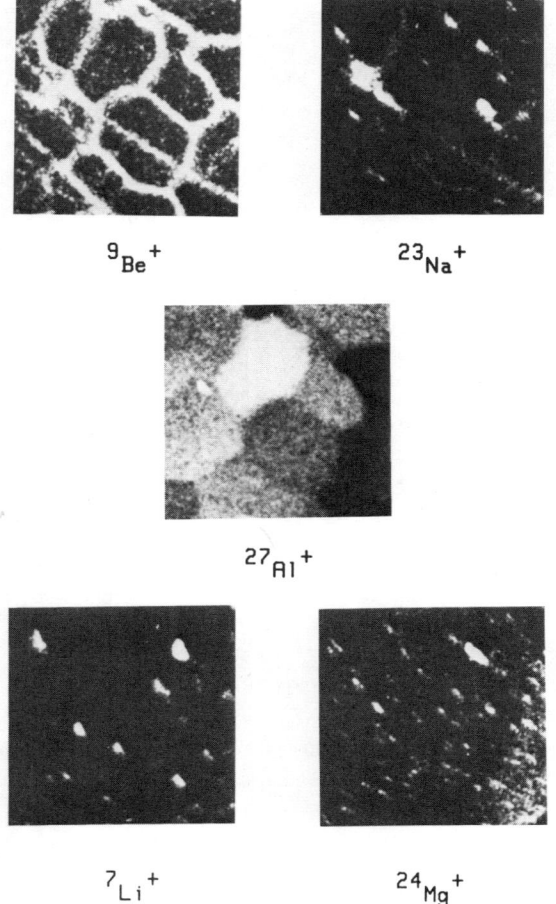

Fig. 5 Ion Micrographs of Li, Na, Be and Mg impurities below the oxide/metal interface (arrow in Fig. 4) for an aluminium sheet covered by a 4.2 ± 0.5 nm oxide layer. Ar^+ bombardment at low energy (2.5 keV) and O_2 jet. Imaged zone is 100 x 100 μm.

The interesting observations may be summarized as follows :

- the existence of a tremendous segregation in the oxide (ox) of alkaline and alkaline earth impurities ; with respect to the bulk (b), the enrichment factor defined by :

$$a = C(ox)/C(b) \propto I(ox)/I(b) \tag{1}$$

is of the order of $10^3 - 10^4$. With Mg, Li, Na and Be present in the bulk at 10, 0.3, 3.7, 0.08 $\mu g \times g^{-1}$ the apparent concentrations in the oxide layer are 1.5, 0.5, 0.3 and 0.08 % respectively. Ion micrographs in the oxide layer revealed an homogeneous distribution at a given depth for the four impurities.

- a continuously decreasing profile for Li in the outermost superficial layers. Below the interface, it is distributed in small randomly distributed nodules.

- a similar profile shape for Be and Mg : a peak with the maximum inside the oxide layer. Below the interface, Be is only segregated at the grain boundaries up to around $Z = 20$ nm while Mg is concentrated in small precipitates not apparently related to the grain boundaries.

- a composite shape for the profile of Na : a sharp decreasing portion at the extreme surface as for Li followed by a peak in the layer as for Be and Mg. Na is partially distributed at the grain boundaries and in small precipitates.

When the primary energy is increased, the depth profile of Li presents a small shift in the tail towards greater depths. The profiles of Na, Be and Mg become broader and more asymetric as illustrated in Fig. 6, in the case of 7 nm thick enriched layer. Note that at this thickness, the transition time of $^{18}O^+$ is roughly equal to those of Be^+ and Mg^+.

When increasing the oxide thickness from 4.2 to 13.2 nm at a given Ep, it is observed that :

- the profile of Li remains unaffected (Fig. 7a)
- the profiles of Be, Mg, Na shift toward greater depths (Fig. 7b for Be) ; the transition time for Be and Mg increases linearly with oxide thickness as illustrated for Be in Fig. 8.
- the transition times of Be and Mg are equal within an uncertainty of 10 % (Fig. 4, 6).

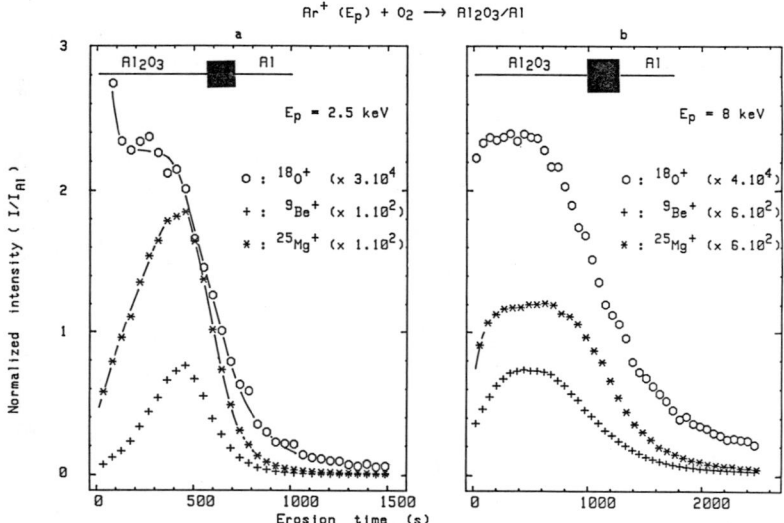

Fig. 6. Normalized depth profiles of the major element $^{18}O^+$ and of the segregated impurities $^9Be^+$ and $^{25}Mg^+$ for an aluminium sheet covered by a 7 ± 0.5 nm oxide layer enriched in oxygen 18. Ar^+ bombardment with an O_2 jet. a : low energy (2.5 keV), b : medium energy (8 keV).

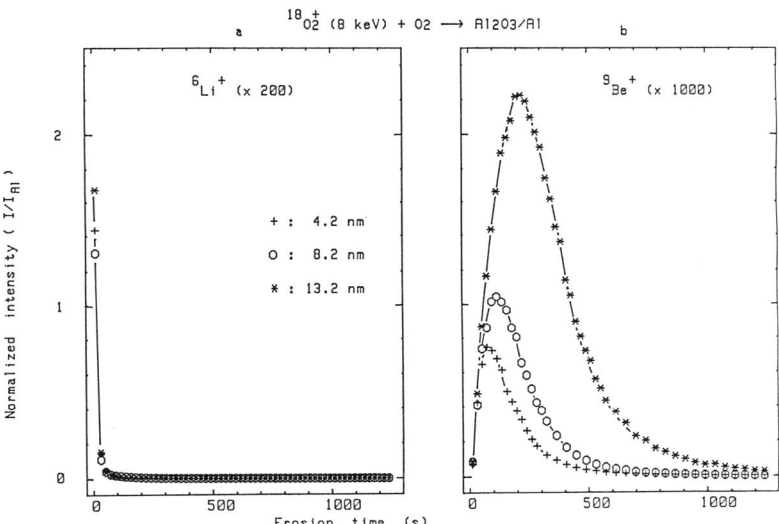

Fig. 7 Normalized depth profiles of segregated Li and Be impurities for an aluminium sheet covered by an oxide layer of increasing thinckness. $^{18}O_2^+$ bombardment at medium energy (8 keV) and O_2 jet.

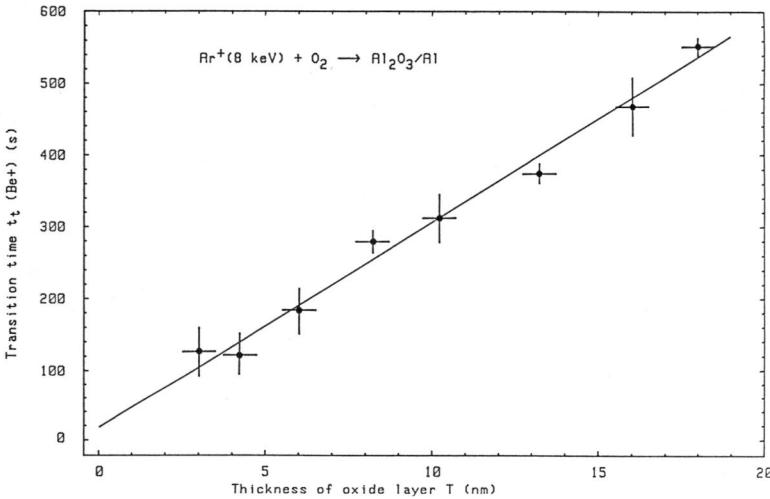

Fig. 8 Variation of the transition time $t_t(Be^+)$ necessary to reach a 50 % reduction in the tail at greater depth with respect to the oxide layer thickness measured by ellipsometry. Ar^+ bombardment at medium energy (8 keV) and O_2 jet.

2.3. Depth profile of implanted oxygen ($^{18}O^+$)

The profile of implanted $^{40}Ar^+$ under Ar^+ bombardment could not be recorded at medium mass resolution. High mass resolution (M/ΔM = 4000) showed the presence at 40 amu of the interferences : $^{40}Ca^+$ - $^{40}K^+$ - ^{24}Mg $^{16}O^+$ - ^{23}Na ^{16}O $^{1}H^+$. Under $^{18}O_2^+$ bombardment, the implantation profile could be recorded and furnished unique information on how oxygen is retained in Al_2O_3 and Al with the presence of O_2^+ sputtering. Fig. 9 gives the implantation profiles for a thin oxide layer (T = 4.2 nm) at different values of (Ep,θ).

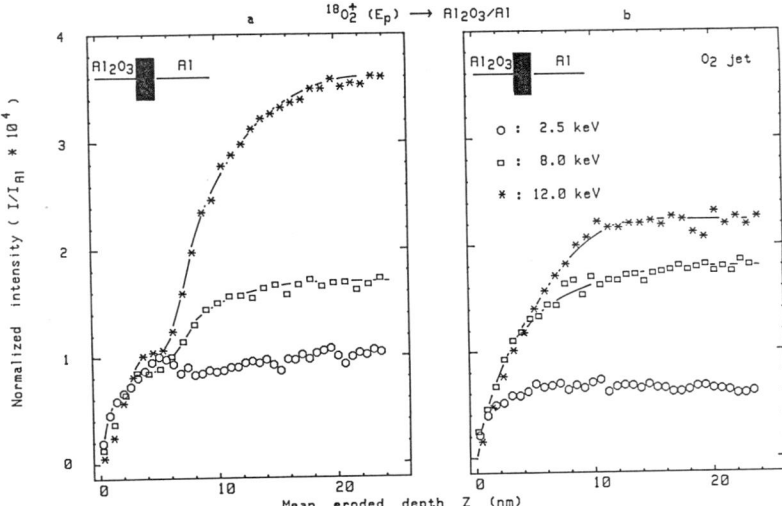

Fig. 9 Depth profile of implanted $^{18}O^+$ for an aluminium sheet covered by a 4.2 ± 0.5 nm oxide layer at increasing energy (decreasing incident angle). Intensities are normalized to the stationary $I(Al^+)$ in the Al bulk.
a : $^{18}O_2^+$ bombardment, b : influence of an O_2 jet.

An increasing Ep (decreasing θ in an ion microscope [9]) leads to :

- increasing the depth Z_s needed to reach a steady-state : about 5 nm at (2.5 keV, 57°) and 20 nm at (12 keV, 36°).
- a variation in the shoulder shape in the interface region.
- increasing the amount of implanted oxygen at the steady-state ; the main variation occurs between 2.5 and 5.0 keV.

Fig. 9 shows that when an O_2 jet is used, the implantation profiles become continuous and the stationary intensity is decreased. A competition seems to occur between ^{16}O and ^{18}O at the surface, at the expense of ^{18}O.

Fig. 10 represents the normalized depth profile of implanted oxygen for oxides of different thickness at 8 keV and with an O_2 jet. Within the experimental scatter, it may be concluded that the curves are not significantly different. A unique implantation curve is hence characteristic of the operating parameters (E_p, θ). The transition depth Z_t corresponding to half the stationary intensity is equal, in these conditions, to 4.6 ± 1.0 nm.

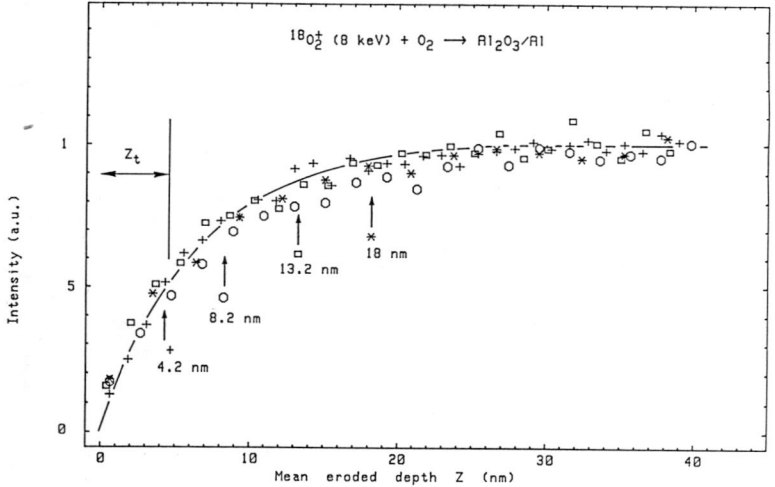

Fig. 10   Depth profile of implanted $^{18}O^+$ for an aluminium sheet covered by an oxide layer of increasing thickness. Intensities are normalized to the stationary values. $^{18}O_2^+$ bombardment at medium energy (8 keV) and O_2 jet.

3. DISCUSSION

3.1. The shape of depth profiles

It is clear that in reality the presented depth profiles represent a distorted picture of the true profiles in the first 20 nm. Since the apparatus fonction f(a) is unknown, deconvolution to obtain the true profiles is not possible [12, 13]. However, the qualitative interpretation of the recorded profiles can be facilitated by reasonable simulation. The convolution of different types of hypothetical true profiles with a gaussian distribution for the resolution function would estimate the relative importance of the different parameters involved in the establishment of depth profiles. Such a simulation is given in Fig. 11 for two typical values of the standard deviation. A broadening effect appears when σ increases from 1 to 2 nm, but the area under the curves (total quantity of matter) remains constant.

Major elements

The profile of oxygen contained in the oxide layer (^{16}O or ^{18}O) may be assimilated to a superficial step function. The computed intensity in Fig. 11 shows the existence of a narrow plateau region before the interface. The slope of the tailing side increases with σ. In these conditions, the transition depth Z_t (transition time t_t) corresponds to the oxide/metal interface.

Under Ar^+ bombardment, the measured profile of $^{18}O^+$ is in qualitative agreement with Fig. 11a (cf. Fig. 1b, 6). Under O_2^+ bombardment, this is not the case and the decreasing profile of O^+ is more difficult to understand (Fig. 2). As the O^+ implantation in the Al_2O_3 layer leads to an excess of oxygen, the establishment of a steady state between the two opposite fluxes does not seem to be reached quickly.

The shape of the profile of the Al^+ matrix ion reflects actually the superficial concentration of oxygen (Fig. 1-2). Without the use of the O_2 jet, the comments presented before for O^+ remain valid for Al^+.

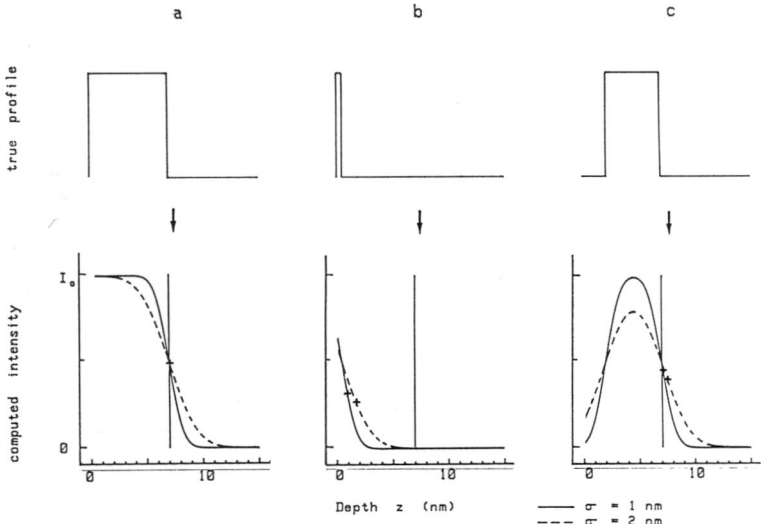

Fig. 11 Computed intensity obtained by convolution of different step function profiles with a gaussian resolution function. Two standard deviation values were chosen : σ = 1 nm (—), σ = 2 nm (--) and the oxide thickness T = 7 nm was chosen. Fig. a simulates the profile of the major element ^{18}O, Fig. b the profile of Li in a superficial layer of 0.5 nm and Fig. c the profile of Be, Mg for a sublayer starting at 2 nm and ending at the oxide/metal interface. In the lower figures, the interface is outlined by a vertical line and transition depth (time) by the + symbol.

Segregated impurities

The experimental profiles suggest that the segregated impurities are not evenly distributed thorough the oxide. It appears that Li is concentrated in a very thin superficial layer while Be and Mg are present in a sublayer, itself located within the oxide layer. The simulated profiles for such distribution are discussed below.

Fig. 11b simulates the profiles of a superficial 0.5 nm thick layer. The initial intensity would be lower than the true one and hence the experimental concentration at $Z = 0$ would be underestimated.

Fig. 11c simulates the profiles of a sublayer starting at 2 nm below the surface and ending at the interface. The broadening in the peaks are in qualitative agreement with the observed depth profiles of Be and Mg (Fig. 6). In addition, the shift in the top position towards the left (initial specimen surface) and the tailing effect at greater depths with increasing energy may be attributed to an asymetry in the actual resolution function. The transition depth (time) does not necessarily correspond to the oxide/metal interface. Practically, the error involved in locating the interface by the transition time of Be^+ and Mg^+ is low and within the experimental scatter ; this explains why $t_t(Be)$ and $t_t(Mg^+)$ vary linearly with oxide thickness (Fig. 8). For the same reason, when T is increased, a higher intensity at the peak top will not necessarily mean a higher concentration (Fig. 7b).

For Na, the experimental profiles suggest the existence of both the two components, i.e. a superficial thin layer as for Li and a sublayer for Be and Mg.

The considerations discussed above are only valid within the oxide layer and in the near interface region since at higher depth (Fig. 4,5), the profile shape depends both on the operating conditions and on the possible heterogeneous distribution (precipitates, grain boundary, enrichment, ...).

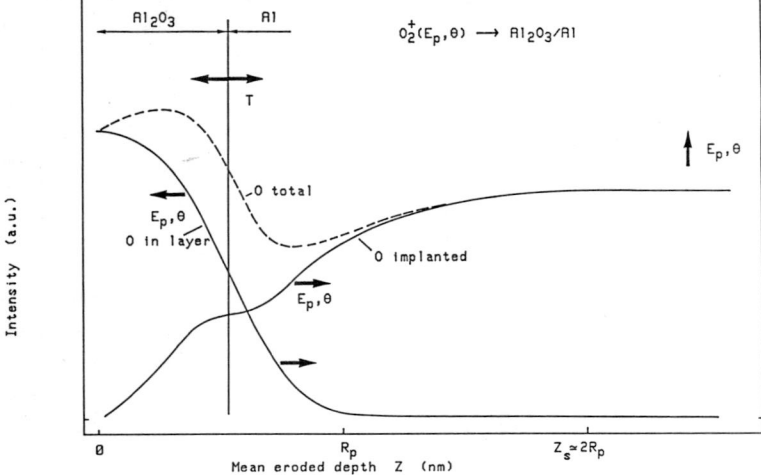

Fig. 12   Outline of the resulting total oxygen profile by summation of oxygen originally present in the oxide layer and oxygen brought by implantation. Evolution with the operating parameters (E_p, θ) and the oxide thickness T. Rp is the range of O^+.

Implanted oxygen. Dynamic oxidation of Al induces by O_2^+ bombardment

During sputtering with O_2^+ the total amount of oxygen available at the surface at each instant is a function of the depth. This is schematically illustrated in Fig. 12. At the beginning of sputtering, oxygen comes essentially from the Al_2O_3 layer. At greater depth, when steady-state sputtering is reached ($Z > Z_s$), the oxygen concentration is imposed by the implantation in Al. However, in the intermediate region generaly comprising the Al_2O_3/ Al interface, the situation is more complicated. In this transient region, the amount of implanted O^+ depends on the depth at which the interface is reached. The AlO_x stoichiometry at every depth is reflected by the total oxygen intensity : ^{16}O for $^{16}O_2^+$ bombardment and ($^{16}O^+$ + $^{18}O^+$) for $^{18}O_2^+$ bombardment. This stoichiometry varies both with the primary parameters (Ep, θ) and the oxide thickness T. When $T > Z_s$, the oxygen depletion is weak, even negligible (Fig. 3b) When $T < Z_s$ (Fig. 2,3a) the oxygen depletion after the interface is important. In this case, the amount of oxygen defined by the implantation is not sufficient to form a stoichiometric oxide probably because the sputtering yield is high. Moreover, when such an understoichiometry is created, a variation occurs in the ionization probability, the ionization yield and the sputtering rate. The conversion of time into depth is therefore not applicable in the interface region ; this contributes to further complicate the interpretation and is probably one of the causes of the observed shoulder in the Al^+ and total oxygen profiles. The existence of a shoulder in the profile of the matrix ion and of $^{16}O^+$ under $^{16}O_2^+$ has already been mentionned for Al [14], as well as for steel [15, 16], Si [17] and Fe [15]. However, the distinction between the oxygen characteristic of the layer and that characteristic of the implantation process was not investigated.

At high energy (Ep > 5 keV, θ < 44°), the stationary oxygen concentration is higher than that necessary to form Al_2O_3. However, in general, the observed shoulder in the profiles is less pronounced although significant at low thickness.

3.2 <u>Surface migration of impurities in aluminium</u>

The enrichement factor (a) defined by relation (1) depends essentially on the sample characteristics : bulk concentration C_b and heat treatment (temperature and time t). It is important to understand why such high values (a = 10^3 - 10^4) are observed. Textor et.al. [2] have interpreted the accumulation of Li and Na at the air/oxide interface and the presence of Mg within the oxide layer near the oxide/metal interface from a purely equilibrium thermodynamic point of view. The differences in the free energy of formation of the different oxides could qualitatively explain the driving force of the segregations. The same argument may be presented for Be which was not mentioned by these authors in their conditions. The tensio-active character of Li and Na could also be of importance in explaining their enrichment in a superficial layer.

Kinetic factors have also to be taken into account. If evaporation does not occur during the heat treatment, the quantity of impurities in the sample remains unchanged. To enrich the oxide layer by a factor of 10^3, it is necessary to pump matter from the bulk reservoir. Rigorously, a depletion below the interface would appear in the depth profiles. This is not the case, at least for the scale of depth investigated in our work (Z < 100 nm). It is worthwhile to compare the depth necessary to reach stationary intensities in our profiles (Z_s = 30-40 nm in Fig. 4) to the distance Z_d that an impurity can travel during the heat treatment : $Z_d = 2(Dt)^{1/2}$ where D is the diffusion coefficient. Taking values of D at 425°C from the literature [18], it can be seen that for t = 2.5 hours, Z_d varies from 9 to 37 µm according to the impurity. The quantity of impurity contained in an oxide layer of thickness T at a mean concentration C_o is proportionnal to $C_o.T$ and in the layer concerned by diffusion to $C_b \times Z_d$. For typical values encountered in our conditions : T = 4 nm, Z_d = 20 µm and $C_o/C_b = 10^3$, one has $(C_o \times T)/(C_b \times Z_d) = 0.2$. In other words, the quantity of impurity segregated in the oxide layer is of the same order of magnitude as the quantity contained in a bulk reservoir extending to a depth 5000 times larger, i.e., approximatively 20 µm. Within this diffusion zone, the impurity is able to be rehomogeneized and while a concentration gradient may result, it is too low to be revealed significantly by SIMS on a 50 - 100 nm depth scale. For Be and Na, the diffusion along the grain boundaries (Fig. 5) aids in the transfer from the bulk to the surface.

4. CONCLUSION

The analytical performances of Ion Microscopy has allowed the detailed chemical characterization of thin thermal oxide layers covering polycrystalline aluminium sheets. The high dynamic range of the technique has made it possible to observe the tremendous superficial segregation ($10^3 - 10^4$) following heat treatment of alkaline (Li, Na) and alkaline earth (Be, Mg) impurities present in the aluminium bulk at concentrations between 10 and 0.08 µgxg-1. The imaging capability has revealed the heterogeneous distribution of these impurities in the metal near the oxide/metal interface, for instance at the grain boundaries for Be and Na. Such information would be very difficult to obtain with other surface analytical techniques.

The interpretation of SIMS depth profiles has been shown to be delicate. The shape of the profiles depends strongly on the operating conditions (incident angle, energy and chemical of the primary beam, O_2 jet) and on the oxide layer thickness. However, reasonable assumptions and the use of simulation have made it possible to deduce the most probable depth distribution of impurities in the oxide/metal interface region.

By bombarding with labelled $^{18}O_2^+$, the distinction of oxygen brought by implantation and oxygen originally present in the oxide layer could be performed and the mechanism of dynamic oxidation of aluminium be investigated.

REFERENCES

(1) K. Wefers, Aluminium, 57, 722(1981).

(2) M. Textor and R. Grauer, Corrosion Science, 23, 41(1983).

(3) A. Csanady, D. Marton, L. Köver and J. Toth, Aluminium, 58, 280(1982).

(4) K.J. Holub and L.J. Matenzio, Appl. Surf. Sci., 9, 22(1981).

(5) E.B. Bas, X.P. Pan, K.J. Rüegg and F. Stucki, Surf. Sci., 138, 172(1984).

(6) B. Goldstein and J. Dresdner, Surf. Sci., 71, 15(1978).

(7) J.M. Gourgout, in "Secondary Ion Mass Spectrometry-SIMS II", Springer Verlag, Berlin, p 286(1979).

(8) J.M. Lang and F. Degrève, Surf. Interface Anal., 7, 53(1985).

(9) F. Degrève and J.M. Lang, Surf. Interface Anal., to be published.

(10) Tran Minh Duc, private communication (1982).

(11) F. Degrève, J. Microsc. Spectrosc. Electron., 6, 223(1981).

(12) S. Hofmann in "Practical Surface Analysis by Auger and X-Ray Photoelectron Spectroscopy", D. Briggs and M.P. Seah Ed., J. Wiley, Chichester, p 141(1983).

(13) H.W. Werner, Surf. Interface Anal., 4, 1(1982).

(14) C.A. Andersen, in "Microprobe Analysis", C.A. Andersen, Ed., J. Wiley (N.Y.), p 535 and 547 (1972).

(15) J.A. Mc Hugh, in "Methods of Surface Analysis", Elsevier Scientific Company, Amsterdam, p 223(1975).

(16) J.C. Joud, R. Faure, R. Namdar-Irani, C. Pichard and p. Poyet, Mém. Et. Sci. Rev. Métal., 235(1982).

(17) R.K. Lewis, J.M. Morabito and J.C.C. Tsai, Appl. Phys. Lett., 23, 260(1973).

(18) Smithell's Metal Reference Book, E.A. Brandes Ed., 6th Edition, Butterworths, London, p 13-55, 56(1983).

SIMS, SAM AND RBS STUDY OF HIGH DOSE OXYGEN
IMPLANTATION INTO SILICON

W.M. LAU*, P. RATNAM** AND C.A.T. SALAMA**
* Surface Science Western, University of Western Ontario,
London, Ontario, Canada N6A 5B7
** Dept. of Electrical Engineering, University of Toronto,
Toronto, Ontario, Canada M5S 1A4

ABSTRACT

Secondary Ion Mass Spectrometry, Scanning Auger Microscopy, and Rutherford Backscattering Spectroscopy have been used to study a buried oxide structure on silicon formed by high dose implantation. All these surface analytical techniques give useful information about the oxygen distribution in the buried oxide structure. The difficulties in these techniques have also been assessed.

Introduction

High dose oxygen implantation into silicon has been found to be a useful procedure of forming a silicon dioxide thin film under a silicon surface [1-4]. The successful fabrication of high performance metal oxide semiconductor devices based on the buried oxide thus formed has been reported [5-7]. Selective ion implantation of high dose oxygen has also been applied to the formation of oxide patterns for very large scale integrated circuits [8]. The advantage of this latter application, as compared to the conventional local oxidation method, is that the lateral diffusion of oxygen is greatly reduced. For the above applications, it is important to examine the distribution of the implanted oxygen accurately. This paper reports on the advantages and problems of using different surface analytical techniques such as Secondary Ion Mass Spectrometry (SIMS), Scanning Auger Microscopy (SAM), and Rutherford Back-scattering Spectroscopy (RBS) for the characterisation of this buried oxide structure.

Experiment

The SIMS analyses were performed with a CAMECA 3f ion microscope equipped with a cesium ion source and a primary ion mass filter. The pressure of the sample chamber was at about 1×10^{-8} torr. The SAM results were obtained with a Perkin Elmer PHI-600 SAM equipped with a differentially pumped argon ion gun. The SAM and SIMS sputtered depths were measured with a Dektak profilometer. The RBS measurements were done with a 1.6MeV He+ beam.
The buried oxide sample was prepared by implanting 1.5×10^{18} atoms/c.c. of oxygen into silicon at 150keV.

Results and discussion

A. SIMS

A cesium ion source was used in this SIMS study because it is suitable for the oxygen detection. Besides, it gives a stable high beam current and a high sputtering rate. Since the presence of cesium on a sample surface can enhance the negative ion yields of many ion species [9], conventionally negative secondary ions are measured when a cesium primary ion source is used. Fig. 1 shows the SIMS depth profiles of O^-, Si^- and SiO_2^- of the as-implanted sample. The sample surface was coated with about 200Å of gold to minimize charging. The SiO_2^- distribution indicates that there is a layer of silicon with low oxygen content under the thin oxide on the sample surface. The SiO_2^- intensity then gradually increases in depth which shows the incoporation of the implanted oxygen in the silicon substrate. The sharp reduction of all ion intensities after about 300sec of sputtering is due to charging of the buried oxide. The ion intensities recover when the charging is reduced by enough ion etching of the oxide. The oxide-silicon interface is reached at about 800sec of sputtering. The interpretation of this profile is very difficult due to this charging problem.

In the CAMECA 3f SIMS system, negative ions are extracted by the negative potential (4500V) on the sample. This negative potential also prevents electrons from reaching the surface. The insulating surface charges up positively due to the gain of

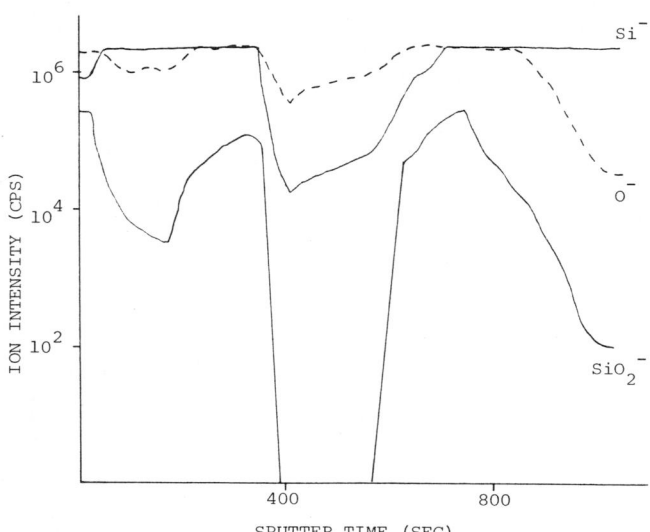

Fig. 1 SIMS profiles of Si^-, O^- and SiO_2^- of an oxygen implanted silicon sample using cesium as the primary ion source

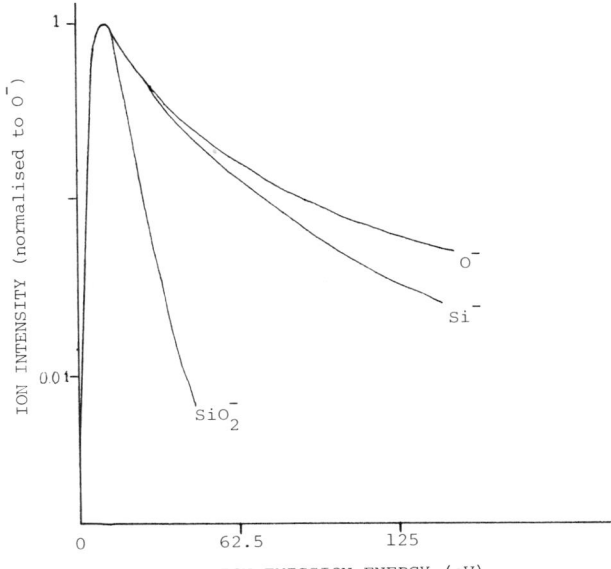

Fig. 2 Energy distribution curves of O^-, Si^- and SiO_2^- of a silicon surface flooded with oxygen under cesium ion bombardment

positive primary ions and loss of secondary electrons. The charging potential reaches a steady state as soon as the leakage current through the sample or sample surface to the conducting sample holder or the presence of some other compensation mechanisms balances out this charging current. The details of these charging and compensation mechanisms have been reported previously [10].

In this SIMS system, secondary ions are energy analysed with an electrostatic analyser (ESA). Under the normal operating condition, ions of near zero emission energy are selected. These ions have about 4500eV of kinetic energy at the ESA due to the extraction voltage. The establishment of a positive charging potential on the sample surface effectively lowers the total extraction potential. Without adjusting the pass energy of the ESA, the detector now measures ions with higher emission energy. The ion intensity drops because of the decreasing population of ions at higher emission energy. Fig. 2 shows the energy distribution curves of O^-, Si^- and SiO_2^- from a silicon surface flooded with oxygen. The different shapes of these three curves clearly explain the fact that due to surface charging the ion intensity reduction in Fig. 1 has the trend of $SiO_2^- \gg Si^- > O^-$. This result indicates that simple ratioing of the ion intensities to a reference ion signal cannot be used as a data correction procedure for surface charging.

Ideally the charging potential can be compensated by applying an offset voltage to the sample. Since energy distribution curves peak at the near zero region, the compensation can be done by scanning the sample voltage (extraction voltage) until the ion intensity is at its maximum level. There is a standard data acquisition procedure in the CAMECA 3f SIMS system for an automatic control of this offset voltage scanning. The application of this procedure has also been reported recently [11]. Fig. 3 shows the depth profile of the same sample measured with this technique. The Si^- ion was used as reference ion species for the voltage scanning. The anomaly due to charging is definitely not completely relieved. The reason is that in practice the automatic offset control only scans a small voltage range (+/- 125V). The actual charging potential near the interface of the buried oxide was measured [10] to be about 300V. The implementation of an automatic offset control routine with a large voltage range is not practical because this will reduce the useful data acquisition time per unit sputter time and hence degrade the depth resolution of the profile. Besides, an adjustment of the ion optics is necessary to compensate for the effect of a high charging potential to the ion trajectory.

Recently the detection of positive secondary ions under cesium ion bombardment has been reported [12,13]. In this detection mode, the positive extraction potential on the sample helps the collection of the secondary electrons from the sample surface and strayed electrons nearby for the compensation of the gain of positive primary ions. An electron flood gun can also be used to

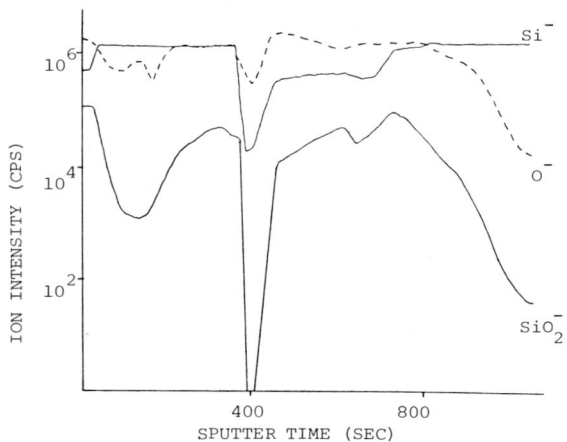

Fig. 3 SIMS profiles of Si^-, O^- and SiO_2^- of an oxygen implanted silicon sample with offset control to compensate the charging effect

neutralize the residual charges. Fig. 4 shows the profiles of Si^+ and O^+ from the sample without gold coating. The oxygen profile agrees well with the calculated profiles reported previously [4]. The profile clearly shows the formation of a buried oxide layer of about 0.15 microns thick beginning at about 0.3 microns from the surface. The asymmetry of the profile shape has also been observed previously [2,4] and been related to the radiation enhanced diffusion of oxygen towards the surface during implantation.

Besides relieving the surface charging problem, it is evident that the detection technique mentioned above also greatly reduces ion intensity due to oxygen background. A detection limit of about 10^{20} atoms/c.c. of oxygen is indicated in Fig. 1 and 3 when O^- is detected. This detection limit reduces to about 10^{18} atoms/c.c. in Fig. 4 when O^+ is monitored. The reason of the low positive ion yield of the adsorbed oxygen is still under investigation.

It is well known that the presence of oxygen can enhance positive ion yields [14]. The enhancement of the Si^+ yield due to the implanted oxygen is evident in Fig. 4. In order to get the oxygen concentration distribution from this depth profile, it is thus relevant to ask if the presence of oxygen enhances the positive ion yield of oxygen. Homma et al. [15] reported that under Ar^+ ion bombardment, the O^+ ion yield is fairly constant from a silicon sample with less than 28% (all percentages are in atomic percents) of oxygen. But the yield increases by a factor of about 16 when the oxygen content increases to 67%. A similar SIMS study was performed using SAM for oxygen concentration calibations. A much smaller self-enhancement of positive ion yield of oxygen was found under Cs^+ ion bombardment. The detailed results will be reported elsewhere.

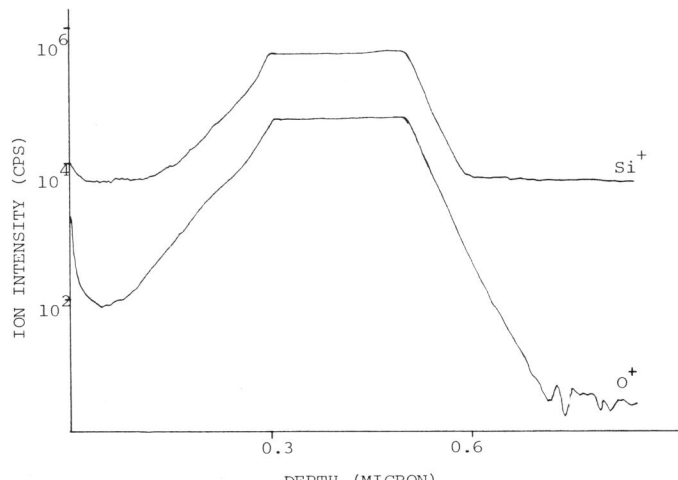

Fig. 4 SIMS depth profiles of Si^+ and O^+ from an oxygen implanted silicon sample using cesium as primary ion source.

The SIMS depth profiles of O^- of a buried oxide under Ar^+ bombardment with an ATOMICA SIMS system have been reported [4]. The ATOMICA system is less susceptible to charging problems compared to the CAMECA system due to the absence of a high voltage on the sample stage. However, a reduction of ion intensity due to charging is still evident in some of these depth profiles. These oxygen depth profiles were obtained by ratioing the O^- signals to the Si^- signals. It is known that the presence of oxygen enhances the Si^- yield as well as the Si^+ yield [14]. There is no report on the self-enhancement of the O^- yield. Hence, this ratioing technique is questionable.

B. SAM

A SAM depth profile of oxygen of the implanted sample is shown in Fig. 5. Since oxygen Auger electron emission is well characterised in silicon oxides, the procedure of getting an

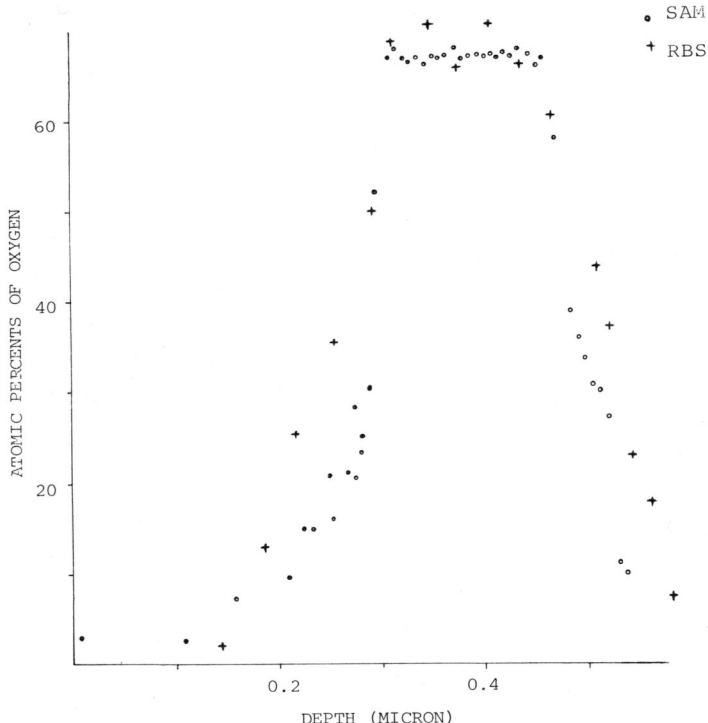

Fig. 5 SAM and RBS depth profile of oxygen implanted in silicon

oxygen profile in this case is routine. A low primary electron beam current density (0.5nA/micron^2) was used, however, to prevent the desorption of oxygen. Another feature of this SAM analysis is that the LMM Auger electron peaks of silicon and silicon dioxide are well separated. The Auger electron spectra of the buried oxide layer unambiguously indicate that the material is actually SiO_2 but not just a physical mixture of oxygen and silicon.

A negative charging potential of about 20V was also observed near the surface silicon - buried oxide interface. Charging, however, is not important in the buried oxide region. If care is not taken, one may get an oxygen profile with a reduction in intensity near this silicon - oxide interface because of the shifting of the oxygen Auger electron peak out of the scanning range in a computer controlled depth profiling procedure. The anomalous abrupt changes in the oxygen Auger depth profiles reported previouly [10, 15] may be due to due this charging effect.

Since the primary electron beam of SAM system can be focused to about 300Å, the SAM technque can be applied to compare the lateral diffusion of oxygen at the edge of an oxide strip formed by local oxidation and by selective ion implantation of oxygen. This idea is demonstrated in Fig. 6 which shows an Auger electron line scan of oxygen across an oxide strip delineated by a silicon nitride mask. The lateral resolution of this line scan is better than 0.1 micron. Hence this technique together with ion beam etching provides useful quantitative information about the extents of lateral diffusion at different depth. The detailed experimental results will be reproted separately.

The practical problem of the SAM analysis is the slow throughput and the limited sensitivity. It took about seven hours for the collection of the data for the depth profile as shown in Fig. 5. The noise background is about 2-3%. A higher beam current or a longer data acquisition time is required to reduce the detection limit.

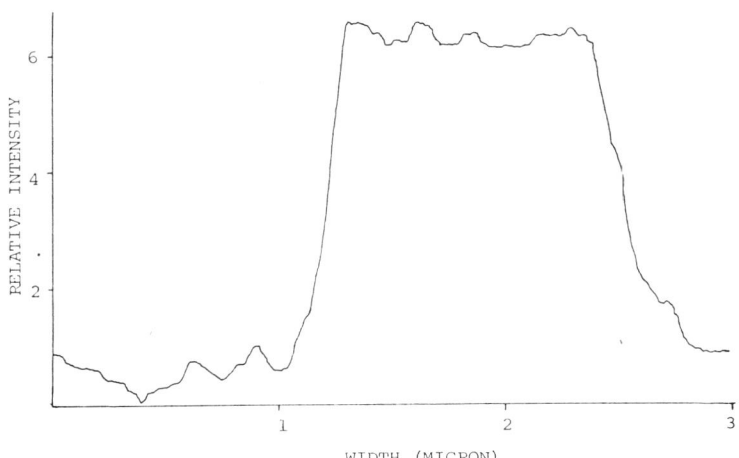

Fig. 6 SAM line scan of oxygen across an oxide strip (1.2 microns)

C. RBS

RBS is another useful technique in the characterisation of a thin film structure. Fig. 7 shows the RBS random spectrum of the buried oxide sample. The reduction of signals near channel 400 is due to the lower atomic density of silicon in the buried oxide. The increase of signals near channel 200 shows the signals directly due to oxygen. The shape of these changes can be used to infer the oxygen distribution. Since the backscattering cross sections and kinematics are well characterised, RBS gives a quantitative results about the oxygen distribution. These results are included in Fig. 5. RBS channelled spectra have also been used to characterise the crystallinity of the silicon film on the buried oxide [4,16]. Similar to SAM, this analysis however has a fairly high detection limit relative to SIMS.

Conclusion

SIMS, SAM and RBS have been used to study a buried oxide structure on silicon formed by high dose implantation. All these surface analytical techniques give useful information about the oxygen distribution in the buried oxide structure. SIMS has the lowest detection limit and highest throughput among these techniques. However, surface charging and matrix dependence of ion yields are often serious problems in SIMS. The detection of positive secondary ions under cesium ion bombardment reduces surface charging and still gives high enough ion yield for the detection of oxygen. It was also found that the signals due to the background oxygen in the SIMS system is greatly reduced under this operation mode as compared to the detection of negative secondary ions. This helps the detection of low concentration of oxygen in the sample. As compared to SIMS, SAM and RBS provide more quantitative results. SAM has the capability of high lateral resolution which will be useful in the study of lateral diffusion of oxygen. However, SAM is the worst technique in terms of throughput.

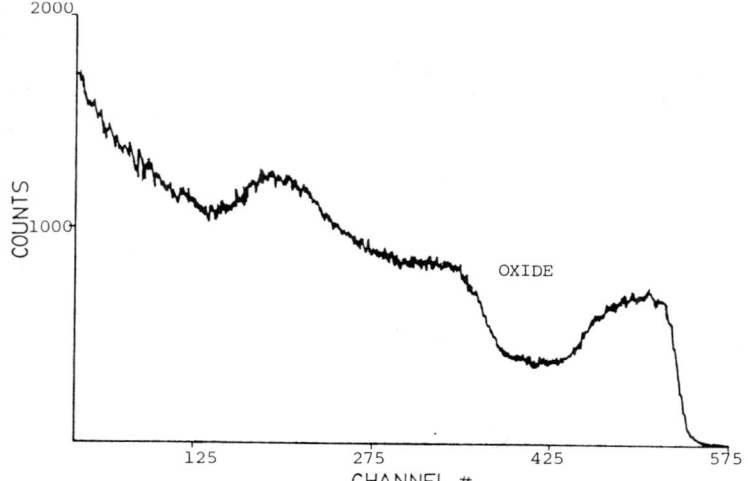

Fig. 7 RBS result of the buried oxide structure

References

1. M.H. Badawi and K.V. Anaud, J. Phys. D, 10(1077)1931.
2. S.S. Gill and I.H. Wilson, Thin Solid Film, 55(1078)435.
3. T. Hayashi, H. Okamoto and Y. Homma, Jpn. J. Appl. Phys., 19 (1980)1005.
4. P.L.F. Hemment, E. Maydell-Ondrusz, K.G. Stevens, J.A. Kilner, and J. Butcher, Vacuum, 34(1984)203.
5. K. Izumi, M. Doken, and H. Ariyoshi, Electron. Lett., 14(1978) 593.
6. H.W. Lam, R.F. Pinizzotto, H.T. Yuan, and D.W. Bellavance, Electron. Lett., 17(1981)356.
7. K. Ohwada, Y. Omura, and E. Sano, IEEE Trans. Electron Devices, ED-28(1981)1084.
8. P. Ratnam and C.A.T. Salama, 2nd Canadian Conf. Semicond. Tech., 1984.
9. M.L. Yu, Phys. Rev. Lett., 40(1978)574.
10. W.M. Lau, N.S. McIntyre, J.B. Metson, D.Cochrane, and J.D. Brown, Surf. Interface Anal., accepted for publication(1985).
11. H. Mutoh and M. Ikeda, SIMS IV, Springer Verlag Series in Chemical Physics #36, 329(1984).
12. W.M. Lau, W. Vandervorst, Canadian J. Phys., accepted for publication (1984).
13. W. Vandervorst, F.R. Shephard, and W.M. Lau, Nucl. Instrum. B, accepted for publication (1985).
14. P. Williams, Surf. Sci., 90(1979)588.
15. Y. Homma, M. Oshima, and T. Hayashi, Jpn. J. Appl. Phys. 21 (1982)890.
16. K. Das, J.B. Butcher, M.C. Wilson, G.R. Booker, D.W. Wellby, P.L.F. Hemment, and K.V. Anand, Inst. Phys. Conf. Ser. 60 (1981)307.

IN SITU ION IMPLANTATION FOR QUANTITATIVE SIMS ANALYSIS

Richard T. Lareau and Peter Williams
Department of Chemistry, Arizona State University
Tempe, AZ 85287

ABSTRACT

The primary ion column of a secondary ion mass spectrometer (Cameca IMS 3f) has been used as an ion implanter to prepare calibrated standards, *in situ*, for quantitative SIMS analysis, with an accuracy better than 10%. The technique has been used to determine oxygen concentrations in contaminated $TiSi_2$ films by implanting a reference level of ^{18}O into a portion of the film.

INTRODUCTION

The major impediment to quantitative analysis using secondary ion mass spectrometry (SIMS) is the difficulty of preparing and calibrating appropriate standards. Standards are necessary due to the extreme sensitivity of sputtered ion yields to the chemistry of the sputtered surface ("matrix effects"), and must in general have very similar, and preferably identical, major element chemistry to the analytical sample. This requirement is fulfilled, and the difficulties of solid-state chemistry overcome, by the use of ion implantation to make standards [1-4]. In principle, implantation allows quantitative incorporation of any species into any substrate. Comparison of the average count rate over the profile integral to the average concentration over the sputtered depth (determined from the implant dose) allows calibration of the instrumental sensitivity for the implant species in the specific matrix sputtered [4]. In the most powerful approach, the implant is superimposed on the intrinsic level to be measured, so that both are measured in one depth profile, with identical ionization efficiency and instrument parameters. It is essential in this approach that the implant be contained entirely in a single matrix, in order that ionization efficiency be constant throughout the depth profile. Despite the power of the technique, implantation as currently practiced suffers from several limitations for SIMS quantification:

1. Implanters are expensive and not universally accessible, particularly if non-standard species (i.e. other than semiconductor dopants) or rare isotopes are to be implanted.

2. Implanters typically operate at rather high beam energies (40-200 keV); implant ranges are such that the implant may not be wholly contained in the outermost layer of a thin film

heterostructure. Implantation at lower energies is possible, but then a major portion of the implant profile may lie within the outer 10-20 nm where ion yields vary strongly as a steady-state level of the primary beam species forms.

3. It is frequently desired to analyze sub-surface films in a thin-film multilayer; because a sizable fraction of the implant dose is retained in the outermost layer (from which it may sputter with a different ionization efficiency) implantation cannot be used directly to quantify such a structure.

We report here a new approach to implant quantitation, using the primary ion column of a secondary ion microanalyzer (Cameca IMS 3f) as an *in situ*, low-energy ion implanter. This approach is shown to be quantitative, convenient, and largely unlimited in the range of elements and isotopes accessible. In addition, the approach allows quantification at any depth in a multilayer structure, and quantification of thin layers (< 100nm).

In situ ion implantation has previously been used by Wittmaack [5] to generate oxygen implants in a SIMS system in a fundamental study of ionization phenomena. The ion microanalyzer used in the present study has considerably improved capabilities over that used by Wittmaack. In particular, the mass resolving power of the primary ion column can be as high as 300. The instrument has been described by Gourgout [6] and the primary mass filter design by LeGoux and Migeon [7]. In particular, the mass filter is designed to accommodate two ion sources simultaneously, so that implantation using one source can rapidly be followed by analysis using the other.

To demonstrate the approach, we chose to analyze oxygen in $TiSi_2$ layers which had previously been analyzed using Auger electron spectroscopy (AES). These samples presented numerous challenges to a conventional implant standard approach. First, the films were rather thin (~ 200 nm), and all had an extended surface oxide layer (some tens of nm). Thus implantation into the original film would deposit an appreciable fraction of the implant in this surface oxide layer, from which it would be sputtered with different (and unquantifiable) ionization efficiency. Second, the oxygen level in some films was thought to be as high as 5 atom %. Superimposition on this level of an ^{16}O implant sufficiently high to be measured above the background would itself alter ion yields; thus, the use of an ^{18}O implant was called for [8], which would be both expensive and inconvenient in a standard implanter.

EXPERIMENTAL

The duoplasmatron ion source of the IMS 3f typically generated up to 1μA of $^{16}O^-$, and the beam therefore contained ~ 2 nA of $^{18}O^-$. Because the analyzed area is typically only ~ 65 μm in diameter (within a 250 x 250 μm² primary beam raster), implantation into a 500 x 500 μm² area was sufficient. In this area, a 2 nA beam implants a dose of ~ 1 X 10^{15} atom·cm^{-2} in only ~ 7 minutes. Thus expensive separated isotopes were not required. Implantation was carried out at an impact energy of 19.5 keV, with the primary ion source at -15 KeV and the sample at +4.5 keV. Negative ion implantation was chosen in order to use the highest atomic ion flux; $^{16}O^+$ constitutes only a 10% component (~ 100 nA) of the positive ion beam. Because the ions impact at about 30° to the surface normal, the range measured normal to the surface corresponds to an effective energy of 16.9 keV. The implanted current was measured in a small Faraday cup built into the sample holder. Because the sample was at elevated potential during implantation, the current was measured before and after implantation. Usually beam stability over the implantation period was better then 3% so that the dose could be calculated from the current and implantation time. Calculation of the dose requires a knowledge of the implanted area; this was set by the primary beam raster calibration, and checked using the secondary ion image of a calibrated test grid, and by measuring sputter craters in an optical microscope and/or by scanning electron microscopy. Because the beam size was rather large (~ 100 μm) only the center of the implanted region received a uniform dose; nevertheless, the dose in this region is precisely given by the ratio of the implanted charge to the nominal rastered area [9]. Because the sample can be precisely aligned with the secondary optics of the ion microscope, location of the center of the implanted area was straightforward even after moving to an unimplanted region to optimize and center the analyzing beam.

The analyzing beam used in the present case was Cs^+, at an impact energy of 12 keV. Cesium was chosen to reduce the signal from background oxygen present in the sample chamber. With the duoplasmatron in operation, the oxygen partial pressure in the sample chamber rises to ~ 1 x 10^{-7}; titanium getters oxygen very efficiently, so that even though $^{16}O_2^+$ could have been used as an analyzing beam, the resulting oxygen background signal was unacceptably high. The cesium beam is oxygen-free and also allows faster erosion rates to reduce the effective oxygen coverage of the sample surface still further. The cesium ion source was left permanently on; switching from implantation to analysis typically took ~ 10 minutes.

RESULTS & DISCUSSION

In order to obtain an estimate of the accuracy of this technique, we have obtained an externally implanted standard of known ^{16}O dose (3.7 x 10^{15} atoms/cm^2 at 70 keV) and compared it to an ^{18}O *in situ* implant (with a dose of 1.3 x 10^{15} atoms/cm^2 at 16.9 keV). The calculated dose refered to *in situ* implant is 4.0 x 10^{15} atoms/cm^2. Therefore the approximate accuracy of the technique appears better than 10%, indicating that our dose calibration is correct, and, in addition that isotope discrimination in the ion emission process is not serious in this case.

A depth profile of an as-received TiSi$_2$ sample with the lowest oxygen level is shown in Figure 1. It can be seen that the surface oxide layer, together with an unexpected high oxygen level near the substrate interface, due apparently to gettering of oxygen by Ti during deposition, leaves only a thin buried layer ~ 100 nm thick with constant low oxygen content. In Figure 2, this sample is shown with a low-energy oxygen implant superimposed; clearly, much of the implant profile is superimposed on the surface oxide, making determination of the implant profile alone quite inaccurate. Figure 3 shows the solution to this problem allowed by *in situ* implantation. This profile was obtained by presputtering the film with the analytical Cs$^+$ beam until the surface oxide was removed. The presputtered region was then implanted and analyzed. This process also allows a steady-state level of Cs$^+$ to be achieved in the surface prior to implantation, so that analysis begins <u>immediately</u> with this level achieved and ion yields are constant throughout the profile. Adsorption of a monolayer of oxygen apparently occurred during the implantation, but this produces a negligible perturbation compared with the surface oxide effects.

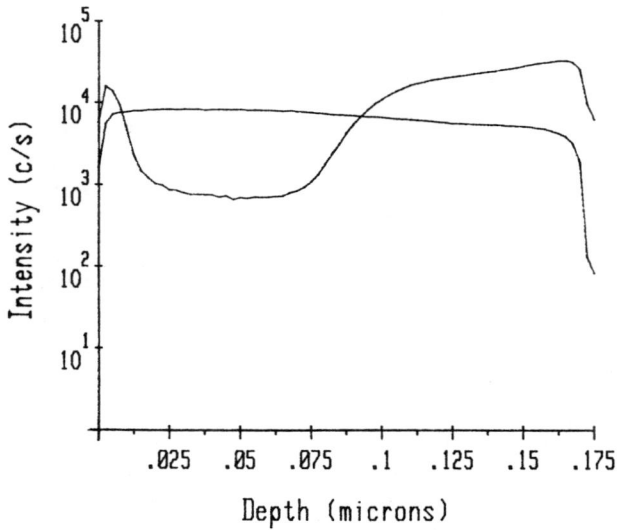

Figure 1: A depth profile of an as-received TiSi$_2$ sample with the lowest oxygen level.

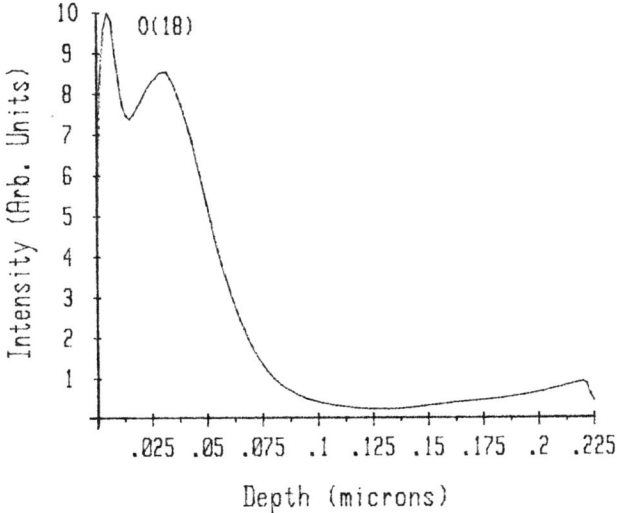

Figure 2: A low-energy ^{18}O implant superimposed on the surface oxide.

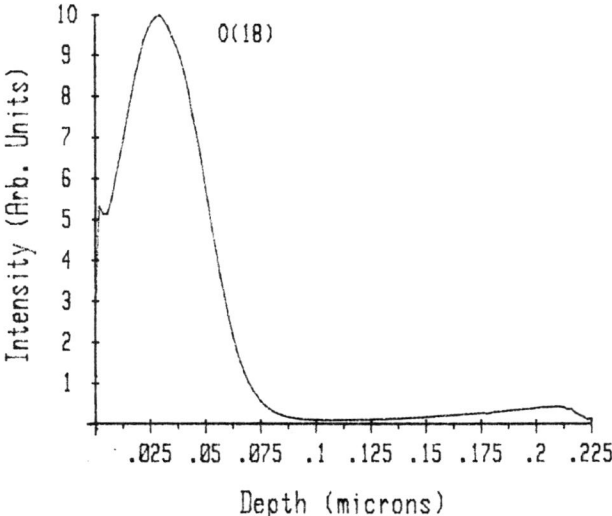

Figure 3: A ^{18}O implant profile obtained after presputtering off the surface oxide layer.

Because all samples showed an instrument-related oxygen background, profiles were obtained at increasing erosion rates until a constant minimum oxygen level was achieved in the center of the film; this was taken to be the intrinsic oxygen level. Even though the films were appreciably thicker than the oxygen projected range (~ 35 nm) it was not possible to use the superimposed implant approach here, because the tail of the implant profile invariably swamped the background level. Instead, an implanted and unimplanted region were compared under optimum conditions for both (high oxygen dose for the implant, with moderate erosion rate; fastest sputtering rate for the unimplanted region). The oxygen signals in both cases were normalized to the silicon signal. Notice that in order for the oxygen background to contribute negligible error to the implant integral, the peak implant concentration was made ~ 100 x the background level. This is only possible with ^{18}O; implantation of such a high level of ^{16}O would severely alter ion yields and render the analysis useless.

The results of the analysis are listed in Table I, together with the results of analyses using AES. The second AES analysis was undertaken after the SIMS results disagreed with the AES analysis supplied with one of the samples. It can be seen that the second analysis is in acceptable agreement with the SIMS results. The major source of error in the SIMS analyses in the present instances lies in the estimate of the raster size. Because the raster drives were in the process of modification during the time this work was performed, internal calibrations were occasionally in error, and external measurements of sputtered crater sizes was needed. We estimate the error in these area measurements to be ~ 5%. Errors in current and crater depth measurements are estimated to be much less than 10%. Rastering of the sample under a stationary beam, with using computer controlled stepping motors promises to be a much more accurate means to achieving controlled raster sizes (and will allow raster sizes larger than the maximum - 500 μ^2 - given by electrical deflection of the beam).

TABLE I (TiSi2 RESULTS)

SAMPLE #	%OXYGEN AES (1)	%OXYGEN AES (2)	%OXYGEN SIMS
1	1%	-	1.2%
2	5%	1%	0.7% 0.8%
3	1%	-	0.8% 0.5%
4	5%	-	7% 7%

CONCLUSION

The *in situ* implantation technique described here offers several major advantages for quantitative SIMS analysis. First, it frees the SIMS analyst from dependence on an external ion implanter. Rare isotopes can be implanted without major expense, and elements detrimental to semiconductor devices (e.g. alkali metals) can be implanted. Second, it offers the possibility of quantifying very thin layers -- < 100 nm -- which need not be surface layers. A wide variety of as-received samples can therefore be analyzed without resorting to special sample preparation. Given appropriate ion sources, any element or stable isotope can be implanted; the low current requirement makes ion source design simple and generally obviates any need for expensive separated isotopes. This appears to be a significant advantage for analysis in minerals and ceramic materials, where frequently the analytical species is not readily available in semiconductor implanters.

This work was supported by the National Science Foundation, Grant DMR 8206028. We gratefully acknowledge the assistance of Mike Warburton (Motorola Inc.) for performing the repeat AES analysis, and Charles Magee (RCA) for the external ion implant standards.

References

[1] Grittins, R.P., Morgan, D.V., Dearnaley, G., *J. Phys.* 5, 1654-63 (1972).
[2] Croset, M., Dieumegard, D., *Corros. Sci.* 16, 703-15 (1976).
[3] Williams, P., *IEEE Trans. Nucl. Sci.* 26, 1809-11 (1979).
[4] Leta, D., Morrison, G.H., *Anal. Chem.* 52, 277-80 (1980).
[5] Wittmaack, K., *Surf. Sci.* 112, 168-180 (1981).
[6] Gourgout, J.-M., Secondary Ion Mass Spectrometry II, ed. by A. Benninghoven, C.A. Evans, Jr., R.A. Powell, R. Shimuzu, H.A. Storms, Springer Series in Chemical Physics, Vol. 9 (Springer, Berlin, Heidelberg, New York, 1979) p. 286-290.
[7] Le Goux, J.J., Migeon, H.N., Secondary Ion Mass Spectrometry III, ed. by A. Benninghoven, C.A. Evans, Jr., R.A. Powell, R. Shimizu, H.A. Storms, Springer Series in Chemical Physics, Vol. 19 (Springer, Berlin, Heidelberg, New York, 1981) p. 52-56.
[8] Williams, P., Stika, K.M., Davies, A., Jackman, T.E., *Nucl. Inst. Meth.* 218, 299-302 (1983).
[9] Any point in the central region of the raster over which the beam passes completely receives a dose which is the integral of the current over the beam profile. For a given total current in the beam, this integral is independent of the beam diameter.

SECONDARY ION MASS SPECTROSCOPY OF CERAMICS

JENIFER A.T. TAYLOR, PAUL F. JOHNSON and VASANTHA R.W. AMARAKOON
New York State College of Ceramics at Alfred University, Alfred, NY 14802

ABSTRACT

Application of SIMS to ceramics is a complicated but rewarding technique for characterization. The variable composition, hardness and insulating nature of these materials render spectra interpretation complex. Basic data reduction procedures from graphical data collection are presented along with spectra from SIMS applied to a glass frit, magnesium sialon, silicon and PTCR barium titanate.

INTRODUCTION

SIMS shows great promise for characterization of ceramics. However, the hardness, variable composition and insulating nature of most of these materials increases the complexity of spectra interpretation. This paper is intended to serve as an introduction for analysts interested in using SIMS on ceramics. The spectra are not to be considered examples of perfect data, whatever that is, but rather as real data collected in the interest of learning more about the instrument and the material at hand. Basic techniques of spectra interpretation are described to illustrate the application of SIMS to this varied group of material. The major advantage of SIMS over other surface analysis techniques is elemental sensitivity; all elements including H can be detected; isotopes can be resolved easily; and instrument parameters, such as the primary ion beam composition and energy, and mode of data collection can be varied to maximize the information represented by the spectra. Sample preparation is relatively simple but all mounting must be done with material that resists decomposition in an ultra-high vacuum while being bombarded by energetic ions. The primary disadvantage of SIMS is the difficulty of interpreting the spectra quantitatively, which is due to the large variation in secondary ion yields.

APPLICATION OF SIMS TO CERAMICS

This paper is a presentation of data from SIMS applied to a number

of different ceramic materials, illustrating some of the techniques used to
interpret spectra as well as the problems involved. All but one of the
analyses were performed using a Leybold-Hereaus Energy Analyzer 10/100 SIMS
which has a minimum beam size of 0.8 mm FWHM. * The raw data were
collected as positive ion yield on an x-y plotter.

Characterizing a Glassy Layer on a Silver Conducting Pad

SIMS of a glassy phase that serves as an insulating layer between
conductive pathways on an alumina substrate proved a rapid and informative
analysis technique. The two samples being compared were supposed to be the
same except for the color in the frit mixed with the silver paste. On
firing, the glassy phase separates from the electroding phase, forming an
insulation between layers of conductive pathways. One of these frits,
which were proprietary preparations being used in a thesis project, proved
to be quite unsatisfactory. The blue frit performed as expected but the
white frit was not wet by solder. SIMS with a low energy beam to aviod
artifacts due to charging or knock-in generated the spectra in Figure 1
which show some interesting differences. The white specimen had strong
titanium and barium peaks while the blue had none and the relative
intensity of aluminum and silica peaks changed. With this information the
composition of the white frit could be altered until adhesion of the solder
was satisfactory. [1]

Analyzing Spectra from Insulators

Sample charging is a serious problem when analyzing insulators, partly
because the effects are not obvious. By acquiring a positive charge, the
sample can deflect the beam, reducing the intensity of detected signal
drastically. The three spectra in Figure 2 are for a magnesium silica
alumina oxynitride containing 2.7% nitrogen. The difference in intensity
between A and B is due entirely to charging of the insulating sample.
Spectra A was flooded with low energy electrons to compensate for the
positive surface charge which developed on this sample after less than a
half minute of bombardment by a 2 keV ion beam. Under a 4 keV beam, the
spectra disappeared completely.

Slight charging may suppress ion clusters peaks more severely than
elemental ions, because molecular ions are generally not as energetic as

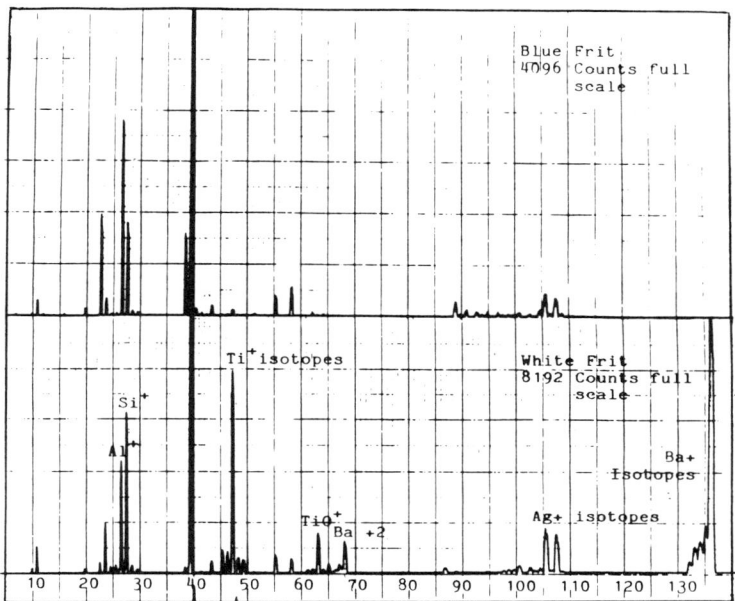

Figure 1. Spectra of a glassy layer on a silver conduction pad, showing the difference in composition between two colors (2ke V Ar$^+$)

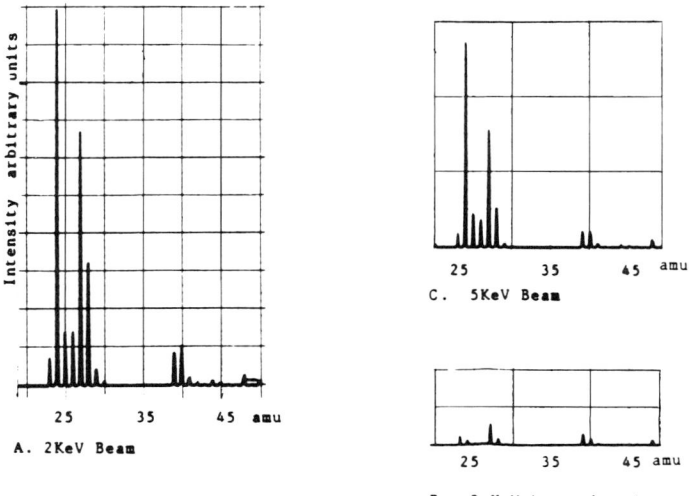

Figure 2. Effect of ion beam energy and flood gun

atomic ions. [2] Such nonlinear suppression may change the relative peak intensities, perhaps facilitating the identification of peaks but the likelihood of an unstable field make interpretation of spectra from charging surfaces a tricky technique. Fluctuations in the ion beam due to unstable surface charging are less easily noticed and can cause misleading peak intensities.

Frequently checking for consistency of major peak heights, such as silicon in this case, can prevent the collection of misleading information. Unstable charging and instrumental variations become apparent if the peak intensity of a matrix element is noted at the beginning, middle and end of each measuring sequence since the concentration of major constituents is not expected to change. For careful work, spectra in which this matrix peak varies by more than 10% from the reference values should be discarded. [3]

Field-enhanced migration of ions released inside a solid insulator can cause misleading depth profiles; a phenomenon aggravated by accelerated diffusion in ion beam damaged lattices. [2] If a specimen containing such mobile ions is subjected to charging the possibility that the surface composition has been changed must be considered. In a test of this scenario, SIMS analysis of the glass represented by spectra A showed no change in peak intensity ratios after being charged for five minutes by irradiation with a 4 keV Ar^+ ion beam. The degree of charging will be a function of the time elapsed and the energy of the primary beam as well as the nature of the sample.

Locating evidence of nitrogen in spectra A of Figure 2 requires careful consideration of each peak. No peaks appear at this intensity above or below those shown. Elemental nitrogen is usually considered too reactive to occur so the absence of a peak at 14 amu is not conclusive. The ratio of peaks at 28, 29, and 30 amu does not correspond to the natural abundance of silicon, which would be 92:5:3 rather than 90:10:3 as in Spectrum B. If the 28 amu peak is entirely due to Si^+, the 29 amu peak is twice as intense as expected. This peak could be N_2-H^+ or $Si(28)-H^+$. Presence of nitrogen as N_2^+ or $Al-H^+$ could also be responsible for part of the peak at 28^+. SiN^+ would be expected at 42 amu, either positive or negative, but no peaks occur. Again, the absence of a cluster ion peak is not conclusive negative evidence. AlN^+ and an isotope of potassium are both found at 41 amu. To check the ratio of K(39) and K(41), another spectrum should be collected at fewer counts full scale. The negative spectra shows no peak that is clearly nitrogen or a nitride. The next step could be to use cesium as a primary beam because nitrogen has a higher cross section for this ion than for argon or oxygen.

Spectrum C in Figure 2 is an example of the effect the energy of the primary beam has on the data collected. This spectrum is the information obtained with a 5 keV primary beam from the same glass represented by spectrum A at 2 keV, both with a flood gun to prevent charging. The major difference is an increase in the 24 to 28 amu peak intensity ratio, from 3.1 to 5.3, possibly due to knock-in of some silicon or suppression of N_2^+. The 24/27 amu peak intensity ratio did not change at all.

Effect of Ion Beam Composition

Comparing spectra from different instruments requires consideration of the variables involved. Changing from argon to oxygen for a primary beam can cause peak intensity ratios to change and different peaks to appear. Figure 3 shows two spectra from the same used silicon wafer as created by primary beams of different composition. Note the difference in the relative intensity of the dominant peaks. Theoretically, more intense oxide peaks should appear in Spectra B from the oxygen primary beam than Spectra A, from the argon beam. The latter should show more significant elemental peaks. Spectra A has a measurable peak at 14 amu which is probably Si^{+2}, while B does not. The SiO^+ peak at 44 amu is about the same relative intensity with respect to Si^+, in both spectra; probably because the existing surface oxidation of the wafer is of much greater magnitude than the oxides due to the oxygen primary beam. The greater intensity of most major peaks in B supports the thesis that an oxygen beam causes higher positive ion yield.

The negative spectra can often provide clues to complement the positive (or vice versa). In this case there were no significant differences beween the negative spectra for argon and that for oxygen primary beams. Sander et al found spectrum for 5 keV argon beam to have dominant peaks at Si^+, Si_2^+, Si_3^+, and Si_4^+ while a 1.2 keV oxygen beam had dominant peaks at Si^-, Si_2^-, and three oxides in the positive spectrum. This work implies that differences between Si spectra were a function of primary beam composition and exposure to gaseous oxygen during bombardment by either argon or oxygen but not beam energy, up to 5 keV. Implantation kinetics could not predict the concentration of near surface oxygen that developed, which seemed to be completely controlled by the reactivity resulting from the damage introduced by sputtering. [4] Careful studies of spectra resulting from different primary beams could become an important

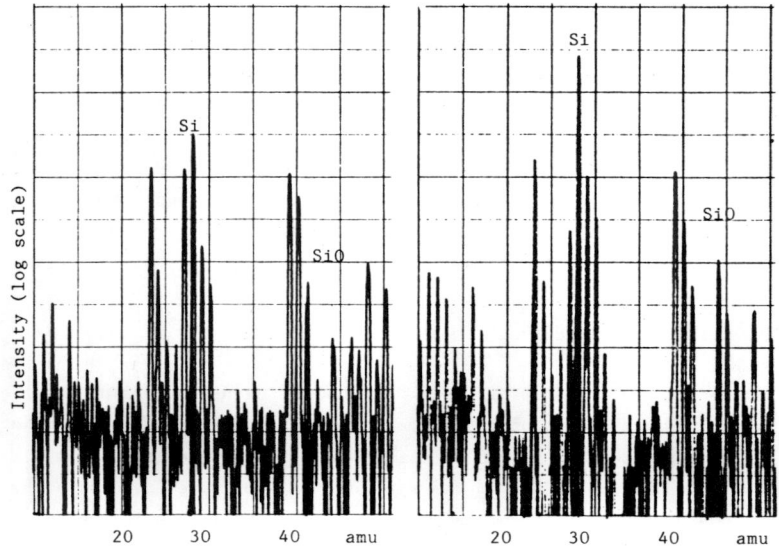

A. Argon primary beam   B. Oxygen primary beam

Figure 3. Effect of ion beam composition on silicon spectrum

source of information about the material being characterized.

Detecting elements from a previous analyses is a common occurrence. The peak at 48 amu in both Spectra A and B is left over from sputtering barium titanate. Such memory effect can be controlled by keeping careful records of the uses of the instrument and running standards to establish the intensity and time dependency of such spurious peaks.

Characterization of Barium Titanate Grain Boundaries

Amarakoon used SIMS to investigate grain boundary compositon in PTCR barium titanate, examining the same basic composition with different additives under identical sputtering conditions. [5] Figure 4 shows the elemental profile for two compositions, whose intended constituents are identical except for 0.2 atom % yttrium, but whose room temperature resistivity differed by a factor of a thousand. The data represented by the composition profile were collected from five spots on an intragranular fracture surface. The major difference between the two profiles is the

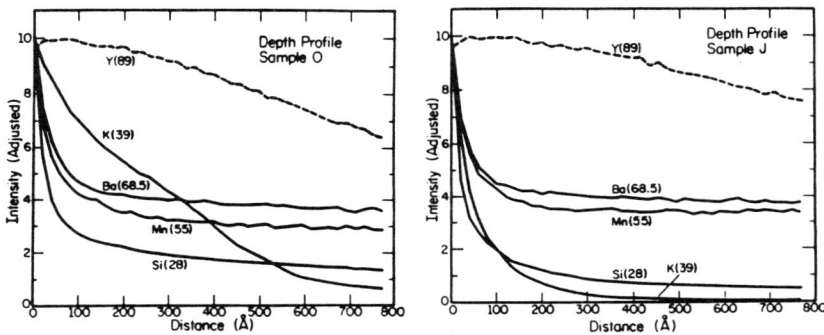

Figure 4. Spectra for similar PTCR barium titanate compositions
(Cs^+ primary ion beam at 9.6 KeV)

distribution of potassium, a contaminant, which has segregated to a layer less than 30 nm thick in sample J and more than 50 nm in sample O.

Potassium acts as an acceptor in PTCR barium titanate so its presence can increase the room temperature resistivity. Using interferometry to measure crater depth, the sputtering rate was determined to be about 0.2 nm/sec. The intensity axis is intended to show only the change in concentration with distance from the grain boundary, not relative intensity, as all profiles were arbitrarily set at ten.

This work is a good example of depth profiling with an ion beam to investigate succeeding layers of the specimen surface. Since all samples were of similar composition, any preferential sputtering or matrix effect should be constant. %

Even qualitative data reduction requires interpretation of spectra that can have mass interference such as that discussed above, between ion clusters and single ions. The detector for a SIMS unit usually reports a charge to mass ratio, with integer resolution. At forty amu, the detector could be reporting magnesium oxide, implanted argon, calcium or an ion cluster of mass 80 and charge +2. Data about expected yields of various oxides are available to facilitate interpreting spectra but material variables can dominate. [6] Separating the probables from the improbables requires checking for isotopes, comparing intensities of related clusters such as the monoxide or dioxide, looking at spectra from standards similar in composition, and increasing the resolution to discriminate between fragments, hydrides and ions. [7] Clusters of ions can form above the surface even when they do not occur in adjacent sites on the surface. [8] Any time hydrogen is present, hydrides of major constituents can appear.

As the examples above illustrate, use of SIMS as an analytical tool for ceramics has great potential. Round robin analyses of the same material by different laboratories are a good solution to the problem of relating data in the literature to individual instrument characteristics. A round robin study is currently being performed on a metal glass. The ASTM Committee E-42 is active in coordinating such cross laboratory work. @

FOOTNOTES

* Leybold-Heraeus Vacuum Products Inc. LAS Group, 5700 Mellon Road, Export, PA 15632 USA Bonner Strasse 504. Postfach 510760 D-5000 Koln 51 FDR.
% Refer to the paper on depth profiling of ceramics elsewhere in this volume for more examples of this technique.
@ C.D. Powell, Chairman ASTM Committee-42 on Surface Analysis, 1916 Race St. Philadelphia, PA 19103.

REFERENCES

1. J.L. Hollenbeck, Unpublished BS Thesis, 1984, New York State College of Ceramics at Alfred University.

2. H.W. Werner, and A.E. Morgan, J Appl. Phys. 47, 1232 (1976).

3. D.E. Newbury, QUANTITATIVE SURFACE ANALYSIS OF MATERIAL ASTM STP 643, N.S. McIntyre, Ed. (ASTM, 1978).

4. P. Sander, U. Kaiser, R. Jede, D. Lipensky, O. Ganschow and A. Benninghoven, Paper E42WeA02 from American Vacuum Society 31st National Symposium Reno, Nevada 1984.

5. V.R.W. Amarakoon, Ph.D Thesis 1984, University of Illinois at Urbana-Champaign (Univ. Microfilms Int. 8422010).

6. W.L. Baun, TECH REP AFML-TR-79-4123 National Technical Information Service, June 1979.

7. S.J.B. Reed, Scanning 3, 119 (1980).

8. G.K. Wehner, METHODS OF SURFACE ANALYSIS, A.W. Czanderna, Ed. (Elsevier Scientific Pub., NY, 1975).

SIMS ANALYSIS OF PURE AND HYDRATED CEMENTS

ERICH NAEGELE AND ULRICH SCHNEIDER
University Gesamthochschule Kassel, Fachbereich Bauingenieurwesen
Mönchebergstrasse 7, D - 3500 Kassel, Federal Republic of Germany

ABSTRACT

SIMS-measurements permit a direct determination of the hydration depth of cement particles from the depth profiles of various elements. Furthermore, surfactants containing an aromatic ring system can be analysed qualitatively and quantitatively using the SIMS-intensities of characteristic aromatic fragments. In addition, a new model for the evaluation of SIMS-spectra of oxidic materials is proposed.

1. INTRODUCTION

The hydration reaction of cement is of fundamental importance in the field of cement and concrete technology. A detailed knowledge of the mechanisms of this very complex reaction system may result in a more sophisticated concrete-mix design respectively.

Amongst the many analytical techniques, which have been applied to study the chemistry of cement, SIMS (Secondary Ion Mass Spectrometry) offers a number of unique advantages. An extremely high depth resolution can be combined with a high accuracy and the possibility of performing comparatively large surface area measurements. Thus it is possible to trace the reaction progress into the interior of the cement particles and to determine and analyse materials which are adsorbed on the cement and its hydration products.

It is the purpose of this paper to show how the hydration reaction proceeds into the interior of a cement particle and how adsorbed species can be detected by SIMS-measurements. This type of analysis comprises also e.g. chemical admixtures and soluble components in the concrete water.

2. EXPERIMENTAL METHODS AND MATERIALS USED

2.1 SIMS

The SIMS-measurements were performed using the Balzers-SIMS-apparatus of the Dornier Systems GmbH laboratories, Friedrichshafen, FRG.

The primary ion-beam were 3 kV Ar^+-ions at a pressure of $8 \cdot 10^{-8}$ Torr and conventional full scale mass spectra were recorded separately for anions and cations, using a Balzers quadrupole mass spectrometer.

As a complete physical theory of the SIMS-process is not at hand, the spectra have been evaluated using a well-established procedure, based on systematic experimental studies on oxidic materials /1-3/. The model is based on three assumptions /3/:

1) The existance of a nonadiabatic dissociation of nascent ion-molecules outside of the surface potential barrier,

2) different energy distribution for surface atoms in homogenous and inhomogenous matrices,

3) the ability of becoming an ion latently exists already in the center-of-mass systems of the particles.

Fig. 1 shows the comparison of experimental data with the model calculations for metal oxides.

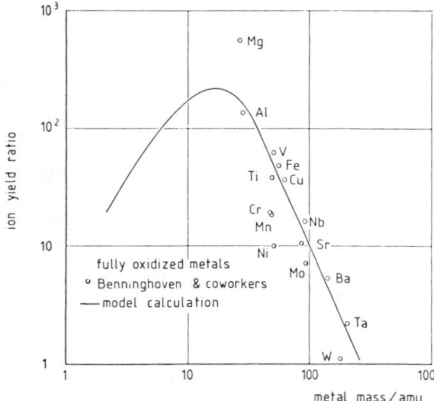

Fig. 1:

Theoretical and experimental ion yield ratios as a function of cation mass for fully oxidized metals /1-3/

During the investigations it turned out that metal element concentrations in an oxidic matrix can be determined quantitatively using the peaks of positive metal and metal oxide ions. At present a determination of the anion concentrations is not possible because similar systematic investigations are not available.

However the emitted molecular particles are always an indication of the types of bonds and of the types of neighbouring atoms. Using molekular peaks of the SIMS-spectra, a determination of the oxidation state is possible.

2.2 Material and Sample Preparation, Test Procedure

A German portland cement PZ 35 F and a German blast furnace slag cement HOZ 35 both with a specified strength of 35 N/mm² after 28 days were selected for the experiments. In addition, a German fly ash approved for concrete-making was included into the research program.
The chemical analyses of the materials are given in Table I. Concentrations of the elements were determined for the first and second monolayers of the pure cements to detect surface hydration of the cement particles during storage. Prior to the SIMS-analysis the cements were freeze-dried. Then samples of both cements were hydrated 10 minutes under water at a water/cement-ratio of 2. The fly ash was hydrated in saturated $Ca(OH)_2$-solution instead of water, to simulate the conditions prevailing in real concretes. Then the hydration was stopped by adding a minimum quantity of sodium dodecylsulphate, and the samples were filtered and freeze-dried before the SIMS-analysis. The freeze-drying is very effective in preventing the cement from further hydration as it is well known that cement hydration ceases completely at temperatures below 4°C /5/.
Further prisms of hardened portland cement paste were made with a water cement ratio of 0,5 according to German standards. Half of the prisms were cured 3 days under water and then stored up to 250 days in air to simulate real conditions. The other half was stored under water for the same period of time. After 250 days alls samples were freeze-dried and analysed.
For all hydrated samples depth profiles were recorded down to a depth of 80 nm from the surface.

Table I: Chemical composition of the cements and the fly ash

	PZ 35 F wt-%	HOZ 35 L wt-%	Fly ash wt-%
Weight loss at 1000°C	1,62	0,7	3,1
CaO	63,20	48,3	1,8
SiO_2	18,40	28,3	52,5
Al_2O_3	3,94	13,1	27,5
Fe_2O_3	3,91	1,5	6,9
MgO	2,40	3,5	2,9
Mn_2O_3	0,39	0,44	-
SO_3	3,08	1,7	0,35
S	-	0,6	-
Na_2O	0,29	0,36	0,45
K_2O	1,61	0,71	4,1
Insoluble Residue	0,73	0,8	
	99,57	100,01	99,60

In a parallel series of tests, several surfactants with different molecular structures were adsorbed onto the portland cement from an aqueous solution of 10^{-4} Mol/l surfactant. The samples were then filtered, freeze-dried and analysed subsequently.

Finally, dodecylbenzenesulfonate was adsorbed on portland cement, the concentrations varying from 0,0028 mMol/g up to 0,83 mMol/g, which is equivalent to 5% up to 300% of a monolayer of surfactant on the surface.

3. RESULTS AND DISCUSSION

3.1 Pure Cements

In Fig. 2 a typical cation SIMS-spectrum of PZ 35 F is shown.

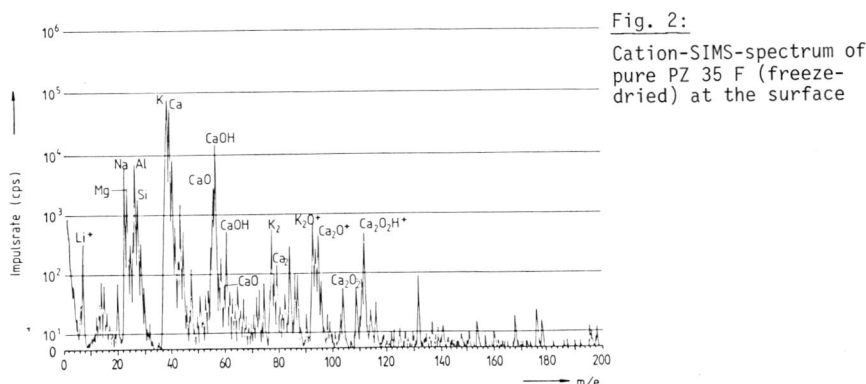

Fig. 2:

Cation-SIMS-spectrum of pure PZ 35 F (freeze-dried) at the surface

Table II gives the concentrations of most of the detected metal elements in portland cement in ion-percent at the surface and in a depth of 0,5 nm from the surface and the theoretical data calculated from the volumetric analysis given in Table I, column 1, assuming homogenous particles.

Elements found in minor quantities are Li and Cs. The concentrations found for Na, Mg, Fe and Al roughly correspond to the data obtained from a volumetric analysis, hence indicating a homogenous distribution of these elements in unhydrated portland cement.

The low concentration of Si in Table II is due to the fact, that Si is emitted as SiO_2^-- and SiO_3^--particles in the SIMS-process. These anionic species cannot be included in the quantitative evaluation procedure, as has been discussed in chapter 2.1. Therefore, the data in Table II represent only the small fraction of Si, that is emitted as Si^+-ions. The true concentration of Si in the cement particles is much higher than the values given in Table II.

The concentration of K^+-ions in the outer layers and their concentration gradient is very high, as can be seen from the values measured immediately at the surface and roughly one atomic layer deeper. Therefore, the enrichment of K^+-ions is limited to about the first seven atomic layers of the cement particles, as can be estimated from the data obtained.

Table II: Composition of the first and second monolayer of portland cement in percent of cations

Element	Experimental Data		Theoretical Values
	Surface	0,5 nm	
Ca	42,8	59,2	67,6
K	44,2	25,6	2,05
Al	5,0	6,9	4,62
Fe	3,5	3,0	2,92
Mg	1,9	2,6	3,59
Na	0,8	0,4	0,56
Si	1,2	2,1	18,36

The high concentration of K in the outer layers of the cement particles agrees well with the data on K^+-concentration in portland cement paste pore solutions reported by Lopez-Flores /4/, who found very high K^+-concentrations up to 0,24 n after one hour of hydration. The concentration did not vary much with the hydration time. After 24 hours the K^+-concentration was 0,26 n. From this it can be concluded, that the K^+-ions dissolve very quickly when the cement comes into contact with water, resulting in a high initial K^+-concentration in the pore solution which remains nearly constant when the further supply of K^+-ions from the cement particles ceases.

The results obtained for blast furnace slag cement are more or less the same as those found for portland cement, except that Ba and Ti, which originate from the blast furnace slag in the blast furnace slag cement are found also in significant quantities.

The results obtained suggest the following type of composition in an unhydrated cement particle:

At the surface alkali ions, especially K^+-ions are drastically enriched compared to the average content measured by volume. In the case of blast furnace slag cement an enrichment of Fe occurs also, which can be explained by the formation of the slag in the blast furnace /5/.

- The hydrates generated by moisture in the atmosphere consist of $Ca(OH)_2$ and KOH but not of $NaOH$. This is supported by the presence of a strong $CaOH^+$-peak in the cation spectrum and a corresponding very strong OH^--peak in the anion spectrum.

- All cations were found in an oxidic environment (partially hydrated at the surface), which contains minor quantities of sulphate-ions, as SO_3^-- and SO_4^--species have also been detected in significant quantities in the anion spectra.

- An analysis of the types of emitted species provides insight into the chemical bonding in unhydrated cement grains. Na, Mg, Mn, Ti and Fe are emitted only as single posititively charged particles.

- Al is mostly emitted as Al^+, but some minor quantities of AlO^- and AlO_2^- have also been detected. Ca is emitted as Ca^+, CaO^+ and $CaOH^+$. K is emitted as K^+ and K_2O^+.

- No anionic species of Ca and K have been as yet detected. Si is mostly emitted as SiO^-, SiO_2^- and SiO_3^-, but a small portion is emitted as Si^+.

- These results show, that Na^+, K^+, Mg^{2+}, Mn^{2+}, Ti^{2+} and Fe^{3+}-ions exist in cement particles, presumably as oxides with ionic bonds. The K_2O^+-species indicates that the alkalies can exist to some extent as free oxides in cement.

- The anionic Si-O-species demonstrate the stability of the silicate groups with their covalent bonds. As only a small portion of Si^+ is emitted it can be concluded that there are mostly mono-or disilicate groups present in cement.

As most of the Al is emitted as Al^+, the bonds in the Al-phases in cement are probably mostly ionic. The large number and high concentrations of the emitted Ca-O-species indicate, that Ca is bound to other units in the lattice moslty by intermediate O-atoms and hence CaO-units exist in the cement particles.

However, these CaO-species must not be interpreted as free oxide. Rather they are incorporated in other structural elements or bound tightly to those. The K_2O, however, exists as free oxide.

3.2 Cements after 10 Minutes of Hydration

3.2.1 Portland Cement

Figure 3 shows the variation of the concentration of some elements with the distance from the surface for the sample hydrated for 10 minutes. In Fig. 4 the ratio of concentrations of SO_3^-- and oxide-ions is shown as a function of the distance from the surface. The same figure can be obtained for the ratio of concentrations of SO_4^-- and oxide-ions. From Fig. 3 it follows that the concentration of Ca shows a marked increase at a depth of about 10 nm from the surface. At the same depth, the concentration of Mg shows a small but significant decrease, and the concentration of Fe becomes constant.

In a depth between 10 and 20 nm the SO_3^-/O^--ratio shows a large concentration gradient as is shown in Fig. 4. The variations of the SO_3^-/O^--ratio at low distances (down to 10 nm) from the surface are not due to the experimental method but indicate a distorted structure in the hydrated sphere of the cement particles. In the deeper layers from 10 to 100 nm from the surface, the SO_3^-/O^--ratio does not vary significantly with the distance from the surface.

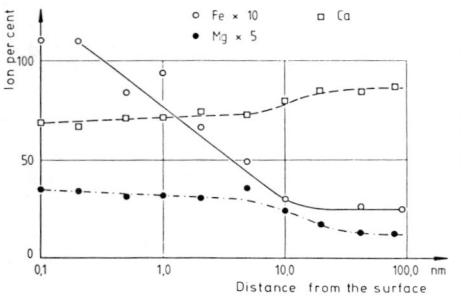

Fig. 3:

Change of metal element concentrations with increasing distance from the surface of partially hydrated portland cement grains (distance on a logarithmic scale)

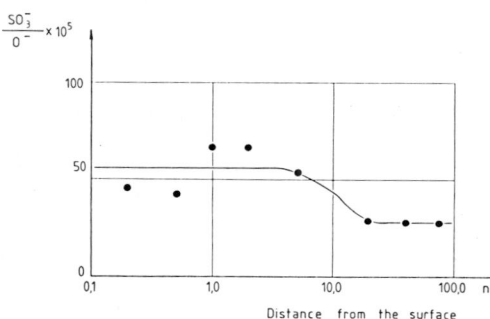

Fig. 4:

Change of SO_3^-/O^--ratio with increasing distance from the surface of partially hydrated portland cement grains (distance on a logarithmic scale)

Hence one can conclude, that the hydration has proceeded to a depth of about 10 - 20 nm into the cement grains during the first ten minutes. In addition, the decrease in the concentration of S-containing species measured by the SO_3^-/O^--ratio clearly shows the formation of phases containing S, i.e. for example ettringite, at the surface of the cement during the first stage of hydration. These components are formed as soon as the hydration starts. Also the K^+-ions are preferably released into the liquid phase. Hence, the concentration of K^+-ions in the lattice of the hydrated samples is much less than in the unhydrated cement (8 ion-percent compared to 44 ion-percent). As Al and Mg are not released so easily, the concentration of these elements in the outer layers of the hydrated sample is higher than in the unhydrated cement. As KOH is liberated easier than $Ca(OH)_2$ from the portland cement in the first minutes of hydration, there is also an enrichment of Ca in the surface of the cement grains in the hydrated samples compared with the unhydrated cement (70 ion-percent to 42,8 ion-percent). From the depth of hydration the degree of hydration can be calculated. If it is assumed that this depth is equal for all cement particles present and if a particle size distribution according to /5/ is used for the cement, a degree of hydration, m, of 0,5 percent is calculated for the sample hydrated for 10 minutes. This agrees very well with results of Locher, Richartz and Sprung /6/, who found for example a decrease in tricalciumaluminate and tricalciumsilicate contents of 0,5 - 2 percent during the first 5 minutes of cement hydration.

3.2.2 Blast Furnace Slag Cement

With the blast furnace slag cement similar results are obtained as with the portland cement. However the hydration has proceeded only about half as far into the interior of the grains as with the portland cement. This agrees with the well known fact, that the hydration reaction is slower with blast furnace slag cement than with portland cement. In addition, the concentration gradients in blast furnace slag cement are not as pronounced as they are in portland cement. This clearly indicates, that the reaction is slower for the blast furnace slag cement. If, though not actually correct, again a grain size distribution according to /5/ is used in addition to the assumptions stated above a degree of hydration of 0,2 percent is obtained for the blast furnace slag cement, which is quite a reasonable value for ten minutes of hydration.

3.2.3 Fly Ash

Fig. 5 shows the depth profiles of K, Al and Fe for the fly-ash, hydrated 10 minutes in saturated $Ca(OH)_2$-solution. The concentrations of Si, Ba and Ti were found not to vary with the distance from the surface.

The results show, that fly ash is able to react with saturated $Ca(OH)_2$-solutions within a period of 10 minutes to a considerable extent. The reactions has proceeded about 5 to 7 nm deep into the particles.

Compared to cement, especially to blast furnace slag cement, this is a surprisingly high value, which clearly demonstrates that fly ash is a puzzolanic material.

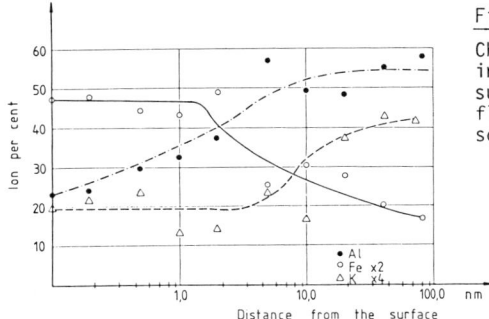

Fig. 5:

Changes of metal elements with increasing distance from the surface for partially hydrated fly ash (distance on a logarithmic scale)

3.3 Portland Cement Pastes

After 250 days of hydration of the portland cement paste the samples show a strong concentration gradient down to 10 nm from the surface of both K^+- and Na^+-ions with increasing concentration towards the surface. This is a result of the drying process before the SIMS-analysis because these ions are the remnants from the pore solution. The calcium ions show a similar concentration gradient but with decreasing concentration towards the surface. All the other ions detected show no variation of concentration with the distance from the surface.

The calcium concentration gradient is a result of the liberation of calcium hydroxide to the surrounding water during the wet curing period. Consequently the concentration of Ca in the near-surface-layers of the specimen stored 250 days in water is significantly less than that of the specimen stored in normal air after 3 days in water.

3.4 Further Applications

It has been shown yet, that SIMS provides an excellent means to analyse cementitous materials of all kinds and to trace the progress of the hydration reaction into the interior of cement or fly ash particles.
This is possible for both, pure and composite cements. High-resolution depth profiles are easily obtained and since the composition of the transitory region between paste and aggregate or paste and reinforcement varies also significantly with the distance from the aggregate or reinforcement respectively, SIMS should,in addition provide an excellent means for studying interfacial bond effects in concrete and reinforced concrete, especially because of the possibility to analyse an interface layer by layer.

4. DETERMINATION OF ADSORBED SURFACTANTS

4.1 Qualitative Measurements

In a first series of tests, a monolayer of p-toluenesulfonate and dodecylbenzenesulfonate was adsorbed on portland cement by hydrating the cement in a surfactant solution. The samples were then analysed on the surface and in a depth of 2 nm below the surface. The spectra obtained for the dodecylbenzenesulfonate showed the typical lines of organic substances and their typical grouping of the lines in the mass spectra, whereas the p-toluenesulfonate yielded a normal spectrum of a 10-minutes hydrated cement. Fig. 6 shows a surface cation SIMS-spectrum of dodecylbenzenesulfonate on portland cement where the groups represent C_2^+-, C_3^+-, C_4^+-, C_5^+- C_6^+-fragments and $\emptyset^+$-, $\emptyset$-C^+-, $\emptyset$-C_2^+-, $\emptyset$-C_3^+-, $\emptyset$-C_4^+- and $\emptyset$-C_5^+-fragments respectively ($\emptyset$ symbolizes the aromatic benzene system). Then the first two layers were abrased and the analysis repeated in a depth of 2 nm below the surface. There, all samples showed the typical spectra of 10 minutes hydrated cement samples. So, the adsorption of surfactant on cement occurs onto the surface within the first two or three layers, in spite of the hydration reaction.

4.2 Quantitative Measurements

In the last series of tests the cement was hydrated in dodecylbenzenesulfonate solutions of 5 different concentrations to obtain different surface concentrations of 0,0028; 0,028; 0,056; 0,17 and 0,83 mMol/g respectively. In Fig. 7 the ratio of some cation-intensities to Ca^+ are plotted against the surfactant concentration, where $\emptyset$ again symbolizes the aromatic ring system. Monolayer coverage occurs above about 0,17 mMol/g.
An extensive analysis of the spectra showed, that for the $\emptyset$-C^+-fragments and the SO_3-species a linear realtionship between the surfactant concentration and the SIMS-intensities is obtained as is shown in Fig. 8. So the $\emptyset$-C^+-peak and the SO_3^--peak may be used for a quantitative determination of such surfactants adsorbed on cement.
However, this method has an upper limit of one monolayer coverage. Concentrations above one monolayer of surfactant cannot be determined quantitatively due to the secondary ion generation process /3/. In this concentration range only qualitative measurements are possible at present, because the SIMS-intensities do not vary any more with the surfactant concentration but become constant.

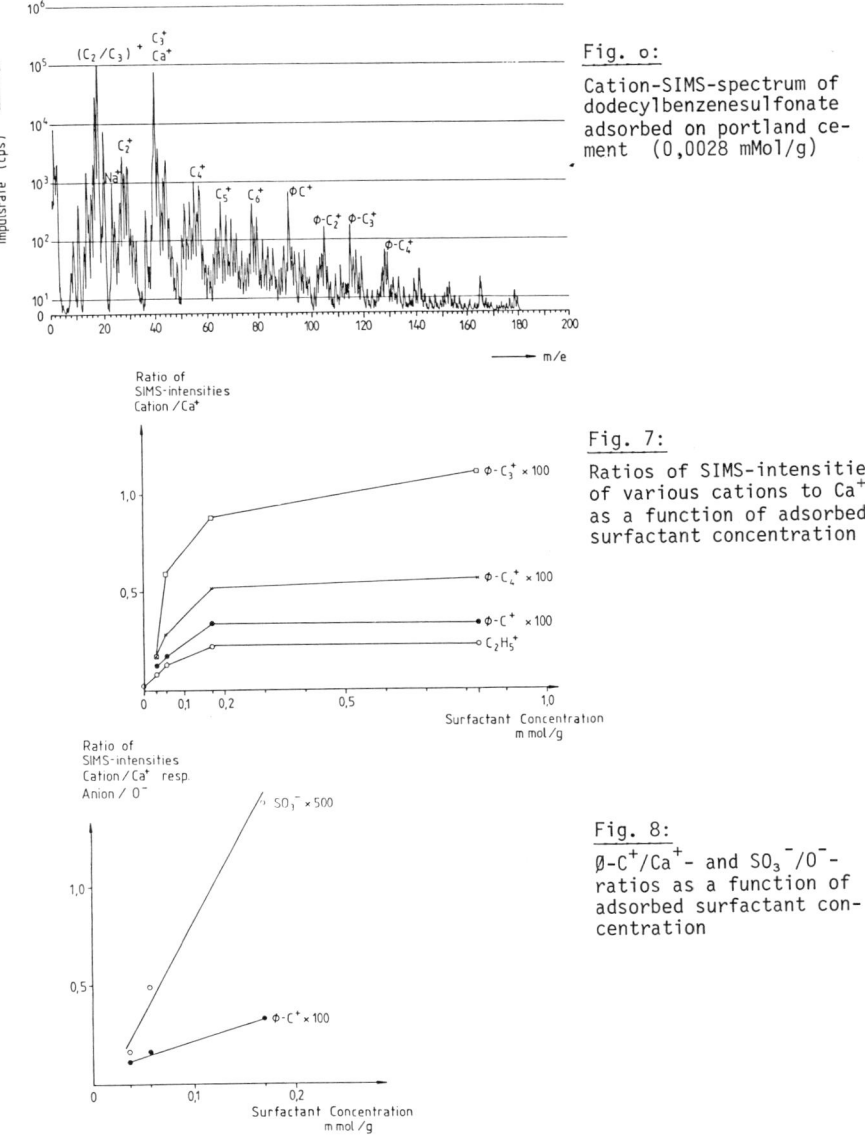

Fig. o:
Cation-SIMS-spectrum of dodecylbenzenesulfonate adsorbed on portland cement (0,0028 mMol/g)

Fig. 7:
Ratios of SIMS-intensities of various cations to Ca^+ as a function of adsorbed surfactant concentration

Fig. 8:
$\phi\text{-}C^+/Ca^+$- and SO_3^-/O^-- ratios as a function of adsorbed surfactant concentration

Hence the adsorption of surfactant can be determined by SIMS, especially for low degrees of surface coverage. These low surface coverages, 5 percent of a monolayer or less are typical for flotation processes and chemical admixtures in concrete.

Thus, SIMS can be used to study the adsorption of flotation reagents on reactive materials as well as for the study of the mechanisms underlying the effects of chemical admixtures in concrete.

5. SUMMARY

In this paper the application of SIMS-measurements in the field of cement and concrete technology is presented. Further a new model for the evaluation of SIMS-spectra of oxidic materials is proposed. Using SIMS, the progress of the hydration of cement can be traced directly by determining depth-profiles of characteristic cations after various times of hydration. It was found in addition, that cements have a surprisingly high content of potassium-ions in their outermost layers, which are easily released when the cement comes into contact with water. After 10 minutes, the hydration has proceeded 10 to 20 nm into the grains of portland cement, 5 to 10 nm into blast furnace slag cement and 5 to 7 nm into fly ash particles. The hydration depth can be detected by significant variations of many metal elements with increasing distance from the surface.

Also, surfactants adsorbed on cement can be analysed qualitatively and quantitatively by SIMS when the surfactant contains an aromatic system and at least 8 - 10 aliphatic C-atoms. For this class of substances, the $\emptyset$-C$^+$-species and, if sulfonic acids or their salts are used, the SO_3^--species in addition, show a linear relationship between the SIMS-intensity and the concentration of surfactant adsorbed on cement, if the surface coverage is less than or at maximum one monolayer. If more than one monolayer of surfactant is adsorbed only qualitative measurements are possible because then the intensities of all emitted species do not vary with the concentration any more. This behaviour is due to the elementary steps in the secondary ion generation process of SIMS.

REFERENCES

/1/ C. Plog, L. Wiedmann, A. Bennighoven
"Empirical Formula for the Calculation of Secondary Ion-yields from oxidized Metal Surfaces and Metal Oxides", Surf.Sci 67 565 (1977).

/2/ C. Plog, W. Gerhard
"Physical Aspects of the Valence Model's Parameter"
Proc. II Int. Conf. on SIMS, Stanford USA, 1979.

/3/ C. Plog, W. Gerhard
"Secondary Ion Emission by Nonadiabatic Dissociation of Nascent Ion Molecules with Energies Depending on Solid Compositions"
Pt. II: Examination of the New Emission Conception by Model Calculations of Metal Yields and qualitative Extension to other kinds of Secondary Ions", Z.Phys B-Condensed Matter, 54 71-86 (1983).

/4/ F. Lopez-Flores
"Flyash and effects of partical cement replacement by flyash"
M.S. Thesis, Joint Highway Research Project, IHRP-82-11
Purdue University, West Lafayette, Indiana, Juli, 13, 1982.

/5/ F. Keil
"Zement"
Springer Verlag, Berlin, Heidelberg, New York, 1971.

/6/ F.W. Locher, W. Richartz, S. Sprung
"Erstarren von Zement"
Zement-Kalk-Gips 29 (1976), S. 435 ff.

APPLICATION OF SIMS DEPTH PROFILING TO CERAMIC MATERIALS

JENIFER A.T. TAYLOR, PAUL F. JOHNSON and VASANTHA R.W. AMARAKOON
New York State College of Ceramics at Alfred University, Alfred, NY 14802

ABSTRACT

Depth profiling is becoming a common method of determining composition gradients for those research facilities that have access to SIMS. The problems with this procedure are briefly discussed but well referenced for those interested in using the technique. Spectra for application of depth profiling to a tantalum pentoxide layer on tantalum, silicon implanted in silica, preferential sputtering of niobium and surface treated lithium alumina silicate glass are presented.

INTRODUCTION

Depth profiling as an analytical technique requires the removal and identification of consecutive surface layers. Removal of material from the surface of the sample is usually accomplished by ion milling or sputtering, during which fragments are eroded with an ion beam. This process is an important part of surface analysis for two reasons: It is widely used both as an analytical tool and in conjunction with other techniques and it causes multitudinous problems in interpretation for the analyst. The basic concept of knocking atoms or ions off the surface of a solid with a high energy beam of ions is straightforward but the mechanism is not. The first section of the paper will be an introduction to some of the typical problems associated with depth profiling of ceramics, along with suggestions from the literature for mitigating the consequent limitations. The second section will be examples of depth profiling used as an analytical technique, including basic data reduction considerations. The work described in this paper was performed using a Leybold-Heraeus Energy Analyzer 10/100 SIMS with minimum beam size of 0.8 mm FWHM. $

ARTIFACTS ASSOCIATED WITH DEPTH PROFILING CERAMIC MATERIALS

The primary beam can knock surface constituents into the matrix or can be implanted in the matrix, both occurrences resulting in an apparent composition that is not necessarily representative of the original. Knock-in is more likely when the atomic mass of the matrix is less than that of the surface constituents. Reducing the energy of the incident ion beam can decrease knock-in and implantation but may enhance preferential sputtering. Preferential sputtering refers to the process of removing some elements from the surface more easily than others, creating a composition gradient which was not present in the original surface. Cones or volcano-shaped protuberances are considered good indications that preferential sputtering has occurred. Zinner has done an extensive review of depth profiling studies including many ceramics and theory-application discrepancies. [1]

Duncan et al working at 5.5 keV with an angle of incidence of 47 degrees to the sample normal concluded that depth profiling with reactive species such as oxygen or cesium rather than argon tends to erode the surface more evenly, up to about two microns. [2] Preferential sputtering can be monitored by analyzing sputter deposited material on a substrate adjacent to the sample. [3]

Depth profiling of polycrystalline ceramic materials is complicated by the insulating nature of many of these materials and by the number of elements present. Under ionic bombardment, insulators develop a surface charge, usually positive, which can cause diffusion of mobile ions and deflection of the beam so the full energy of the primary ions is not incident on the sample. Such charging of insulators can usually be successfully controlled with a low energy electron flood gun, considered standard equipment on most SIMS units.

Depth profiling with SIMS can seriously affect subsequent AES or XPS data. The crystal lattice near the surface can be rendered amorphous, changing the chemical nature of surface atoms including breaking of bonds and forming of new bonds. Thomas found the Si 2p peaks in silica spread from 1.8 ev to 2.7 ev at FWHM during one minute of sputtering with 1 keV helium ions. [4] If the analyst were depth profiling through a thin film looking for small changes in this line as an indication of changes in the silicon bonding, such broadening could have obscured the information. Hofman illustrates with XPS the reduction of niobium pentoxide to metal on sputtering. [5]

Accurate determination of the amount of material removed is difficult and subject to much discussion. The change in erosion rate due to specific surface conditions and interfaces near the surface such as is found in grain boundary regions causes uncertainty in standard rates. [6] The most common procedure is to measure the ion dose which is then compared with literature values to ascertain probable amount of material removed. Measuring the crater depth and relating to the total elapsed time is the most direct method, requiring the use of a profilometer or interferometry.

Standards are available which will allow the quantification of the sputter rate on layers of known thickness. Anodically grown tantalum pentoxide on tantalum is available in two thicknesses. # [7] The US National Bureau of Standards is offering a reference material composed of layered nickel and chromium thin films intended to be used to calibrate sputtering rates in surface analysis as well as to monitor ion beam stability. @ [8] This reference material shows mixing at interfaces clearly, but may behave differently from ceramics. Generally, recognizing the possible artifacts and using other analytical techniques before and after sputtering to ascertain how the surface has been affected are the best precautions against misleading information. Standards of similar composition help reduce the variables.

Despite all these complications, depth profiling is commonly used with apparent success. Clegg et al undertook a study of depth profiling of boron in silicon, comparing analyses of the same material by several different instruments. They found that the profile width data showed agreement to within 10% over a large concentration range. System configuration had a significant effect on the peak to background range with quadropole-based raster-scanning instruments being sensitive to the quality of the primary beam while imaging ion microscope systems have memory effects. This group of analysts recommends analysis of identical samples prepared by ion implantation to facilitate comparison of results between laboratories. [9]

Detailed discussion of depth profiling with SIMS can be found in the references given, each of which lists more articles on the topic. Depth profiling of glass is addressed in Reference 10. The rest of this paper will be concerned specifically with application of the technique to ceramics.

APPLICATION OF DEPTH PROFILING TO CERAMIC MATERIALS

The mode of data collection is an important consideration. The instrument parameters such as intensity scale, scan speed and response time can affect the data collected and should be established before serious analyses begins. The vertical axis indicating intensity can be either log or linear scale. Using the log scale ensures that no peaks will exceed the maximum number of counts that can be accommodated by the chart paper or computer program. A linear scale has a larger difference in height between peaks making interpretation easier. Figure 1 shows a linear and log version of the same information, collected from an anodic tantalum oxide surface on tantalum, sputtered for two hours. Comparing peak intensity is easier in linear form because the difference between peak heights is greater, but information is lost about those peaks whose intensity is less than the threshold value for the linear count. Collecting the data first on log to see all peaks and then in linear to compare specific peak intensities is one way to circumvent this problem. The peak intensity ratios for Ta/TaO in Figure 1 show a difference due to data collection mode that might be mitigated by computerized data reduction.

Electronically gating the signal so data is only collected when the beam is passing through the middle of its range is another way to modify a SIMS spectrum. The advantage of this procedure is it reduces the number of secondary ions collected from the edges of the crater created by the ion beam. Generally the preferred information about the composition comes from the deepest layer of material, at the bottom of the crater.

Figure 2 shows log spectra of plasma-implanted silicon in vitreous silica, both gated and ungated modes. The silica peak at 60 amu is not visible on the input gated mode after two hours of sputtering. After six hours the silica peak has begun to show up in the input gated signal as the crater bottom approached the vitreous silica substrate. The initial Si/SiO peak intensity ratio was 49, dropping to 20 after six hours of sputtering,

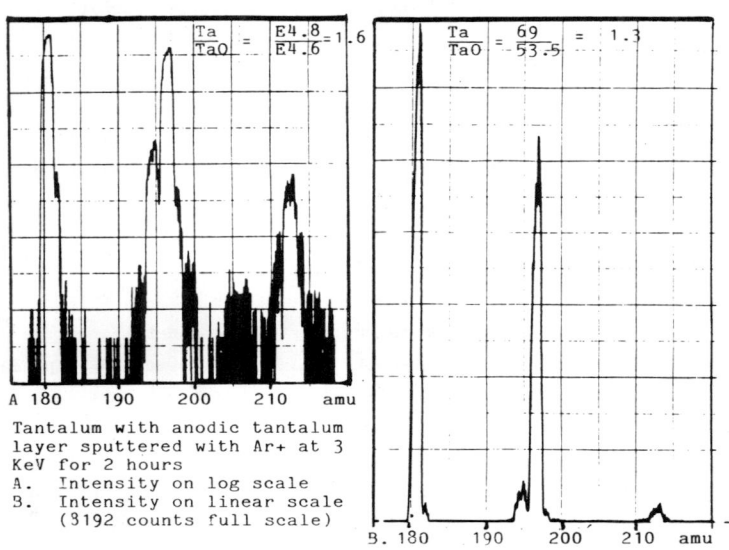

Tantalum with anodic tantalum layer sputtered with Ar+ at 3 KeV for 2 hours
A. Intensity on log scale
B. Intensity on linear scale (3192 counts full scale)

Figure 1

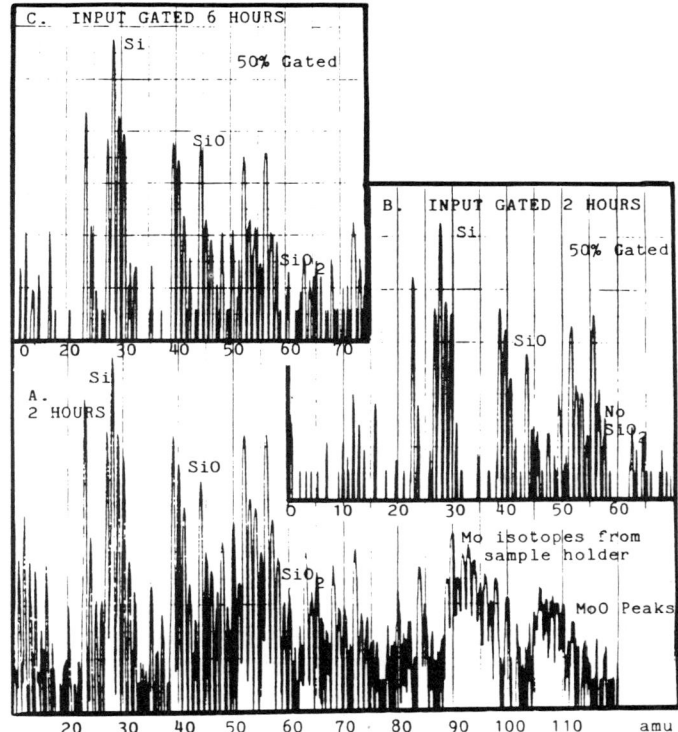

Depth profiling of Si thin film on SiO$_2$ showing log spectra after 2 and 6 hours of sputtering

FIGURE 2

both numbers calculated from linear spectra. Spectrum A shows clearly peaks at 52 amu for Cr and 56 amu for Fe, probably tertiary ions sputtered from the steel sample rod. If the peak at 56 amu were there without the peak at 52 amu, Si$_2^+$ would be the more likely species. The molybdenum isotopes from the sample holder appear between 90 and 100 amu followed by the associated MoO peaks.

Depth profiling for such long periods is not recommended because of the atomic mixing that occurs under ion bombardment. Using a low energy, static SIMS mitigates this problem but in any case, careful documentation of suface character with other techniques such as electron optics, XPS, AES and ISS is necessary to identify artifacts. Magee et al address these concerns for SIMS work on a silica-silicon interface, layers on gallium-arsenide, and chlorine and boron implanted in silica. [11]

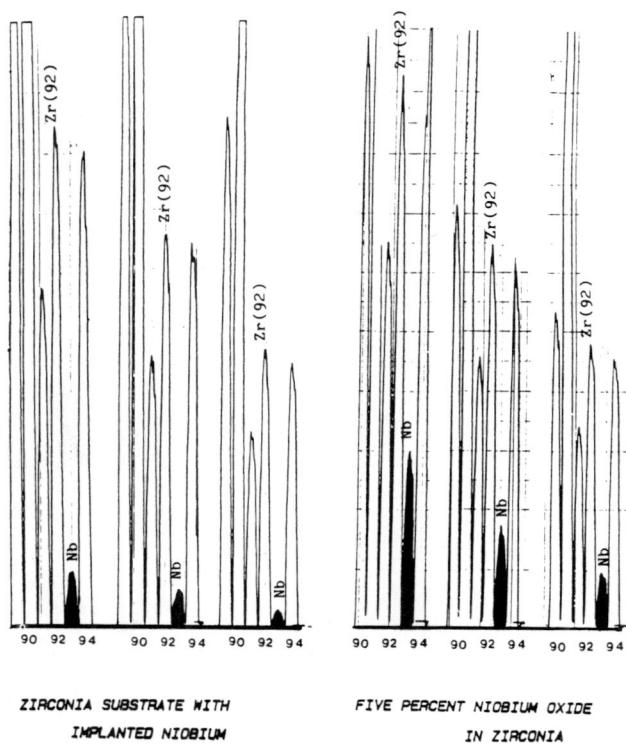

ZIRCONIA SUBSTRATE WITH IMPLANTED NIOBIUM

FIVE PERCENT NIOBIUM OXIDE IN ZIRCONIA

FIGURE 3

Preferential sputtering is a constant concern, especially in ceramics because of the number and diverse properties of elements present. Figure 3 shows limited spectra for niobium implanted in yttria stabilized zirconia with major isotopes of Zr at 90, 91, 92, and 94 amu; Y at 89 amu, and a small peak for Nb at 93 amu. Depth profiling with SIMS was used to ascertain the thickness of the layer of implanted Nb. The spectra seemed to indicate a decreasing concentration of niobium since the peak intensity ratio of Nb to Zr(92) dropped from 0.114 to 0.059 during 30 minutes of sputtering with argon. [12] However, on sputtering under identical conditions a standard containing five percent niobium oxide homogeneously dispersed in zirconia, a similar decrease in the ratio was seen. One explanation for the changing peak intensity ratio for the standard is that niobium had segregated to the surface during sintering, not likely since the standard was sintered rapidly in plasma. The most probable explanation is that niobium was being preferentially removed, causing the surface to become depleted in niobium so the Nb/Zr(92) ratio decreased. This same phenomena could occur in the niobium implanted sample, meaning the changing ratio is not an indication of layer thickness. The rate of decreasing niobium concentration with sputtering might be extracted from this spectra if it were compared with the decrease in peak intensity ratio of standards with similar concentrations of niobium.

Depth profiling treated glass proved a valuable technique for analyzing the effect of various solutions on the composition of the surface. Figure 4 Spectrum A is a standard untreated specimen showing fairly even levels of the measured ions for the duration of the sputtered layer, except for potassium. Spectrum A establishes the composition profile to be expected from this glass when sputtered. Spectrum B shows the same glass after a surface treatment that significantly decreased the concentration of several of the measured ions. The concentration level of

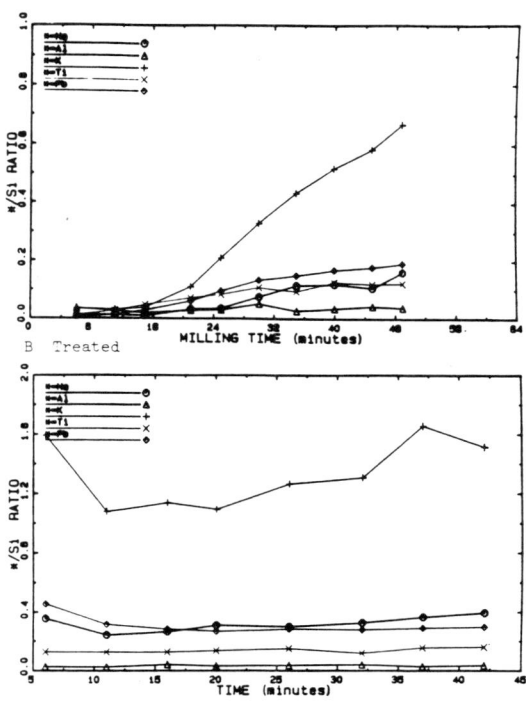

FIGURE 4 Concentration Profiles for Various Ions in Treated and Untreated Specimens

the depleted ions does not attain the bulk value for the peak intensity ratio with silicon of 1.5 to 2 after 48 minutes of sputtering, except for potassium. [13]

Depth profiling of ceramics with SIMS is becoming a common analytical technique for determining composition gradients. Cross laboratory comparisons, such as that described by Clegg et al, should lead to improved reproducibility while the growing data base of information about SIMS of ceramic materials facilitates interpretation.

FOOTNOTES

$ Leybold-Heraeus Vacuum Products Inc. LAS Group, 5700 Mellon Road, Export, PA 15632 USA Bonner Strasse 504. Postfach 510760 D-5000 Koln 51 FDR
# The anodic tantalum pentoxide on tantalum, designed for use with argon in the energy range of 0.5 to 3 keV, can be ordered from Division of Materials Applications, National Physical Laboratory, Teddington, Middlesex TW11 OLW as Certified Reference Material NPL No. 57B83 BCR No. 261
@ Available from J. Fine, National Bureau of Standards, Gaithersburg, MD 20899

REFERENCES

1. Zinner, E. Scanning 3 (1980) 57.

2. S. Duncan, R. Smith, D.E. Sykes and J.M. Walls, Vacuum 34 (1984) 145.

3. G.K. Wehner, METHODS OF SURFACE ANALYSIS A.W. Czanderna Ed (Elsevier Scientific Pub, NY, 1975) p 5.

4. J.H. Thomas, APPLIED ELECTRON SPECTROSCOPY H. Windawa and F. Ho, Eds. (John Wiley and Sons, NY, 1982) p 33.

5. S. Hofman, SECONDARY ION MASS SPECTROSCOPY SIMS III, A. Benninghoven, J. Giber, J. Laszlo, M. Riedel, H.W. Werner, Eds. (Springer-Verlag, New York, 1982) p 186.

6. S. Hofman, SECONDARY ION MASS SPECTROSCOPY SIMS III, A. Benninghoven, J. Giber, J. Laszlo, M. Riedel, H.W. Werner, Eds. (Springer-Verlag, New York, 1982) p 186.

7. C.P. Hunt, M.T. Anthony and M.P. Seah, Surface and Interface Analysis 2 (1984) 92.

8. J. Fine of NBS, Paper E42FrM08 American Vacuum Society 31st National Symposium, Reno, Nevada, 4-7 Dec. 1984 .

9. J.B. Clegg, A.E. Morgan, H.A.M. DeGrefte, F. Simondet, A. Huber, G. Blackmore, M.G. Dowsett, D.E. Sykes, C.W. Magee and V.R. Deline, Surface and Interface Analysis 6 (1984) 162.

10. H. Bach and F.G.K. Bauke, J Amer. Cer. Soc. 65, 527 (1982).

11. C.W. Magee, R.E. Honig and C.A. Evans, Jr., SECONDARY ION MASS SPECTROSCOPY SIMS III, A. Benninghoven, J. Giber, J. Laszlo, M. Riedel, H.W. Werner, Eds. (Springer-Verlag, New York, 1982) p 172.

12. G.W. Bieniecki, M.S. Thesis, 1984, New York State College of Ceramics at Alfred Univ.

13. T. Hobbs, Unpublished work New York State College of Ceramics at Alfred Univ. 1985.

ULTRASENSITIVE ELEMENTAL ANALYSIS OF MATERIALS USING
SPUTTER INITIATED RESONANCE IONIZATION SPECTROSCOPY

J.E. PARKS, D.W. BEEKMAN, H.W. SCHMITT, and M.T. SPAAR
Atom Sciences, Inc., 114 Ridgeway Center, Oak Ridge, Tennessee 37830

ABSTRACT

Sputter Initiated Resonance Ionization Spectroscopy (SIRIS) is a technique being developed by Atom Sciences, Inc. to perform ultrasensitive elemental analysis of materials. SIRIS uses sputtering to atomize a solid sample and resonance ionization (RIS) to selectively ionize an element of interest. The SIRIS technique is capable of detecting impurities at the 0.1 ppb level ($5 \times 10^{12}/cm^3$) in a routine analysis time of 5 minutes. RIS and the SIRIS technique are briefly reviewed. We report a detection efficiency for SIRIS of 2 ppb sensitivity and recent results are given for standard well-characterized samples of boron-doped silicon. Current progress is described for the development of depth profiling and the analysis of silicon in gallium arsenide. The SIRIS detection of silicon in standard reference materials certified by NBS is presented.

INTRODUCTION

Sputter Initiated Resonance Ionization Spectroscopy (SIRIS) is a technique which provides a new technology for the ultrasensitive analysis of trace elements in solid materials and has commercial potential for a wide variety of applications. SIRIS is element sensitive and is free from molecular interferences, isobaric effects, and other ambiguities. This technique has the potential of providing ultrasensitive analyses to the sub-ppb level routinely in reasonable measurement times. The development of SIRIS at Atom Sciences was begun late in 1980 and this work led to a patent for the technique and apparatus granted April 10, 1984, patent #4,442,354. As the commercial developer of SIRIS, Atom Sciences has pursued the development with the objective of providing SIRIS both as an analysis service and as an analytical instrument.

The SIRIS technique is based on a laser ionization technique developed at the Oak Ridge National Laboratory by Hurst and his co-workers [1]. This technique is called Resonance Ionization Spectroscopy (RIS) and has been adopted by many to study basic physical processes and to develop analytical detection techniques. This technique is also sometimes referred to as multiphoton ionization. The RIS technique has been well documented, but the essence of RIS consists of one or more resonantly excited transitions by the absorption of one or more photons of the proper wavelength or energy and then followed by absorption of a final photon energetic enough to cause ionization. This concept for creating an ionized atom of a preselected element has been shown to be both selective and efficient. Only the atoms and all of the atoms of the preselected element which are present within the region of space probed by the laser beam can be ionized with commercially available laser systems.

The SIRIS concept is illustrated in Figure 1. RIS requires that the

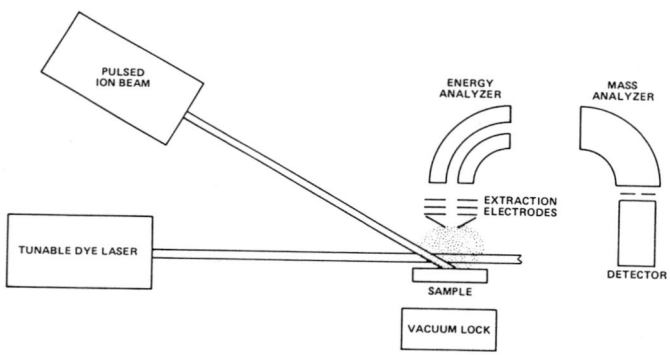

Fig. 1. Schematic diagram of SIRIS apparatus.

atoms to be probed be free from each other and interact only with the photons of light. The atoms must be in the gaseous state and therefore, in order to analyze a solid, the solid must be vaporized. SIRIS uses sputtering to accomplish this. A pulse of ions from an energetic ion beam impinges upon the sample located in vacuum and sputters atoms representative of the constituents of the solid sample into the gaseous state. Secondary ions which are produced in the sputtering process are first rejected by electrostatic fields, energy discrimination, time discrimination, or a combination of these, thus leaving a neutral atom cloud to be probed by the laser. Then the neutral atom cloud is probed by the RIS process to resonantly excite and ionize just the atoms of the selected element. After a trace element has been ionized, it is then extracted and counted in a standard particle detector. If isotopic information is needed or if verification of the ionized element is desired, a mass analyzer with an energy filter may be inserted prior to detection.

The apparatus designed and built for SIRIS at Atom Sciences has been described in detail elsewhere [2]. The basic elements of the system are illustrated in Figure 1 and consist of a pulsed argon ion beam, target chamber, pulsed dye laser system, and double-focusing magnetic mass spectrometer, all in a system capable of ultrahigh vacuum. The device is controlled by a microprocessor and includes a computer-based data acquisition system. Winograd, et al. [3] and Pellin et al. [4] have reported on similar approaches to SIRIS and have described their apparatus elsewhere. Recently, Donahue and his co-workers [5] have reported the modification of a SIMS instrument to do SIRIS work.

The present instrument used by Atom Sciences produces a 50 μA beam on target in a pulse of approximately 1 μs duration synchronized with the laser 30 times per second. The diameter of the laser beam is approximately 1 mm and is centered about 3 mm from the surface of the sample. There are several factors which affect the sensitivity of the SIRIS measurements. These include the geometry, timing, extraction efficiency, mass spectrometer transmission, sputtering yield, and RIS efficiency. The geometrical factor depends on the solid angle the laser beam intercepts and is normalized to the total angle over which the sputtering takes place and the directional distribution of the sputtered particles. The timing factor depends on the energy distribution of the sputtered atoms and the time at which the laser is pulsed to probe the sensitive region. At some instant of time, the slow particles will not have reached the sensitive volume and the fast ones will have passed on through.

Using typical values for the key parameters, calculations show that it is reasonable to expect geometry factors of 10% and timing factors of 15%. The extraction of RIS produced ions can be estimated to be at least 50% while the transmission of a magnetic mass spectrometer for properly selected ions is as high as 80%. Since all sputtered neutral atoms are not in their lowest energy state, it is also reasonable to expect a 50% loss due to the fact that the RIS technique can't detect an atom unless it is in a specific energy state.

Therefore assuming a sputtering yield of 5 for a 50 μA beam of 1 μs duration, 1.6×10^9 particles will be sputtered during each pulse. Of these, 0.3% (10% x 15% x 50% x 50% x 80%) or 4.8×10^6 will be detected. Since the laser operates at 30 Hz, there will be 9000 laser pulses in 5 minutes and 4.3×10^{10} atoms will be probed. In order to count 100 atoms in 5 minutes, the sample would have to have 1 part in 4.3×10^8 or 2 ppb. Under the same conditions with a sample of gallium arsenide, we find that we can count 2×10^6 ^{69}Ga ions per laser pulse, which corresponds to 6×10^6 ions per pulse for a monoisotopic element in elemental form. In the standard 5 minutes counting time with a pulse repetition rate of 30 Hz, 5.4×10^{10} counts would be obtained from a pure sample, or the 100 counts required for 10% counting precision would be obtained from a sample with $100/5.4 \times 10^{10} = 2 \times 10^{-9}$ fraction of the element in question. Previously, we reported [6] that this 2 ppb detection limit for gallium inferred above from measurement on a pure sample (really, 50% Ga in GaAs) was confirmed by measurements on a dilute sample, silicon doped with 0.5 ppm gallium.

RESULTS

Measurements at Atom Sciences using the SIRIS technique have concentrated on the analysis of well-characterized materials for impurities found in the bulk. Silicon samples have been analyzed for gallium, indium, and boron. Certified reference materials of various steel samples obtained from the National Bureau of Standards have also been analyzed and reported for aluminum, vanadium, and boron [7]. We report here some more recent results for the analysis of boron in silicon and steel and then for silicon in steel and other materials. The RIS schemes used for ionizing boron and silicon are similar and are three color schemes involving two excitation steps followed by a photoionization step. These required two dye lasers synchronized with each other to produce the correct wavelengths.

A set of five standard reference materials from the National Bureau of Standards consisting of steel with boron concentrations certified at 110, 25, 9, 5, and 1.3 ppm were analyzed for boron by SIRIS. A typical mass scan (for the 1.3 ppm sample) is shown in Figure 2. The ordinate represents averages of the detector output for 300 laser pulses (10 sec) at each magnetic field setting. Figure 3 show the ^{11}B peak heights from the different samples plotted against the NBS certified values of B concentration, that is, the ordinate and abscissa were set equal for that point. The straight line is a least squares fit to the data.

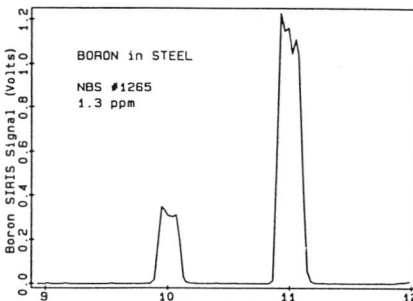
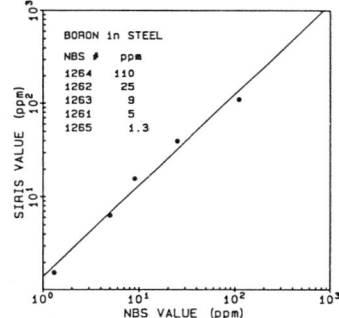

Fig. 2. Mass spectrometer magnetic field scan showing SIRIS signal for boron isotopes at 10 and 11 amu.

Fig. 3. Correlation plot showing SIRIS values vs. NBS values for boron concentrations in five NBS standard stainless steel samples.

Since we wished to study the applicability of SIRIS to samples of lower concentration, we obtained a set of silicon wafers doped in the bulk with boron at levels of 15.4, 3.8, 0.045, 0.017, and 0.0053 ppm as measured by resistivity. These samples, grown and characterized by Wacker, were obtained from the Materials Characterization Laboratory of Tektronix, Inc. Figure 4 shows the SIRIS results plotted against the concentrations quoted

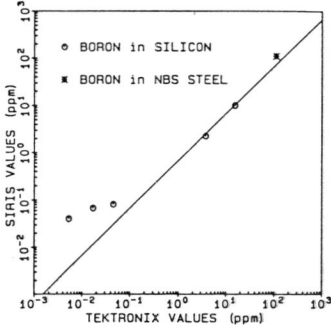

Fig. 4. Correlation plot showing SIRIS values vs. resistivity values for boron concentrations in silicon.

by the supplier (from resistivity). The NBS certified sample with the highest B content was remeasured along with the silicon samples, and it was again used to interrelate the ordinate and abscissa scales. The concordance of the steel sample with the two silicons highest in boron suggests that no large matrix effects occur with boron in these two materials. The SIRIS results for the three samples lowest in B lie above the least squares line fitting the steel and the two highest silicons. We attribute this to contamination of these samples by sputtering of boron from the stainless steel extraction electrodes, which presumably contain boron at levels much higher than those of the more dilute silicon samples. The sputtering may be by the argon ion beam itself, or by secondary ions sputtered from the sample. This interpretation was confirmed by remeasuring the same set of samples with the sample position altered to place it farther from the extraction electrodes, the presumed source of contamination. In fact, the results shown in Figure 4 are from the second experiment. In the first, the three lowest points were between 10 and 100 times higher on the ordinate scale than in the experiment shown, so that reducing the probability of contamination did markedly reduce the signal produced by the low-boron samples. The stainless electrodes were later replaced with ones made of tantalum and the measurements shown in Figure 5 show a slight improvement. This illustrates one of the problems that arises in the measurement of very low levels of a substance by any technique.

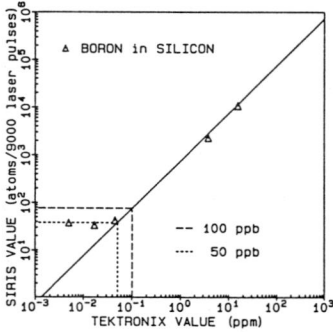

Fig. 5. Correlation plot showing SIRIS values vs. resistivity values for boron concentrations in silicon after replacing steel electrodes with tantalum electrodes.

The present SIRIS apparatus does not include quality depth profiling capability and that is the subject of the present development effort. The object of this work is to demonstrate that SIRIS can be used to measure the concentration of silicon as a function of depth in gallium arsenide. To do this, we have developed a three-color RIS scheme for silicon similar to our boron scheme. As a first step we have demonstrated the detection of silicon in the same steel samples shown in Figure 3. In addition, samples of niobium and tungsten were also analyzed for silicon and the results compared to the values supplied by Alfa Products of Morton Thiokol, Inc. The correlation of the SIRIS results with the NBS and Alfa results is shown in Figure 6. The correlation of the SIRIS results with the NBS values are excellent. These results also correlate well with the Alfa results, although the Alfa results are not certified and the silicon is detected from different matrices. The next step is to detect silicon in the bulk of some epitaxially grown gallium arsenide samples which have been well characterized.

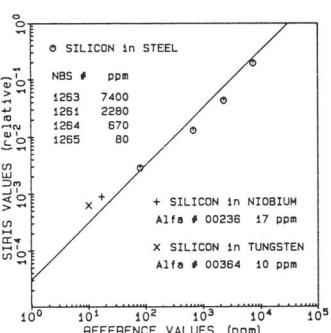

Fig. 6. Correlation plot showing SIRIS values vs. NBS values for silicon in steel.

The last step of the demonstration will be to decrease the diameter of the ion beam and provide SIRIS with electronic rastering so that a larger size crater can be milled to the desired depth. Ideally, the crater diameter should be ten times the ion beam diameter. For a 1 mm diameter crater this will require that the ion beam diameter of the present system be reduced from about 2 mm to 0.1 mm. The modifications to accomplish this are now in progress.

SUMMARY

Development of the SIRIS technique and apparatus has continued at Atom Sciences with improvements in the efficiency and sensitivity of the process. Sensitivities at the sub-ppm level have been demonstrated for a number of elements and detection efficiencies have been demonstrated that will make possible the detection of elements at the few ppb level. Elimination of background contamination generated from the system remains a problem to be solved. The development of depth profiling capability has begun and will be a valuable addition to the SIRIS technology. SIRIS has the exciting potential for ultrasensitive detection of elemental substances and other materials in the sub-ppb range with little ambiguity of interpretation of results.

ACKNOWLEDGEMENTS

The authors appreciate the cooperation of K. K. Smith and G. E. McGuire of the Materials Characterization Laboratory of Tektronix, Inc. and the National Bureau of Standards in supplying well-characterized samples.

The partial support of this work by the Avionics Laboratory, Air Force Wright Aeronautical Laboratories, Aeronautical Systems Division (AFSC), United States Air Force, Wright-Patterson AFB, Ohio 45433 is greatly appreciated.

REFERENCES

1. G. S. Hurst, M. G. Payne, S. D. Kramer, and J. P. Young, Rev. Mod. Phys. $\underline{51}$, (1979) 767.
2. J. E. Parks, H. W. Schmitt, G. S. Hurst, and W. M. Fairbank, Jr., Thin Solid Films $\underline{8}$, (1983) 69.
3. F. M. Kimock, J. P. Baxter, D. L. Pappas, P. H. Kobrin, and N. Winograd, Anal. Chem. $\underline{56}$, (1984) 2782.
4. M. J. Pellin, C. E. Young, W. F. Calaway, and D. M. Gruen, Surface Sci. $\underline{144}$, (1984) 619.
5. D. L. Donohue, W. H. Christie, D. E. Goeringer, and H. S. McKown, Anal. Chem. (accepted for publication).

6. J. E. Parks, H. W. Schmitt, G. S. Hurst, and W. M. Fairbank, Jr., in Resonance Ionization Spectroscopy 1984, edited by G. S. Hurst and M. G. Payne, Institute of Physics Conference Series Number 71, (The Institute of Physics, Bristol, UK, 1984), pp. 167-174.

7. J. E. Parks, H. W. Schmitt, G. S. Hurst, and W. M. Fairbank, Jr., Laser-based Ultrasensitive Spectroscopy and Detection V, Richard A. Keller, Editor, Proc. SPIE 426, (1983) 32.

MICROFOCUSSED ION BEAMS FOR
SURFACE ANALYSIS AND DEPTH PROFILING

DAVID R KINGHAM, P VOHRALIK, D FATHERS, A R WAUGH AND A R BAYLY

VG Scientific Ltd, The Birches Industrial Estate,
Imberhorne Lane, East Grinstead, Sussex, RH19 1UB

ABSTRACT

SIMS microprobe analysis using a liquid metal ion source (LMIS) for the primary ion probe can give fast, sub-micron resolution, chemical imaging of matrix and trace elements. The high brightness and small source size of the LMIS is vital for this high performance. Using a new 30kV gallium ion micro-probe a spatial resolution of 50nm can be achieved with a probe current density in excess of 1 A/cm^2. A newly configured SIMS instrument is described which incorporates either a gallium or caesium ion probe as well as a duoplasmatron for ultimate sensitivity and depth profiling performance. For small area, high spatial resolution, depth profiling a raster-scanned ion micro-probe is ideal. A quadrupole mass spectrometer gives a mass range of 1-800 or 2-1200 amu. Examples are presented of up to 50nm resolution SIMS imaging with both positive and negative secondary ions and up to 50nm resolution ion induced secondary electron imaging of a wide variety of samples including superconductors, steels, alloy fractures, optical fibres, integrated circuits, catalysts, polymers and biological specimens. The recent addition of framestore to the system has enhanced the imaging capability, allowing secondary electron and several different secondary ion images of a sample to be overlaid in contrasting colours.

INTRODUCTION

SIMS microprobe analysis can now be readily achieved using a Liquid Metal Ion Source (LMIS) to produce the primary ion probe. These sources have high brightness and small source size and permit the formation of high intensity sub-micron ion probes with energies from a few keV up to 30keV. The use of such ion probes to extend the imaging capability of Secondary Ion Mass Spectrometry well into the submicron range has recently been demonstrated [1, 2]. Ease of operation and full UHV compatibility of a gallium microprobe (VG Scientific MIG100) have allowed it to be included in both dedicated SIMS and multi-technique surface analysis instruments, e.g. in combination with X-ray Photoelectron Spectroscopy (XPS) and Auger Electron Spectroscopy (AES). Applications of SIMS microanalysis have been demonstrated on a wide variety of samples including superconductors, steels, alloys, optical fibres, integrated circuits, catalysts, polymers and biological specimens. SIMS has proved to be especially useful when AES analysis is difficult for reasons such as surface charging, electron induced sputtering or migration, low sensitivity and excessive analysis time. Particular advantages of SIMS are its good to excellent sensitivity for all elements and the short time required to acquire an image, typically only a minute.

In this paper we report on three further developments on the VG Scientific "SIMSLAB" which have significantly enhanced its imaging capabilities.

(i)   50nm Probe Size

By using the gallium LMIS on a 30kV electrostatic column (MIG300) a probe size down to 50nm can be achieved. This improvement takes the SIMS imaging capability to the limits of resolution attained in Auger imaging.

ii)   Caesium Liquid Metal Ion Source

A caesium LMIS has been used in a 10kV electrostatic column. Although the caesium source is less convenient to use, it does offer a clear advantage for SIMS of certain samples because it enhances the negative ionisation efficiencies for many chemical species.

iii) Digital Imaging and Framestore System

A full digital imaging system based on a DEC PDP-11 computer for system control, data acquisition and data archiving and a framestore for monitoring acquisition and for data display has been installed on the SIMSLAB. This system offers particular advantages for an inherently consumptive technique, such as SIMS, where data cannot be regenerated. Using the framestore it is possible to simultaneously acquire, view and store data in a manner which is simply not possible using conventional analogue techniques. Permanent storage of the data in a form where it can be rapidly recalled as an image is achieved by transfer to disk storage. The system also provides the capability to manipulate and combine images, to annotate and produce hardcopy and to acquire three dimensional data which may be analysed post hoc to give SIMS images, depth profiles, linescans and unconventional data such as image sections [3].

SIMSLAB APPARATUS AND DATASYSTEM

The results shown later in this paper come from SIMSLAB systems of slightly different configurations. This is a significant strength of SIMSLAB, that components can easily be added or removed to suit particular applications. All configurations share the MM12-12S SIMS quadrupole analyser with a mass range of 1-800 or 2-1200 amu and, to neutralise surface charging of insulating samples, a LEG31 electron gun.

Technical details of the ion probes and SIMS system have been published elsewhere [1, 2]. Briefly, the probe current range at 10kV is <50pA to >100nA with corresponding probe spot sizes from 200nm to 50µm. The less favourable physical characteristics of caesium have imposed a number of restrictions on its convenience as a LMIS compared to gallium which is simple to use, has a long lifetime and is fully UHV compatible. In particular the high reactivity of caesium requires that the source is filled and wetted in vacuum in-situ; the high mobility of liquid caesium imposes limitations on the source design and positioning and the high vapour pressure of liquid caesium results in some evaporation and inability to bake out the whole system after source initialisation. Nevertheless, the caesium source has produced a probe with characteristics similar to gallium with a small reduction in current due to the lower source brightness. It is also comparable to the gallium source in stability during day to day operation; but its long term reliability is not as good as gallium and source recharging is required more frequently.

The new 30kV gallium probe has achieved the expected improvement in probe size and current density, with image resolution of about 50nm at a probe current of 50pA and a probe current density of about 2 A/cm^2.

The images shown here were acquired using a digitally scanned ion probe. SIMS ion images were recorded using pulse counted, mass analysed, secondary ion detection. In some cases each detected ion was recorded in an image as a bright point and photographic integration of pulses in individual pixels led to the development of grey level contrast. Other images were acquired using the newly developed framestore computer system.

The major hardware components of the framestore system are a DEC LSI-11 computer with a 10Mbyte Winchester disk system (5Mbyte fixed, 5Mbyte removable), a 512 by 512 pixel by 8 bit framestore, a control keyboard and special interfaces for machine control. System software enables acquisition of ion induced SEM and SIMS images, linescans and depth profiles as well as mass spectra and ion energy distributions. Special attention has been paid to the acquisition modes and to data manipulation for hardcopy.

The two outstanding advantages which are gained by using a framestore are, firstly, to store data automatically as it is generated, and secondly to be able to decouple the acquisition and photographic (hardcopy) processes. The first of these is of particular importance in SIMS since it is a consumptive technique. The second feature allows flexible acquisition modes such as frame integration, averaging, smoothing or edge enhancement.

All the SIMS images (using the framestore) presented here were acquired in a frame averaging mode. In this mode, as successive frames are added the signal remains at a constant level ie, a constant brightness on the screen, whereas the noise diminishes steadily. The acquisition may be terminated when the desired signal to noise ratio has been achieved. For each image acquisition the mass window and target bias may be preset via a menu; the sensitivity (gain) and dwell time may be varied as the beam scans - then the acquisition may be switched to the averaging mode. Typical acquisition times are one minute per frame.

Once acquired, an image may be stored on disk as a byte packed binary file. Similarly, a previously archived image may be rapidly recalled for display. These processes take only 2 seconds. In normal operation all images are 256 by 256 pixels by 8 bits. This allows the framestore to be 'quandranted' to provide 4 independent displays which may be viewed or copied either separately or together.

A number of image processing, image manipulation and image annotation routines have been written to enable simple interactive control over the final hardcopy. These routines require one or more specified source quadrants for input and will output the result to a specified destination quadrant. Processing options include two dimensional derivative, smoothing, gradient, edge enhancement, intensity scaling and software zoom and rotation. Images may also be arithmetically or logically combined. Intensity transformations may be effected by using different pre-programmed colour scales, by digital manipulation of contrast and brightness and by colour slicing.

It is a great advantage in analytical imaging, including SIMS imaging, to be able to manipulate images numerically. Two illustrations of this are as follows. The first is the ability to overlay independent images, for example, SIMS images at different selected masses or a SIMS image with a reference image of ion induced secondary electrons. This is especially important because the images may appear to be very different; for example a SIMS image of a trace element or contaminant may contain little or no information on the physical nature of the surface. The overlay is therefore a direct and visually striking method of determining the distribution of chemical species in relation to physical features. The second is the ability to use a reference image to remove unwanted information. The most obvious example of this is the elimination (or at least the reduction of) topographical detail in order to leave only chemical information. This can be achieved by ratioing the SIMS image to the SEM image or total ion image (the total ion image is obtained by operating the quadrupole analyser as a high pass mass filter). In our experience the total ion image is to be preferred since detector effects are identical to the normal SIMS image.

The combination of the high brightness LMIS and the digital framestore can therefore provide high resolution chemical images capable of simultaneously showing the distribution of a number of chemical species and their disposition relative to surface micro-features.

SIMS IMAGING RESULTS

The images shown here demonstrate high spatial resolution; digital acquisition, display and manipulation; fast mapping of matrix elements; analysis of electrical insulators including polymers, powders and semi-conductors; and include applications where SIMS microscopy can provide a unique analysis capability and produce data which complements that from other analysis techniques.

The example in Fig 1 shows analysis of an optical fibre. This is a sample where surface charging is severe for electron and ion probes so that a neutralisation technique must be used. The resultant experimental difficulties have proven to be more tractable, in our experience, using either the gallium or caesium ion probes with SIMS analysis, than using an electron probe with Auger analysis. In this example the dopant concentrations are within the detection range of Auger ($>0.1\%$) and happen to be relatively low sensitivity elements (Ge and P) in SIMS.

323

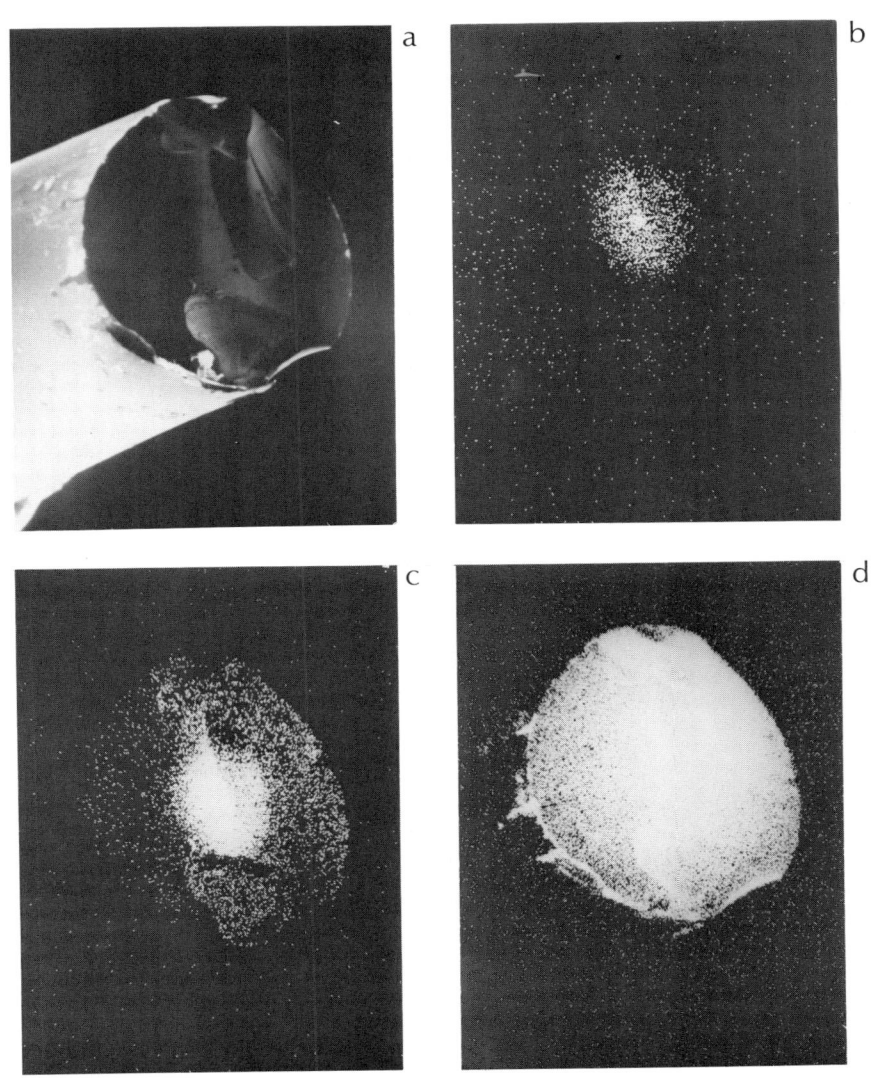

FIGURE 1 : A 125μm diameter optical fibre. (a) Ion induced SEM, 20s exposure; (b) P⁺ core dopant 1280s exposure; (c) Ge⁺ core dopant, 1280s exposure; (d) O⁻ matrix element, 80s exposure.

Phosphorus doping in silicon matrices is also known to present problems in SIMS due to the mass interference ^{30}SiH which coincides with ^{31}P. Even so, in this sample the ^{31}P concentration sufficiently high (0.3%) and the vacuum sufficiently clean that the interference is insignificant. It should be noted that sample preparation for this analysis was rudimentary, as it was for all the results shown. A single fibre was attached to a standard sample stub with silver paste and a fresh surface formed by snapping the fibre. No further preparation was involved and the fibre tip was overhanging the stub so that analysis was performed without a backing material present (which could lead to cross contamination). The images were obtained using a 10kV gallium probe at a current of 0.5nA.

Figure 2 illustrates the crystallographic contrast available from an ion induced secondary electron image, in this case of a weld edge in stainless steel. Though the origin of this enhanced contrast is not yet fully understood our investigations indicate that ion beam cleaning of the surface, which occurs automatically during imaging, may be a significant factor.

Figure 3 is also an ion induced secondary electron image, illustrating the depth of field obtainable. This image with a field width of 1mm allows the probe to be focussed onto a specific micro-area for high sensitivity analysis.

Biological specimens are also amenable to ion microprobe analysis and a potential application of finely focussed ion beams is in the microdissection of such specimens. Figure 4 shows an ion induced SEM of blood cells with three 0.5μm holes drilled into one cell.

A composite superconductor consisting of hexagonal arrays of Nb_3Sn filaments embedded in a bronze (Cu/Sn) matrix is shown in fig 5. The irregularly shaped filaments are some 5μm across. The surface was exposed to pure oxygen at 10^{-6} mbar during the acquisition of these images.

A unique application of the gallium probe is shown in fig 6. The sample is a single polymer fibre which has been treated to produce a fluoride coating. Surface charging had to be neutralised for the optical fibre, but another problem which confronts Auger analysis is the strong desorption of fluorine by an electron probe. Even the relatively low energy electron neutralising beam used in this SIMS analysis presented some difficulties since fluorine ions were also desorbed from metallic surfaces such as the specimen stub and so spurious ESD (electron stimulated desorption) signals would arise if the electron beam was not accurately positioned. Once positioned properly ESD contributed only a weak background, as mentioned by Briggs [4] in his work on static SIMS characterisation of polymers. Fluorine mapping on an untreated control fibre showed that the fluorine level was below the ESD background.

Catalysis is another important area amenable to SIMS microanalysis. An auto exhaust catalytic cleaner, seen in fig 7, was analysed with the MIG300 gallium probe operated at 27kV. The problem was to determine whether zinc and phosphorus, known to be responsible for poisoning the catalyst, were associated or not. The images of $^{64}Zn^+$ and $^{31}P^+$ show correlations which are consistent with these elements being chemically associated. With such samples topographical contrast can be a significant

FIGURE 2 :   A weld edge in stainless steel. Micrograph
             obtained using a MIG100 gallium microprobe at
             10kV.

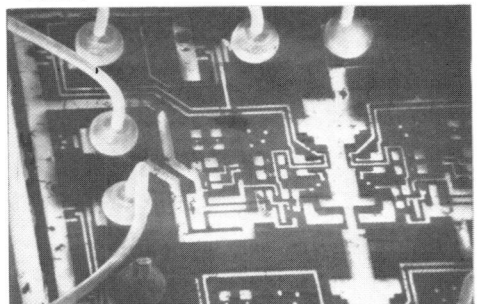

FIGURE 3 :   An integrated circuit and its wire bonding.
             Micrograph obtained using MIG100 gallium micro-
             probe at 10kV with 0.3nA probe current and 20s
             exposure.

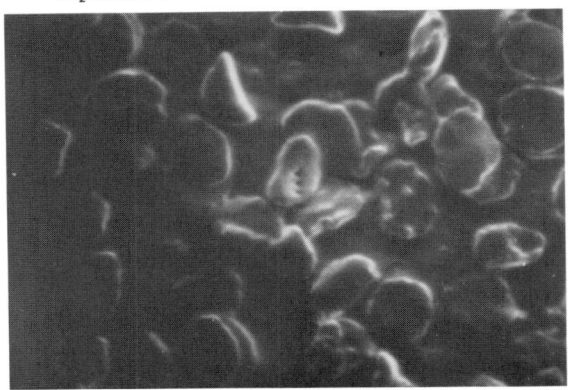

FIGURE 4 :   Ion-induced SEM of human blood cells. Probe
             0.1nA, 10kV Ga. Holes drilled by static beam in
             10s.

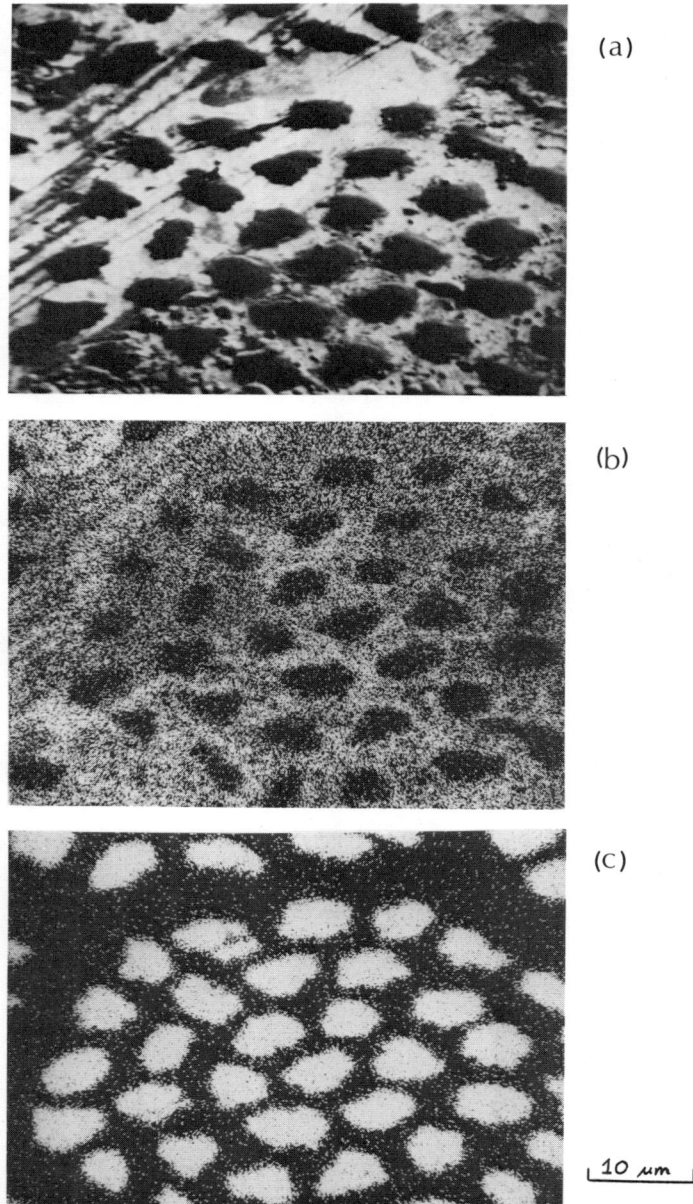

FIGURE 5 : Composite superconductor. (a) ion-induced SEM; (b) $^{63}Cu^+$ SIMS Image, 320s exposure; (c) NbO^+ SIMS image, 80s exposure.

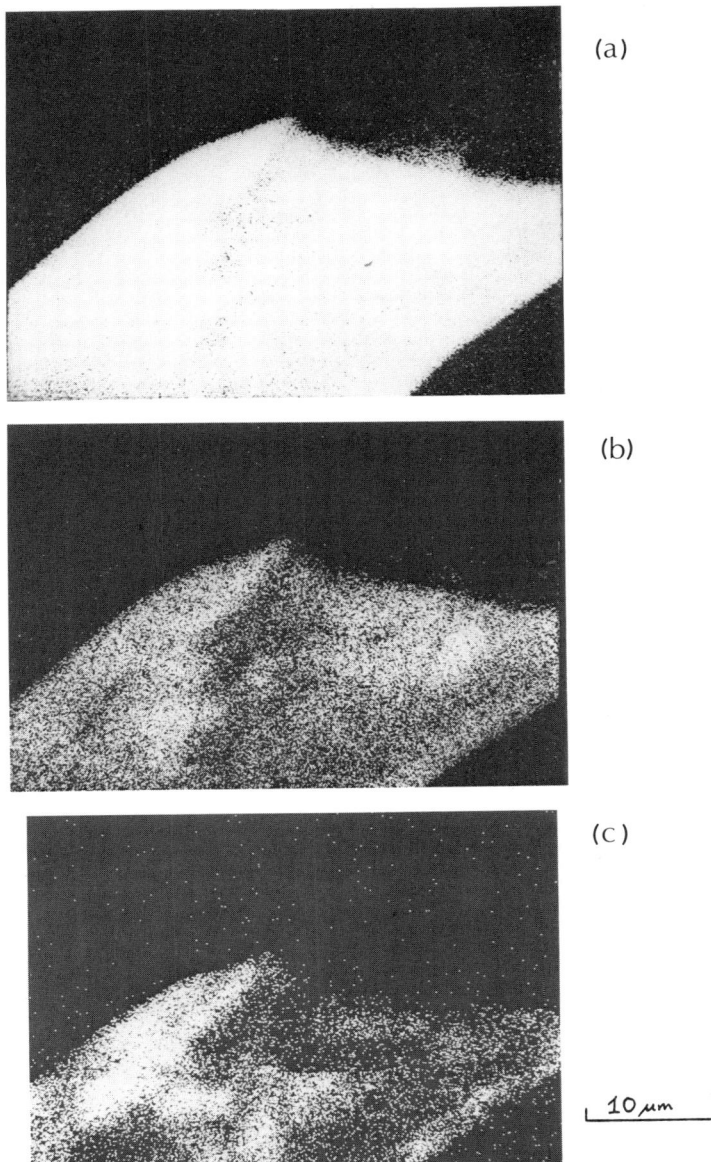

FIGURE 6 : A polymer fibre with a fluoride surface treatment imaged with a 0.1nA, 20kV Ga probe. (a) total negative ion image, 80s exposure; (b) C_2^- image, 160s exposure; (c) F^- image, 160s exposure.

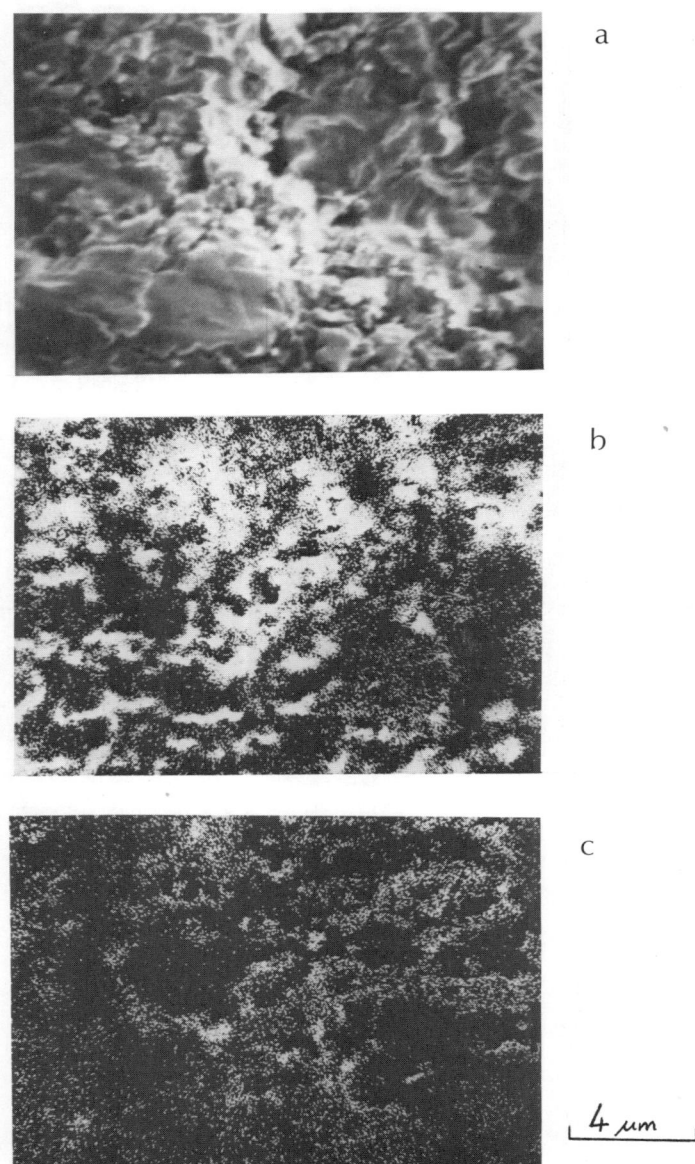

FIGURE 7 : An auto exhaust catalytic cleaner (powder supported catalyst) imaged with a 27kV Ga probe. (a) ion induced SEM, 0.1nA, 20s exposure; (b) Zn^+ image, 0.2nA, 320s exposure; (c) P^+ image, 0.2nA, 320s exposure.

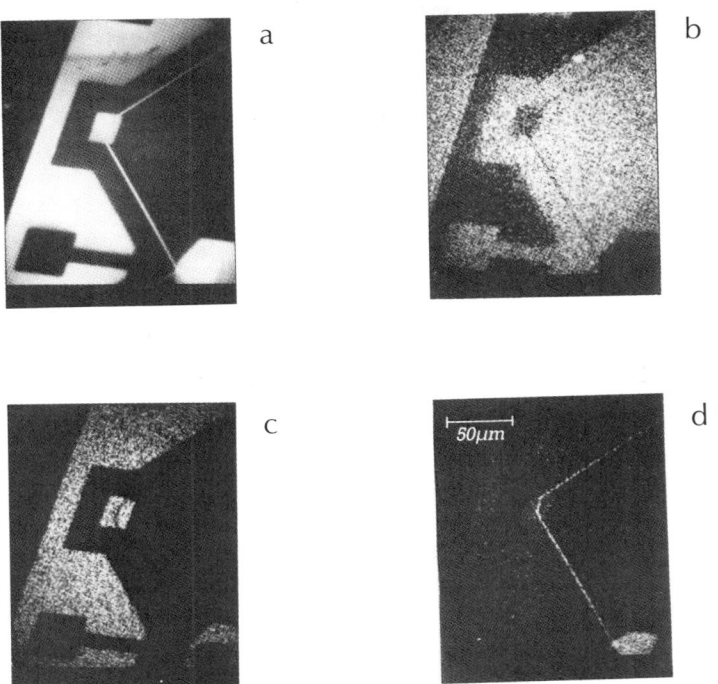

FIGURE 8 : A GaAs FET integrated circuit imaged with a 0.1nA, 5kV, Cs^+ probe. (a) ion induced SEM, 20s exposure; (b) Au^- image, 320s exposure (c) Cr^+ image, 240s exposure; (d) Al^+ image, 240s exposure.

cause of image contrast. However, such contrast is expected to be essentially the same for all SIMS signals, so that a more detailed comparison of images, acquired by computer, would help to distinguish topography from true concentration variations.
Fig 8 shows an example of the high negative ion sensitivities achieved with the caesium probe in the $^{197}Au_2^-$ map from the gold track of a GaAs integrated circuit. The $^{52}Cr^+$ and $^{27}Al^+$ maps also show good sensitivities for positive ions probably due to natural oxide layers on those tracks.

Unfortunately, the black and white SIMS and ion-induced SEM images shown here do not illustrate the powerful imaging capability of our new framestore. In particular the ability to overlay independent SIMS and SEM images in different colours is a direct and visually striking method of determining the relative distributions of different chemical species and their relation to physical features.

Levi-Setti et al [5] at the University of Chicago have also demonstrated an impressive Scanning Ion Microscopy capability and the relevance of work on focussed ion beams for semiconductor lithography has been discussed by Prewett [6].

CONCLUSION

Fast, high resolution, chemical imaging can now be attained with SIMS microscopy on the VG Scientific SIMSLAB using a primary ion probe from a high brightness liquid metal ion source. The 30kV primary ion column has extended the spatial resolution to 50nm for both SIMS and ion-induced SEM images. The caesium source has produced probe characteristics similar to gallium and extends the range of elements which may be analysed at the sub-micron level. The technique incorporates an electron flood gun to neutralise surface charging and allow analysis of insulators, including polymers, where Auger analysis may not be suitable. The addition of a digital framestore to the system greatly enhances the imaging capability and allows all data to be stored as it is acquired, for subsequent manipulation and image processing.

REFERENCES

[1] A.R. Bayly, A.R. Waugh and K. Anderson
    Nucl. Instrum. Meth. 218 375 (1983)

[2] A.R. Waugh, A.R. Bayly and K. Anderson
    Vacuum 34 103 (1984)

[3] F.G. Rudenauer, Surface and Interface Analysis 6 132
    (1984)

[4] D. Briggs, Surface and Interface Analysis 5 113 (1983)

[5] R. Levi-Setti, P.H. La Marche, K. Lan and Y.L. Wang, Proc.
    SPIE, 394 (1984)

[6] P.D. Prewett, Vacuum 34 931 (1984)

SOME APPLICATIONS OF SIMS AND SSMS IN MATERIALS CHARACTERIZATION

J. VERLINDEN, R. VLAEMINCK*, F. ADAMS AND R. GIJBELS
University of Antwerp, U.I.A., Dept. of Chemistry, Universiteitsplein 1,
B-2610 Antwerp-Wilrijk, Belgium.
*Bell Telephone Mfg.Co., Gasmeterlaan 106, B-9000 Gent, Belgium

ABSTRACT

Some applications of spark source mass spectrometry are given showing that this technique, often considered as suitable for bulk analysis only, can also be used for the determination of impurities in thin layers and for in-depth profiling in rather thick samples.
Secondary ion mass spectrometry, in addition to its applications in surface- and in-depth analysis, can successfully be applied to quantitative analysis when using the matrix ion species ratio as an environment sensitive indicator, or when using the isotope dilution method.

INTRODUCTION

Spark source mass spectrometry (SSMS) and secondary ion mass spectrometry (SIMS) are two powerful trace element analytical techniques with wide, but typically different-application areas. Whereas SIMS in its various configurations is well-known for surface-, in-depth- and lateral materials characterization [1], SSMS is mostly applied to qualitative and quantitative bulk analysis of a large variety of materials [2]. In recent years there is, however, an increased interest in SSMS for layer- and in-depth analysis [3]. Indeed, this technique may have some advantages over the typical surface analysis techniques when a microvolume has to be checked for all elements simultaneously, when very low (sub-ppm) concentrations are to be determined, when better than semi-quantitative results are required and when the surface roughness prevents the application of other techniques. This is especially so when relatively thick layers (well in excess of 1 μm) are to be studied.
In our laboratory we have both techniques available. SSMS is mostly used for bulk analysis of metals and high-purity semiconductors, and SIMS for the characterization of micro-electronics materials. The combined application of both techniques has been found quite valuable for solving some problems, as will be illustrated below.

EXPERIMENTAL

The spark source mass spectrometer is a JEOL JMS-01 BM-2 double focusing instrument with electrical detection equipment. For the analysis of layers photoplate detection (Ilford Q2) was used. For in-depth profiling the electrical detection (ED) system was used in the magnetic peak-switching mode with a Hall probe magnetic field monitor, thus allowing coverage of the entire mass range at constant accelerating voltage. The spark-gap width was kept constant with an automatic spark-gap controller. Some experimental parameters are given in Table 1.
The secondary ion mass spectrometer is a CAMECA IMS-300 instrument with electrostatic sector attachment.

Table 1: Experimental spark parameters

Sample	GaAs	Si	Gold	Re	Ni-Cu	Mg-Al
Spark voltage (kV)	2.4	2.8	1.5	3.2	2.4	3.8
Repetition frequency (kHz)	3	3	1	3	3	3
Pulse width (μs)	20	20	20	20	20	20
Counter probe:						
material	Au	Si	Au	n.a.	Cu	Al
dimensions (mm)	0.5($\emptyset$)	0.6 x 1.0	0.5($\emptyset$)	n.a.	2.0 x 1.7	1.0 x 2.3
Detector	Plate	Plate	E.D.	Plate	E.D.	E.D.

n.a. = not applicable

The rhenium filaments were investigated using a 6.0 keV Ar^+ primary ion beam rastered over a square area measuring ca. 500 μm on side and a primary ion current of 0.3 μA. The Si and GaAs layers were bombarded with a 5.5 keV O_2^+ beam of 400 μm diameter and a primary ion current of 3 μA. All measurements were carried out at oxygen saturation above the sample surface.

RESULTS

1. Bulk and surface analysis of 25 μm thick rhenium filaments

    Rhenium filaments are frequently used as substrate in thermal ionisation mass spectrometers. For high precision isotope ratio measurements it is necessary that the filaments do not give rise to spurious signals from interfering impurities.
    We have compared rhenium filaments from various manufacturers, all with the same nominal specifications: 25 μm thick, 700 μm wide and 99.99 % pure. The results of SSMS analyses of the different filaments are shown in Table 2. It is clear that the relative contents of some impurities differ considerably from one filament to another. The element iron, used as internal standard for SSMS analysis, was quantitatively determined by isotope dilution (ID-SSMS) and ID-SIMS) following a procedure described elsewhere [4] and the results are given in Table 3. In addition, taking sample 2 as standard for Fe in rhenium, also the MISR (matrix ion species ratio) procedure [5] taking ReO^+/Re^+ as environment sensitive ratio was attempted; the results are also listed in Table 3. The precisions (r.s.d.) obtained with ID-SIMS, ID-SSMS (photoplate detection) and MISR were 5 %, 12 % and 10 % respectively; all the results are in satisfactory agreement. Since the MISR method is the easiest and fastest, it was used for further Fe determinations.
    A study was made on the effect of heat treatment on the bulk and surface purity of the filaments. Therefore, the samples were annealed in vacuum (10^{-5} torr) at 1600°C for 15, 30 and 60 minutes respectively. SSMS analysis of the treated samples did not reveal depletion of any of the impurities within experimental error (30 %), except for sample 3 where the elements Na, Mg, Al, Si, Cl, K and Ca were found at smaller concentrations after a 15 minutes anneal, as compared to the untreated sample; this is shown in Table 4. Longer annealing did not result in any further purification, suggesting that the depletion was due to the removal of surface contaminants. This could be demonstrated by SIMS in-depth profiling: a surface enrichment by a factor of 100 or more was observed for the elements mentioned above on the untreated sample. The contaminants were largely removed after a 15 minutes anneal [4].

Table 2: Analysis of rhenium filaments by SSMS (ppmw).

Element	Sample 1	Sample 2	Sample 3	Sample 4
Na	1.4	1.0	5.6	0.47
Mg	0.61	0.51	4.1	1.2
Al	1.5	1.1	4.5	0.58
Si	4.7	2.0	14	0.93
P	1.9	0.76	5.1	1.2
S	3.3	0.31	4.2	0.31
Cl	0.51	0.08	1.8	0.03
K	5.8	3.1	7.5	0.54
Ca	7.7	1.8	7.0	5.2
V	< 0.03	0.11	< 0.03	0.02
Cr	5.7	5.3	4.1	6.3
Mn	0.60	0.23	0.18	0.46
Fe	67	48	40	32
Co	0.41	1.1	0.21	0.16
Ni	10	8.3	2.4	1.9
Cu	4.0	0.38	1.7	0.20
Zn	0.21	< 0.06	0.31	0.23
Sr	< 0.04	0.25	< 0.04	0.04
Mo	67	130	30	48
Rh	< 0.03	< 0.03	< 0.03	77
In	7.3	< 0.04	< 0.04	0.04
Sn	14	1.0	0.76	0.15
Ba	0.43	0.96	< 0.06	0.06
W	12	31	1250	12
Pb	0.83	0.17	< 0.1	< 0.1

Table 3: Determination of Fe in Re by ID-SIMS, ID-SSMS and MISR (ppmw).

Sample	ID-SIMS	ID-SSMS	MISR
1	75	67	81
2	49	48	= 48
3	34	40	36
4	38	32	27

Table 4: Analysis of untreated and annealed rhenium filaments (sample 3) by SSMS (ppmw).

Element	Untreated	15' 1600°C	30' 1600°C	60' 1600°C
Na	5.6	0.19	0.14	0.16
Mg	4.1	0.27	0.21	0.23
Al	4.5	0.37	0.27	0.17
Si	14	0.39	0.11	0.71
Cl	1.8	0.05	0.03	0.14
K	7.5	1.3	0.97	1.2
Ca	7.0	1.9	1.2	2.1
Fe	40	38	32	33

In general no differences were found in surface concentrations between samples treated at 1600°C for 15 and 60 minutes respectively. A 15 minutes anneal at 1600°C is thus effective in removing surface contaminants. A temperature of 1600°C is however not effective in purification of rhenium filaments in the bulk; for this purpose temperatures around 2000°C should be used [6].

2. Analysis of layers

For this application SSMS may have some advantages over other surface analysis techniques, or at least it can yield important additional information, when layers having a thickness ≥ 1 μm are to be analysed for all elements and when at least semi-quantitative results are required.

In SIMS elemental sensitivities may differ by 4 orders of magnitude making quantification difficult, especially when no standards are available. The SSMS relative sensitivities are more uniform (usually within a factor of 3). Especially in the analysis of semiconductor materials high sensitivity is of most importance. SIMS detection limits for elements in Si and GaAs, taken from the literature [7-20] are shown in Table 5. Many elements can be detected at concentration levels below 0.05 ppma; this is however not the case for F, Si, S, Cl, Zn, Ge, Se, Sn and Te in GaAs and for F, Cl and Sb in Si, as well as for other elements not reported in the literature. For the determination of such elements SSMS may be advantageous. For comparison, detection limits obtained by SSMS for elements in Si and in GaAs are listed in Table 6 [21]. Most elements can be determined with a detection limit of 0.01 ppma, this is especially so for bulk analysis. For the analysis of thin layers by SSMS, detection limits depend on the amount of material available. The volume v (μm^3) of material consumed in order to detect an impurity at a concentration c (ppma) is given by

$$v = \frac{A.n.10^{20}}{N.I.d.c.Y} \quad (1)$$

where A is the atomic weight of the matrix, n the number of ions necessary to have a just detectable (visible) line on the ion-sensitive emulsion (n ∼ 3.10^3), N is Avogadro's number, I the abundance of the isotope measured (%), d the matrix density (g/cm^3) and Y the useful ion yield (number of atoms consumed in order to deliver one ion to the detector). In the following some examples will be given of the application of SSMS and SIMS in layer analysis.

Table 5: Detection limits of SIMS for elements in silicon and gallium arsenide.

Element	SILICON DL (ppma)	Ref.	GALLIUM ARSENIDE DL (ppma)	Ref.
H			100	17
Li			0.001	20
Be	0.01	8	0.002	17
			0.4	18
B	0.01	9	0.01	17
	0.002	10		
C	0.01-0.1	7	0.01-0.05	7
O	0.1-0.5	7	0.01	20
	24	11		
F	0.1-0.5	7	0.1-1	7
Na	0.001-0.01	7	0.001	20
Mg	0.01-0.05	7	0.01	20
Al	0.001-0.01	7	0.001	20
			0.2	17
Si			0.5	17
			4	19
			20	20
P	0.1	10		
	0.02	12		
S			1	17
			10	20
Cl	0.1-0.5	7	0.05-0.1	7
K	0.001-0.01	7	0.001	20
Ca	0.01-0.05	7	0.01	20
Ti	0.01-0.05	7	0.001	20
V			0.001	20
Cr	0.06	13	0.002	18
			0.12	19
Mn			0.002	17
Fe	0.01-0.1	7	0.005	20
Co			0.005	20
Ni			0.005	20
Cu	0.01-0.05	7	0.2	17
			0.001	20
Zn			0.2	17
			40	19
			10	20
Ga	0.01-0.1	7		
Ge			0.5	17
As	0.4	9		
	0.06	14		
	0.01	15		
Se			1.6	18
			18	18
			20	20
Sn			0.5	17
			0.2	19
Sb	0.1-1	7		
	4.0	16		
Te			1-5	7

Table 6: Detection limits of SSMS for elements in silicon and gallium arsenide.

DL (ppma)	SILICON	GALLIUM ARSENIDE
0.1	Mn, Co, Sr, Ce	Na, Ge, Sc, Sr, Ti
0.01	Be, Na, Mg, Al, P, S, Cl, K, Ca, Sc, Ti, V, Cr, Ni, Br, Rb, Zr, Mo, Ru, Pb, Cd, In, Cs, Pr, Nd, Tu, Yb, Ta, Au, Hg	Li, Be, Mg, Si, S, Cl, K, Ca, Sc, V, Cr, Ni, Br, Rb, Mo, Ru, Rh, Pd, Cd, Sn, Te, Cs, Ba, Pr, Nd, Sm, Gd, Dy, Er, Y, Nb
0.001	Li, B, F, Cu, Zn, Ga As, Se, Zr, Nb, Rh, Ag, Sn, Br, Sb, Te, I, Ba, La, Sm, Eu, Gd, Tb, Dy, Pt, Tl, Pb, Bi, Th, U, Ho, Er, Lu, Hf, W, Re, Os, Ir	B, F, Al, P, Ti, Mn, Fe, Co, Cu, Zn, Y, Nb, Ag, Zr, Sb, I, La, Lu, Eu, Tb, Ho, Tu, W, Re, Os, Ir, Pt, Au, Hg, Te, Pb, Bi, Th, U

2.1. Analysis of 2.5 μm contaminated Si layers on a pure Si substrate

Spreading resistance measurements revealed a contamination (5.10^{15} at/cm^3) in a 2.5 μm Si layer, the substrate was pure (10^{14} at/cm^3) silicon. SIMS analysis revealed the presence of Al in low, just detectable, concentration in the layer. Since this element is used for alignement of the apparatus and thus may give rise to memory effects, and because other less sensitive elements may be present at even higher concentrations than aluminium (see table 5), SSMS was also used for this study. The sample was therefore sparked against a counter electrode, cut from the sample itself, following the x-y scan method [22]. Under our experimental SSMS conditions, a value of 10^{-7} was determined for Y [23]. In principle a surface of 5 x 5 mm^2 would thus be sufficient to reach the 0.01 ppma level (eq. 1). Since more material was available the layer was not sparked to the full depth but over an area of ca. 2 mm^2, until a charge of 2.10^{-7} C was collected on the ion-sensitive plate. The analysis confirmed the presence of Al at a concentration of about 0.06 ppma. In addition also the elements Cr and B were detected at concentrations near 0.01 ppma (near detection limit). No other elements, except the typical surface contaminants Na, K and Ca, also observed in the SIMS spectrum, were found by SSMS.

2.2. Analysis of 3 μm LPE GaAs layers on pure GaAs substrate

A similar contamination problem, though much more pronounced, was reported for an LPE (liquid phase epitaxy) GaAs layer. The SIMS spectrum of GaAs showed the presence of Al and Si contaminants. The same sample was also sparked against a pure Au counter electrode. Various SSMS exposures were collected up to a maximum charge of 10^{-8} C. In this way the sub-ppm level was reached. The same contaminants as revealed by SIMS were observed at concentrations of 75 ppma Al and 125 ppma Si. All other elements were present below 0.5 ppma. The uncertainty is within a factor of 3 or better. More accurate results (< 30 %) can be obtained even if only one standard is available, since the SSMS intensities are linearly proportional to concentration over 5 orders of magnitude [24]. Quantification methods for elements in GaAs by SIMS are described in literature [25] giving results accurate within a factor of 2. From our experiments, the SIMS detection limits for Al and Si in GaAs were estimated to be 0.03 ppma and 0.4 ppma respectively.

2.3. Determination of C in 3 μm thick gold layers

A comparative study on the determination of C in gold coatings (deposited from cyanide solution) on a nickel substrate was made by SSMS and charged particle activation analysis (CPAA). The layers used for SSMS analysis were 3 μm thick, those for CPAA 50 μm. For both analyses a standard of gold containing 0.128 wt % C was made available. The results obtained with both techniques are given in Table 7, showing that good accuracies (<10 %) can be obtained by SSMS. The carbon concentrations in this application are relatively high. For the determination of the elements C, N and O by SSMS, detection limits of 0.1 ppma or better can be obtained [26], when special care is taken to improve the quality of the vacuum.

Table 7: Determinations of C in a gold layer with SSMS and CPAA.

Sample	SSMS (wt%)	CPAA (wt%)
1	0.073	0.076
2	0.061	0.065
3	0.092	0.084

3. Lateral- and in-depth profiling

SSMS has various applications for in-depth analysis when the profile extends over 10 μm or more, although also profiles were obtained in samples of less than 1 μm thickness [3]. Limits of lateral and depth resolution of SIMS and SSMS are compared with those of some other techniques in Figure 1. "Depth resolutions" (crater depths) of 1 μm for metals and 0.1 μm for insulators and lateral resolutions of 10-100 μm can be obtained by SSMS [3].

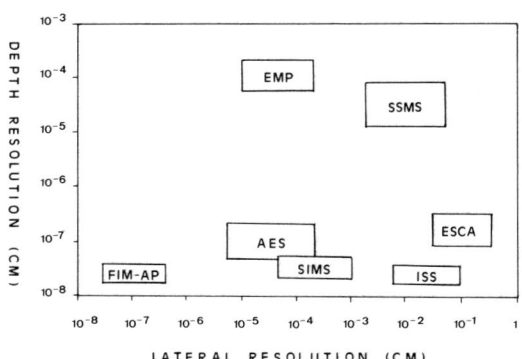

Figure 1: Typical values of depth resolution and lateral resolution for SIMS, SSMS and some surface analysis techniques.

For in-depth analysis the material consumption in the spark source should accurately be controlled; detailed studies of the parameters responsible for sample consumption were reported [27]. Control of sparking parameters can be used to vary erosion rates over roughly two orders of magnitude. Figure 2 shows an in-depth profile measured by SSMS for a diffusion profile extending over some mm. It concerns the diffusion of Mg in an Al-AlMg diffusion couple heated at 462°C for 525 hours. The results obtained by SSMS by controlled spark erosion through the junction are compared with those obtained with SIMS (ion probe) on a lateral cut through the sample [27]. Figure 3 shows a diffusion profile of Ni in Cu obtained after in-depth analysis with SIMS and SSMS. The sample was high-purity copper on which a 40 nm Ni-layer had been deposited. The sample was heated in vacuum at 900°C for 27 hours and diffusion occurred over some ten micrometers [28]. The SSMS results shown in Figs. 2 and 3 were obtained using respectively a pure Al and Cu counter electrode of the same dimensions as the sample electrode and by sparking in a plane-to-plane geometry.

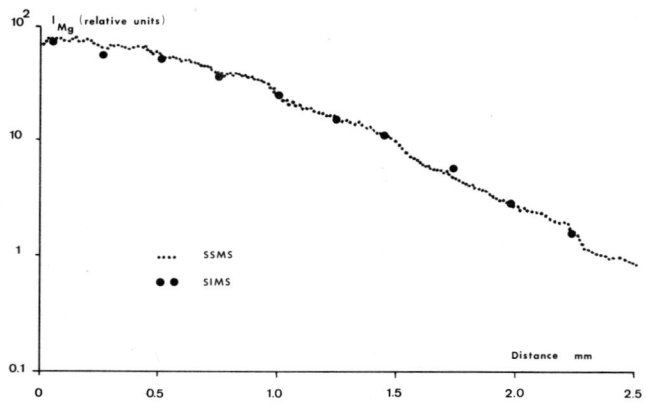

Figure 2: Diffusion profile of magnesium in an Al-AlMg diffusion couple
(. SSMS in-depth, ○ SIMS laterally)

Our results show that SSMS can successfully be applied to in-depth analysis when profiles extend over depths of some micrometers to some millimeters. Concentration profiles of various elements in silicon and gallium arsenide were also measured by SSMS for total erosion depths between 2 and 20 μm [3].

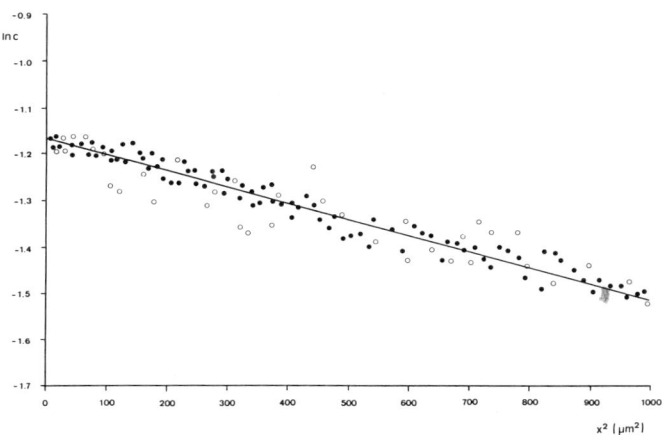

Figure 3: Diffusion profile of nickel in copper (○ SSMS in-depth, o SIMS in-depth)

CONCLUSION

Some applications are described, which show the potentialities of SSMS for layer and in-depth analysis. For layer analysis SSMS is especially useful when measuring low concentrations of elements with relatively high detection limits in SIMS. For in-depth analysis SSMS is useful when profiles extend over some micrometers or more.

As to SIMS, in addition to its powerful possibilities for surface- and in-depth analysis, it can also be used for quantitative analysis when standards are available: the MISR method yields accuracies better than 20 %.

REFERENCES

1. H.W. Werner and A.P. von Rosenstiel, J. Phys., 45 (1984) C2-103.
2. J.R. Bacon and A.M. Ure, Analyst 109 (1984) 1229.
3. V.I. Derzhiev, G.I. Ramendik, V. Liebich and H. Mai, Int. J. Mass Spectrom. Ion Phys. 32 (1980) 345.
4. J. Verlinden, R. Gijbels, H. Silvester and P. De Bièvre, Comission of the European Communities, Physical Sciences, Report EUR 9653 EN (1985).
5. J.D. Ganjei, D.P. Leta and G.H. Morrison, Anal. Chem. 50 (1978) 285.
6. J. Verlinden, R. Gijbels, H. Silvester and P. De Bièvre, J. Microsc. Spectrosc. Electron, 7 (1982) 291.
7. A.M. Huber and M. Moulin, J. Radioanal. Chem. 18 (1972) 75.
8. R.G. Wilson and J. Comas, J. Appl. Phys. 53 (1982) 3003.
9. W. Vandervorst, Ph.D. Thesis, K.U.L. Leuven, Belgium (1983).
10. M. Grasserbauer, G. Stingeder, E. Guerrero, H. Pötzl, R. Tielert and H. Ryssel, in SIMS III, A. Benninghoven, J. Giber, J. Laszlo, M. Riedel, H.W. Werner eds., Springer 1982, p. 321.
11. T.J. Magee, Appl. Phys. Lett. 39 (1981) 260.

12. J.M. Gourgout, in SIMS II, A. Benninghoven, C.A. Evans, R.A. Powell, R. Shimizu, H.A. Storms eds., Springer 1979, p. 286.
13. R.G. Wilson, P.K. Vasudev, D.M. Jamba, C.A. Evans and V.R. Deline, Appl. Phys. Lett. 30 (1977) 559.
14. P. Williams and C.A. Evans, Appl. Phys. Lett. 30 (1977) 559
15. R.G. Wilson, J. Appl. Phys. 52 (1981) 3985.
16. W.K. Chu, Rad. Eff. 47 (1980)1.
17. K. Wittmaack, Appl. Phys. Lett. 52 (1981) 527.
18. M. Gauneau, A. Rupert and R.P. Favennec, in SIMS III, A. Benninghoven, J. Giber, J. Laszlo, M. Riedel, H.W. Werner eds., Springer 1982, p. 342.
19. W. Gerigk and M. Maier, in SIMS II, A. Benninghoven, C.A. Evans, R.A. Powell, R. Shimizu, H.A. Storms eds., Springer 1979, p. 64.
20. A. Huber, G. Morillot and P. Merenda, J. Microsc. Spectrosc. Electron, 4(1979)493.
21. Yu. A. Karpov and I.P. Almarin, J. Anal. Chem. USSR, 34 (1979) 1085.
22. R.K. Skogerboe, in A.J. Ahearn (Ed.), Trace Analysis by Mass Spectrometry, Academic Press, New York and London (1972).
23. K. Swenters, unpublished results
24. E. Van Hoye, F. Adams and R. Gijbels, Talanta 23(1976) 789.
25. A.E. Morgan and J.B. Clegg, Spectrochim. Acta, 35B(1980) 281.
26. J.B. Clegg and E.J. Millett, Philips tech. Rev. 34 (1974) 344.
27. J. Van Puymbroeck, J. Verlinden, K. Swenters and R. Gijbels, Talanta, 31 (1984) 177.
28. R. Gijbels, J. Verlinden and K. Swenters, JEOL News, 20A (1984) 5.

CHARACTERIZATION OF THIN METAL FILMS BY SIMS, AUGER, AND TEM

B. K. FURMAN, J. P. BENEDICT, K. L. GRANATO, AND R. M. PRESTIPINO,*
AND D. Y. SHIH**
* IBM Corporation, Poughkeepsie, NY 12602
** IBM Corporation, East Fishkill, Hopewell Junction, NY 12533

ABSTRACT

Secondary ion mass spectrometry (SIMS), Auger spectroscopy, and transmission electron microscopy (TEM) were applied to study Au, Cu, Ti, Ti-W, Cr, Al-Cu, and Al metal films deposited on both Si and ceramic substrates.
SIMS analysis of as-deposited metal films was used to characterize impurity levels, both metallic and gaseous, incorporated during deposition. The results revealed that the levels varied under generally accepted deposition conditions. As-deposited and annealed films were examined with SIMS, Auger, and TEM to study interdiffusion, grain growth, and impurity segregation as a function of processing conditions. Metallic impurities were observed to modify Au/Ti interdiffusion. Large variations in residual H, C, O, and N were observed in as-deposited Al and Al-Cu films.
Hydrogen, incorporated during deposition of Ti films, was observed to redistribute after thermal annealing in N_2 or thermal cycling in forming gas (N_2 - 10% H_2). Samples thermally cycled in forming gas absorbed additional H into the Ti layer. SIMS/ion imaging was used to study the incorporation and segregation of H. Differences in H behavior were observed to be dependent upon metal structure, composition, and substrate material.

INTRODUCTION

The use of thin metal films in semiconductor device and packaging applications is widely established [1-5]. As computer technologies become more complex, faster, and more reliable, the need to control and understand all aspects of thin film materials properties becomes increasingly important. One area of primary interest is the role of metallic and gaseous impurities, incorporated into metal films during deposition or subsequent processing, on the thin film properties. Complete characterization of a thin film system often requires an arsenal of analytical tools for surface and materials analysis. In general, no single technique will provide complete characterization; the application of multiple techniques can usually provide a better understanding of the system being studied. This paper discusses three analytical techniques used to characterize several thin metal film systems: Secondary ion mass spectrometry (SIMS) [6], Auger electron spectrometry (Auger) [7], and transmission electron microscopy (TEM) [8].
In this study, Cr, Ti-W, Al-Cu, Al, Ti, Cu, and Au thin films with various processing were analyzed. TEM and Auger analysis were used to study interdiffusion, grain growth, and impurity segregation. SIMS analysis, however, was the major technique used because no sample preparation was necessary and the time required for in-depth profiling was much less than

with Auger. Additionally, SIMS was the only one of the applied techniques [9] that could realistically study the absorption of H as a function of processing. SIMS/ion imaging provides lateral information on H and hydride formation as shown by Williams et al. [10] for Niobium Hydride. This study demonstrates the importance of SIMS/ion imaging as a technique to characterize impurity distributions in metal films.

EXPERIMENTAL

The metal films were prepared by a number of deposition techniques under varying conditions, as described in Table I. Evaporation was carried out in a Varian 3120 three hearth electron-beam evaporator with a liquid N_2 trapped diffusion pump system. Metal films were deposited sequentially on a thermally grown SiO_2 layer on a Si wafer. The pressure during evaporation was maintained at mid-10^{-7} torr. The typical evaporation rate was 1 nm/sec and the substrate temperature was controlled at 150° C throughout deposition.

Sputter deposited films were prepared by dc Magnetron sputtering. The initial chamber pressures were 3-5 x 10^{-7} torr. The Al and Ti-W films were deposited at 1.5 kW and 0.67 Pa Ar pressure. The typical deposition rates were 1.5 nm/sec.

A tube furnace was used for N_2 annealing at 400° C for 1 hour. Thermal cycling experiments were carried out using a belt furnace with a peak temperature of 350° C for no more than 5-6 min. in forming gas. The samples were processed repeatedly to obtain 30, 60, or 100 min. anneals.

TABLE I SAMPLE MATRIX

METAL	SUBSTRATE	THICKNESS (UM)	DEPOSITION
Au/Ti	SiO_2/Si	0.3/1	Evaporated
Au/Ti	SiO_2/Si	0.5/0.8	"
Au/Ti	Ceramic	0.3/1	"
Ti	Ceramic	0.5	"
Cu/Ti	SiO_2/Si	0.5/0.3	"
Cr	SiO_2/Si	0.2	"
Al-Cu	SiO_2/Si	2	Evaporated
Al	SiO_2/Si	1-2	Sputtered
Al-Cu	SiO_2/Si	2	"
Al-Cu/Ti-W	SiO_2/Si	1/0.5	"

For SIMS analysis, a CAMECA ims 3F ion microscope was used. This instrument allows in-depth profiling as well as lateral ion imaging. Table II lists the conditions used for the various modes of analysis.

The instrument used for Auger measurements was a Physical Electronics (Phi) Model 595 Scanning Auger Spectrometer. A 3 keV electron beam with a 25 nA beam current was used for the analysis. The samples were sputtered at 10 nm/min. calibrated by sputtering through a known thickness (120nm) Au film using 2 KeV Ar+ ion bombardment.

TEM analysis was done with a Phillips 420T transmission electron microscope. Bright and dark field micrographs as well as selected area diffraction patterns were generated using a tungsten filament and 120 keV accelerating potential. Two methods of sample preparation for the TEM were used. The first involved peeling the metal film from the substrate on which it was deposited and thinning, using an ion mill, until an electron transparent region was obtained. The second method was to study the electron transparent region of the sample formed by ion sputtering during SIMS analysis.

TABLE II SIMS ANALYTICAL CONDITIONS

	Impurity Analysis	
	Gaseous	Metallic
Primary ion species	Cs^+	O_2^+
Energy	10KeV	12.5KeV
Current	1-2µA	.5-1uA
Spectroscopy	Negative	Positive
Analysis area (diameter)		
Ion imaging	150um	150um
Depth profiling	30um	30um
Species detected	H, C, O, F N(AlN, TiN) Au	Ti, Cr, Cu, Al, Si, Au
Sputter Rates	50-100 A/sec.	5-25 A/sec.
Chamber pressure	1×10^{-8} torr LN_2 trap	2×10^{-7} torr

RESULTS

Auger analysis of as-deposited and annealed samples was used to study surface composition and Ti out-diffusion through the Au in the Au/Ti films. Figure 1 represents three Auger surface surveys from an as-deposited film, and two annealed samples. The as-deposited sample showed only Au present on the surface. Although the two annealed samples were prepared and processed under "identical" conditions, differences in surface composition were observed. One sample indicated high levels of Ti and O on the surface as well as Au. The other sample showed the presence of possible Cr contamination but limited Ti or O. Differences in Ti out-diffusion between the samples were shown by Auger in-depth profiles (Figure 2). The as-deposited sample showed only Au present over the depth analyzed (about 50 nm). Both annealed samples indicated surface oxides limited to the top 10 nm layer.

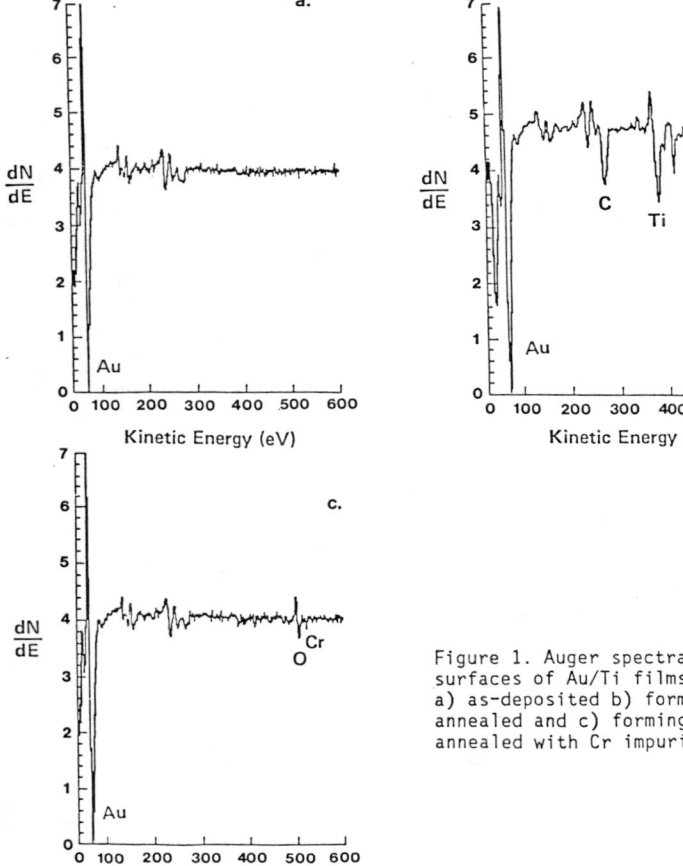

Figure 1. Auger spectra of Au surfaces of Au/Ti films on Si for a) as-deposited b) forming gas annealed and c) forming gas annealed with Cr impurity.

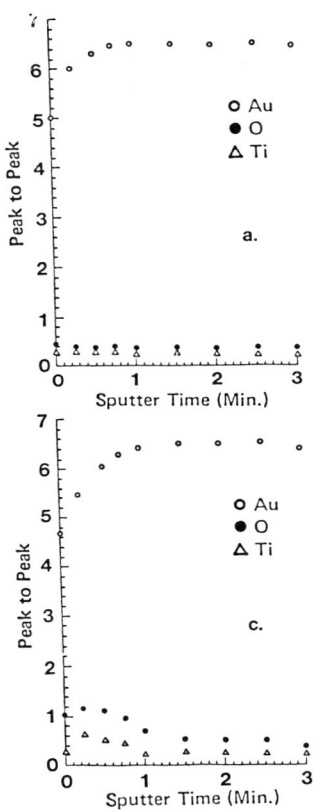
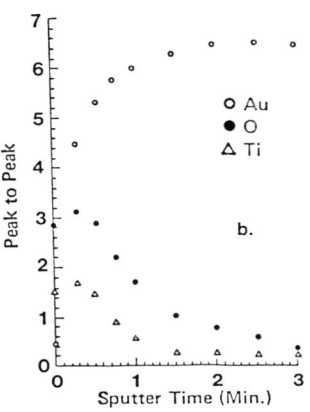

Figure 2. Auger depth profiles of Au/Ti films on Si for a) as-deposited b) forming gas annealed and c) forming gas annealed with Cr impurity.

Auger analysis was also used to characterize gaseous contaminates (C, and O) in both as-deposited and annealed samples. It is well established that small amounts of interstitial impurities such as C and O can alter the material properties of Ti [11], and interdiffusion and alloy formation in Au/Ti [12] and Au/Ti-W [13] thin-film systems. Auger analysis indicated that all the samples contained low levels of C and O in the Ti. Cr was detected at the Au/Ti interface in the as-deposited and one of the annealed samples. The other annealed sample, which showed Ti outdiffusion, contained no detectable Cr.

SIMS analysis (Figure 3) of these samples confirmed the presence of high Cr contaminant levels co-deposited with the Ti on one annealed sample. The in-depth distribution of Cr in this sample was very non-uniform and appeared segregated into three specific regions: the Au/Ti interface, the Ti/Si interface and a region about 400 nm below the Au/Ti interface. SIMS/ion imaging of the lateral Cr distribution indicated a very uniform distribution. TEM analysis of these samples indicated only minimal grain growth after annealing, in agreement with results previously reported by Tisone and Drobek [14], and no detectable Cr segregation. A precise identification of the Cr location and phase was not possible, because of the small (about 10nm) grain size of the Ti. Microdiffraction and small spot EDX indicate that Cr is present in all Ti regions analyzed. Both SIMS and Auger confirm that the Cr preferentially segregates to the Au surface [15].

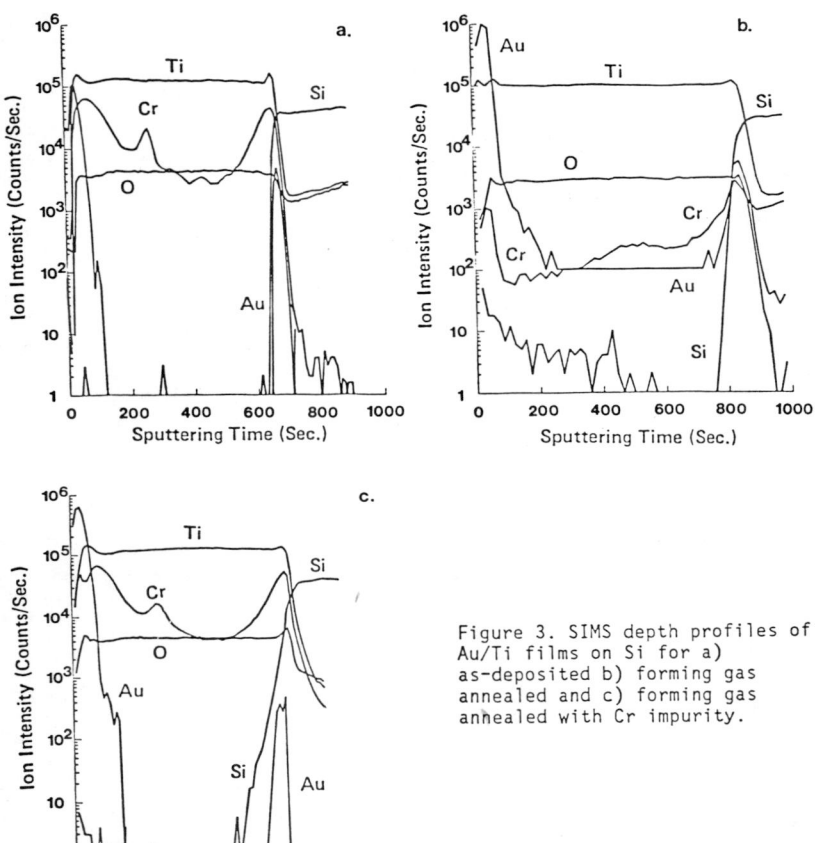

Figure 3. SIMS depth profiles of Au/Ti films on Si for a) as-deposited b) forming gas annealed and c) forming gas annealed with Cr impurity.

The SIMS analysis confirmed uniformly low O and C levels. Slight increases in N were observed after annealing in N_2 or forming gas. The most significant results are shown in Figures 4-8. Hydrogen present in the as-deposited Ti films was observed to redistribute to the Au/Ti interface after annealing (Figure 4). Samples annealed in forming gas also absorbed additional H and exhibited H segregation (Figure 5 & 6).

Similar results were observed in other Au/Ti samples as shown in Figure 7. Hydrogen absorption was observed on samples with different thicknesses of Au/Ti and on a sample with only Ti. Samples with Ti layers thicker than 400 nm show that the H absorption does not penetrate the entire thickness of the layer. The H in-depth distributions were similar in all these Au/Ti samples regardless of the presence of Cr in the Ti. The lateral distributions, however, were vastly different. In addition, one Au/Ti sample experiencing very poor adhesion showed a similar in-depth distribution of H, but the density of these H rich areas was very low.

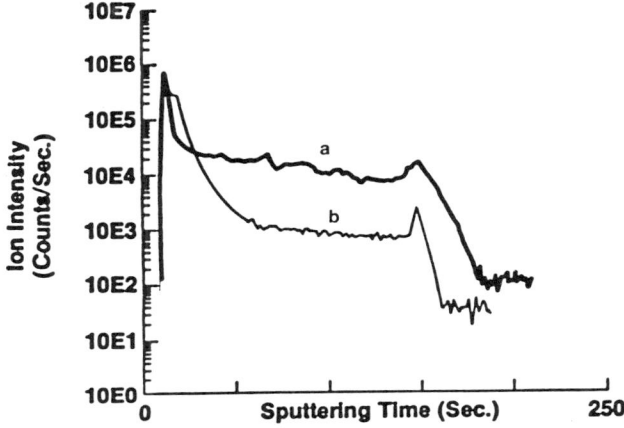

Figure 4. SIMS in-depth profiles of H in Au/Ti films on Si for a) as-deposited and b) annealed in N_2.

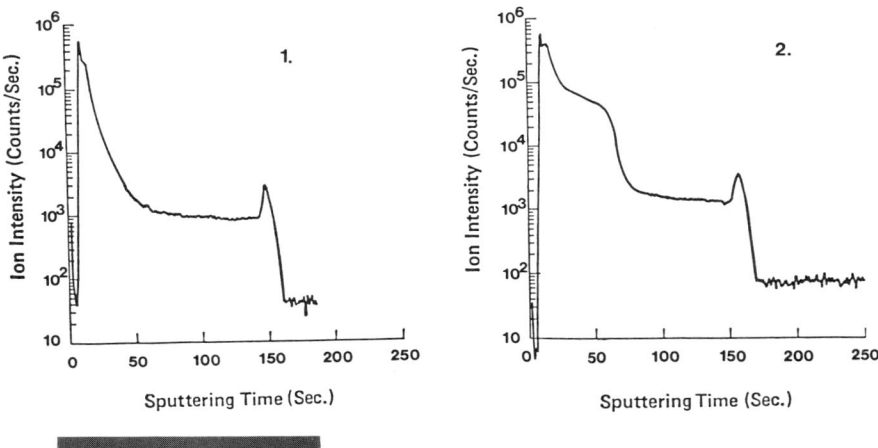

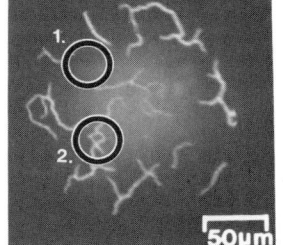

Figure 5. SIMS ion micrograph of lateral H distribution recorded between 25 and 50 sec. sputtering time (top 400 nm of Ti). Sample is Au/Ti film annealed in forming gas. SIMS in-depth profiles obtained from regions 1) and 2) are also shown.

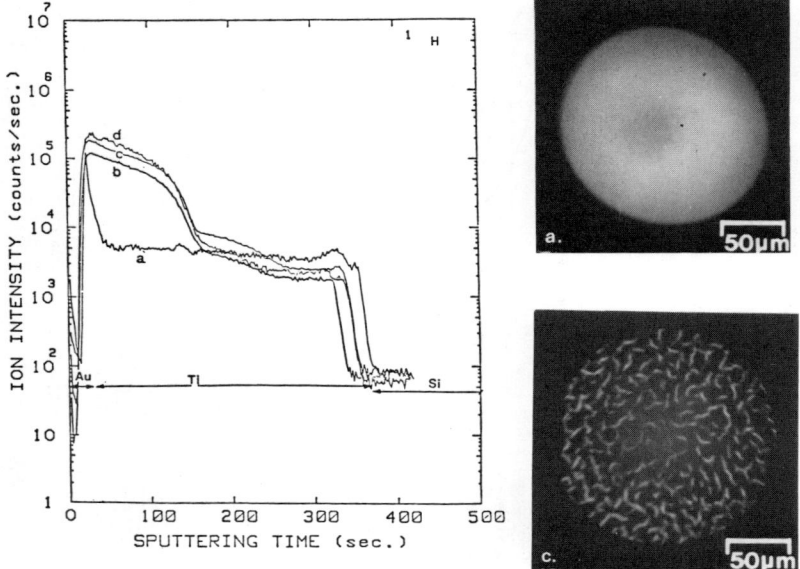

Figure 6. SIMS in-depth and lateral distributions of H in Au/Ti films on Si for a) as-deposited and b) 30 min. c) 60 min. d) 100 min. anneal in forming gas. Ion micrographs were recorded in the top 400 nm of the Ti layer.

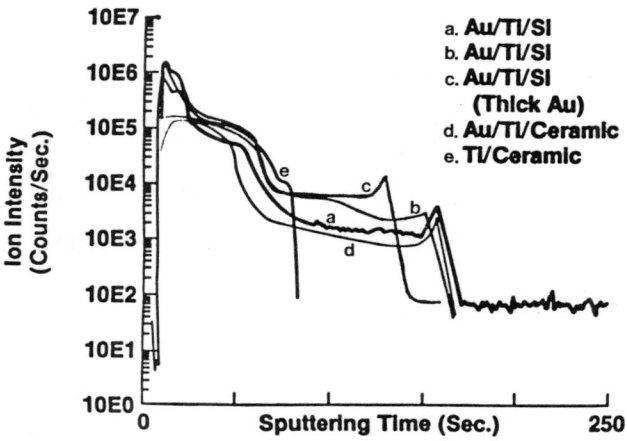

Figure 7: SIMS in-depth profiles of H for five Au/Ti or Ti-only samples on Si or ceramic substrates.

Figure 8 displays ion micrographs of the lateral H distribution in the samples above. A dramatic difference in H "decoration" is observed between similar samples processed under identical conditions. Figure 9 shows the absorption of H into a Cu/Ti sample as a function of annealing in forming gas. In this case, the distribution of H both in-depth and laterally is different from that observed in the Au/Ti or Ti only sample. TEM analysis indicates that the Ti grain size for all samples containing Ti is similar. Additionally, TEM observed large structures that were similar in shape and distribution to those shown to be H rich by ion imaging. The presence of TiH_2 lines in the electron diffraction patterns in conjunction with the SIMS results and the reported low solid solubility of H in Ti [16] indicates the presence of titanium hydride formation.

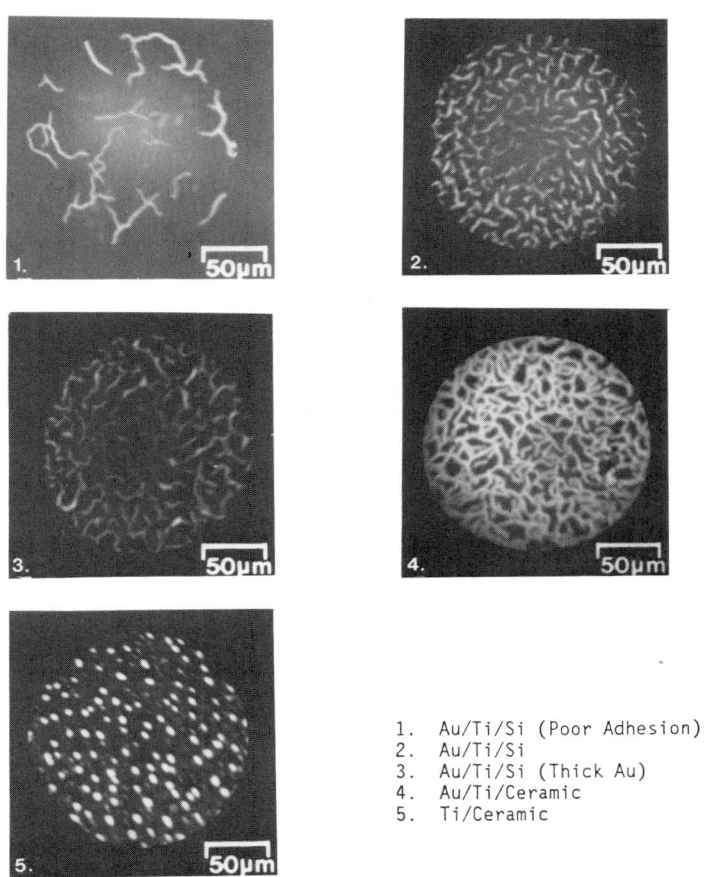

1. Au/Ti/Si (Poor Adhesion)
2. Au/Ti/Si
3. Au/Ti/Si (Thick Au)
4. Au/Ti/Ceramic
5. Ti/Ceramic

Figure 8: Ion micrographs of H in samples shown in Figure 7.

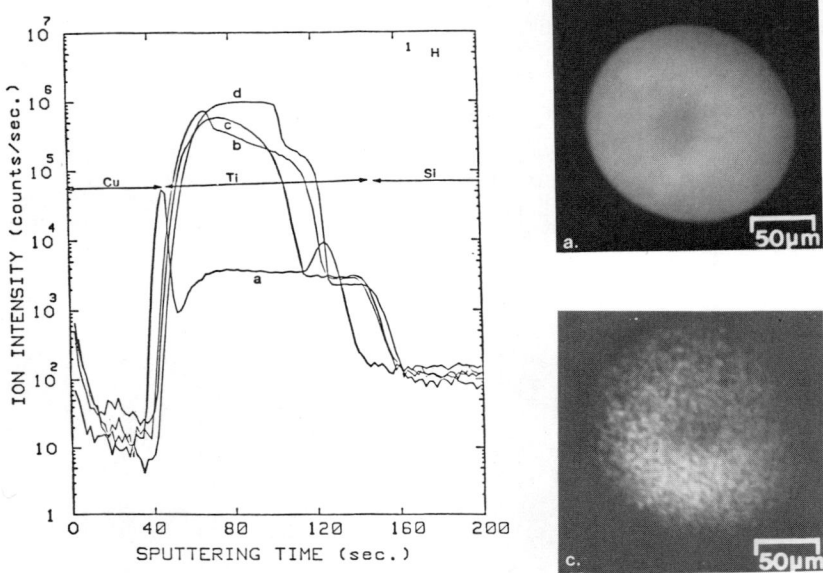

Figure 9: SIMS in-depth and lateral distributions of H in Cu/Ti films on Si for a) as-deposited and b) 30 min. c) 60 min. d) 100 min. anneal in forming gas. Ion micrographs were recorded in the top 400 nm of the Ti layer.

Kirchheim [17] has proposed results on Pd which show that internal stresses strongly affect H solubility, mobility, and hydride formation. Further work to characterize the role of internal stress within the films may provide further understanding of the differences in H distribution observed in this study. Additionally, it is important to understand the role hydride formation may have on the materials properties of Ti thin films.

Analysis of other thin metal films (Cr, Al, Al-Cu, Ti-W) for gaseous impurities indicates no H absorption during annealing in forming gas. In fact, H levels decreased in Cr films as shown in Figure 10. The results did show large differences in impurity levels in as-deposited films under "clean" deposition conditions. These results are summarized in Table III. The influence of gaseous contaminants on grain size, electromigration and hillock growth has been reported by Reimer [18] for electron beam evaporated Al films. Additionally, N and H have been reported [19,20] to adversely affect the reliability of sputtered Al-Si metal films.

TABLE III SIMS ANALYSIS OF GASEOUS IMPURITIES IN BLANKET FILMS

SAMPLE DESCRIPTION	ION INTENSITY (COUNTS/SEC.)*			
	H	C	O	N(AlN)
EVAPORATED Al-Cu				
AS DEPOSITED	300	700	8000	25
400°C 1 HR. N2	300	600	8000	50
SPUTTERED Al-Cu				
AS DEPOSITED	300	800	1300	200
400°C 1HR. N_2	300	800	1500	200
350°C 1 HR. FG	300	800	1400	200
SPUTTERED Al				
AS DEPOSITED	1×10^3	2×10^4	5×10^4	1.2×10^5
AS DEPOSITED	800	1.5×10^4	1.5×10^4	8×10^2
AS DEPOSITED	200	1.5×10^4	2×10^4	5×10^2
Al-Cu				
AS DEPOSITED	300	1.5×10^4	1.5×10^4	-
400°C 1 HR. FG	300	1.5×10^4	1.5×10^4	-
TiW				
AS DEPOSITED	4×10^3	1.5×10^4	1×10^5	-
400°C 1 HR. FG	4×10^3	1.5×10^4	1×10^5	-

* All counts normalized to Al_3^- at 5×10^5 cps

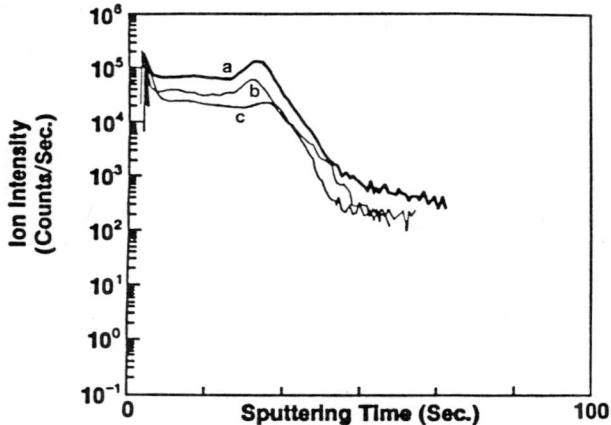

Figure 10: SIMS in-depth profiles of H in Cr films on Si for a) as-deposited and b) 30 min. c) 60 min. d) 100 min. anneal in forming gas.

CONCLUSIONS

These results illustrate the importance of SIMS/ion microscopy as the major viable technique to study H redistribution and absorption in Ti thin metal films as a function of deposition and processing. Hydrogen co-deposited with Ti was observed to redistribute when annealed in N_2 or in forming gas. Additionally, Ti was observed to getter H from the forming gas into specific structures and then to form hydrides during processing. TEM measured Ti grain size to be much smaller (about 100X) than the H rich structures observed by SIMS. These results indicate that H decoration is not simple grain boundary segregation.

The SIMS and Auger results demonstrate that metallic impurities such as Cr can affect out-diffusion and surface composition in Au/Ti films. It has also been reported that the annealing ambient can affect the interdiffusion of Au/Ti [21,22]. This may ultimately affect solderability and/or bonding onto the Au surfaces during subsequent processing. Additionally it illustrates the importance of interrelating the results obtained from multiple techniques in thin film characterization. Although Auger analysis provides excellent surface composition and depth profiling of major constituents in very thin films (<200nm), SIMS can provide additional information on trace impurities as well as in-depth analysis of thicker films.

This study demonstrates the value of SIMS/ion microscopy, in general, for the analysis of gaseous impurities in thin metal films. The results indicate impurity levels of H, C, O, N, and F can vary dramatically in samples deposited under similar conditions. This fact may also help to explain why similar films thought to be "identical" initially, behave differently when subjected to reliability and functionality testing. Further work to correlate the effect, if any, of these impurities on the materials properties and/or reliability of thin metal films is required. This would provide unique information to allow better understanding of deposition, processing, and ultimately, testing of thin metal films in semiconductor and packaging applications.

REFERENCES

1. M. Wittmer, J. Vac. Sci. Technol. A. 2, 273 (1984).
2. K. N. Tu, J. Vac. Sci. Technol. A. 2, 216 (1984).
3. J. M. Morabito, J. H. Thomas III and N. G. Lesh, IEEE Trans on Parts, Hybrids, and Packaging. PHP-11, 253 (1975).
4. C. W. Ho, D. A. Chance, C. H. Bajorek, and R. E. Acosta, IBM J. Res. Develop. 26, 286 (1982).
5. T. W. Orent and R. A. Wagner, J. Vac. Sci. Technol. B. 3, 844 (1983).
6. R. J. Blattner, and C. A. Evans, Jr., SEM, Inc., ed. by O. M. Johari. 1, 55 (Chicago, IL, 1980).
7. J. W. Morabito, Thin Solid Films. 19, 21 (1973).
8. G. Thomas and M. J. Goringe, *Transmission Electron Microscopy of Materials* (J. Wiley and Sons, New York, 1979).
9. J. F. Ziegler, et al., Nucl. Instrum. Methods. 149, 19 (1978).
10. P. Williams, C. A. Evans, M. L. Grossbeck, and H. K. Birnbaum, Anal. Chem. 49, 1399 (1977).
11. N. E. Paton and R. A. Spurling, Metall. Trans. 7A, 1769 (1976).
12. G. W. B. Ashwell and R. Heckingbottom, J. Electrochem. Soc. 128, 649 (1981).
13. J. E. Baker, R. J. Blattner, S. Nadel, C. A. Evans, and R. S. Nowick, Thin Solid Films. 69, 53 (1980).
14. T. C. Tisone and J. Drobek, J. Vac. Sci. Technol. 9, 271 (1971).
15. P. H. Holloway, Appl. Surf. Anal. ASTM STP 699, 5 (1980).
16. H. Numakura and M. Koiwa, Actu. Metall. 32, 1799 (1984).
17. R. Kirchheim, Proc. Matls. Res. Soc., 252 (Boston, MA, 1984).
18. J. D. Reimer, J. Vac. Sci. Technol. A. 2, 242 (1984).
19. J. Klema, R. E. Pyle, and E. Domangue, Proc. Inter. Rel. Phys. Sym., 1.1-1 (Las Vegas, NV, 1984).
20. G. Nelson, Y. Guan, G. Fitzgibbon, J. Curry, R. Muollo, and A. Thomas, Proc. Inter. Rel. Phys. Sym., 1.2-1 (Las Vegas, NV, 1984).
21. J. M. Poate, P. A. Turner, W. J. DeBonte, and J. Yahalom, J. of Appl. Phy. 46, 4275 (1975).
22. C. Chang, Appl. Phys. Lett. 38, 860 (1981).

TITANIUM SILICIDE FORMATION ON HEAVILY DOPED ARSENIC-IMPLANTED SILICON

S.L. DOWBEN*, D.W. MARSH,** G.A. SMITH,** N. LEWIS,** T.P. CHOW,** AND W. KATZ***
*Present address: Cornell University, Ithaca, NY 14853
**General Electric Corporate Research and Development, PO Box 8, Schenectady, NY 12301
***General Electric Knolls Atomic Power Laboratory, Schenectady, NY

INTRODUCTION

Every new generation of metal/oxide/semiconductor (MOS) technology has achieved higher densities and switching speeds. In order to match these characteristics of MOS circuits, a metallization which has a low resistivity, has electrical and chemical stability, can withstand high-temperature processing and can be manufactured relatively easily and reliably is needed. These requirements make the refractory metals a suitable if not ideal choice [1,2]. However, there has been some question as to the reliability of processing during silicide formation when using refractory metals. When the metallization is used to form self-aligned silicide structures over heavily doped source and drain regions, it is crucial to understand the subsequent behavior of the dopant during the processing period. Whereas others have studied different aspects of dopant redistribution [3-8], we report in this paper a systematic study of the electrical, structural, and elemental properties of titanium silicide formation on arsenic implanted silicon as a function of implanted dose and processing temperature.

EXPERIMENTAL

Varying doses ($5 \times 10^{14}/cm^2$ to $6 \times 10^{15}/cm^2$) of $^{75}As^+$ at 150 keV were implanted into 300-500 ohm-cm, n-type (100) silicon substrates. All implants were then activated before Ti deposition at 900°C for 30 min in a N_2 ambient. The sheet resistance after activation varied from 8.3 to 7.0 ohms/square for the implants $5 \times 10^{14}/cm^2$ and $6 \times 10^{15}/cm^2$, respectively. After a dip in 10% HF to remove the native surface oxide, titanium was e-beam evaporated onto silicon substrates at room temperature to a thickness of approximately 1000Å. The background pressure before evaporation was 1×10^{-6} Torr or lower. The heat treatment was performed in H_2 at 650°C to 900°C for times ranging from 10 to 60 minutes. To minimize contamination (mainly oxygen and nitrogen) from the ambient, the film was covered with a clean silicon wafer during annealing.

After annealing, the elemental and structural properties were studied using backscattering spectrometry (RBS), secondary ion mass spectrometry (SIMS), x-ray diffraction (XRD), and transmission electron microscopy (TEM). The electrical sheet resistance of the films was measured with a four-point probe. The RBS was done on a General Ionex Tandetron using a 3.0 MeV He^+ beam and a scattering angle of 120°. The SIMS was done on a Cameca IMS-3f using a 10 keV Cs^+ primary beam and negative secondary ion spectrometry. The SIMS arsenic profiles shown were subtracted from unimplanted standards in order to eliminate a spectral overlap due to $^{75}(TiSi)^+$. Therefore, the arsenic profiles represent the real arsenic concentration present in the films. The TEM was performed on a Hitachi H-600 operated at 100 keV and the x-ray data was taken using a Read camera with a glancing angle of 15° and exposed to Cr (K alpha) radiation for 16 hours.

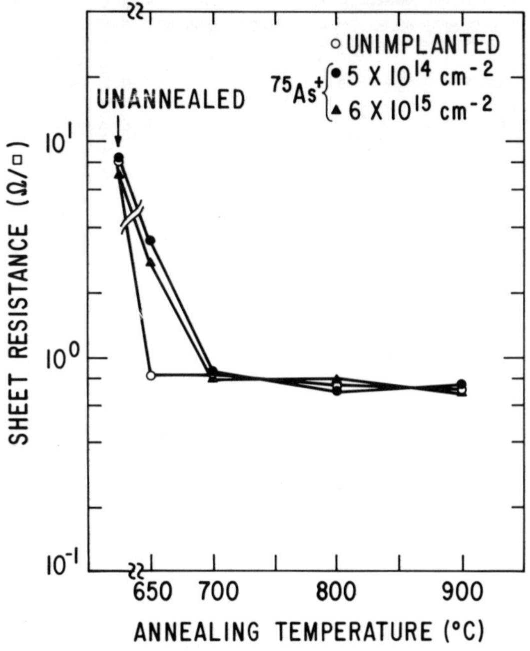

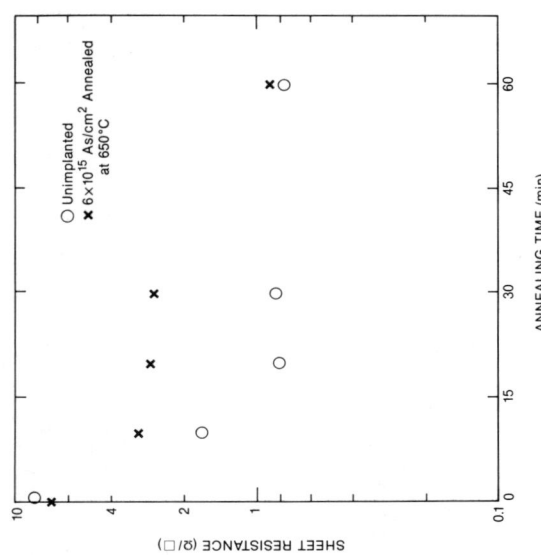

Figure 1. Sheet resistance versus temperature for samples annealed for 30 min.

Figure 2. Sheet resistance versus time for an annealing temperature of 650°C.

RESULTS

The sheet resistance versus temperature is shown in Figure 1 for the implanted and unimplanted samples. All of the samples had an initial sheet resistance of approximately 8 ohms/square and 0.8 ohms/square for annealing temperatures of 700°C and above. The only difference in the results is the implanted samples do not reach steady state until 700°C while the unimplanted samples have reached it by a temperature of 650°C. It is interesting to note that this behavior was found for p-type dopants as well [9]. The sheet resistance versus time is shown in Figure 2 for the $6 \times 10^{15}/cm^2$ dose and unimplanted samples. Initially, the implanted and unimplanted samples show sheet resistance values of 7.0 and 8.0 ohms/square, respectively. After a 20 minute anneal, the unimplanted sample has already approached a constant value of 0.8 ohms/square while the sheet resistance of the implanted sample is still declining. After 60 minutes, both samples have reached a similar sheet resistance value of 0.9 and 0.8 ohms/square for the implanted and unimplanted samples, respectively.

Figure 3a shows the RBS spectra of Ti films deposited on 150 keV, $6 \times 10^{15}/cm^2$ arsenic implanted silicon wafers at various annealing temperatures. All of the annealed samples show a very similar profile except for the 650°C anneal. The surface region of the 650°C anneal appears to be a titanium-rich layer on top of titanium silicide. The silicide was calculated to be $TiSi_2$. The region above 2.2 MeV has been expanded in Figure 3b in order to see the arsenic profiles more clearly. While the arsenic profiles are relatively constant throughout the silicide layer for the 700°C, 800°C, and 900°C annealed samples, the 650°C annealed sample is distinctly different. Here the arsenic profile shows a peak near the surface; in fact it appears the arsenic has piled up at the $Ti/TiSi_2$ interface. This pileup could not be seen in the $5 \times 10^{14}/cm^2$ dose sample since the concentration was too low to be detected by RBS.

Since the main differences in structure occur between the 650°C and 700°C anneal, these samples were studied in more detail by observing the titanium silicide formation as a function of time for each of these temperatures (Figures 4 and 5). Again, the region above 2.2 MeV has been expanded to show the arsenic profiles more clearly. One can see that even for the longest annealing time of 60 minutes for the 650°C anneal, the arsenic peak still remains concentrated at the surface. At 700°C, the arsenic is distributed rather uniformly with no peak at the interface, even for the 10 minute anneal. Therefore, the arsenic segregation to the surface is primarily a function of temperature, possibly signifying the formation of an additional phase.

It is apparent from the SIMS data as seen in Figures 6 and 7 that this pileup does indeed exist for the $5 \times 10^{14}/cm^2$ dose sample. The high concentration of arsenic near the surface of the 650°C annealed sample in Figure 6 is in agreement with the RBS data seen in Figure 3. It can be clearly seen that the profiles of the $5 \times 10^{14}/cm^2$ implanted samples (Figure 7) are similar to the $6 \times 10^{15}/cm^2$ implanted samples (Figure 6). Therefore, it appears that arsenic implanted silicon (at least for doses greater than or equal to $5 \times 10^{14}/cm^2$) will retard silicide formation at 650°C and so a higher anneal temperature must be used.

In fact it appears from the RBS data that more than one phase may be present in the 650°C annealed sample. Therefore, these samples were analyzed using x-ray diffraction. Table 1 lists the different phases found from the diffraction study. Note that for anneals of 700°C or higher only the $TiSi_2$ phase is found for either the implanted or unimplanted samples.

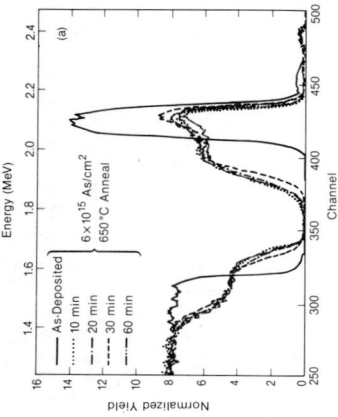
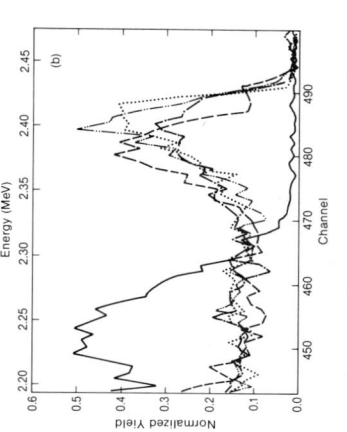

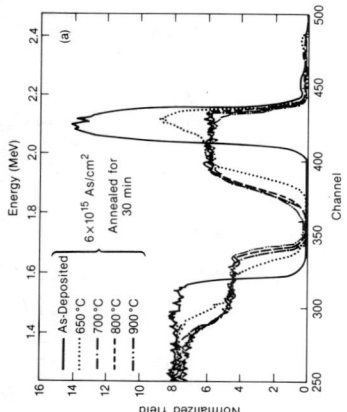

Figure 3. RBS spectra of titanium silicide formation versus temperature: (a) entire energy region and (b) energy region above 2.2 MeV.

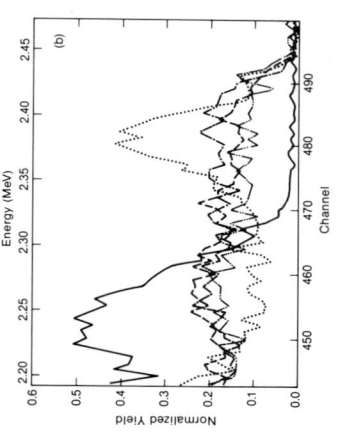

Figure 4. RBS spectra of titanium silicide formation versus time at an annealing temperature of 650°C: (a) entire energy region and (b) energy region above 2.2 MeV.

359

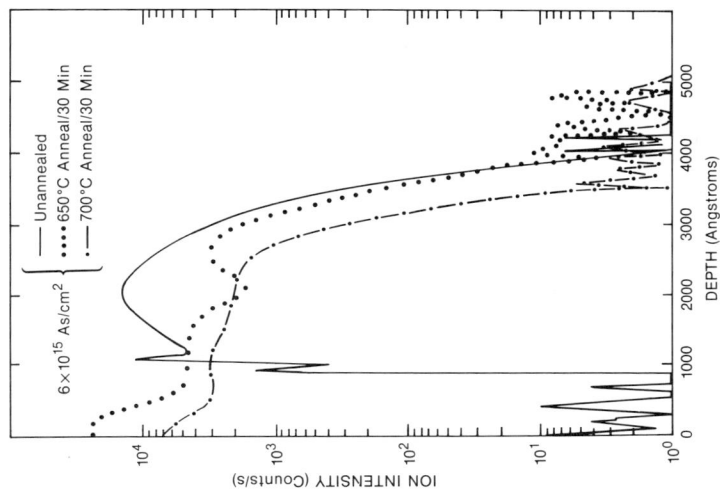

Figure 6. SIMS profiles of arsenic distribution for 6×10^{15} cm^{-2} implanted samples.

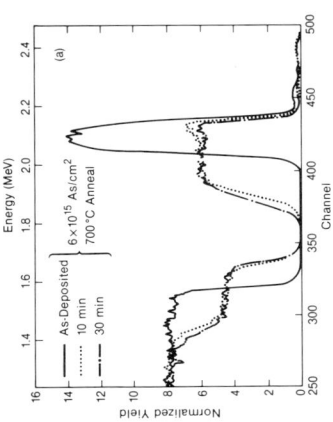

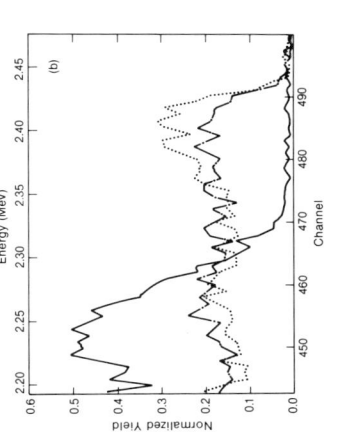

Figure 5. RBS spectra of titanium silicide formation versus time and an annealing temperature of 700° C: (a) entire energy region and (b) energy region above 2.2 MeV.

However, the 650°C sample shows phases of $TiSi_2$, Ti_5Si_3, and possibly Ti_5As_3 to be present for the $6x10^{15}/cm^2$ implanted sample. The arsenide phase was detected from a diffraction line at 1.77A of noticeable intensity which could not be attributed to diffraction lines of Ti_5Si_3 or $TiSi_2$. However, since some of the major diffraction lines of Ti_5As_3 overlap with the major diffraction lines of $TiSi_2$ and Ti_5Si_3 [10-12], a positive identification is not possible. Therefore, the titanium and arsenic rich region near the surface seen in the RBS and SIMS results for the implanted samples annealed at 650°C is due to Ti_5Si_3 and possibly Ti_5As_3.

Finally, the sample morphology was studied using cross-sectional TEM. The results as seen in Figure 8 show the grain size of the 700°C anneal sample to be larger than the 650°C sample (only the $6x10^{15}/cm^2$ implanted samples were analyzed). In addition, the 650°C anneal sample had a higher number of faulted grains than the 700°C anneal sample. Both types of samples contained a number of small (~200A) particles particularly in the fault-free grains. However, the results from x-ray spectroscopy and electron diffraction were inconclusive in trying to identify the particles.

DISCUSSION

Reviews have already appeared on dopant redistribution in refractory metal silicides [13,14] in general. All low temperature studies (700°C or less) show no redistribution of arsenic (i.e., no snowplow effect) during titanium silicide formation. In fact, for these low temperatures the arsenic is seen to initially segregate to the moving metal/silicide interface and continue to diffuse outward to the surface even after silicide formation is completed. Our results show that arsenic may have formed a Ti_5As_3 phase which appears to be unstable at 700°C or higher. While our study found incomplete silicide formation to occur at 650°C, other studies have found the opposite result at or near this temperature [3,5,6]. It is known that arsenic concentrations greater than $1x10^{21}/cm^3$ [8] and other impurities such as residual gases in the Ti deposition chamber [15] may affect silicide growth. Since the arsenic concentrations in our study was approximately $8x10^{20}/cm^3$ or less, other contaminants may have been responsible for the discrepancy in results. It is interesting to note that while the n-type dopants exhibit a snowplow effect in the near noble metal silicides, they do not show any such effect in the refractory metal silicides [3-8,13,14]. On the other hand the p-type dopants do not show any snowplow effect in either the near noble or refractory metal silicides [9,16-19]. This seems to suggest that there must be a chemical effect involved (such as the formation of a stable or unstable phase as found in this study) and not just the presence or nonpresence of defects as has been suggested by others [3].

CONCLUSION

The electrical, structural, and elemental character of titanium silicide formation was investigated for arsenic implanted silicon substrates. The arsenic diffuses toward the surface for all of the temperatures and implant doses studied except for the 650°C anneals. At this temperature arsenic was found to segregate to the surface. In fact, x-ray diffraction identified this high arsenic concentration near the surface as a possible Ti_5As_3 phase which was not present for any of the other annealing temperatures studied.

Note: It has been brought to the authors' attention that the titanium arsenide phase identified in the 650°C annealed samples could actually be a metastable phase of $TiSi_2$. Beyers and Sinclair identified a metastable

Table 1. A list of phases present for the $6 \times 10^{15} \text{cm}^{-2}$ arsenic-implanted samples annealed at various temperatures for 30 min.

SAMPLE	PHASE(S)
UNANNEALED	- - -
650°C, 30 MIN	$TiSi_2, Ti_5Si_3, Ti_5As_3(?)$
700°C, 30 MIN	$TiSi_2$
800°C, 30MIN	$TiSi_2$
900°C, 30 MIN	$TiSi_2$

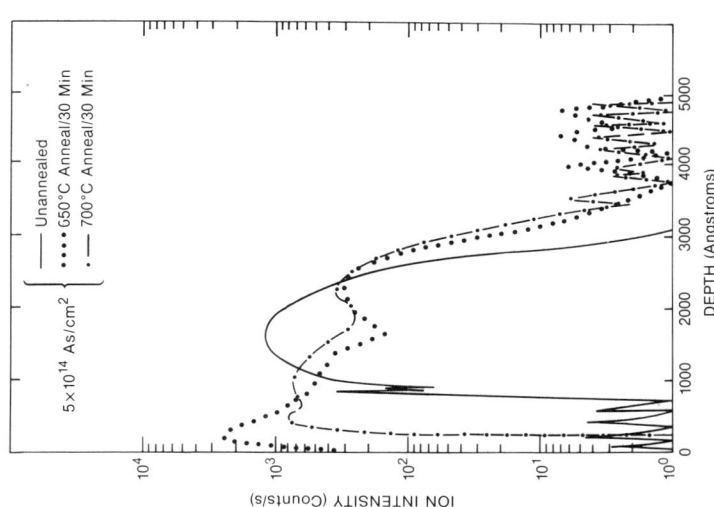

Figure 7. SIMS profiles of arsenic distribution for $5 \times 10^{14} \text{cm}^{-2}$ implanted samples.

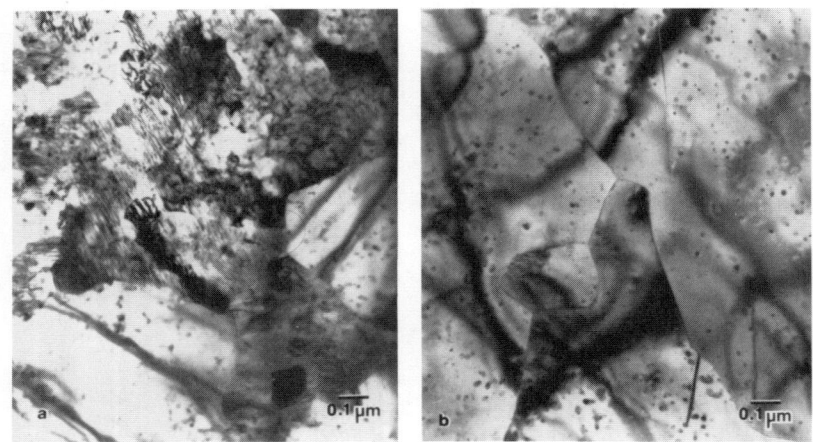

Figure 8. TEM micrograph of $6 \times 10^{15} cm^{-2}$ arsenic implanted samples: (a) 650°C anneal and (b) 700°C anneal. Both samples were annealed for 30 min.

phase of TiSi$_2$ (C49 or ZrSi$_2$ structure) which appears for films formed by reacting Ti with a Si substrate or by codeposition [20].

ACKNOWLEDGEMENTS

The authors have used the program RUMP from Cornell University to produce the graphs used in this article. The authors would also like to thank Dr. James W. Mayer and his group at Cornell University for the use of his accelerator on which the RBS data was taken.

REFERENCES

1. S. P. Murarka, Silicides for VLSI Applications (Academic Press, New York; 1983).

2. T. P. Chow and A. J. Steckl, IEEE Trans. Electron Devices, ED-30 (1983) 1480.

3. M. Wittmer and K. N. Tu, Phys. Rev. B29 (1984) 2010.

4. J. Amano, P. Merchant, and T. Koch, Appl. Phys. Lett. 44 (1984) 744.

5. M. Wittmer, C.-Y. Ting, and K. N. Tu, Thin Solid Films 104 (1984) 191.

6. M. Ostling, C. S. Petersson, C. Chatfield, H. Norstrom, F. Runovc, R. Buchta, and P. Wiklund, Thin Solid Films 110 (1984) 281.

7. P. Revesz, J. Gyimesi, and E. Zsoldos, J. Appl. Phys. 54 (1984) 1860.

8. H. K. Park, J. Sachitano, M. McPherson, T. Yamaguchi, and G. Lehman, J. Vac. Sci Technol. A2 (1984) 264.

9. T. P. Chow, W. Katz and G. Smith, Appl. Phys. Lett. 46 (1985) 41.

10. Standard Powder Diffraction Pattern 18-1395.

11. Standard Powder Diffraction Pattern 31-1405.

12. Standard Powder Diffraction Pattern 29-1362.

13. C. B. Cooper III, R. A. Powell, and R. Chow, J. Vac. Sci. Technol. B2 (1984) 718.

14. C.-D. Lien and M-A. Nicolet, J. Vac. Sci. Technol. B2 (1984) 738.

15. L. S. Hung, J. Gyulai, J. W. Mayer, S. S. Lau, and M-A. Nicolet, J. Appl. Phys. 54 (1983) 5076.

16. P. Pan, N. Hsieh, H. J. Geipel, Jr., and G. J. Slusser, J. Appl. Lett. 53 (1982) 3059.

17. F. Jahnel, J. Biersack, B. L. Crowder, F. M. d'Heurle, D. Fink, R. D. Isaac, C. J. Lucchese, and C. S. Petersson, J. Appl. Lett. 53 (1982) 7372.

18. C.-Y. Wei, W. Katz, and G. Smith, Thin Solid Films 104 (1983) 203.

19. T. P. Chow, W. Katz, and G. Smith, J. Electrochem. Soc. (in press).

20. Robert Beyers and Robert Sinclair, J. Appl. Phys. 57 (1985) 5240.

THE CHARACTERIZATION OF INTENTIONAL DOPANTS IN HgCdTe USING
SIMS, HALL-EFFECT, AND C-V MEASUREMENTS

L.E. LAPIDES, R.L. WHITNEY, AND C.A. CROSSON
Santa Barbara Research Center, Goleta, CA 93117.

ABSTRACT

The properties of selected dopants in liquid-phase epitaxial (LPE) layers of HgCdTe have been studied using secondary ion mass spectrometry (SIMS), Hall-effect, and capacitance-voltage (C-V) measurements. The layers were grown from Hg-rich melts on {111}-oriented CdTe and CdZnTe single-crystal substrates. Diodes, for the C-V measurements, were homojunctions formed by ion implantation or heterojunctions formed by the growth of a second layer on the base layer. Dopant concentration distributions in both single- and double-layer structures were characterized by SIMS and C-V measurements. The dopant profiles measured by SIMS were quantified using relative sensitivity factors calculated from ion implanted impurity profiles measured on standard reference samples. Using specialized SIMS techniques, such as molecular ion spectrometry, As concentrations as low as 2×10^{15} cm^{-3} have been measured. In the HgCdTe:In/HgCdTe:As system minimal dopant interdiffusion is observed in SIMS profiles. The growth of the second layer has insignificant effect on the As distribution in the base layer, and C-V data indicate that the electrical properties change only slightly. Carrier types and concentrations were determined by Hall effect and C-V measurements. Good agreement between dopant concentrations and carrier concentrations was observed, indicating 100% activation of the dopant atoms, for all dopants studied. Examples of implant calibration profiles, dopant concentration distributions, carrier concentration vs temperature measurements, and $1/C^2$ vs V data are presented, along with graphs and tables comparing dopant profiles with electrical properties.

INTRODUCTION

The ternary alloy $Hg_{(1-x)}Cd_xTe$ is an important semiconductor for photovoltaic infrared detector fabrication. By changing the alloy composition the bandgap can be varied from 0 to 1.6 eV, allowing fabrication of photovoltaic devices which are tuned to the individual IR atmospheric windows. The demand for large area detector arrays led to the development of epitaxial HgCdTe growth on CdTe substrates and later on CdZnTe substrates. The standard approach to detector array fabrication has been to grow a p-type epitaxial HgCdTe absorbing layer and ion implant to form a p-n junction. The carrier concentration of the p-type layers is determined by the Hg vacancy concentration, which is adjusted by annealing the layers. Improved control over the electrical properties can be achieved by intentional impurity doping. In addition, p-n junctions can be formed by growing an appropriately doped layer on the absorbing layer. To evaluate a given dopant, its segregation coefficient, activation coefficient, distribution after diode formation, and electrical properties after diode formation must be determined. In this study the majority of data presented will relate to the evaluation of As as a p-type dopant and In as a n-type dopant in HgCdTe.

The segregation coefficient for a particular dopant is the ratio of the concentration of the dopant in an epitaxial layer to the concentration in the melt. The activation coefficient is the ratio of the carrier concentration in the layer due to the dopant to the dopant concentration in the epilayer. Once both are known, the carrier concentration in a layer can be controlled by determining how much of the dopant to put in the melt. The dopant distribution and electrical properties after diode formation, whether by implant or the growth of another layer, are important for p-n junction placement and diode performance. The preference, of course, is for stable dopants whose distribution and electrical properties remain constant before and after diode formation. The choice of dopant can also affect other properties, such as minority carrier lifetime; however, the evaluation of the lifetime and other affected properties involves extensive analysis which is beyond the scope of this work.

Segregation coefficients and the electrical properties of dopants in liquid phase epitaxial layers of HgCdTe have been studied by Kalisher [1]. Radford and Jones have characterized diodes formed by ion implantation into, and the growth of a second layer on, an As-doped HgCdTe epitaxial layer [2]. In this study the distribution and electrical properties of intentional dopants in HgCdTe epitaxial layers are analyzed, and the effect of diode formation upon these properties discussed. After a brief review of the sample preparation procedures and the characterization techniques, the results of the SIMS, Hall-effect, and C-V measurements are compared. These results are then discussed with respect to dopant activation coefficients, distribution, and electrical properties.

EXPERIMENTAL PROCEDURES

The HgCdTe epitaxial layers were grown from Hg-rich infinite-melt systems using vertical liquid phase epitaxy. The substrates were single crystal {111}-oriented CdTe and CdZnTe. The dopants were introduced into the melt in elemental form. Diode formation occurred by B^+ implantation or by the growth of a second "cap" layer. The implant-formed diodes were delineated by implantation through a mask while the double layer diodes were defined by a mesa etch. In all cases the base layer was p type, with the implant or the cap layer forming the n-type region. After base layer growth the layer was cleaved into two parts: one was used for diode fabrication and the other was cleaved in two parts again. One base layer sample was used for SIMS analysis and the other for Hall-effect measurements. Similarly, the piece used for diode fabrication was divided into two parts, one for diodes and one for SIMS.

The SIMS analyses were performed using a Cameca IMS-3f ion microanalyzer interfaced with a HP9825 computer and a HP9872A graphics plotter. The primary beam species was O_2^+ when positive secondary ions were studied, and Cs^+ was used when negative secondary ions were analyzed. The secondary ions were detected using an electron multiplier. Following the SIMS analysis the sputtered crater depth was measured using a Sloan Technology Dektak surface profilometer.

The method used to quantify the SIMS data relates the secondary ion intensity to the concentration of the element profiled through the following equation:

$$C_i = (I_i/I_m)RSF_i(m). \tag{1}$$

Here C_i is the concentration of element i, I_i is the secondary ion intensity of i, I_m is the secondary ion intensity of the matrix element m (usually ^{125}Te), and $RSF_i(m)$ is the relative sensitivity factor of i with respect to m. The RSFs are determined by analyzing samples that have been ion implanted with the element of interest. In certain cases where implant calibration samples for an element have not yet been fabricated the RSF can be calculated. The calculation is performed using empirical equations which relate RSF to ionization potential for positive secondary ions and RSF to electron affinity for negative secondary ions [3].

Several of the dopants studied are difficult to analyze because of low ion yields and background signals at the masses of interest. The molecular ion spectrometry technique was used to overcome these obstacles. With this method a molecular ion composed of the element of interest and either a matrix element or a primary beam element is analyzed instead of the elemental ion. These molecular ions are formed during the sputtering process, and do not exist in the sample. Because of this, a RSF for an element which relates the ion intensity of the molecular ion analyzed to the concentration of the element can still be determined from an implant calibration sample. This is illustrated by Figure 1, which shows an As implant profiled using $^{205}TeAs^-$ instead of $^{75}As^-$. In the case of As analysis, the use of the TeAs molecular ion results in a factor of 10 increase in sensitivity over As^- detection [4].

Hall-effect measurements were used to determine the majority carrier type and concentration. Samples were fabricated in the van der Pauw configuration

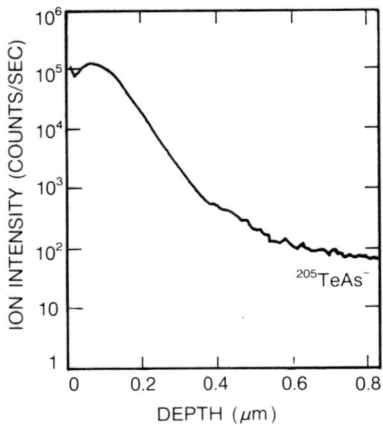

Figure 1. Profile of As implant (200 keV, 4×10^{14} cm^{-2}) using TeAs$^-$ secondary ions. For this experiment $RSF_{As(TeAs)}(^{125}Te) = 8.2 \times 10^{21}$ cm^{-3}.

and tested from 30K to 300K. The net carrier concentration was determined by extrapolating the $1/eR_h$ vs $1/T$ curve to $1/T = 0$.

Capacitance-voltage measurements used a PAR 410 C-V Plotter for 1 MHz C-V measurements and a PAR Model 124 Lock-in Amplifier for measurements at lower frequencies. The measurement apparatus was calibrated before each set of measurements using a standard air capacitor. Diodes of various areas were fabricated, and measured at 80 kHz, 200 kHz, and 1 MHz. Data from a specific frequency were chosen for each sample by determining the frequency at which the capacitance measurements came the closest to scaling with diode area.

RESULTS

SIMS profiles of intentionally As-doped epitaxial layers were performed using both Cs$^+$ and O$_2^+$ primary ions. The detection limit for As analysis using As$^+$ detection is ~2×10^{16} cm^{-3}; therefore, for good signal-to-noise ratio or for layers with low As doping TeAs$^-$ secondary ion analysis was used. When analyzing HgCdTe:In/HgCdTe:As double layer heterostructures maximum accuracy was achieved by analyzing the layer twice, once with O$_2^+$ primary ions and In$^+$ detection and once with Cs$^+$ primary ions and TeAs$^-$ detection. The depth profiles were then overlaid to determine the extent of interdiffusion occurring during the growth of the second layer (Figure 2). This procedure was validated by repeating the analysis for a highly doped sample and profiling In$^+$ and As$^+$ simultaneously. The difference in the As depth scale between the simultaneous profile and the overlay was approximately 0.1 µm. The interdiffusion region between the two layers for the samples studied was found to range from 0.25 to 0.5 µm.

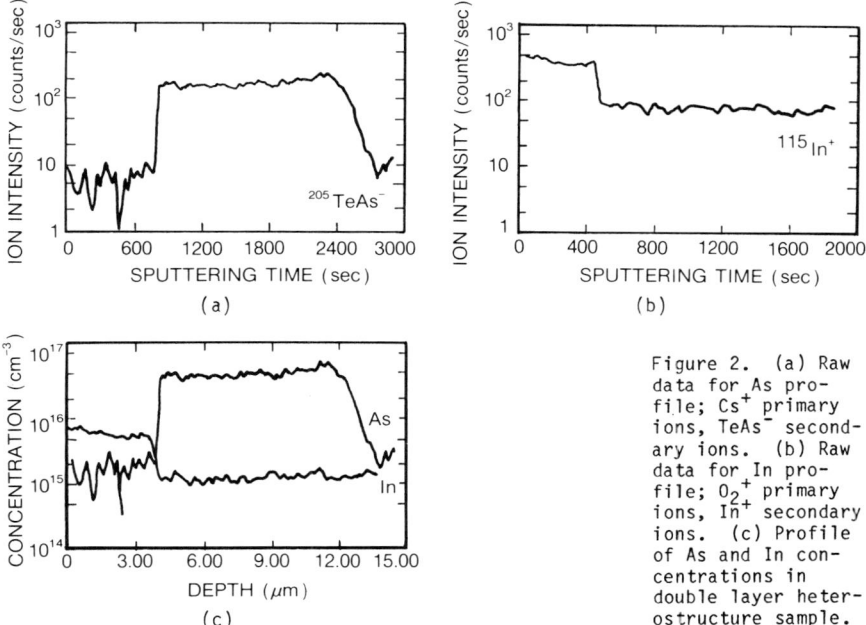

Figure 2. (a) Raw data for As profile; Cs^+ primary ions, $TeAs^-$ secondary ions. (b) Raw data for In profile; O_2^+ primary ions, In^+ secondary ions. (c) Profile of As and In concentrations in double layer heterostructure sample.

Hall-effect measurements on n-type and high p-type samples are relatively straightforward, producing "classical" R_h vs $1/T$ curves (Figure 3). The measurement of low doped p-type samples is complicated by mixed conduction effects because of the high mobility of the electrons. For these samples a computer model developed by Lou and Frye [5] is used to fit the data, with N_A as a parameter. Hall-effect measurements of N_D and SIMS measurements of [In] are in agreement to ±13% (see Figure 4a). Figure 4b shows a plot of N_A vs [As]; this also shows good agreement between Hall-effect and SIMS measurements (±11%). In addition, data on Al-doped epitaxial layers also show good agreement between N_D and [Al].

Capacitance-voltage measurements in HgCdTe are not straightforward due to factors such as series resistance, bias-dependent resistance of nonohmic contacts, and surface effects which can increase the effective area of the diode. The resistance in series with the parallel capacitance and conductance of the diode can cause a change in the measured capacitance, as the measurement equipment assumes only the parallel capacitance and conductance. Schematic diagrams of the actual and equivalent circuits are given in Figure 5. The effective capacitance is given by [6],

$$C' = C/[(rG + 1)^2 + \omega^2 r^2 C^2], \tag{2}$$

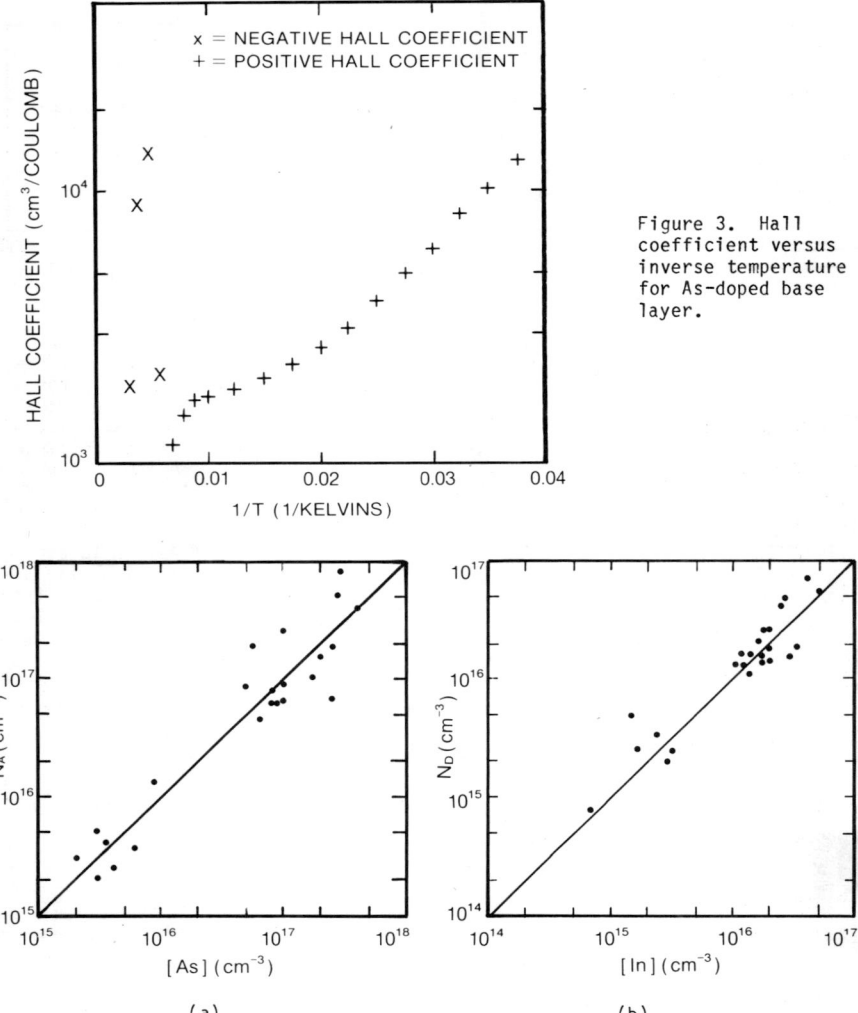

Figure 3. Hall coefficient versus inverse temperature for As-doped base layer.

Figure 4. Carrier concentration versus impurity dopant concentration, indicating 100% electrical activation of dopant atoms. (a) N_D vs [In]. (b) N_A vs [As].

where C is the diode capacitance, G is the conductance, r is the series resistance, and ω is the measurement frequency. The conflict which arises during measurement is in the determination of the measurement frequency to be used. This is a result of the desire to increase frequency to (1) increase the signal-to-noise ratio and (2) to keep ω > 1/τ, where τ is the trap

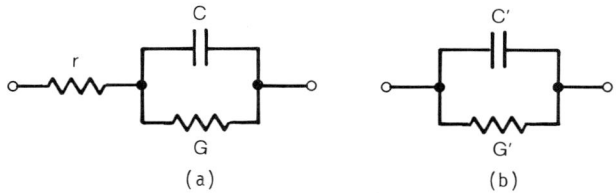

Figure 5. (a) Actual circuit and (b) equivalent circuit measured by C-V test equipment.

relaxation time, and the conflicting need to decrease frequency to decrease the effect of the series resistance on the measurement [7].

Because of the large junction resistance of the diodes the approximation $rG \ll 1$ can be used. Equation (2) then becomes

$$C' = C/[1 + \omega^2 r^2 C^2]. \tag{3}$$

If we assume that the capacitance C is proportional to $[V_{bi} - V_a]^{1/2}$, where V_{bi} is the built-in potential of the p-n junction and V_a is the applied voltage, then the term $\omega^2 r^2 C^2$ in (3) just adds a constant to the $1/C'^2$ vs V_a plot:

$$1/C'^2 = [2(V_{bi} - V_a)/A^2 e \varepsilon_0 \varepsilon N_B] + 2\omega^2 r^2. \tag{4}$$

Here A is the area of the diode, and N_B the reduced carrier concentration $N_B = N_A N_D/(N_A + N_D)$. It is assumed here that the dielectric constant ε is the same in both layers of the DLHS diodes. Since N_B is derived from the slope of the curve, there is no change in the expression for N_B:

$$N_B = -2[A^2 e \varepsilon_0 \varepsilon (d(1/C'^2)/dV_a)]^{-1}. \tag{5}$$

The assumption that $1/C^2$ is proportional to $(V_{bi} - V_a)$ is called the abrupt junction approximation because the expression for C as a function of V is derived from the assumption of a perfectly abrupt electrical junction [8]. The measured C-V curves were analyzed in two ways: first, using the abrupt junction calculation, N_B was calculated as a function of the applied voltage. Since the depletion region thickness W is given by

$$W(V_a) = \varepsilon_0 \varepsilon A/C(V_a), \tag{6}$$

N_B was plotted as a function of W. These plots showed N_B to be almost constant with respect to depletion region thickness, supporting the hypothesis of a nearly abrupt junction (see Figure 6).

The second method of analysis consisted of assuming a relationship of the form

$$C = B(V_{bi} - V_a)^\alpha, \tag{7}$$

where B and α are constants. If the abrupt junction approximation holds, then (7) is equivalent to (4), with $\alpha = -1/2$. In practice, α was determined by a

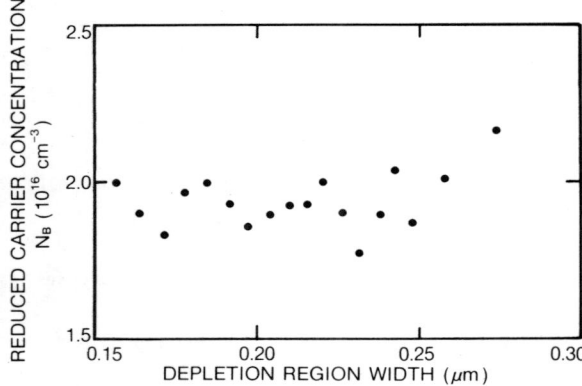

Figure 6. Reduced carrier concentration N_B versus depletion region width W for double layer heterostructure diode.

least squares fit of the curve

$$V_a = \alpha[d(\ln C')/dV_a]^{-1} + V_{bi}. \tag{8}$$

For the double layer heterostructures that were studied, α ranged from -1/2.17 to -1/2.38 (see Figure 7).

The carrier concentration of the n-type cap layer in the double layer heterostructure samples can be determined by a Hall-effect measurement on the single layer witness sample grown simultaneously with the cap layer. Using the value of N_B from C-V measurements, a value for the base layer carrier concentration can be calculated. In Figure 8 N_A values from Hall-effect measurements on base layers are compared with N_A values from C-V data.

C-V measurements were also performed on the homojunction diodes. In this case the objective was to evaluate the abruptness of the junction by comparing measured zero bias depletion width with the value calculated using the abrupt junction approximation. For base layer doping of $N_A = 10^{17}$ cm^{-3} and 180 keV, 5×10^{12} cm^{-2} B$^+$ implant ($N_D \sim 10^{17}$ cm^{-3}), the calculated depletion region width is 0.130 μm. Measured depletion widths ranged from 0.142 to 0.151 μm, indicating a nearly abrupt junction.

DISCUSSION

In analyzing the results the objectives of this study should be reiterated: to characterize the distribution and electrical properties of intentional dopants in HgCdTe epitaxial layers, and the effect of diode formation processes upon these properties. Prior to diode formation information about the electrical activity of the dopant and its distribution is desired. The SIMS measurements on As-doped base layers have shown uniform distribution of the dopant through the depth of the epitaxial layer. A comparison of SIMS and Hall-effect measurements, as presented in Figure 4, shows that there is good

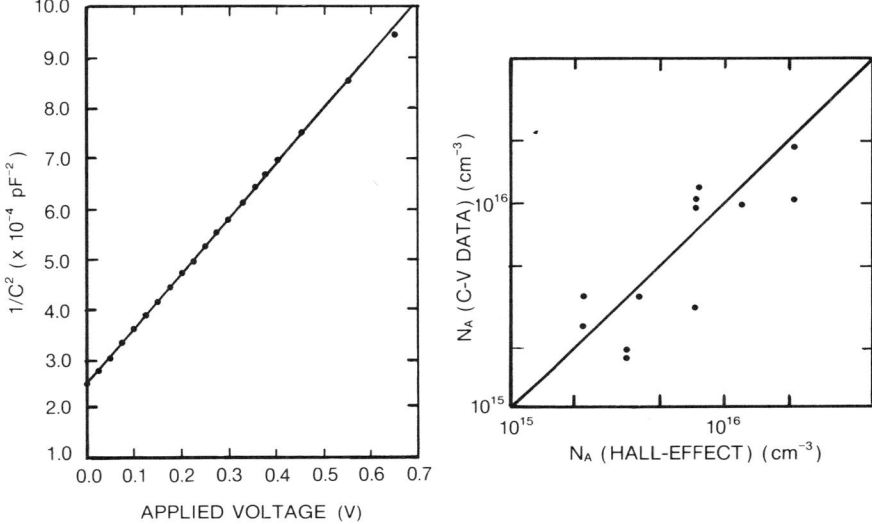

Figure 7. $1/C^2$ versus applied voltage for a double layer heterostructure diode. The best fit exponential to this data is $\alpha = -1/2.22$.

Figure 8. N_A from Hall-effect data on base layers versus N_A calculated from C-V results on double layer heterostructure diodes.

agreement between the concentration of the dopant atoms in the layers and the carrier concentration of the layers. This indicates that the dopants evaluated contribute a single active carrier per atom, or that the activation coefficient is equal to 1.

After diodes have been fabricated on the layers the same information is again needed: the electrical properties of the material and the dopant distribution. Dopant distributions were evaluated using SIMS and C-V measurements. For the heterostructure diodes SIMS measurements show an interdiffusion region of less than 0.5 μm. Depletion region widths from C-V tests are about 0.2 μm, and analysis of the C-V curves indicates a nearly abrupt p-n junction. Also, the net carrier concentration at the edge of the depletion region remains almost constant as the depletion region is enlarged. These data taken together show that while there is some interdiffusion of dopants during the growth of the cap layer, the dopant distribution is relatively stable with respect to the diode formation process. The same conclusion is reached for the homojunction diodes, where the C-V data show a nearly abrupt junction, as expected for an ion-implant-formed electrical junction. This is in contrast to the results of Bubulac et al. [9], who found graded p-n junctions formed by ion implantation into HgCdTe epitaxial layers. Their results, however, may be due to the larger implant dose used or due to the material produced when HgCdTe epitaxial technology was relatively immature.

The electrical properties of the dopants are evaluated by comparing Hall-effect data taken prior to diode formation and post-diode-formation C-V results. The agreement between N_B values from Hall-effect and C-V measurements presented in Figure 9 shows that for heterostructure diodes the electrical properties are minimally affected by the growth of the cap layer. This is confirmed by a comparison of the SIMS data on double layer samples and the value for N_A calculated from C-V results. This comparison shows that the activation coefficient of As in base layers after the growth of the cap layer is still 1. This comparison is difficult to make for the implanted diodes because the doping from the implant is damage related [9,10].

CONCLUSION

Selected dopants in HgCdTe epitaxial layers have been characterized with the overall goals of control and stability of the diode formation process. The layers were characterized before and after the diode formation procedure using SIMS, Hall-effect, and C-V measurements. The specialization of these techniques to the study of HgCdTe was described. Electrical activation coefficients were evaluated with the control of the electrical properties of the layers in mind. The analysis of the data showed that in both implant-formed diodes and heterostructure diodes the dopant distributions were relatively stable. The results also show the electrical stability of the heterostructure diodes.

ACKNOWLEDGMENTS

The authors would like to thank P.E. Herning, M.H. Kalisher, J.M. Myrosznyk, and R.F. Risser for epitaxial layer growth, H.P. Bevans, M.S. Langell, S. Tallarico, and P.S. Villa for device processing, D.H. Patterson for Hall-effect measurements, and M.E. Boyd for C-V testing. The SIMS analyses were performed at Charles Evans and Associates, San Mateo, CA. This work was jointly sponsored by the Defense Advanced Research Projects Administration under contract F04701-82-C-0029, monitored by AF/STC, and by the Army Night Vision and Electro-Optics Laboratory under contract DAAK70-83-C-0006.

REFERENCES

1. M.H. Kalisher, Proceedings of the Sixth American Conference on Crystal Growth, Atlantic City, NJ, July, 1984; to be published.

2. W.A. Radford and C.E. Jones, Proceedings of the IRIS Detector Specialty Conference, Seattle, WA, August, 1984; to be published.

3. L.E. Lapides, G.L. Whiteman, and R.G. Wilson, Thin Films and Interfaces II, J.E.E. Baglin, D.R. Campbell, and W.K. Chu, eds. (North Holland, New York, 1984).

4. L.E. Lapides, Surf. Interface Anal.; to be published.

5. L.F. Lou and W.H. Frye, J. Appl. Phys. $\underline{56}$ 2253 (1984).

6. A.M. Goodman, J. Appl. Phys. $\underline{34}$ 329 (1963).

7. J.D. Wiley and G.L. Miller, IEEE Trans. Electron Dev. $\underline{ED\text{-}22}$ 265 (1975).

8. S.M. Sze, Physics of Semiconductor Devices, 2nd edition (John Wiley and Sons, New York, 1981), p. 79.

9. L.O. Bubulac, W.E. Tennant, S.H. Shin, C.C. Wang, M. Lanir, E.R. Gertner, and E.D. Marshall, Jpn. J. Appl. Phys. Suppl. $\underline{19\text{-}1}$ 495 (1980).

10. A. Kolodny and I. Kidron, IEEE Trans. Electron Dev. $\underline{ED\text{-}27}$ 37 (1980).

PART IV

High-Energy Methods and Materials Characterization Using Photon Beams

HYDROGEN MEASUREMENT
OF THIN FILM SILICON:HYDROGEN ALLOY FILMS
TECHNIQUE COMPARISON

GARY A. POLLOCK
ARCO Solar, Inc., P.O. Box 2105, Chatsworth, CA 91313

ABSTRACT

The hydrogen content in thin film silicon:hydrogen alloy material has been measured by five techniques: ^{15}N nuclear resonance reaction analysis, secondary ion mass spectrometry, infrared spectroscopy, nuclear elastic scattering spectroscopy, and Rutherford backscattering spectroscopy. NRA and SIMS data agree most closely in this study, although SIMS data from three of four laboratories are considerably scattered. The hydrogen content of TFS ranges from 18% at a deposition temperature of 150°C to 10% at 250°C. It varies little, if any, with thickness. Some advantages and disadvantages of the five techniques are discussed.

INTRODUCTION

This paper compares five techniques for measuring hydrogen in thin-film silicon:hydrogen (TFS) alloy films. The techniques are: ^{15}N nuclear resonance reaction analysis (NRA); secondary ion mass spectrometry (SIMS); infrared (IR) spectroscopy; nuclear elastic scattering (NES); and Rutherford backscattering spectrometry (RBS).

TFS films are important materials for photovoltaic power sources, and the hydrogen content of the films is important to their photovoltaic properties. Therefore, selecting the appropriate technique for analysis is important, and understanding the accuracy of the various techniques available is necessary.

I will discuss the preparation of a set of TFS film samples, the procedures used to analyze the samples by each of the five techniques, the results of the analyses, and some advantages and disadvantages of these techniques.

In a previous study, Ziegler et al focused on depth profiling techniques for hydrogen measurement in ion implanted, preamorphized crystalline silicon using NRA, NES, and SIMS [1]. Even though the emphasis of this paper is on total hydrogen content, these techniques are still applicable. Infrared spectroscopy was used in an early study of TFS-like material to determine the main groups of spectral bands associated with silicon-hydrogen bonds and to estimate the hydrogen content of this material [2].

EXPERIMENTAL PROCEDURE

Sample Preparation

TFS samples are deposited using RF glow discharge deposition of silane. The films are deposited on crystalline silicon (c-Si) substrates for the NRA, SIMS, IR, and RBS techniques, on aluminum foil substrates for NES, and on glass (7059) substrates for film thickness measurement. The film thickness is measured with a Dektak II surface profilometer.

The TFS films were deposited at three substrate temperatures: 150, 200, and 250°C, and in two thickness ranges: thin (0.1-0.3 μm) and thick (~3-4 μm) to provide samples for this study.

Each laboratory participating in this study received a sample cut from each of the six c-Si substrates (except as noted for NES).

Analytical Methods

The NRA method [3] basically uses $^{15}N^+$ ions at or above the 6.385 MeV resonance energy for the reaction $^1H + ^{15}N \longrightarrow ^4He + ^{12}C$ + gamma (4.43 MeV). By using 6.385 MeV ^{15}N, the yield of the 4.43 MeV gamma rays is proportional to the amount of hydrogen present at the surface. By increasing the ^{15}N energy, a depth profile of hydrogen can be obtained. The calibration for the hydrogen content in this case is based on the total integrated hydrogen per unit area using hydrogen implanted into sapphire. The accuracy of this method is essentially limited only by the accuracy for measuring the ion current during ion implantation.

SIMS measurements were made using either a Cameca IMS-3F secondary ion mass spectrometer with a cesium ion source or a secondary ion quadrapole mass spectrometer (SIQMS) instrument. The SIMS method [4] consists of using a primary ion source to sputter remove sample material, a fraction of which are secondary ions. Detection of the mass of these secondary ions allows one to depth profile various species present in the sample. The calibration of the hydrogen content by this method is based on hydrogen ion implants in crystalline silicon. Each laboratory participating in this experiment used its own standard.

IR spectroscopy measurements were made using a Nicolet Model MX-1 FTIR spectrophotometer. In this method, the background subtracted area of the IR absorption peak at 640 cm^{-1} in TFS is integrated to determine hydrogen content after the method of Brodsky et al [2] and Fang et al [5]. This peak corresponds to the Si-H wagging mode. The calibration of the IR method is based on theoretically and empirically determined oscillator strengths and the integrated area of the absorption peak. IR spectroscopy thus detects the bound hydrogen in the sample but does not provide depth profile data.

NES measurements were made at a cyclotron facility using 12 MeV protons for scattering. The method uses proton-proton forward scattering in a non-coincidence configuration [6]. Calibration for hydrogen content is based on a thin polystyrene foil reference standard. The total hydrogen content in the sample is measured, but profiling data are not provided.

RBS measurements were made using 2 MeV He$^+$ ions. The method does not measure hydrogen directly, but deduces the hydrogen content from the change in backscattered ion yield between the silicon in the TFS film and the silicon in the c-Si substrate [7].

Laboratories performing the measurements were: ARCO Solar (IR); California Institute of Technology (RBS); Charles Evans & Associates (SIMS); Jet Propulsion Laboratory (SIMS); Princeton University (NES); RCA (SIMS); Solar Energy Research Institute (SIMS); and State University of New York at Albany (NRA).

RESULTS

Table I reports the total hydrogen content in each of the six films as measured by the five techniques: NRA, SIMS, IR, NES, and RBS. Figure 1 shows graphically the data in Table I. The NRA results appear to be the most consistent in this study.

Table I. Hydrogen content (atom %) in TFS films as measured by five different techniques. ± = typical instrument error.

Sample Deposition Parameter	NRA	SIMS				IR	NES	RBS
		1	2	3	4			
150 °C								
0.13 μm	18.2 ± 0.2	15.8 ± 0.3	23.0 ± 0.6	14.3 ± 0.3	18.6 ± 0.2	18.0 ± 0.5	19.4 ± 0.5	20 ± 3
4.11 μm	17.1 ± 0.2	14.5 ± 0.3	25.0 ± 0.6	14.0 ± 0.3	14.4 ± 0.2	12.0 ± 0.5	12.6 ± 0.5	
200 °C								
0.31 μm	13.0 ± 0.2	14.5 ± 0.3	13.0 ± 0.6	9.7 ± 0.3	13.4 ± 0.2	9.6 ± 0.5	10.9 ± 0.5	21 ± 4
2.93 μm	12.6 ± 0.2	9.3 ± 0.3	14.0 ± 0.6	8.4 ± 0.3	10.9 ± 0.2	5.6 ± 0.5	10.9 ± 0.5	
250 °C								
0.11 μm	9.8 ± 0.2	8.5 ± 0.3	10.7 ± 0.6	7.6 ± 0.3	10.0 ± 0.2	9.5 ± 0.5	13.2 ± 0.5	40 ± 5
3.11 μm	9.6 ± 0.2	8.0 ± 0.3	4.3 ± 0.6	7.2 ± 0.3	8.6 ± 0.2	5.8 ± 0.5	8.3 ± 0.5	

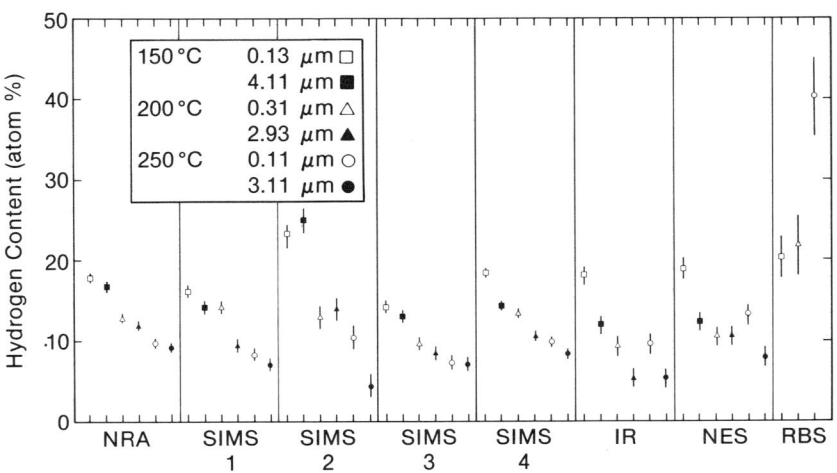

Fig. 1. Graphic plot of hydrogen content data listed in Table I.

The hydrogen content is given in atomic percent based on an assumed silicon density of 5×10^{22} atoms/cm^3.[1] Depth profiles of the hydrogen by NRA and by SIMS indicate that the in-depth distribution is uniform, so there should be no discrepancy in the hydrogen content measured by the profiling and non-profiling techniques. As an example, Fig. 2 shows the hydrogen depth profile measured by NRA on the 200°C, 0.31 μm sample.

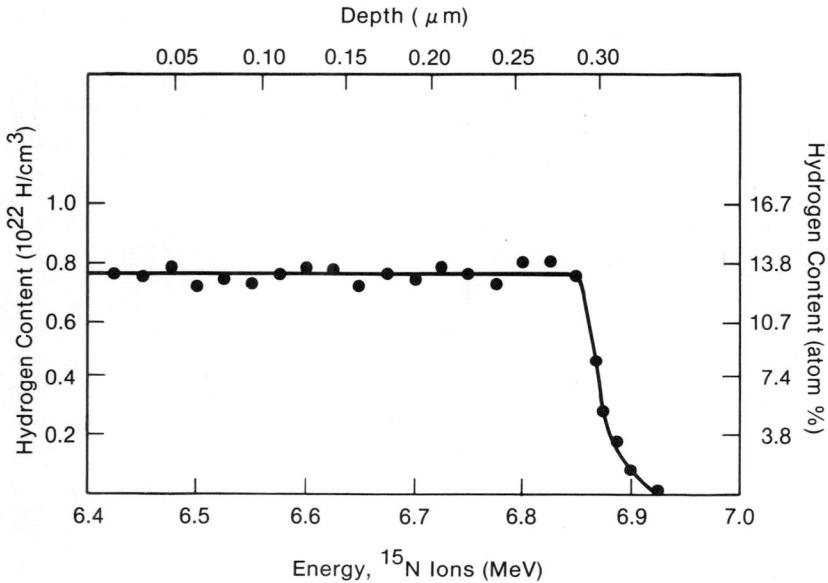

Fig. 2. Depth profile of 200°C, 0.31 μm TFS sample measured by NRA.

The NRA results show that hydrogen content is dependent on substrate temperature. As the temperature is increased, the amount of hydrogen incorporated into the TFS decreases. Similar temperature dependence has been reported for glow discharge of a-Si:H measured by IR and RBS [7]. The other techniques generally show this trend but with much more scatter.

The data from SIMS Lab 4 most closely match the NRA data. The data from Labs 1 and 3 report lower hydrogen content. The Lab 2 data are more scattered than those from the other three labs. Repetitive measurements by Lab 2 on all six samples indicate the precision was ±6% for all but the 200°C, 0.31 m sample, which was 13%. The precision for other SIMS labs was generally ± 2-3%, except again for that one sample. There were no obvious reasons for the greater variation in data from Lab 2.

---

[1]The NRA data are independent of actual film density. The density of one film was measured by NRA using the total energy loss in the film and the Dektak measured film thickness. The density was 97% of that of c-Si.

The NRA data indicate there is little or no thickness dependence in the hydrogen content, yet the IR data tend to indicate that there is a dependence. S. Hasegawa has reported the existence of a thickness dependence in the hydrogen content for glow discharge a-Si:H films deposited at 250°C, measured by IR spectroscopy [8]. This apparent discrepancy between these techniques may be due to presence of "non-bonded" hydrogen in TFS.

The NES hydrogen data are also more scattered than the NRA data. The hydrogen content measured by NES is lower than that determined by NRA for the thicker film and higher than that determined by NRA for two of the three thin films. The reason for overestimating the hydrogen content by NES in the thinner films may be due to the lower signal-to-noise (S/N) ratio for such samples. The difference in S/N ratio can be seen in Fig. 3. Figure 3a shows the hydrogen peak for the thin 150°C film as measured by NES; Fig. 3b is a similar plot for the thick film.

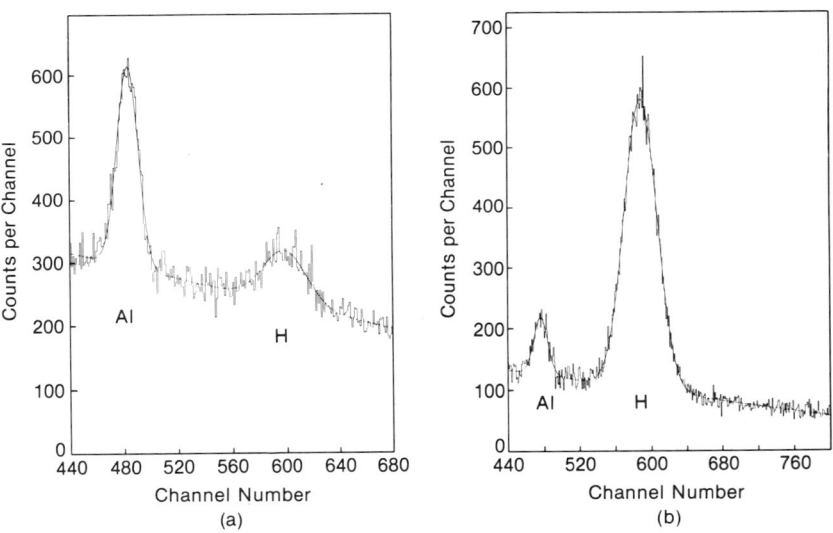

Fig. 3. Determination of the hydrogen yield by NES for T_s = 150°C TFS films. (a) 0.11 µm thick. (b) 4.11 µm thick.

The RBS data are significantly outside the range of the other techniques. This technique does not measure hydrogen directly and other impurities such as carbon, oxygen, and nitrogen, if present, must be accounted for in order to correctly deduce the hydrogen content. SIMS data from Lab 4 for these other impurities indicate that oxygen may be present in the range of 10^{19}-10^{20} atoms/cm^3. Although the correction for such levels of oxygen are insignificant, they would cause even further discrepancy in the comparison of the RBS results with the other results.

DISCUSSION OF TECHNIQUES

The NRA measures hydrogen directly and can provide depth profile information. The calibration can be very precise, limited only by the accuracy with which ion current can be measured during ion implantation of the standards. It is the most quantitative method known for determining hydrogen content in thin films [3].

NRA requires use of a large accelerator and high energy ions. Such facilities are limited in number and under the circumstance of high oxygen content the high ion energy has caused loss of hydrogen from the sample [9].

SIMS is the most sensitive analytical technique for surface and thin film elemental characterization [10]. SIMS can provide data on the distribution of hydrogen in TFS films as well as the distribution of other elements in the films, such as dopants and contaminant impurities. The depth distribution of hydrogen and other elements can be non-uniform in multilayer thin films, such as in a typical TFS solar cell structure, so that the ability to depth profile can be important.

The quantitative analysis by SIMS is very complex. There are matrix effects; that is, significant variations occur in the ion yields for specific ions when the composition of the film changes by about 1% or more. Although studies of the matrix effect, such as [10], have given valuable insight into the nature of this effect, it must be taken into account for quantitative analysis by SIMS. Standards are necessary for such analysis. The matrix of the standard must be very similar to that of the film being analyzed. As stated above, hydrogen ion implanted silicon was used as the standard in this study.

An earlier study did use an a-SiH$_x$ film for the SIMS standard where NRA was used to determine its hydrogen content [11]. In so doing, the SIMS and NRA results were in very good agreement. The comparison of ion implanted standards using c-Si and TFS substrates for quantitative analysis of hydrogen and other elemental species by SIMS is the subject of another study under way jointly by the Jet Propulsion Laboratory and ARCO Solar.

IR spectroscopy is among the most commonly used techniques for hydrogen measurement in TFS type films [11]. It is relatively simple to perform. It does not allow measurement of the hydrogen depth distribution and it measures only bound hydrogen. The calibration method for measurement of hydrogen has been questioned because of possible variations in the oscillator strength caused by the local silicon, hydrogen environment [12]. The interpretation of the presence of non-bonded hydrogen could be due to failure to properly account for these oscillator strength variations.

The NES technique used in this study (non-coincidence mode) can determine the content of all constituents in thin film silicon alloys such as a-Si:H and a-Si:Ge:H [6]. It is not a profiling type technique. Using the coincidence mode, the NES technique improves the S/N ratio for the hydrogen peak significantly over the non-coincidence mode as well as reducing the background noise [12]. This improves the sensitivity and precision of the technique. The coincidence mode eliminates the simultaneous determination of other constituents of the film.

As stated above, RBS does not directly measure hydrogen content. Other impurities in the film must be accounted for to use the method, so some other analytical technique, such as SIMS, must be used in conjunction with it [7]. Therefore, there is no real advantage in using this technique.

CONCLUSIONS AND COMMENTS

Although any one of several techniques can measure the hydrogen content of thin film silicon:hydrogen alloy films, there is considerable scatter in the data from the various techniques and even between different labs using the same technique. However, some specific conclusions can be drawn from this study and the direction future studies should take can be indicated.

1. The hydrogen content of TFS depends on substrate temperature, decreasing from 18% at 150°C to 10% at 250°C.

2. The total hydrogen content of TFS has little or no thickness dependence.

3. The NRA technique could be used to establish a secondary standard for calibration of other techniques.

4. There are reasons for using more than one technique: determining total versus bound hydrogen, measuring other constituents along with hydrogen, and depth profiling.

Further studies of the techniques for hydrogen measurement in TFS films should investigate:

1. Reasons for scatter in results by different techniques.

2. How to provide reliable calibration standards.

3. Whether there is truly significant non-bonded hydrogen in TFS material.

ACKNOWLEDGMENTS

Grateful acknowledgment is extended to Dr. W.A. Lanford for the ^{15}N NRA data; and to Craig Hopkins at Charles Evans & Associates, Dr. L.L. Kazmerski at SERI, and Dr. C.W. Magee at RCA for the SIMS analyses. Special thanks go to Dr. William Hamilton at JPL for his SIMS data and for continuing studies and also to Reinhard Schwarz at Princeton University for the NES data. Thank you to Frank So and Dr. Marc Nicolet at Cal Tech for the RBS data and our continued association.

ARCO Solar contributors include Vic Grosvenor, who prepared the samples, Philip Yan, who conducted the IR measurements, and Kim Mitchell, who provided technical advice. Dorothy Houk and Mari Kristiansen assisted in preparing and editing this manuscript.

REFERENCES

1. J.F. Ziegler et al, Nuc. Inst. and Meth., <u>149</u>, 19-39 (1978).

2. M.H. Brodsky, M. Cardona and J.J. Cuomo, Phys. Rev. B, <u>16</u>, 3556-71 (1977).

3. W. Lanford, Solar Cells, <u>2</u>, 351-63 (1980).

4. G.J. Clark et al, App. Phys. Lett., <u>31</u>, 582 (1977).

5. C.J. Fang, et al, Journal of Non-Crystalline Solids, <u>35 & 36</u>, 255-260 (1980).

6. R. Schwarz, <u>Technical Digest of the International PVSEC-1</u>, P-II-38, Kobe, Japan, Nov. 1984.

7. K. Kubota et al, Nuc. Inst. and Meth., <u>168</u>, 211-15 (1980).

8. S. Hasegawa and Y. Imai, Philos. Mag. B, <u>46</u>, 239-51 (1982).

9. M. Fallavier et al, App. Phys. Lett., <u>39</u>, 490-92 (1981).

10. V.K. Deline, "Secondary Ion mass spectrometry: SIMS II," in <u>Proceedings of 2nd Int. Conf. on Secondary Ion Mass Spectrometry (SIMS II)</u>, (Springer-Verlag, 1979), p. 48.

11. R.C. Ross, I.S.T. Tsong and R. Messier, J. Vac. Sci. Technol., <u>20</u>, 406-09 (1982).

12. Noboru Fukaka et al, Jp. J. App. Phys., <u>21</u>, L532-34 (1982).

HYDROGENATION DURING THERMAL NITRIDATION OF SiO_2

A.E.T. KUIPER*, F.H.P.M. HABRAKEN** AND JAMES T. CHEN***
* Philips Research Laboratories, 5600 JA Eindhoven, The Netherlands
** Technical Physics Dept.,State University,3508 TA Utrecht,The Netherlands
*** Philips Research Laboratories, Sunnyvale, CA 94086, U.S.A.

ABSTRACT

The incorporation of nitrogen and hydrogen during nitridation of SiO_2 was studied over the temperature range of 800-1000°C and for ammonia pressures of 1, 5 and 10 atm. The nitrogen content of the nitrided films was determined with Rutherford backscattering spectrometry. Nitrogen in-depth profiles were obtained applying Auger analysis combined with ion-sputtering. Hydrogen profiles in the films were measured using nuclear reaction analysis. Both the nitrogen and hydrogen incorporation were found to increase with temperature in this range. A higher ammonia pressure primarily increases nitridation of the bulk of the oxide films. Depending on the nitridation conditions up to 10 at.% of hydrogen may be incorporated. As distinct from the nitrogen profiles, the hydrogen in-depth profiles are essentially flat. The concentration of hydrogen in the films, however, was always found to be smaller than that of nitrogen : measured H/N ratios varied between 0.09 and 0.85, the smaller values being obtained for the thinner oxides and higher nitridation temperatures. The model previously postulated to explain the nitrogen incorporation during atmospheric nitridation of SiO_2 proves to be valid at higher pressures as well. By considering the role of OH as reaction product of the nitridation process, the hydrogen results can be accommodated within the same concept. The model predicts a low H/N incorporation ratio for a thin surface and interface layer and a substantially larger ratio for the bulk of the film. If this prediction is correct, which seems to be indicated by the etch-rate behaviour of the nitrided oxides, then this would have considerable importance for the electrical properties of this material.

INTRODUCTION

Most reports on thermal nitridation of thin oxide films mention the possible application pf these films as ultrathin stable gate dielectrics in submicron MOS devices or as radiation-hardened insulators in MOSFETS. These potential uses arise from the higher dielectric strength, higher density and higher chemical stability of the material as compared to untreated silicon dioxide. This explains the growing interest in the process of nitridation, and hence the increasing amount of data published both on the analysis of nitrided oxides and the electrical characterization of such films.
We have reported earlier, as have many others, that the kinetics of atmospheric nitridation of silicon dioxide can be characterized by a rapid, reaction-controlled initial nitridation of both the surface and the interface region of the oxide, followed by a much slower diffusion-limited nitridation of the bulk of the oxide [1]. The rate of this second part depends also on the thickness of the oxide film, i.e. thicker oxides take up more nitrogen. The slowing-down of the reaction arises from the formation of a nitride or oxynitride layer at the surface, which reduces the transport of reactive molecules to the inner film, and from the inability of the inner oxide film to carry off all oxygen-containing reaction products. The latter is in contrast to the situation at the interface, where the nitridation can occur at a higher rate - at least during the start of the process - since

reaction products will be consumed in a a subsequent oxidation reaction with the underlying silicon substrate. This model accounts for the observations, made by various researchers, that nitrogen piles up at the surface and near the interface, and that the amorphous film thickness increases with nitridation time.

We have shown that during nitridation in NH_3 substantial amounts of hydrogen are incorporated in the oxide; depending on the temperature at which nitridation is carried out, 1 to 3 at% H is built in [2]. Apart from the incorporation of nitrogen in the oxide, it is most likely that these amounts of hydrogen will also affect the electrical properties of the material. The frequently reported high trapping density, observed in nitrided oxides, may equally well be related to hydrogen as to nitrogen.

For this reason we have continued our study of thermally nitrided oxides, paying attention to both nitrogen and hydrogen.
Nitrogen contents in the films were measured quantitatively with Rutherford-Backscattering Spectrometry (RBS). Elemental in-depth profiles were obtained with Auger Electron Spectroscopy (AES), applying ion-sputtering, for Si, O and N, and with Nuclear Reaction Analysis (NRA) for H.

In this paper we report on the effect of high-pressure nitridation on the kinetics of the process and on the nitrogen profile that develops in the oxide films. Additionally we present data on the attendant incorporation of hydrogen and compare these with the figures for atmospheric nitridation. The implications of hydrogenation, occurring during nitridation of oxides, for the interpretation of etch rate and oxidation resistance measurements will briefly be indicated. A relation between trap density and bulk hydrogen content, inferred from the measured variations in $^H/N$ incorporation ratio, will be postulated.

EXPERIMENTAL

Nitridation experiments at pressures between 1 and 10 atmospheres of pure ammonia were performed in a Gasonics high-pressure system. 100 mm Si substrates, oxidized previously to grow 80, 300 or 800 Å of SiO_2, were loaded into this system while purging N_2. The process tube was then evacuated to reduce the moisture content of the reactor. Next, the system was pressurized by injection of liquid ammonia. There is no refreshment of the reaction gas during an experiment. Upon termination of the nitridation cycle the system was vented and the wafers were unloaded in a nitrogen environment.

The 300 Å and 800 Å-thick oxides were nitrided at 1000°C in 10 atm NH_3 for various reaction times. As for the 80 Å-thick oxides, the temperature (800°C, 900°C or 1000°C) and the ammonia pressure (1, 5 or 10 atm) were also varied.

We employed nuclear reaction analysis to obtain hydrogen in-depth profiles, making use of the resonant nuclear reaction $^1H(^{15}N,\alpha\gamma)^{12}C$. The geometry of the set-up, together with the applied energies and beam currents, have been published in a previous paper [2]. Unlike the conditions in our earlier work, the base pressure in the NRA analysis chamber is now in the 10^{-9} Torr region, which strongly reduces the contribution to the measured hydrogen profile arising from water or hydrocarbons adsorbed at the surface of the samples. This UHV set-up is very useful when measuring thin films or low hydrogen contents.

RESULTS

1. Nitrogen incorporation

In figure 1 are plotted the RBS results for nitrogen, measured in all samples that were nitrided at 1000°C in 10 atm NH_3. We interpret the kinetic

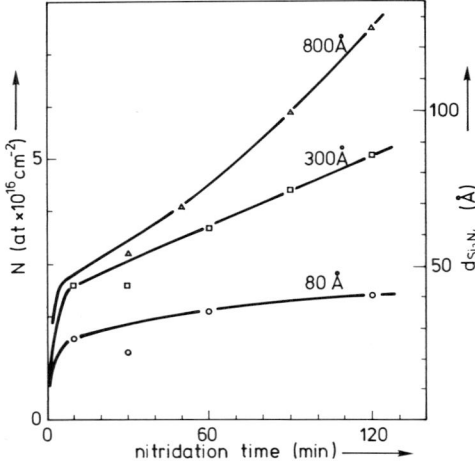

Figure 1:

Nitrogen contents, measured with RBS, in oxide films after nitridation at 1000°C in 10 atm NH$_3$. The oxide thickness is indicated.

data in figure 1 in terms of a rapid surface and interface reaction occurring in the first 10 minutes, followed by a slower diffusion-controlled process. Similar reaction kinetics has been found for nitridation in 1 atm NH$_3$ [1]. The main difference with the atmospheric results is that the nitrogen content in the films does not appear to saturate so quickly, at least not for the thicker oxides, resulting in appreciably higher nitrogen concentrations (or larger equivalent nitride thicknesses), even at a lower temperature. In our concept of nitridation, as summarized in the introduction, this behaviour suggests that the increase of the ammonia pressure leads to an enhanced nitridation of the inner part of the oxide film, which then accounts for the dependence of the effect on oxide thickness.

This explanation is furthr substantiated by the Auger results. An example is given in fig7ure 2 for a 300 Å-thick oxide film. In the last profile of this series bulk nitridation has proceeded so far that the interface pile-up of nitrogen is no longer distinguishable. In similar profiles

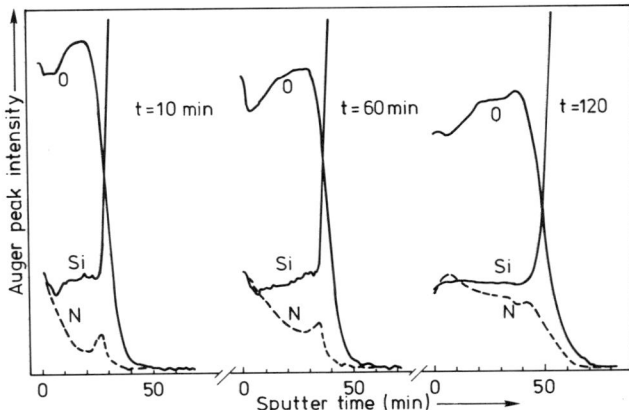

Figure 2 : Auger in-depth profiles of 300Å-thick oxide films, nitrided at 1000°C in 10 atm NH$_3$ for different reaction times t.

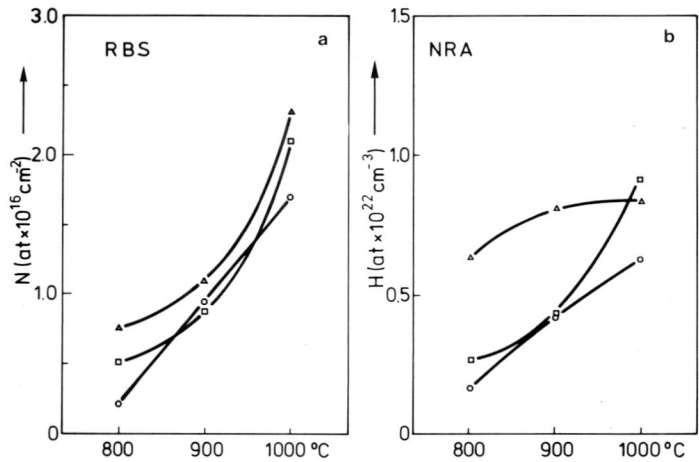

Figure 3 : Nitrogen contents, measured with RBS (a), and hydrogen contents derived from NRA (b), in 80Å-thick oxides after 1 h nitridation at various ammonia pressures: O = 1 atm NH_3, △ = 5 atm NH_3, □ = 10 atm NH_3.

of oxides nitrided at atmospheric pressure this interface pile-up could still be distinguished after a reaction time of 120 min [1].

Another cross-section of the RBS results is obtained by plotting all nitrogen data for one oxide thickness and reaction time versus nitridation temperature. This is done in figure 3a for the 80 Å-thick oxides after 1 h of reaction. It clearly shows that the temperature is a far more important parameter than the ammonia pressure. It is somewhat surprising that the curve for 10 atm NH_3 falls below that for 5 atm NH_3. Comparison of the Auger profiles showed that the 10 atm samples had suffered from surface oxidation. The difference in nitrogen content between the 5 and 10 atm samples amounts to 2×10^{15} at.cm^{-2} (fig. 3a) or one monolayer. The relative effects of small deviations among the experimental conditions will be large for these 80 Å-thick oxide films. Therefore, we will not attempt to interpret the deviating 10 atm curve in terms of the nitridation mechanism, but instead stress the importance of the temperature over the ammonia pressure.

2. Hydrogen incorporation

For quantitative hydrogen determinations we rely on NRA measurements. Hydrogen in-depth profiles were obtained by collecting a series of data points, each point from a fresh spot on the sample and using a different energy of the incoming ^{15}N ion beam. With the beam energy the probing depth is increased simultaneously. The atomic percentage of hydrogen in the film is calculated using an LPCVD nitride film with known H content as a calibration standard.

Figure 4 shows an NRA hydrogen profile measured for a 300 Å-thick nitrided oxide, which is characteristic of all samples analyzed in that the profile is flat. This is in agreement with the profiles we have published before for oxides of different thicknesses nitrided at 800 and 1000°C in 1 atm. NH_3 [2]. Even in the thinnest oxides a plateau in the hydrogen profile could be distinguished. We ascribe the higher concentration at the surface to adsorbed molecules (a monolayer or less). We have estimated the depth resolution of the NRA technique when applied to samples of this kind to be 30 Å.

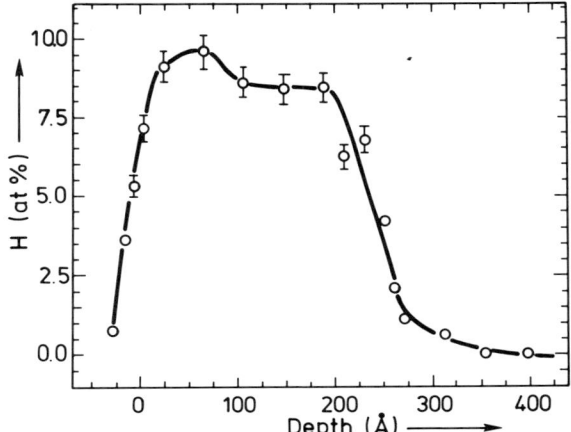

Figure 4 : Hydrogen profile, determined by NRA, in a 300Å-thick oxide film after 2 h nitridation at 1000°C in 10 atm NH_3.

The homogeneous distribution of hydrogen in the nitrided oxide, which is different from the nitrogen distribution, indicates that, although the amount of hydrogen incorporated is related to the nitrogen content, there is no such evident correlation with the position of the nitrogen atoms in the oxide film.

All measured NRA plateau or bulk values for the hydrogen content fall between 2.4 and 10 at%. From these values a plot similar to that of figure 3a may be constructed for the hydrogen incorporation: figure 3b. The conversion from at.% to concentration values is accomplished using the film composition as determined from RBS.

Although there is some similarity between figures 3a and 3b, the relation between N and H incorporation during nitridation does not appear to be straightforward.

To further elucidate this we have derived H/N ratios by combining the NRA and RBS results. In the calculation of H/N it is tacitly assumed that the H content is negligibly low. The error introduced in this way is about equal to the hydrogen content, i.e. some 10% or less.

In figure 5 H/N ratios are compiled for 80Å-thick oxides nitrided for 1 hr at different temperatures and ammonia pressures. Added is a data point at 1160°C, which was derived from the results of a former publication [2]. The trend that emerges from this figure is that the number of hydrogen atoms incorporated per nitrogen atom decreases with decreasing pressure and increasing temperature. We recall that both the amount of nitrogen and the amount of hydrogen increase with temperature (up to 1000°C) and pressure (up to 5 atm) : fig. 3.

We have shown earlier that at still higher temperatures (1160°C) the hydrogen content is found to be lower again [2]. The temperature dependence of the hydrogen incorporation may be governed by outdiffusion of hydrogen (or H-containing species), causing both the decrease of H/N ratio when the nitridation temperature is raised, and the reduction of the hydrogen content itself above 1000-1100°C. However, the way in which nitrogen is incorporated in the film structure may also play a role; we will return to this point in the discussion section.

The H/N ratio also exhibits some thickness-dependent behaviour, al-

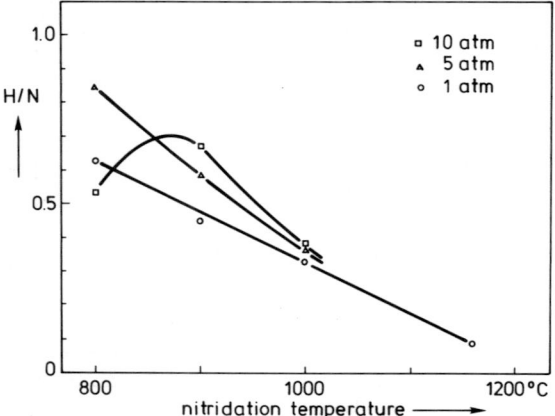

Figure 5 : H/N ratios derived for 80Å-thick oxides after nitridation at different NH_3 pressures for 1 h.

though far less pronounced than with the temperatureas parameter. Thicker films have a somewhat higher H/N ratio, e.g. the combination 1000°C-10 atm NH_3-1h yields H/N=0.38 for 80 Å SiO_2 and 0.48 for 300 Å SiO_2. This suggests that relatively more hydrogen is incorporated in the bulk of the film than near the surface or the interface.

DISCUSSION

The results presented support the concept of oxide nitridation as outlined in the introduction. Higher ammonia pressures bring about an enhancement of the rate of the diffusion-controlled part of the nitridation process, resulting in a larger amount of nitrogen incorporation in the bulk of the oxide than occurs with atmospheric nitridation. As a consequence, high-pressure nitridation is more effective for thicker oxides. A higher ammonia pressure, however, also affects the hydrogen content of the nitrided oxides. Figures 3 and 5 show that temperature is a more important parameter than ammonia pressure and that the nitrogen and hydrogen incorporation are somehow correlated. This again is in support of the proposed nitridation mechanism in which NHx has been postulated as the reactive species [1].

The N-H correlation makes it meaningful to consider the process dependence of the H/N ratio. This ratio increases with higher ammonia pressure, lower temperature and thicker oxide films.

Since the hydrogen and nitrogen in-depth profiles are not found to be similar (the hydrogen profiles are essentially flat), at least part of the hydrogen atoms (or H-containing species) that are incorporated in the oxide during nitridation must have some mobility, as opposed to the incorporated nitrogen atoms. Furthermore, all H/N ratios calculated are smaller than 1, varying from 0.85 at 800°C to 0.09 at 1160°C (figure 5). This we are inclined to interpret as follows.

The process of nitridation of SiO_2 must be conceived as an equilibrium between nitridation and oxidation [1]. Everywhere in the oxide film a competition takes place between nitriding NH_x groups (which will produce OH_y groups) and oxidizing OH_y fragments (which will liberate NH_x groups again). For simplicity we assume x and y equal to one. While nitridation of the surface is occurring the OH products formed can easily escape as long as the formation of a diffusion-limiting nitride layer is not completed. At the same time, the interface region will be nitrided, with the OH groups being consumed in the oxidation reaction of the underlying silicon. Hydrogen, released in this latter reaction, can easily desorb from the material into the ambient. After a short time the nitridation at both interfaces is considerably slowed down as a result of the formed nitride layers, which constitute diffusion barriers for both NH and OH. At this point the nitridation of the bulk has not proceeded very far, since the higher concentration of OH (desorption from the bulk is more difficult than from the surface) has prevented the equilibrium from shifting as far as at the interfaces. The process is now in its (out)diffusion-controlled regime. A further increase of the NH concentration (by raising the ammonia pressure) will both shift the equilibrium so as to incorporate more nitrogen atoms and increase the OH concentration, and hence the hydrogen content.

At considerably higher temperatures SiO_2 can be converted completely into Si_3N_4, due to the enhancement of the outdiffusion of OH. An increase of the nitridation temperature will therefore shift the equilibrium favourably; because of the enhanced outdiffusion of OH this will not result in a proportional increase of the hydrogen content, whence the decrease of the H/N ratio at higher temperatures.

In view of the foregoing it is most likely that hydrogen, measured in the nitrided films, is incorporated as OH or NH, and possibly also as SiH at nitridation temperatures lower than 1000°C. This implies that wherever in the film hydrogen is detected, the coordination of the atoms deviates from that of a completely cross-linked amorphous network as in SiO_2 or Si_3N_4. The concept of hydrogen incorporation, as outlined above, would suggest that in the surface and interface regions nitrogen atoms are completely coordinated with silicon, such that on a molecular scale Si_3N_4 and/or Si_2ON_2 is formed. Some AES and XPS data have been published that support this supposition [3,4]. Unfortunately, because of the depth resolution of NRA, which is limited to 30 Å, the picture of very small H contents at both interfaces and relatively high H contents in the bulk cannot be substantiated directly from the measured hydrogen profiles. Nevertheless with this cocept it can at least be understood why the H/N ratio increases with oxide film thickness. Further, it may also be helpful in interpreting the chemical and electrical behaviour of nitrided oxides.

For example, the hydrogen incorporated during nitridation also affects the etch rate of the film. This may be derived from the etch rate measurements shown in figure 6. Nitridation in 1 atm NH_3 results in a reduced etch rate of a surface and interface layer of the film. This behaviour has been observed before and has been related directly to the larger nitrogen concentration in these regions. However, a very similar etch characteristic is measured. However, a very similar etch characteristic is measured after nitridation at 10 atm, although this study shows that upon such treatment the nitrogen content in the bulk of the film approaches that of the surface and interface regions (figure 2). Therefore, the higher etch rate of the bulk of the nitrided oxide film has to be related to the mode of hydrogen incorporation (NH, OH, SiH). Wet-etching may be envisaged as a process in which atomic bonds are broken by chemical reactions that proceed at a solid-liquid interface. If the etchant is reactive towards different atomic bonds, the measured etch rate will be an average rate, weighed over the relative abundancies of the bonds involved. The persistence of the higher etch rate of the bulk of oxide films upon ongoing nitridation may hence be ascribed to a growing concentration of e.g. SiOH groups having a large etch rate, which balances the simultaneously increasing concentration of etch-resistant Si-N bonds. Similar reasoning may apply to the thickness-dependent oxidation resistance of nitrided oxides.

In various publications a high trapping density has been reported for thermally nitrided oxides. The relatively large hydrogen concentrations in these films, as evidenced by the results presented, may very well be related to

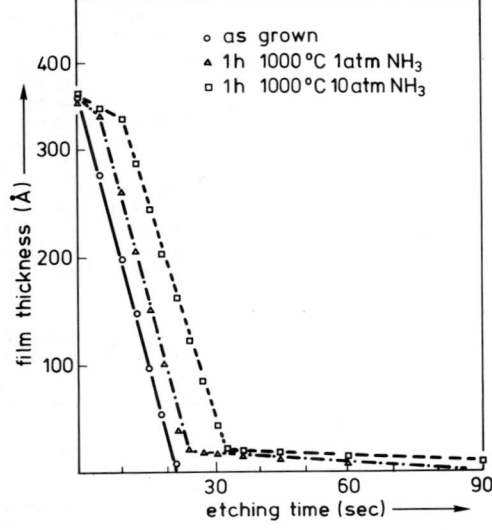

Figure 6:

Effect of 1 h nitridation at 1000°C on the etch rate behaviour of silicon oxide in 1:10 HF at 20°C.

this observation. Lai et al. have found evidence that the trapping density is related to OH groups in the film [5]. For the device applicability of nitrided oxides a further study of the incorporation of hydrogen is at least as important as that of nitrogen. Especially the mechanism of bulk nitridation, where the larger H/N ratios seem to originate from, deserves a closer examination.

REFERENCES

1. F.H.P.M. Habraken, A.E.T. Kuiper, Y. Tamminga and J.B. Theeten, J. Appl. Phys. 53, 6996 (1982)

2. F.H.P.M. Habraken, E.J. Evers and A.E.T. Kuiper, Appl. Phys. Lett. 44, 62 (1984)

3. R.P. Vasquez, M.H. Hecht, F.J. Grunthaner and M.L. Naiman, Appl. Phys. Lett. 44, 969 (1984)

4. T.W. Ekstedt, S.S. Wong, Y.E. Strausser, J. Amano, S.H. Kwan and H.R. Grinolds, in Insulating Films on Semiconductors (Ed. J.F. Verwey and D.R. Wolters), North-Holland 1983, 189

5. S.K. Lai, D.W. Dong and A. Hartstein, J. Electrochem Soc. 129, 2042 (1982)

GROWTH AND COMPOSITION OF LPCVD
SILICON OXYNITRIDE FILMS

F.H.P.M. HABRAKEN[*] AND A.E.T. KUIPER[**]

[*] Technical Physics Department, Utrecht State University, P.O. Box 80.000, 3508 TA Utrecht, The Netherlands

[**] Philips Research Laboratories, 5600 JA Eindhoven, The Netherlands

ABSTRACT

Silicon oxynitride films with various O/N concentration ratios were deposited in a Low Pressure Chemical Vapour Deposition process from SiH_2Cl_2, N_2O and NH_3 or ND_3. The resulting films were analysed with respect to their chemical compositions using a number of high energy ion beam methods. The results of the analysis are related to the growth kinetics. It is suggested that the SiH_2Cl_2 decomposes more difficult on growth surfaces which contain more oxygen. The effect is attributed to the larger electronegativity of oxygen compared to nitrogen.

INTRODUCTION

Silicon nitride films (Si_3N_4) are applied nowadays in electronic devices which are based on metal-nitride-oxide silicon (MNOS) structures. Furthermore the material is widely used during integrated circuit processing as oxidation and diffusion mask. Although a lot of effort has been made to grow oxygen free nitride films, the controlled incorporation of oxygen into silicon nitride films to grow silicon oxynitride films may be advantageous. In this way, for instance, the di-electric constant, internal and interfacial stress and refractive index of the material can be varied. However, there is no unambiguous way of growing silicon oxynitride films. If we confine ourselves to chemical vapour deposition processes, various gases, gas phase compositions, temperatures and total pressures may be used. In advance it is not obvious that these different processes results in the same material with respect to all relevant properties, also not if films with the same overall stoichiometry (for instance oxygen/nitrogen concentration ratio) are compared.

In this paper we present results obtained in the study of the growth and composition of silicon oxynitride films, deposited by low pressure chemical vapour deposition (LPCVD) [1]. The reactant gases were SiH_2Cl_2, NH_3 and N_2O, the temperature was 820°C and the total pressure amounted to 200 m Torr. LPCVD is preferred above CVD at atmospheric pressures (APCVD) because larger batch sizes are allowed and a better uniformity is obtained. The choice for the particular reactant gas mixture is prompted by the fact that it represents a simple extension of a widely used deposition process of Si_3N_4. Parameter in our investigations was the N_2O/NH_3 gas phase concentration ratio. As analysing techniques we used mainly high-energy ion beam methods. Special attention will be paid to the hydrogen and chlorine contents in the films. It will be suggested that there is a relationship between the film composition and the deposition kinetics.

EXPERIMENTAL

Oxynitride films were grown in a standard LPCVD reactor normally used for silicon nitride deposition. The sum of the N_2O and NH_3 gas flows was kept constant at 75 sccm. The SiH_2Cl_2 gas flow was 25 sccm at all N_2O/NH_3 gas phase ratios [1]. Some samples were grown with ND_3 instead of NH_3 to reveal the origin of the incorporated hydrogen.

The oxygen, nitrogen, silicon and chlorine contents of the films were measured with Rutherford Backscattering Spectrometry using 2 MeV He$^+$ particles. Details of the procedure and experimental set up are given in Ref. 2 Depth profiles of oxygen, nitrogen and silicon were obtained by means of Auger electron spectroscopy in combination of 2 keV Ar ion sputtering [1,2]. Hydrogen concentration profiles were determined by means of Nuclear Reaction Analysis with the reaction $^1H(^{15}N,\ )^{12}$. Hydrogen profiling using the reaciton $^1H(^{15}N,\ )^{12}C$ is impeded in films which also contain deuterium because of the reaction $^2D(^{15}N,n\ )^{16}O$ [3]. Therefore, in the deuterated films the D and H content was measured using Elastic Recoil Detection (ERD), where we made use of 28 MeV ^{28}Si primary beam. The recoiled particles were detected at an angle of 35° with the primary beam. Scattered and recoiled heavy particles were stopped in a 9 m Mylar absorber foil in front of a 150 m totally depleted silicon detector. This technique provides us also with values for the O/N concentration ratio in the films. These values appeared to be in excellent agreement with those deduced from RBS measurements. Details of the procedures and equipment used in NRA and ERD will be given elsewhere [4].

RESULTS AND DISCUSSION

At first we consider the growth rate of the material. Fig. 1 shows the growth rate as a function of the N_2O/NH_3 gas phase input ratio. For the Si_3N_4 it is about 58 Å/min but it declines strongly with increasing N_2O/NH_3 ratio [1]. RBS spectra, obtained under channeling conditions, reveal that the amount of silicon in the deposited films does not exceed the amount which is necessary to compensate the oxygen and nitrogen bonds.

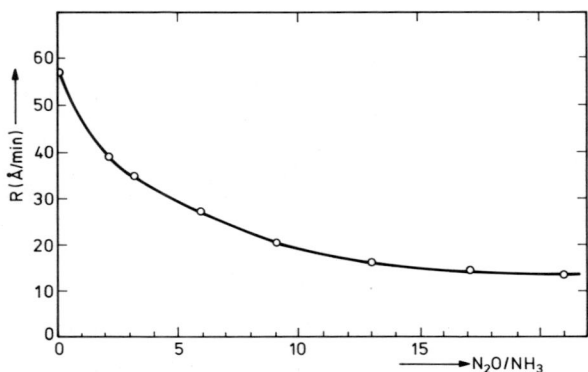

Fig. 1 Growth rate of silicon oxynitride as a function of the gas phase input ratio $N_2O/\ NH_3$.

An Auger lineshape study by van Oostrom et al [5] indicates that these materials consist of a mixture of silicon oxide and nitride on an atomic scale with no detectable amount of free silicon present.

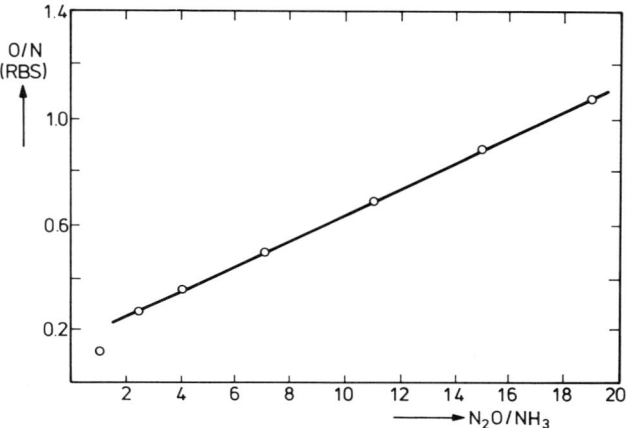

Fig. 2 O/N concentration ratio in the films, measured with RBS, as a function of the gas phase ratio N_2O/NH_3

Fig. 2 shows the O/N concentration ratio in the films as a function of the N_2O/NH_3 gas phase ratio . As expected, O/N increases with N_2O/NH_3, but the NH_3 gas flow has to be strongly reduced to the advantage of the N_2O flow in order to introduce oxygen in the films. Apparently N_2O has a much lower reactivity towards the growth surface that NH_3.

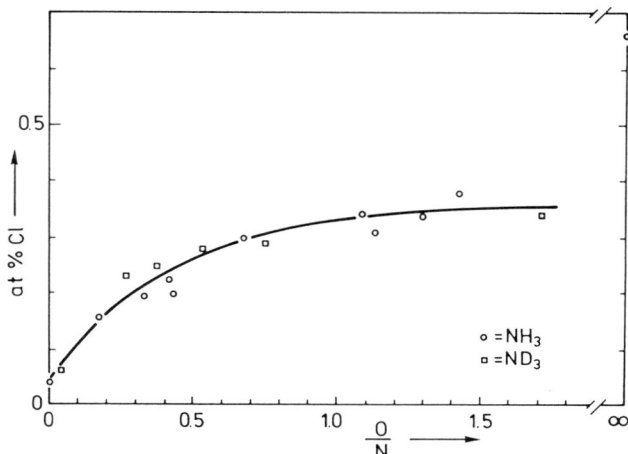

Fig. 3. Chlorine concentration in silicon oxynitride films as a function of O/N.

The chlorine concentration in the films increases with increasing N_2O/NH_3 gas phase ratio, or stated differently, with increasing O/N ratio in the film, as revealed by RBS measurements (fig. 3). Its concentration varies between 0.05 and 0.65 at .% going from Si_3N_4 towards SiO_2. Since LPCVD is

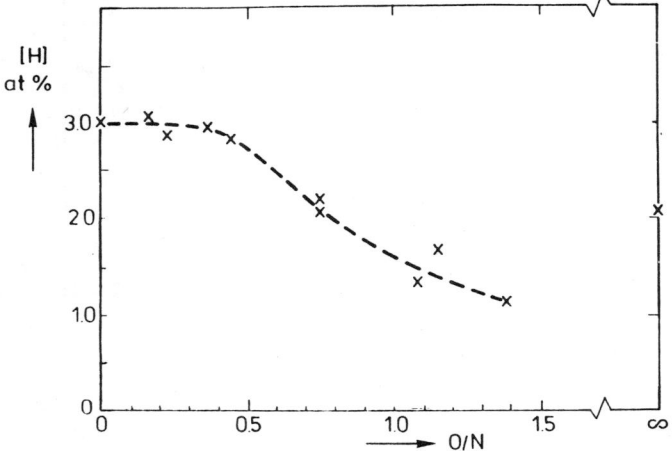

Fig. 4   Hydrogen concentration in silicon oxynitride films as a function of O/N.

believed to be a surface reaction controlled process, the surface itself is one of the reactants. The chemical composition of the surface is at the moment the best described by the O/N ratio, which is the reason for the way of plotting the analysis data.

The Cl data are obtained in the films of various thicknesses, but its concentration scatter around one and the same curve, indicating that the Cl is for the largest part homogeneously distributed with respect to the depth of the oxynitride films.

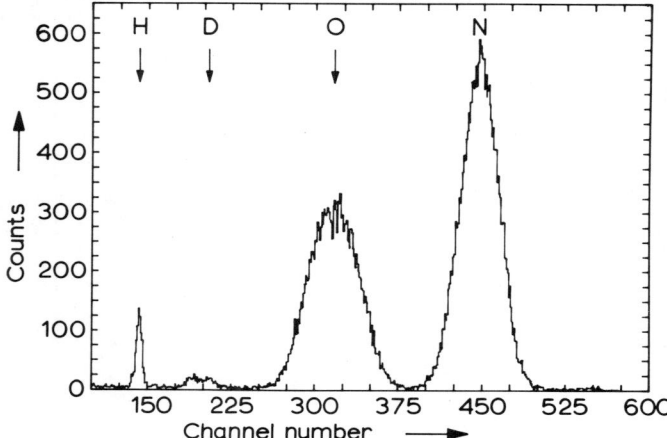

Fig. 5   ERD spectrum of a 200 Å silicon oxynitride film with O/N = 0.53.

A further element of interest is hydrogen. Fig. 4 shows the hydrogen concentration as a function of O/N, measured with NRA. For O/N < 0.4 it is constant at 3 at .%, but for larger O/N ratios it decreases, until somewhere between O/N = 1 and   it must increase again. The situation for hydrogen is a bit more complicated because it may originate both from NH_3 and from SiH_2Cl_2 giving rise to N-H and Si-H bonds [6,7]. Therefore layers were grown from SiH_2Cl_2 and ND_3.

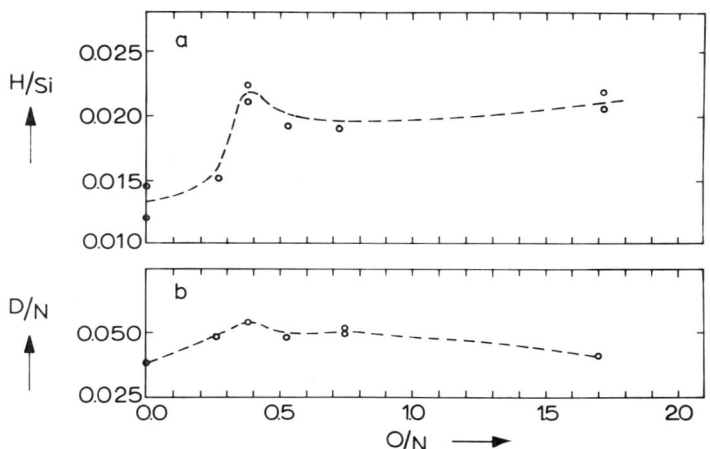

Fig. 6   Amount of D relative to N and amount of H related to Si in as grown silicon oxynitride films as a function of O/N.

The deuterated films were analysed by means of ERD. A typical ERD spectrum is given in Fig. 5. The relative elemental concentrations were deduced from the spectra by correcting the number of counts in the peaks with the relative cross sections for recoiling of the considered elements. We assumed the scattering process in our experimental situation to be completely of the Rutherford type, although this may not be true for the recoiling of D. However, a deviation of the cross section from the Rutherford value will only change the magnitude of the D/N concentration ratio but not its behavior as a function of O/N, the H/N ratios deduced from the ERD spectra were converted into a H/Si concentration ratio using the RBS data.

Fig. 6 shows the D/N and H/Si concentration ratios in the oxynitride films as a function of O/N. The H/Si ratios contain also the hydrogen of the possibly present hydrogen surface layer, which cannot be completely resolved in ERD from the bulk hydrogen, because the thickness of the deuterated oxynitride films amounted to 20-30 nm only. Despite of this an increase in H/Si with O/N is clearly present. For O/N > 2, H/Si increases further up to a value of about 0.06 at O/N =   , as is derived from fig. 4. D/N shows a more complicated behavior as a function of O/N: initially it increases, but for O/N > 0.4 it tends to decrease.

Isotope exchange is supposed to have a negligible effect on the hydrogen incorporation as is discussed on the basis of anneal experiments in ref. 4. With this in mind, it follows from the analyses that the concentration of the elements which are introduced via SiH_2Cl_2 (H,Cl) increases with increasing O/N concentration ratio in the films. As well for Cl as for H the increase is the strongest in the region of low O/N ratios (cf. figs. 3 and 6).

In the same region a strong decrease of the growth rate occurs (fig. 1). From growth experiments, in which the flow of SiH_2Cl_2 was varied, it is deduced that the decomposition of SiH_2Cl_2 is the growth rate limiting step [8]. Therefore we believe that the increase of the H and Cl both originating from SiH_2Cl_2, as well as the decline of the growth rate with increasing O/N ratio are both appearances of the same phenomenon, i.e. a more difficult decomposition of SiH_2CL_2 on surfaces, which contain more oxygen.

As for the origin of this retardation of larger O/N ratios we suggest that this is due to bonding of the silicon atoms of incompletely decomposed dichlorosilane moleclues to one or more oxygen atoms. The larger electronegativity of oxygen compared to nitrogen causes a strengthening of Si-H bonds and a shift in infrared absorption frequency to higher frequencies [9,10]. Similar phenomena have been observed in the LPCVD of SiO_x mixtures from SiH_4 and N_2O [11]. The increase of the Cl concentration can be readily understood within the same picture. The same holds true for the increase of the D/N ratio for O/N < 0.4. Here we consider the increase of the electronegativity of the groups to which the incompletely decomposed ND_3 molecules are bound during the surface reaction. In this case the electronegative element oxygen resides in the next-nearest neighbor position. The decrease for O/N > 0.4 is not yet understood at the moment.

The strengthening of Si-H bonds may be beneficial for the electrical performance of these films in MNOS like structures, as is discussed in ref. 4.

ACKNOWLEDGEMENTS

The investigations have been supported in part by the Commission of the European Communities within the framework of ESPRIT.

REFERENCES

1) A.E.T. Kuiper, S.W. Koo, F.H.P.M. Habraken and Y. Tamminga, J. Vac. Sci. & Technol. B1, 62 (1983).

2) F.H.P.M. Habraken, A.E.T. Kuiper, A. van Oostrom, Y. Tamminga and J.B. Theeten, J. Appl. Phys. 53, 404 (1982).

3) J.L. Weil and K.W. Jones, Phys. Rev. 112 1975 (1958)

4) F.H.P.M. Habraken, R.H.G. Tijhaar, W.F. van der Weg, A.E.T. Kuiper and M.F.C. Willemsen, to be published.

5) A. van Oostrom, L. Augustus, F.H.P.M. Habraken and A.E.T. Kuiper, J. Vac. Sci. Technol. 20, 953 (1982).

6) H.J. Stein, J. Electron. Mater. 5, 161 (1976).

7) H.E. Maes and J. Remmerie, Proc. Silicon Nitride Thim Insulating Films, ECS 83-8, 73 (1983).

8. J. Remmerie and H.E. Maes, to be published.
M.J.H. Kemper, S.W. Koo, Y. Tamminga and M.F.C. Willemsen, Private Communication.

9) J.A. Schaefer, D. Frankel, F. Stucki, W. Gopfel and G.J. Lapeyre, Surf. Sci. 139, L209 (1984).

10) G. Lucovsky, Solid State Comm. 29, 571 (1979).

11) B. Verstegen, F.H.P.M. Habraken, W.F. van der Berg, J. Holsbrink and J. Snijder, J. Appl. Phys., 57, 2766 (1985).

MeV HELIUM MICROBEAM ANALYSIS: APPLICATIONS TO SEMICONDUCTOR STRUCTURES.

R.A. BROWN*, J.C. McCALLUM*, C.D. McKENZIE* and J.S. WILLIAMS**.
*School of Physics, University of Melbourne, Parkville 3052, Australia.
** Microelectronics Technology Centre, R.M.I.T. Melbourne 3000, Australia.

ABSTRACT

We have employed an MeV He^+ microbeam for the analysis of micron-scale semiconductor structures. The analysis combines a unique beam scanning and mapping capability with powerful ion beam techniques such as particle-induced x-ray emission, Rutherford backscattering and ion channeling. These analysis capabilities are applied to the measurement of dopant profiles in polycrystalline resistors, identification of micro-alloying in metal-GaAs contacts and damage/atom location in individual polycrystalline silicon grains. The latter application utilises the newly-developed technique of channeling contrast microscopy.

INTRODUCTION

The Melbourne University microprobe facility offers a powerful and unique means of probing micron-sized structures in single-crystal substrates by combining a scanned He^+ microbeam with ion beam techniques including high resolution Rutherford backscattering and channeling. This analysis method can be used to study damage and impurity distributions in semiconductors with a lateral resolution of better than 5 μm and an in-depth resolution of better than 100Å. A unique data storage and mapping facility allows near-surface features to be imaged according to secondary-electron emission, elemental mass, concentration and depth. This information is obtained by Particle-Induced X-ray Emission (PIXE) or Rutherford Backscattering (RBS)/channeling spectrometry. We have recently developed a novel imaging technique termed channeling contrast microscopy [1] which has considerable potential for the analysis of microscopic variations in semiconductors. Channeling contrast has been applied successfully to study locally laser-annealed regions (< 30 μm) in ion implanted Si and to the imaging of individual grains in polycrystalline Si. In this paper the various imaging and data acquisition facilities of the Melbourne microprobe are discussed and some examples of profitable analyses are given.

EXPERIMENTAL DETAILS

Details of the performance and uses of the Melbourne microprobe are reviewed elsewhere [2,3,4]. A schematic of the microprobe is shown in Figure 1. A monoenergetic beam of ions from a 5U Pelletron is used to illuminate a circular object aperture, O, of 5 - 300 μm in diameter. The transmitted beam is further restricted by a divergence-limiting aperture, A, used to minimise aberrations and to ensure a beam divergence of $<0.3°$ for channeling studies. The remaining beam is focussed to a spot on target by a set of four magnetic quadrupole lenses, Q, arranged as a 'Russian quadruplet'. At present helium microbeams of <3 μm diameter with beam currents up to 1 nA are routinely produced; higher currents are available with larger beam diameters (e.g. 10 nA for 10 μm). A set of deflection coils, C_{1-4}, driven independently in the X and Y directions by triangular waveform generators, is used to scan the focussed beam uniformly over the target. Scans greater than 1 x 1 mm^2 can be achieved.

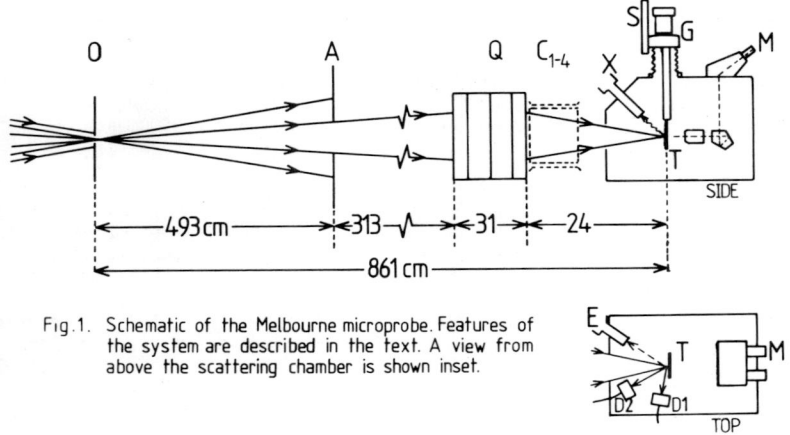

Fig.1. Schematic of the Melbourne microprobe. Features of the system are described in the text. A view from above the scattering chamber is shown inset.

The target, T, is incorporated in the scattering chamber and supported from an x-y translation stage, S, mounted above the chamber. A two-axis goniometer, G, is also mounted on the micrometer stage to facilitate alignment of targets for channeling. Nuclear-reaction products and/or backscattered particles are collected by surface-barrier detectors, D1 and D2, which can be externally positioned at desired scattering angles to provide either high depth resolution or optimum mass resolution. Particle-induced x-rays are collected by the Si(Li) x-ray detector, E. The rear-viewing microscope, M, facilitates target positioning and allows the beam diameter to be measured when the beam impinges upon thin glass cover slips mounted in the specimen plane.

MAPPING AND DATA STORAGE CAPABILITIES

Since secondary-electrons are emitted from a surface under ion bombardment, secondary-electron images of surface features can be generated using scanned ion microbeams in much the same way as images obtained in a scanning electron microscope. Such ion-induced secondary-electron images are routinely used to locate regions of interest on a sample. An example of an ion-induced secondary-electron map is shown in Fig. 2b, taken from a particular part of an integrated circuit containing aluminium contacts, polysilicon interconnects and surrounding SiO_2 regions. An optical micrograph of the same portion of the circuit is shown for comparison in Fig. 2a.

Having located regions of interest, the signals from the various x-ray and particle detectors can be examined. It is possible to display in real-time either i) integrated x-ray and particle energy spectra from the scanned area, or ii) two-dimensional maps of the scanned area, analogous to the secondary-electron images, but with the storage oscilloscope intensity representing the yield within a selected energy region of the collected x-ray or particle spectrum. An example of an x-ray (PIXE) map is shown in Fig. 2c, taken from the same region of the integrated circuit as shown in Figs. 2a and 2b. For the PIXE map, the energy window corresponds to the K_α x-ray line from aluminium. Thus, the bright regions on the PIXE map correspond to aluminium contacts and interconnects.

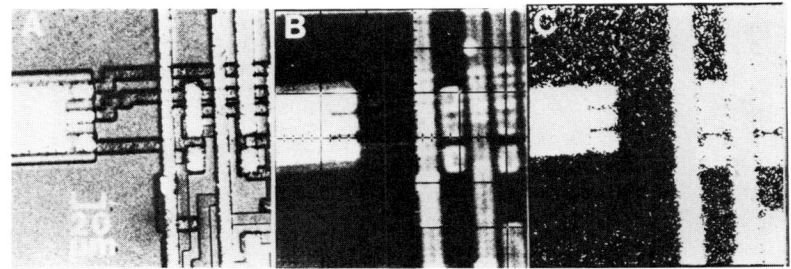

Fig.2 a) optical micrograph, b) secondary-electron map, c) PIXE map of part of an integrated circuit containing Al contacts, polysilicon interconnects and surrounding SiO_2 regions.

Fig. 3a illustrates a map generated from a Rutherford backscattering spectrum of an antimony-implanted polysilicon resistor. The light region on the map indicates the implanted resistor body and was imaged by setting an energy window about the antimony signal in the RBS spectrum. Fig. 3b shows for comparison a PIXE map of the same resistor with the energy window set on aluminium K_α x-rays to image the contact pads.

In addition to real-time mapping and display of integrated energy spectra, all data is stored using an on-line PDP 11/40 computer. Upon detection of a particle or x-ray, the energy (pulse height) and x,y scan voltages are digitised and stored on magnetic tape. Following accumulation, the stored data can be sorted into a three-dimensional array with indices of position (x,y) and energy. This allows i) two-dimensional maps to be produced using the yield in any selected energy window, or ii) energy spectra to be produced from any selected area within the map. As we illustrate in the following section, this powerful data storage capability is particularly useful for the measurement of composition and damage profiles in selected micron-scale regions of semiconductor structures.

SELECTED APPLICATIONS

Composition vs. Depth Profiling

To illustrate selected-area concentration vs. depth profiling, we make use of the ion-implanted polysilicon resistor shown in Fig. 3. Prior to deposition of the aluminium metallisation for the resistor contacts and interconnections, a further shallow antimony implant was made into the contact regions to increase the near-surface carrier concentration and facilitate direct ohmic contacting to deposited metals. Fig. 4 shows the selected-area RBS spectra obtained from the resistor body (open circles) and the contact pads (open triangles). The Sb profile in the resistor body exhibits the expected Gaussian distribution for a single 100 KeV Sb^+ implant with a range of ~500Å. In addition, the profile from the contact region shows a high near-surface Sb concentration as a result of the second shallow Sb implant.

Micro-alloying

A second microbeam application we have pursued is the analysis of laterally non-uniform metal-semiconductor contacts. A particular system of

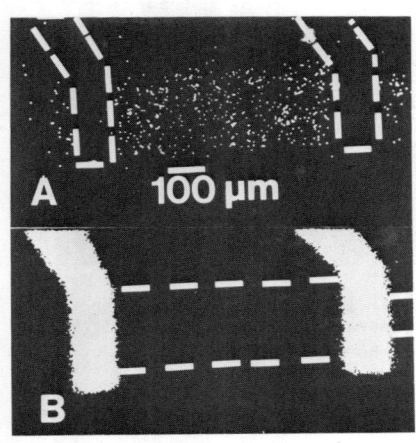
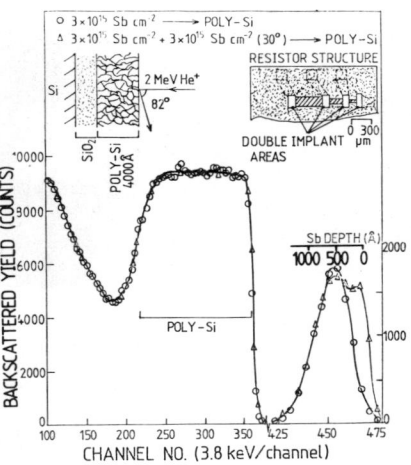

Fig.3 a) RBS map of Sb distribution across an ion-implanted polysilicon resistor.
b) PIXE map of Al in the same resistor, indicating the aluminium contact pads.

Fig.4 Selected-area RBS spectra obtained from the resistor body (open circles) and the contact pads (open triangles) of the resistor structure shown in Fig. 3.

interest has been Sn migration into GaAs for the source and drain contacts of FET devices [5]. Initial electrical measurements indicated promising results for ohmic contacts and RBS/ion channeling measurements (using a large 1mm beam spot on large Sn-migrated areas) suggested the incorporation of up to 2 atomic % Sn in crystalline GaAs. This latter result was difficult to reconcile in view of the expected microstructure following the alloying of molten Sn with GaAs. We consequently investigated the lateral uniformity of Sn migration using the helium microbeam. Fig. 5 is a PIXE map of a migrated contact region on a FET device indicating (bright) Sn-rich regions, typically < 20 μm in size. Fig. 6 shows an RBS spectrum taken from within one of these regions. Both the GaAs and Sn regions of the spectrum provide evidence for local micro-alloying of Sn with GaAs (up to 20 atomic % Sn) over a depth of >2000 Å. The Sn content in regions surrounding the Sn-rich alloys was estimated to be 0.1 atomic %.

Channeling Contrast Microscopy

Channeling contrast microscopy has provided a unique way of examining impurity and damage distributions within individual grains of ion-implanted polysilicon. For this analysis a commercially produced polysilicon material with grain sizes ranging from a few microns to approximately 1 mm was implanted with Sb^+ ions and annealed at 650°C for 30 mins. Channeling contrast analysis was carried out by aligning the sample so that the He^+ beam (unscanned) was channeled in a particular grain of interest. The beam was then scanned over a region containing the aligned grain.

A channeling contrast map can be produced by imaging the backscattered yield from an energy window set within the Si part of the spectrum; those

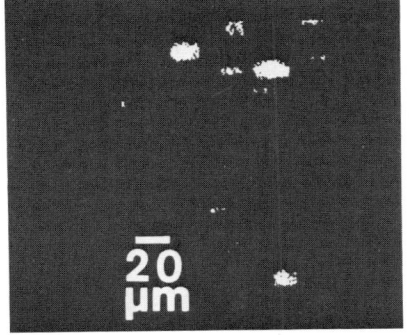
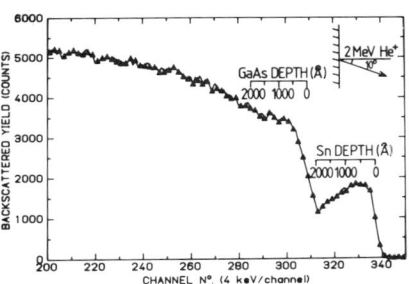

Fig. 5 PIXE map of Sn across a contact region of a GaAs FET device, indicating strong localisation of Sn in regions <20 µm in size.

Fig. 6 RBS spectrum taken from within a Sn-rich region of Fig. 5, indicating the possibility of local micro-alloying of Sn with GaAs over a depth ~2000 Å.

scanned areas (grains) in which the beam is well channeled will give rise to a low yield (density) of backscattered particles. Such a map is illustrated in Fig. 7b. The clearly defined dark/bright regions are indicative of channeling contrast from adjacent grains. For comparison, Fig. 7a shows an optical micrograph of the same region on the sample. In addition to the well-channeled grains (imaged as a dark band from lower left to upper right in Fig. 7b), a partially channeled grain (arrowed) is indicated. This does not appear to correspond to any grain observable in the optical micrograph of Fig. 7a and may indicate that it is a subsurface grain. The feature running diagonally across the lower half of the scan region is a scribe track which was used as a location marker. As expected, channeling is poor in the vicinity of this scratch. In Fig. 7c we show a map obtained by imaging the Sb yield. Two features are of interest: i) the Sb yield is lower (darker region) in the well channeled grains, indicating

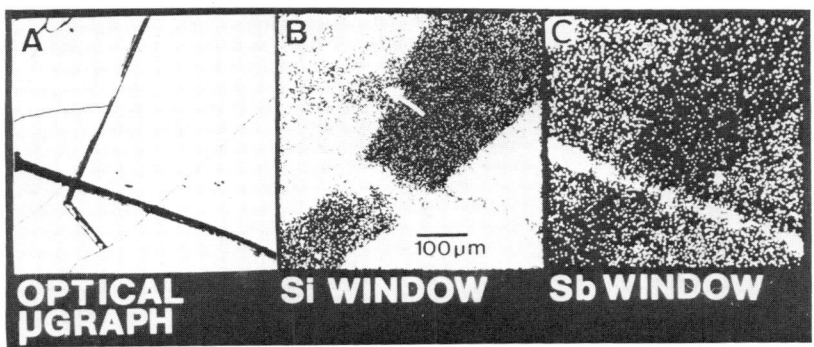

Fig. 7 a) optical micrograph of a 500 x 500 µm^2 region of polysilicon, showing individual grains.
  b) Channeling contrast map of the same region as in a).
  c) Map taken from the Sb yield. See text for further details.

substitutionality of Sb, and ii) the scratch appears bright, indicating segregation of Sb to this region. Selected-area RBS/channeling spectra from the maps of Fig. 7 are shown in Fig. 8. The spectrum from within the well-channeled grain (regions 3 and 4 in the inset) indicates a silicon minimum yield of 25% in comparison with a 'random' spectrum from the same area of surrounding random grains. Similar comparison of the Sb yield indicates 70% substitutionality of Sb in the well-channeled grain. Interestingly, the non-Gaussian shape of the Sb profiles indicates redistribution of Sb during annealing.

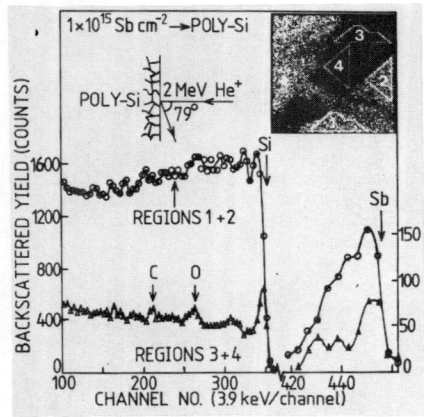

Fig. 8 Channeled microbeam spectra from selected regions of equal area inside and outside the well-channeled grain as indicated in the inset.

CONCLUSION

We have demonstrated that a helium microprobe combining several analysis techniques, including secondary-electron imaging, PIXE x-ray detection, Rutherford backscattering and channeling together with sophisticated mapping and data storage capabilities constitutes a powerful method for probing micron-scale structures in semiconductors. The low-divergence property of the Melbourne University microprobe allows imaging using a newly developed technique of channeling contrast microscopy. This technique has exciting prospects for the analysis of semiconductor structures.

The Australian Research Grants Committee and the Commonwealth Special Research Centres Scheme are acknowledged for financial support of this project.

REFERENCES

[1] J.C. McCallum, C.D.McKenzie, M.A. Lucas, K.G. Rossiter, K.T. Short and J.S. Williams, *Appl. Phys. Lett.* 42, 827 (1983)
[2] G.J.F. Legge, D.N. Jamieson, P.M.J. O'Brien and A.P. Mazzolini, *Nucl. Instr. & Meth.* 197, 85 (1982)
[3] J.C. McCallum, C.D. McKenzie and J.S. Williams, *IEEE Trans. NS-30*, 1228 (1983)
[4] G.J.F. Legge, *Nucl. Inst. & Meth.* B3, 561 (1984)
[5] H.B. Harrison, S.T. Johnson, B. Cornish, F.M. Adams, K.T. Short and J.S. Williams, *Proc. Mat. Res. Soc.* 13, 193 (1983)

CONCENTRATION MEASUREMENTS AND DEPTH PROFILING OF
PHOSPHORUS AND BORON BY MEANS OF
(p,γ) RESONANT REACTIONS

M.J.M. PRUPPERS[*], F. ZIJDERHAND[*], F.H.P.M. HABRAKEN AND W.F. van der WEG
Department of Technical Physics, [*]Department of Nuclear Physics,
State University Utrecht, P.O. Box 80.000, 3508 TA Utrecht, The Netherlands

ABSTRACT

Small concentrations of phosphorus and boron in thin films have been measured by detection of proton induced prompt γ-rays. This technique is almost free of interference from other elements in the film or the substrate. Furthermore it is possible to obtain depth information without sputtering. The detection limit of both elements is around 100 ppm; depth resolution at shallow depths in silicon is about 50 nm in the case of boron, whereas it is better than 10 nm in the case of phosphorus.

As an application of this method, the incorporation of phosphorus and boron in thin films of hydrogenated amorphous silicon is measured. The films were prepared in a glow discharge reactor by mixing PH_3 or B_2H_6 with the SiH_4/H_2 gas mixture.

INTRODUCTION

Atomic composition of and impurity concentrations in thin films can be determined with a wide variety of techniques. Several of these techniques make use of energetic ion beams [1], like Rutherford backscattering (RBS) [2] and elastic recoil detection (ERD) [3] which use energy analysis of backscattered and recoiled particles, respectively. In nuclear reaction analysis (NRA) products of particle induced nuclear reactions are detected. This technique has been used to determine concentration profiles of for instance hydrogen [4] and aluminum [5].

We investigated the use of one particular set of nuclear reactions: proton induced γ-ray emission - (p,γ) reactions - for the determination of small concentrations ($\leq$ 1 at.%) in thin films of doped hydrogenated amorphous silicon (a-Si:H:P,B).

The basic principle of the method is the fact that the γ-ray yield of the nuclear reaction is proportional to the concentration of the capturing nucleus. The method is isotope sensitive, so there is almost no interference from other elements in the film or the substrate. This in contrast with RBS, especially in the case of P and B in a Si-matrix. Moreover it provides depth information without thinning the film due to the energy loss of the protons as they penetrate into the film.

In several cases it is even possible to determine the concentration ratio of two elements in the same film when the yield of two different reactions is measured. The use of an external reference can then be avoided.

In this paper the principles of the method and its advantages and limitations are discussed. Both the use of narrow resonances as well as the low energy wing of a very broad resonance are described. Some results of analysis of a-Si:H:P,B films are presented.

EXPERIMENTAL

The experimental setup is shown schematically in figure 1. The proton beam was provided by the Utrecht 3MV Van de Graaff accelerator. The energy-spread of the beam was lower than 200 eV. The samples were mounted in a small vacuum chamber (pressure $\sim 10^{-6}$ Torr) at an angle of $45°$ with respect to the incident beam direction. To measure the collected proton charge

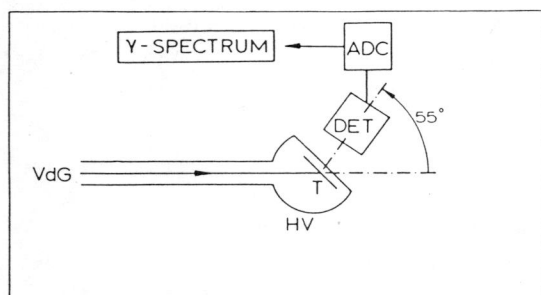

Figure 1.
The experimental setup; VdG = van de Graaff accelerator, HV = high vacuum chamber, T = target, DET = Ge(Li)-detector, ADC = analog to digital converter.

deposited on the sample the target holder was connected with a current integrator. The chamber was electrically insulated from pumps and accelerator tube and connected with the target holder to avoid influence of secundary electron emission on the current integration. Typical beam currents used were about 40 μA while the total dose on each sample varied from 6 to 300 mC, depending on the concentration of the measured element. The effects of heating of the sample by the proton beam were minimized by water cooling at the back of the sample.

The γ-rays were detected with a coaxial 126 cm^3 Ge(Li)-detector which was positioned at an angle of $55°$ with respect to the incident beam direction and at a distance of about 1 cm. The resolution of this detector was better than 3 keV at a γ-energy of 1.33 MeV. Spectra of 8192 channels were recorded with a PDP 11/34 computer. A ^{60}Co radioactive source was used to correct for the deadtime of the detection system at high counting rates.

The a-Si:H:P,B films were deposited on tantalum substrates in a conventional capacitively coupled glow discharge reactor [6]. The gas flow consisted of 10 sccm SiH_4 and 15 sccm H_2. Pressure, rf power and substrate temperature were fixed at 0.25 mbar, 20 W and 250 °C, respectively. The total gasflow remained 25 sccm while replacing different amounts of H_2 by PH_3 or B_2H_6, diluted in H_2. In this way the P/Si or the B/Si ratio in the gas phase was varied from 10^{-2} to 10^{-4}.

THEORY

By capturing a proton a target nucleus $^A_Z X$ is transmutated into a compound nucleus $^{A+1}_{Z+1} Y$. A and Z are the mass number and the atomic number, respectively of the nucleus. Only within a small energy interval Γ around certain proton energies E_r the reaction can take place. Γ and E_r are called the resonance width and the resonance energy, respectively. The resonances in the proton energy are related to the energies of excited states of the compound nucleus.

After capturing a proton the compound nucleus decays from the resonance level to lower levels or the ground level by particle or γ-ray emission. Let us only consider the γ-decay of the compound nucleus, or, when a particle is emitted, the residual nucleus. An example of the latter case is $^{15}N(p,\alpha\gamma)^{12}C$ [4]. Some recent compilations of the properties of many nuclei, including the γ-decay, are given in ref. [7].

The width Γ of the resonance, defined as the FWHM of the cross-section peak at the resonance energy, can be used to obtain information about the depth distribution of the nucleus under study. When Γ is small compared with the target thickness and the energy spread of the incident proton beam the depth at which the reaction actually takes place depends on the energy of the incident protons. When this energy is larger than the resonance energy the protons will reach this resonance energy at a certain depth because they loose energy in the film. In other words the yield of γ-rays with energy E_γ per unit charge versus proton energy, the so called yield curve, resembles the concentration profile versus depth.

Thin target yield

In the case of a narrow resonance and a thin film ($\Gamma \leq$ the total energy loss in the film) the total area O under the measured yield curve is:

$$O \propto Q_p \cdot N_{A_X} \cdot d.e.b. \int_{-\infty}^{+\infty} \sigma(E).dE \qquad (1)$$

where Q_p is the total proton dose per measuring point, N_{A_X} the atomic density of the capturing nucleus $^A X$, d the geometrical thickness of the film, e the relative detector efficiency and b the branching ratio i.e. the fraction of all compound nuclei that decays by emitting a γ-ray of energy E_γ. σ(E), the cross-section for capturing a proton and emitting a γ-ray, can be described by [8]:

$$\sigma(E) = \sigma_r \cdot \frac{(1/4) \cdot \Gamma^2}{(E-E_r)^2 + (1/4) \cdot \Gamma^2} \qquad (2)$$

The value of σ(E) at $E=E_r$ is given by

$$\sigma_r = \lambda^2 \cdot \omega\gamma/\pi\Gamma \qquad (3)$$

where λ is the wavelength of the incident proton,

$$\lambda^2 = \left(\frac{m_p + m_t}{m_t}\right)^2 \cdot \frac{h^2}{2m_p \cdot E_r} \qquad (4)$$

and

$$\omega\gamma = \frac{2J+1}{(2i+1) \cdot (2I+1)} \cdot \frac{\Gamma_p \cdot \Gamma_\gamma}{\Gamma} \equiv \frac{S_{res}}{(2i+1) \cdot (2I+1)} \qquad (5)$$

m_p and m_t are the masses of a proton and the target nucleus, respectively; J the resonance spin (the angular momentum of the compound nucleus in its resonance state), i the proton spin, I the target spin, Γ_p and Γ_γ the energy widths for formation and decay of the compound nucleus and S_{res} the resonance strength.

Thick target yield

In the case of a thick target (Γ << total energy loss in the film) with a uniform concentration of element $^A X$ the relation between O and the step-height H_{A_X} of the yield curve is given by

$$O = H_{A_X} \cdot d \cdot (dE/dx) \qquad (6)$$

where dE/dx is the stopping power of the target material.

Finally, the concentration N_X of element X is thus proportional to:

$$N_X \propto \frac{(2I + 1)}{\left(\frac{m_p + m_t}{m_t}\right)^2 \cdot a} \cdot \frac{E_r \cdot (dE/dx)}{e \cdot b \cdot S_{res}} \cdot \frac{H_{A_X}}{Q_p}, \qquad (7)$$

a is the natural abundance of isotope $^A X$. Thus, by measuring the step-height of two different reactions $^A X(p,\gamma)$ and $^B Y(p,\gamma)$ the concentration ratio N_X/N_Y of the two elements X and Y can be calculated provided that all the constants in eq. (7) are known.

In the case of a very broad resonance ($\Gamma \gg d.dE/dx$) the energy E_γ of the emitted γ-rays is proportional to the actual proton energy E'_p (neglecting Doppler effects and momentum of the γ-photon):

$$E_\gamma = Q + \frac{m_t}{m_p + m_t} \cdot E'_p \tag{8}$$

Q is the rest energy difference between compound nucleus after and proton plus target nucleus before the reaction. The actual proton energy E'_p depends on the depth at which the proton is captured. In other words, at a constant energy of the incident protons, a complete depth profile appears in the γ-spectrum [9].

Detection limit

The sensitivity of the method can be improved by maximization of (cf. eq. (7)):

$$\frac{\left(\frac{m_p + m_t}{m_t}\right)^2 \cdot a}{(2I + 1)} \cdot \frac{S_{res}}{E_r \cdot dE/dx} \cdot e \cdot b. \tag{9}$$

This can be achieved by first choosing a suitable isotope for which the first part of eq. (9) is as high as possible. The second part of eq. (9) can be maximized by choice of a strong resonance (which must be narrow when a good depth resolution is required) at low proton energy. Finally a γ-transition for which e.b is as large as possible can be chosen. The latter choice sometimes depends on the presence of γ-rays originating from other resonances, from reactions of protons with other elements in the film or the substrate or from the natural background.

Depth resolution

The depth resolution Δx depends on the resonance width Γ, the energy spread of the proton beam Γ_b, the energy straggling Γ_s (a function which increases with depth) and the stopping power dE/dx. Δx can be approximated by:

$$\Delta x = \frac{(\Gamma^2 + \Gamma_b^2 + \Gamma_s^2)^{\frac{1}{2}}}{dE/dx} \tag{10}$$

At shallow depth the straggling can be neglected, while at greater depth it dominates the depth resolution. In the case of a broad resonance also the energy resolution of the detection system must be taken into account.

The reactions with narrow resonances we used for profiling of P, B and Si are listed in Table I together with their properties. Most of the data

Table I   Narrow resonances

Element	P	B	Si
Isotope	^{31}P	^{11}B	^{30}Si
a(%)	100	80	3.1
Reaction	$^{31}P(p,\gamma)^{32}S$	$^{11}B(p,\gamma)^{12}C$	$^{30}Si(p,\gamma)^{31}P$
E_r (keV)	811	1438	163	620
E_γ (MeV)	7.42	3.64 / 1.62^a	4.44	7.90
Γ (eV)	< 420	45±20	5700±400	68±9
S_{res} (eV)	1.00±0.08^b	4.9±1.5^c	0.35±0.07^d	3.9±0.2^b
dE/dx in Si (eV·nm^{-1})^e	46	33	105	55

Remarks:
a. The 1.62 MeV γ-ray is more suitable for relative concentration measurements and profiling than the 3.64 MeV γ-ray.
b. From ref. [11].
c. Calculated with $S_{res,811}$ and $S_{res,1438}$ from [7] and $S_{res,811}$ from [11].
d. Calculated from Γ_p, Γ_γ and Γ from [7].
e. From ref. [12].

are from ref. [7]. Of the $^{28}Si(p,\gamma)^{29}P$ reaction we used the low energy wing, near 1460 keV (at this energy σ(E) is only a slowly varying function of energy), of the 52 keV broad resonance at 1652 keV [10]. At a fixed proton energy of 1460 keV not only a complete Si depth profile but also the step-height of the ^{31}P reaction at the 1438 keV resonance can be measured (provided that the equivalent film thickness is larger than 22 keV).

RESULTS AND DISCUSSION

The incorporation of P and B in thin films of hydrogenated amorphous silicon as a function of the P/Si- and B/Si-ratio in the gas phase is measured. Depth profiles of P and Si in an a-Si:H:P film with 1.05% PH_3 in the gas phase are presented in figure 2. For the P-profile the 811 keV resonance was used. The concentration of P appears to be uniform troughout the entire film. Using eq. (7) the P concentration defined as $N_P/(N_P + N_{Si})$ was determined at 1.17 ± 0.22 at.% (neglecting the presence of H).

The P concentration in the film versus the gas phase ratio is plotted

Figure 2.
Depth profiles of P and Si in an a-Si:H:P film containing 1.05 at.% P. Thickness about 220 nm.

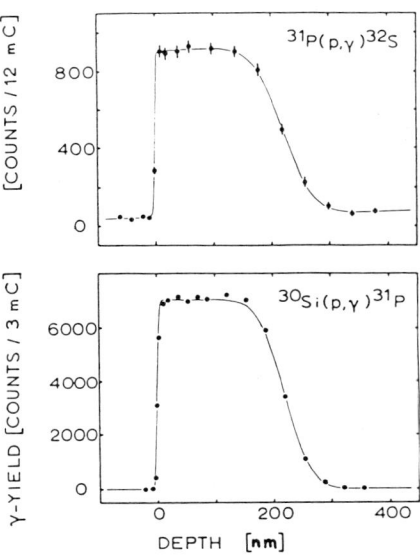

in figure 3. These results are obtained with the 1438 keV resonance of $^{31}P(p,\gamma)^{32}S$. Leidich et al. [13] proposed a power law to describe the relation between the P concentration in the film and in the gas phase:

$$\frac{P}{P+Si}\bigg|_{film} = \left(\frac{P}{P+Si}\bigg|_{gas}\right)^{\gamma} \qquad (11)$$

When fitting this relation to the results of figure 3 we obtained $\gamma = 0.98 \pm 0.03$. Leidich et al. have measured $\gamma = 0.82$ whereas Thomas III [14] has found $\gamma > 1$. It is likely that the exponent γ in eq. (11) is

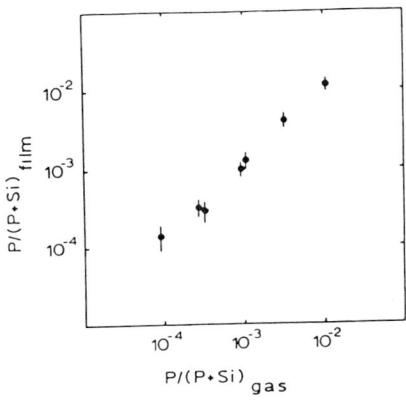

Figure 3.
Incorporation of P in a-Si:H:P films versus gas phase doping ratio.

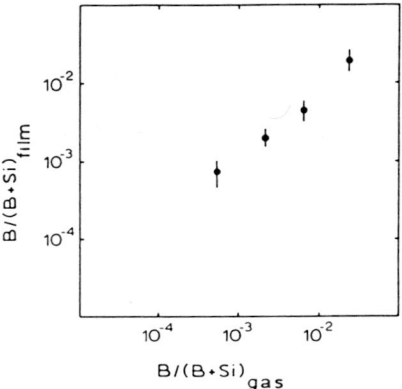

Figure 4.
Incorporation of B in a-Si:H:B films versus gas phase doping ratio.

sensitive to the conditions during the growth of the films. Preliminary results show that variation of the rf power of the glow discharge influences the exponent γ. The implications of these results for the understanding of the physics and chemistry of amorphous silicon growth and the relation to electrical and optical properties of the films will be published elsewhere [15].

Figure 4 presents the results for B doped films. The exponent γ is also close to unity for these samples: γ = 1.05 ± 0.08.

In figure 5 an example of a γ-yield curve obtained with the ^{28}Si reaction is presented. The beam energy was fixed at 1460 keV. The shape of the cross-section which was used to fit the experimental data (the solid line in figure 5) was taken from ref. [10]. The depth scale was calculated using eq. (8).

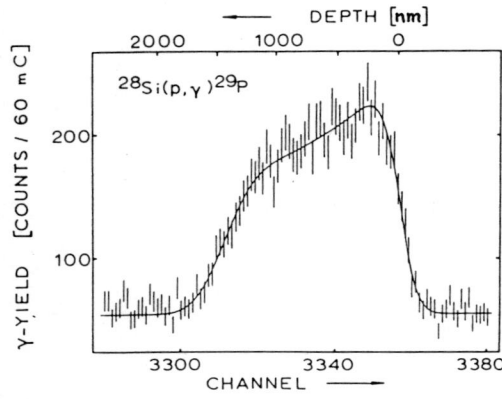

Figure 5.
Si profile obtained with the reaction ^{28}Si(p,γ)^{29}P for a 1.5 μm thick a-Si:H:P film.

From the figures 3 and 4 it can be concluded that the detection limit of the method is about 100 ppm for both P and B. Measuring times per measuring point for the samples with the lower concentrations are about one hour.

Depth resolution at shallow depths in Si appeared to be about 50 nm in the case of B. This can be improved by using smaller angles between the incident beam direction and the sample surface. In the case of P and Si the depth resolution near the surface is better than 10 nm.

CONCLUSIONS

By using the technique discussed in this paper it is possible to measure absolute concentrations of P and B in a Si-matrix with good sensitivity and depth resolution. There is no need for calibration standards. The absolute accuracy of the concentration measurements depends on the accuracy of the nuclear physics constants. One of the limitations of the technique is the need for high beam currents.

Nuclear reaction analysis with (p,γ) reactions turns out to be a powerful method to study the physics and chemistry of amorphous silicon growth.

REFERENCES

[1] W.K. Chu, J.W. Mayer, M.-A. Nicolet, T.M. Buck, G. Amsel and F. Eisen, Thin Solid Films 17, 1 (1973).

[2] W.K. Chu, J.W. Mayer and M.-A. Nicolet, Backscattering Spectrometry, Academic Press (1978).

[3] P.M. Read, C.J. Sofield, M.C. Franks, G.B. Scott and M.J. Thwaites, Thin Solid Films 110, 251 (1983); S. Nagata, S. Yamaguchi, Y. Fujino, Y. Hori, N. Sugiyama and K. Kamada, Nucl. Instr. and Meth. B6, 533 (1985).

[4] W.A. Lanford, Solar Cells 2, 351 (1980).

[5] J.S. Rosner, P.M.S. Lesser, F.H. Pollak and J.M. Woodall, J. Vac. Sci. Technol. 19, 584 (1981); M. Erola, J. Keinonen, H.J. Whitlow, A. Anttila and M. Hautala, Thin Solid Films 115, 125 (1984).

[6] J. Bezemer, M.J.M. Pruppers, F.H.P.M. Habraken and C.W. Hooyman in "Poly-micro-crystalline and amorphous semiconductors", MRS-Europe, Strasbourg, June 5th - 8th, 445 (1984).

[7] P.M. Endt and C. van der Leun, Nucl. Phys. A310, 1 (1978) (for A=21-44); F. Ajzenberg-Selove, Nucl. Phys. A300 (1978) (A=18-20), A320 (1979) (A=5-10), A336 (1980) (A=11-12), A360 (1981) (A=13-15) and A375 (1982) (A=16-17), all pag. 1.

[8] P.M. Endt and M. Demeur, Nuclear Reactions, vol. 1, North Holland (1959).

[9] W. Rudolph, C. Bauer, P. Gippner and K. Hohmuth, Nucl. Instr. and Meth. 191, 373 (1981).
[10] F. Terrasi, A. Brondi, P. Cuzzocrea, R. Moro and M. Romano, Nucl. Phys. A324, 1 (1979).
[11] D.G. Sargood, Phys. Reports 93, 61 (1982).
[12] J.F. Ziegler, Handbook of Stopping Cross-sections for Energetic Ions in all Elements, Pergamon Press (1980).
[13] D. Leidich, E. Linhart, E. Niemann, H.W. Grueninger, R. Fischer and R.R. Zeijfang, J. Non. Cryst. Sol. 59 & 60, 613 (1983).
[14] J.H. Thomas III, J. Vac. Sci. Technol. 17, 1306 (1980).
[15] M.J.M. Pruppers, J. Bezemer, F.H.P.M. Habraken and W.F. van der Weg, to be published.

MATERIALS CHARACTERIZATION WITH INTENSE POSITRON BEAMS

I.J. ROSENBERG, R.H. HOWELL, M.J. FLUSS, and P. MEYER
Lawrence Livermore National Laboratory, Livermore, CA   94550

ABSTRACT

   We have developed an apparatus that provides a high flux, low energy, monoenergetic positron beam to investigate various processes which occur when a positron beam impinges on a metal surface, including annihilation at the surface, trapping in vacancies, and the emission of both fast and thermally desorbed positronium. We report here the first angular correlation of annihilation gamma-rays measurements and positronium time of flight experiments at a material surface. We applied a simple free electron model which explains the general trend of the data but differences due to the surface specific properties and the deviation from a free electron metal are evident.

INTRODUCTION

   The similarities and differences between electrons and positrons have made positrons an important tool in the investigation of the bulk properties of materials (e.g., vacancy defects and electron momentum distributions). More recently, the development of monoenergetic, low energy positron beams has extended the possibilities of applying these techniques to surfaces, and introduced techniques specific to surfaces. Sensitivity to surface contamination, phase, and defect conditions has been demonstrated for positronium production and positron emission. Defects near the surface have been profiled and several experiments have been performed substituting positrons for electrons including diffraction and energy loss spectroscopy [1]. While we have found that positron beams are highly sensitive to changing surface conditions, the interaction of the positron as it leaves the bulk and either attaches itself to the surface or is ejected is still not adequately understood. The experiments presented in this paper investigate these interactions, looking at both the ejected positronium using time of flight techniques, and positrons attached to the surface using angular correlation measurements of the annihilation radiation.
   Unlike the first low energy beams applied to materials research, which were derived from radioactive sources, our positrons are generated in a Bremstrahling target at the end of the LLNL 100 MeV electron linac [2]. These positrons impinge on an annealed tungsten moderator, resulting in a monoenergetic positron beam which is then electrostatically accelerated and magnetically guided and focused onto a target mounted in an ultrahigh vacuum (UHV) chamber located in a low radiation background environment. Depending on the appropriate mode for the investigation chosen, the beam energy is set between 0.5 to 20 keV, the pulse duration between 10 nanoseconds and 3 microseconds, and the frequency between 300 and 1440 pulses per second. Intensities range up to 10^6 positrons per pulse. The vacuum chamber is a standard UHV design with a base pressure of 2×10^{-10} torr; it is equipped to monitor, clean, and anneal the sample by Auger analysis, ion sputtering, and resistance heating.
   Using this beam, we have measured the electron-positron momentum distribution of positrons trapped at, and positronium (Ps, the bound electron-positron pair) emitted from, the surface of metal samples using two dimensional (2D) ACAR. We have also measured the energy distribution of Ps ejected from metal surfaces by measuring the time-of-flight of Ps decaying in front of a well collimated detector (Ps-TOF) in front of the sample.

Background Discussion

A low energy positron impinging on a surface may either diffract away from, or penetrate into the material. Positrons that penetrate will quickly thermalize and will then either diffuse to the surface, become trapped at a void or vacancy, or annihilate within the bulk. For low energy positrons in materials with few trapping sites, the return to the surface is the most likely case. At the surface, the positron will either be re-emitted, or combine with an electron and either annihilate there or be emitted as Ps.

Two surface emission positronium channels have been experimentally observed: the first, resulting from expelled, slow positrons that pick up an electron at the surface, yields energetic Ps; and the second, a thermally activated process where bound positron-electron pairs are desorbed at elevated temperatures yielding low energy Ps. The maximum energy of a Ps atom emitted from a surface (Φ_{Ps}) under the fast mechanism is the energy obtained from the negative of the Ps work function, which is given by the Ps binding energy ($\frac{1}{2}R_H$) minus the sum of the positron and electron work functions. Using a free electron model [3], one can obtain an expression for the Ps density in momentum space:

$$d^3N/dK^3 \quad \alpha \quad K_z(K_z^2 - K_\parallel^2 + 2k_f^2 - 4\Phi_{Ps})^{-\frac{1}{2}} \qquad (1)$$

where K refers to Ps momentum, k_f is the Fermi momentum for the metal in question, and the subscripts z and $\parallel$ refer to the directions perpendicular and parallel to the sample. This expression is then integrated taking into account the geometry of the particular experiment to provide expressions that can compared to the Ps data.

ACAR MEASUREMENTS

A two dimensional momentum distribution of the electrons in a material may be obtained by measuring the angular deviation of the two annihilation gamma-rays from anticollinearity. Our 2D-ACAR distributions were measured using a system similar to that of West [4] using position sensitive Anger cameras. Each camera consists of a 13 mm by 400 mm NaI crystal connected to an array of 37 phototubes yielding a position resolution (FWHM) of -8 mm. The detectors were placed 13.7 m away from the sample resulting in an angular resolution of 0.9 mrad. Valid gamma-ray events were selected by requiring a coincidence between the two Anger camera detectors and the positron pulse from the linac. Positions and relative detection time for each event were stored on tape and the angular correlation distributions were calculated and corrected by the geometric efficiency matrix after the experiment.

The positron beam was produced in 3 microsecond wide pulses at 900 per second. Only 5% of the available positron intensity was used due to saturation in the Anger cameras. Even so, the high intensity of the positrons in the beam pulse resulted in accidental backgrounds and loss of events due to pileup rejection larger than those in systems using random positron sources. Pile-up events were rejected by taking the first event in the energy window of each detector for each beam pulse. Accidental coincidences were minimized during the analysis of the data by setting a narrow time window on the valid events. Events that satisfied all constraints but were outside the time window were made into accidental angular correlation spectra that were then normalized and subtracted from the data as background.

The sample in these measurements [5] was a copper single crystal oriented with the [121] axis along the beam direction and the [111] axis along the line joining the camera centers. The surface was cleaned by Argon sputtering and there was less than 10% surface contamination at the end of a 24 h run.

Perspective drawings of the smoothed 2D-ACAR spectra obtained with 18 keV and 740 eV positrons are shown in Fig. 1. The positron beam is moving in the positive p_z direction as it strikes the surface. For the 740 eV run the sample was biased to attract re-emitted positrons; for 18 keV it was set at zero to enhance the bulk nature of the data. The background from accidental coincidences that was subtracted from both angular distributions is shown below the 740 eV spectrum. Approximately 5×10^5 counts are shown in each of these spectra, representing ~24 h of data collection.

The data accumulated at 18 keV are similar to those reported by Berko and co-workers [5] for positrons annihilating in the bulk of an annealed Cu single crystal. In comparison, the 740 eV spectrum is much narrower overall, with an asymmetric shift toward the negative p_z direction due to the annihilation in flight of energetic Ps. Careful inspection of the 18 keV spectrum also reveals a small asymmetric Ps peak which masks the neck usually evident at zero momentum in this crystal orientation. The energetic Ps component was stripped from the underlying spectrum by subtracting the spectrum for positive p_z from

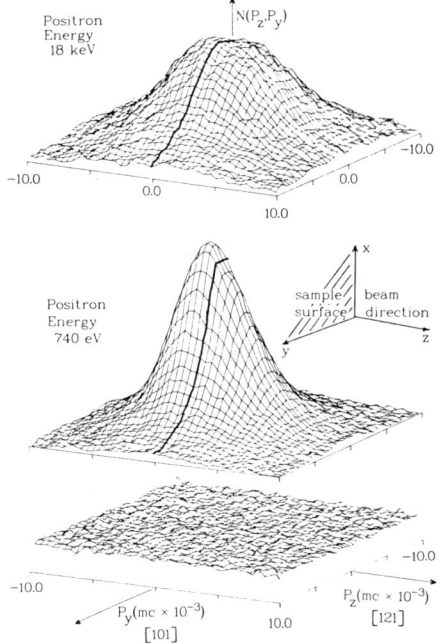

Fig. 1. The 2D-ACAR spectra for positron beams of 18 keV and 740 eV impinging on a single crystal of copper. The sample surface orientation and a typical background spectrum are also shown. The heavy contour at $p_z=0$ emphasizes the asymmetry of the low-energy data.

that for negative p_z. This procedure has the effect of clamping the asymmetric distribution to zero at zero momentum, introducing a significant error only within 0.5 mrad of zero due to the resolution function. Applying the 0.9 mrad resolution to the data yields the maximum Ps momentum, $4.5 \pm 0.25 \times 10^{-3}$ mc, corresponding to a work function energy of -2.6 ± 0.15 eV in reasonable agreement with the value of -2.4 eV [6] obtained from electron and positron work function values. The most probable Ps momentum is found at $p_z = 2.5 \times 10^{-3}$, $p_y = 0.0$ mc, corresponding to a forward energy of 0.80 eV. In the time-of-flight experiment described below the maximum Ps energy was found to be 2.6 ± 0.3 eV in reasonable agreement with the present 2D-ACAR results. The yield of energetic Ps in the 740 eV spectra was 4% of the total counts for para-Ps implying 14% total Ps if a statistical distribution of spins is assumed.

Separating the parallel momentum into x and y components in eqn. (1) and then integrating along the line between the cameras yields the equation:

$$d^2N/dK_x dK_z \propto K_z \sin^{-1}[(4\Phi_{Ps} - K_x^2 - K_z^2)/(K_z^2 - K_x^2 + 2k_f^2 - 4\Phi_{Ps})]^{\frac{1}{2}} \quad (2)$$

The contours computed from this model, assuming no momentum dependence in the Ps formation interaction and a free electron distribution for the electrons, are compared to the Ps data at the top of Fig. 2. The forward shapes of the contours in the data are qualitatively described by the model. However, significant deviations from the model contours occur for the position of the most probable momentum which is less in the data than in the model, indicating that the simple assumptions used in deriving the model contours may be inadequate.

Because the Ps momentum is strongly shifted toward the vacuum, the distribution of momenta into the sample is nearly Ps free. These data are shown with equally spaced contours for both energies at the bottom of Fig. 2. The narrowing of the pair-momentum distribution of positrons in the surface state compared to the bulk can be seen to be the result of a loss of strength of high momentum components from core electrons and a narrowing of the peak in the low momentum part of the distribution from the valence band. A similar narrowing of 1D-ACAR data for positrons trapped in vacancies and vacancy clusters has been observed [7].

The width of the underlying momentum distribution from the 740 eV data is anisotropic with p_y narrower than p_z (6.6 ± 0.2 mrad versus 8.0 ± 0.2 mrad FWHM). Such anisotropy was predicted in the momentum distribution calculated using an independent particle model with a positron trapped in the surface image charge potential of aluminum [8]. There is also a

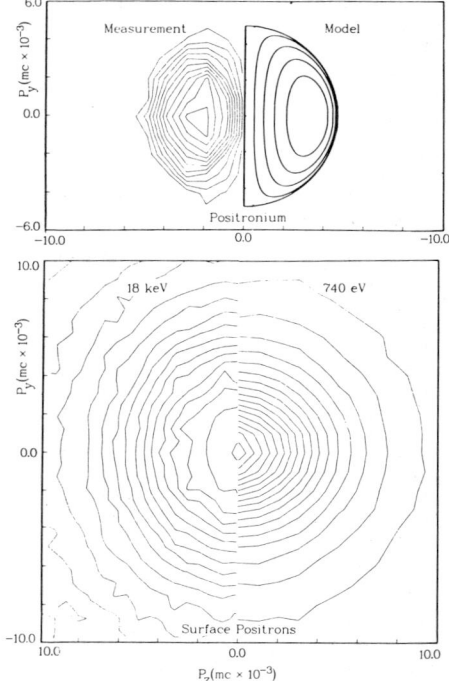

Fig. 2. Top, contour maps comparing measured and computed momentum distributions for positronium ejected from the copper surface. Bottom, positive p_z contours for the 18 keV and 740 eV beams with no energetic positronium contribution.

strong inferred possibility from the low amount of free Ps found that a localized or trapped Ps contribution may be contained in the broad part of the distribution. In the direction parallel to the surface, delocalization would lead to a narrowing of the momentum distribution so that either picture is consistent with our data.

Ps-TOF MEASUREMENTS

A technique that takes advantage of the pulsed positron beam is time of flight measurements of the Ps velocity. We use a detector, similar to that reported by Mills [9], to detect annihilating Ps atoms. The detector is placed a fixed distance from the sample, and is shielded such that it can only detect annihilation gamma rays from those Ps atoms decaying in a narrow slice parallel to the sample. Fig. 3 shows data, corrected by the triplet lifetime for decay in flight, obtained with a Ps flight path of 11.5 ± 0.5 cm and 15 ns time resolution on the same sample as the 2D-ACAR experiment. These spectra were each obtained in 30 min. The shape of the low temperature distribution is determined by the electron energy distribution at the surface and other factors in the electron-positron interaction along with the measurement geometry and scattered Ps intensity. The low energy Ps resulting

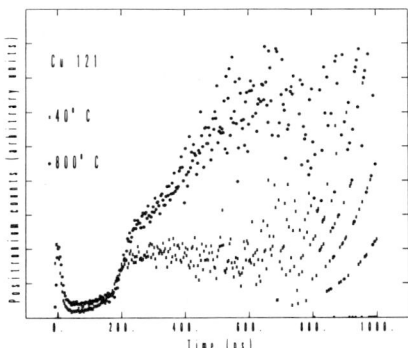

Fig. 3. Positronium time-of-flight distributions for clean copper at room temperature and 800°C.

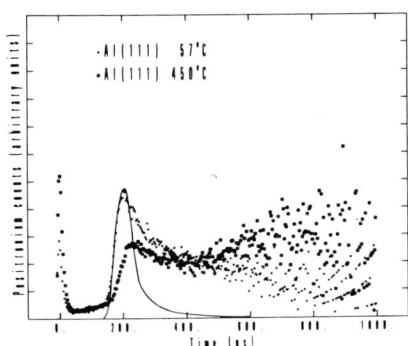

Fig. 4. Positronium time-of-flight distributions for clean aluminum. The solid line represents a nearly free electron model.

from thermal desorption is clearly seen as an additional component in the high temperature data in Fig. 3. In Fig. 4 we present data taken using an Al(111) sample. As a nearly free electron metal, it should be a better test of our model than was the Cu. Here the integration of eqn. (1) must be done over two dimensions (i.e., both parallel components), and then be converted to time space. Due to the geometry of the vacuum system, Ps emitted at angles greater than β are not detected. The result is:

$$dN/dK_z \propto t^{-3} \times \begin{cases} [(1-\tan^2\beta)\tau+o]^{\frac{1}{2}} - [\tau+o]^{\frac{1}{2}} & \Phi_{Ps}\cos^2\beta > \tau \\ [2\tau+o-\Phi_{Ps}]^{\frac{1}{2}} - [\tau+o]^{\frac{1}{2}} & \tau > \Phi_{Ps}\cos^2\beta, \Phi_{Ps} > \tau \end{cases} \quad (3)$$

Where $\sqrt{\tau}$ is the distance between the sample and the detection region divided by the elapsed time and o is the Fermi energy minus Φ_{Ps}. As can be seen by comparing the data with the theory, the theory does a good job predicting the behavior of the energetic Ps, but does not account for the slower atoms to the right of the peak. One can also see a well known effect of heating Al; the equilibrium vacancy concentration increases, which reduces the flux of positrons back to the surface, thereby reducing the yield of Ps. However, one can still see by the shape of the curve that the ratio of thermal to energetic Ps has increased at the higher temperature.

ACKNOWLEDGMENTS

The assistance of M. Connor and L. Bernardez in collecting the data, A. Coombs in developing the instrumentation, and J. Kimbrough in developing data acquisition software and hardware is gratefully acknowledged. Work performed under the auspices of the U.S. Department of Energy by the Lawrence Livermore National Laboratory under contract number W-7405-ENG-48.

REFERENCES

1. For a comprehensive review of positron beam measurements, see Positron Solid-State Physics, edited by W. Brandt and A. Dupasquier, (North

Holland, New York, 1983) and Positron Annihilation, edited by Paul G. Coleman, Suresh C. Sharma, and Leonard M. Diana (North Holland, New York, 1982).
2. R.H. Howell, R.A. Alvarez, and M. Stanek, Appl. Phys. Lett. $\underline{40}$, 751 (1982).
3. J. B. Pendry, in Positron Solid-State Physics, edited by W. Brandt and A. Dupasquier, (North Holland, New York, 1983), p. 408, and private communication.
4. R.N. West, J. Mayers, and P.A. Walters, J. Phys. E $\underline{14}$, 478 (1981).
5. R.H. Howell, P. Meyer, I.J. Rosenberg, and M.J. Fluss, Phys. Rev. Lett. $\underline{54}$, 1698 (1985).
6. S. Berko, M. Haghgooie, and J.J. Mader, Phys. Lett. $\underline{63A}$, 335 (1977).
7. O. Sueoka, J. Phys. Soc. Jpn. $\underline{36}$, 464 (1974).
8. A. Alam, P.A. Walters, R.N. West, and J.D. McGervey, J. Phys. F $\underline{14}$, 761 (1984) and R.N. West (private communication).
9. A.P. Mills, Jr., L. Pfeiffer, and P.M. Platzman, Phys. Rev. Lett. $\underline{51}$, 1085 (1983).

EFFECTS OF AMBIENT GAS ON THE OUT-DIFFUSION OF NICKEL AND COPPER THROUGH THIN GOLD FILMS

R. K. LEWIS, S.K. RAY, and K. SESHAN

IBM Z/41C, Route 52, Hopewell Junction, NY   12533

ABSTRACT

The low temperature (300-400°C) out-diffusion of an underlying metal through a thin film overlayer of another metal has been studied in ambients of N_2 and forming gas. Surface oxidation of this out-diffused metal can interfere with the attachment of terminating components used in semiconductor devices. AES depth profile analysis has shown that the out-diffusion of Ni through Au is suppressed in a forming gas ambient. It has also been shown that the extent of this out diffusion can be modified considerably by the presence of Cu.

INTRODUCTION

Composites of thick and thin metallizations are frequently utilized for attaching the terminating components used in advanced semiconductor devices and packages.[1,2] During the fabrication of these structures, multiple thermal excursions in the temperature range 200-400°C are often necessary. As a result, diffusion of metal from an underlying layer through an overlayer of metal to the top surface takes place during these heating cycles. Accumulation of these metals along with their oxides can degrade the joining behavior of these metal film composites. It is important, then, to study the extent of such diffusion effects, the mechanism of the process and any effect, resulting from ambient gas or other process conditions, this may have on the out-diffusion in metal film structures. In recent years, a number of papers have been published on effects of ambient gas on interdiffusion in thin film couples.[3-6] In this paper, the effects of thermal exposures in nitrogen and nitrogen mixed with a few percent of hydrogen (forming gas) on the out-diffusion of Ni through Au thin films deposited on Mo and Cu surfaces is presented. It has been found that the presence of hydrogen in nitrogen suppresses the out-diffusion.

EXPERIMENTAL PROCEDURE

The first type of thin metal composite was prepared using a Mo coated surface plated with 5μm of nickel followed by a 60 to 80nm layer of gold using conventional electroless metal plating procedures. A short (five to ten minute) excursion at a high temperature (550 to 600°C) in a hydrogen atmosphere was used to produce a gold nickel solid solution in the top layer. X-ray diffraction has shown that a solid solution of 18-20 at.% nickel in gold is achieved by this thermal excursion. The second type of composite was a sputtered thin films of Au and Ni with nominal thicknesses of 120nm and 2μm respectively deposited on Cu. Both composites were annealed in controlled atmospheres of N_2 and forming gas. The gases and the annealing times used are listed in Table I. Auger depth profile analysis was done on each sample.

TABLE I
Metal composites studied

Sample	Fig. No.	Annealing Temperature	Time-Ambient Gas*
	1	"as received"	
Au-Ni-Mo	2	400°C	1 hr. in N_2*
Composite	3	400°C	1 hr. in F.G.
Au-Ni-Cu	4	"as received"	
Composite	5	350°C	1 hr. in N_2*
	6	350°C	1 hr. in F. G.

* All gases contained less than 10ppm O_2.

AUGER ANALYSIS

The Auger analysis was carried out with a Physical Electronics Model 590 Auger electron spectrometer. The depth profiles of the Au-Ni-Mo and Au-Ni-Cu composites are shown in Figs. 1, 2, and 3 and Figs. 4, 5 and 6 respectively. Argon was used for the sputtering ion beam. For these analyses, a combined slow (10nm/min. on Ta_2O_5) and fast (120nm/min. on Ta_2O_5) sputtering rate was used. The slow rate was produced by a large raster, then in the same crater, a fast rate was obtained using a small raster. This procedure allows the depth profiles to be connected directly along their abcissas as shown. The relative concentration values plotted for the ordinate were the Auger peak to peak intensities divided by their respective elemental Auger sensitivity factors found in the Physical Electronics handbook.

RESULTS AND DISCUSSION

The abient effects[7] on the annealing of Au-Ni composites have been reviewed in detail. The surface Au/Ni ration shown on the depth profile plots was determined by averaging the Au and Ni values over the first few nanometers and taking their ratio. This figure has been used by us as an accurate predictor of wettability for 60/40 Sn/Pb solder. It has been shown that a Au/Ni ratio of five or greater provides good solder wettability whereas a ratio significantly less than five generally results in poor wettability. The heat treatment of the Mo-Ni-Au composite in N_2 for one hour at 400°C produced extensive outdiffusion as shown in Fig. 2. The Au/Ni ratio decreased to 0.2 from an initial value of about five. The Au/Ni ratio, after a similar annealing step in forming gas was about six, as shown in Fig. 3, indicating no significant degradation of the surface had occurred. The solder wettability of the sample heated in N_2 degraded significantly. In contrast, the wettability of the sample remained excellent after the annealing was done in forming gas.

On the other hand the heat treatment of the Au-Ni-Cu composite at 350°C in N_2 did not produce the extensive Ni outdiffusion found with the Au-Ni-Mo composite as shown in Fig. 5. This can be attributed to the existence of a pronounced oxide barrier at the Ni/Au interface as indicated by the Auger oxygen depth profile. The surface properties did not degrade with this composite which is consistent with the high surface Au/Ni ratio observed. The heat treatment of the Cu-Ni-Au composite in forming gas on the other hand did show limited Ni out diffusion together with Cu as shown in Fig. 6. We attribute this

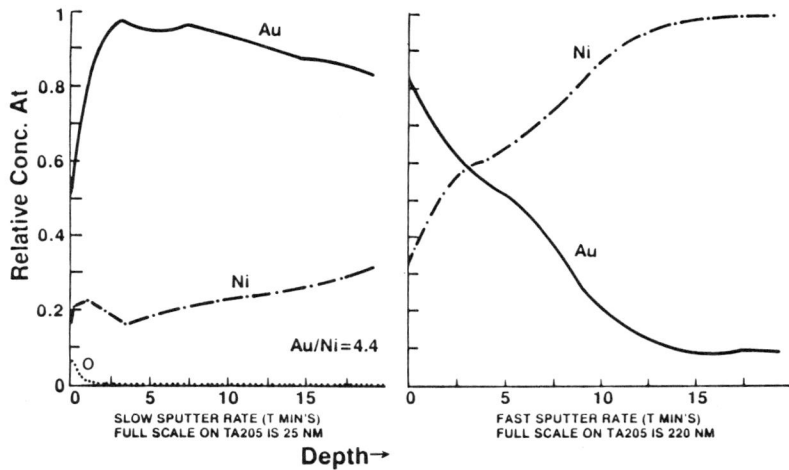

Fig.1 Au-Ni-Mo Composite "As Received"

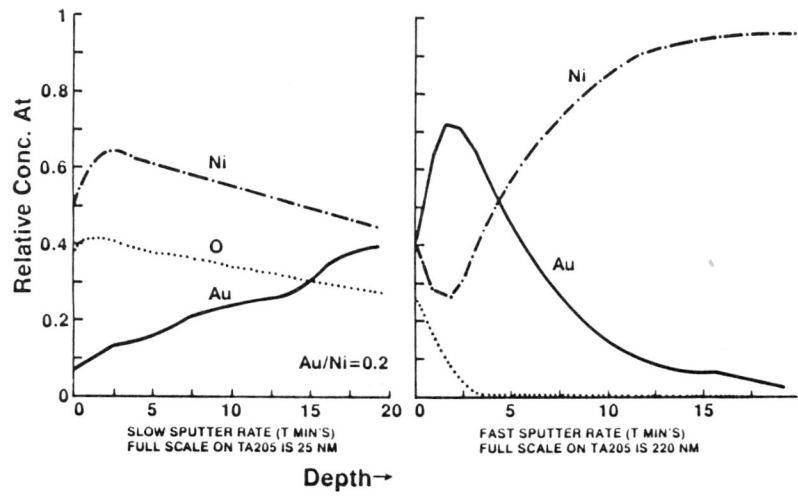

Fig.2 Au-Ni-Mo Composite Annealed 400C 1Hr In N2

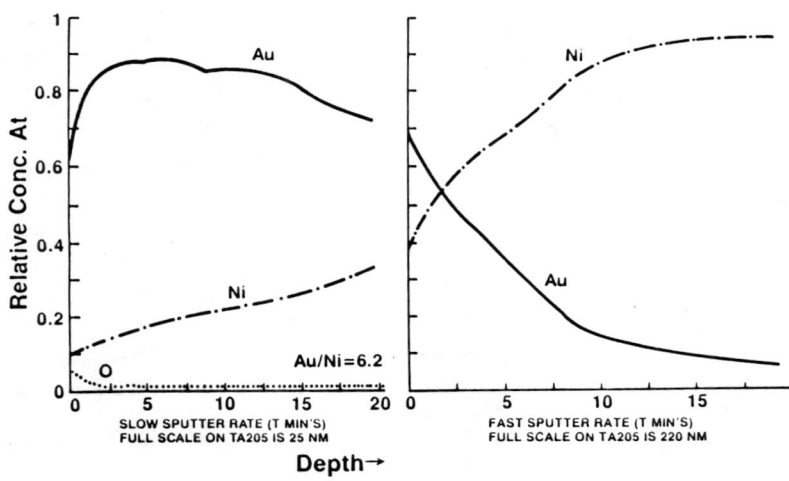

Fig.3 Au-Ni-Mo Composite Annealed 400C 1 Hr In F.G.

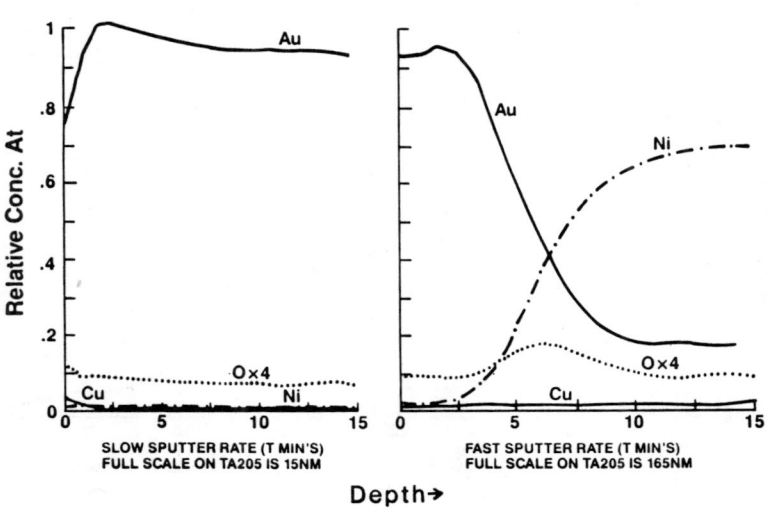

Fig.4 Au-Ni-Cu Composite "As Received"

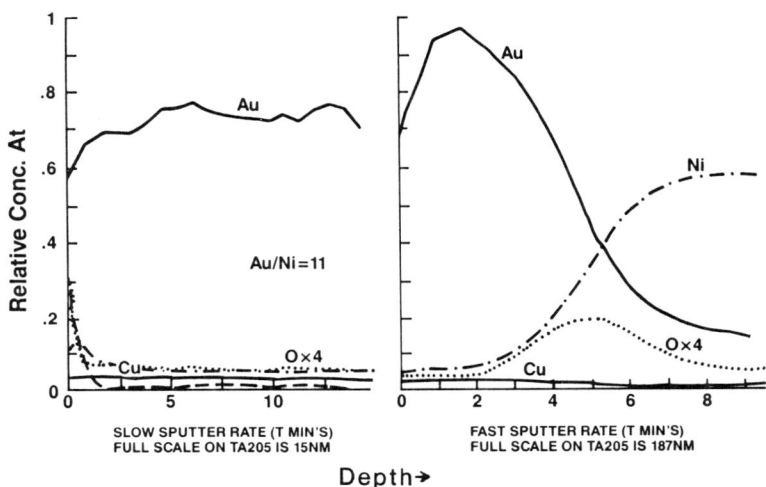

Fig.5 Au-Ni-Cu Composite Annealed 350C 1Hr In N2

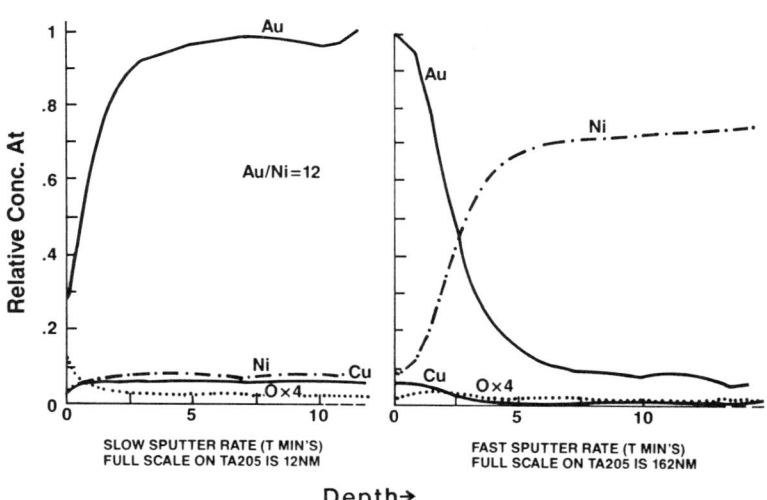

Fig.6 Au-Ni-Cu Composite Annealed 350C 1Hr F.G.

out-diffusion to the fact that the NiO barrier at the Ni/Au interface was reduced during annealing in the forming gas ambient as indicated in the oxygen profile. The absence of this barrier allowed the Cu to react with the Au and Ni to promote the rapid grain boundary diffusion of Ni and Cu into the Au forming a Au-Ni-Cu ternary solution. The presence of the reducing ambient also prevented more extensive out-diffusion and oxidation at the surface of the Au layer. The ineffectiveness of the thick (2.0μm) Ni film as a diffusion barrier for Cu is consistent with a grain boundary diffusion model.[8] Because Ni has a much higher melting temperature than Cu this model predicts that Cu diffusion in the Ni layer follows C-type kinetics. In C-type kinetics lattice diffusion is considered negligible and significant atomic transport occurs only within the grain boundaries.

CONCLUSION

These studies have shown that there is a strong influence of the ambient gas used during annealing at 300 to 400°C on the out-diffusion and oxidation at the surface of thin films of Au and Ni deposited on Mo and Cu. With the Au-Ni-Mo composite heating in nitrogen with up to 10ppm oxygen produced extensive Ni out diffusion through the Au and oxidation which severely degrades surface properties. One possible explanation is that the out-diffusion is aided by the oxidation at the surface of the diffusing species. Heating in a reducing ambient of forming gas on the other hand suppresses the Ni out-diffusion and no significant oxidation of Ni occurs. However, with the Au-Ni-Cu composites, annealing in N_2 did not cause extensive out-diffusion because of the NiO barrier formed at the Au/Ni interface. Heating in the reducing ambient of forming gas supressed this barrier which then allowed Ni and Cu to react with the Au. The presence of the Cu appears to promote the out-diffusion of Ni into the Au. Because no significant surface oxidation took place in forming gas, the Ni and Cu out-diffusion was limited to a few percent in the Au film and the surface properties of the Au were not degraded.

ACKNOWLEDGEMENTS

We wish to thank T. Tang for his technical assistance, and V. Marcotte and H. Wildman for their helpful discussions.

REFERENCES

1. C.W. Ho, D.A. Chance, C.H. Bajorek and R.E. Acosta, IBM J. Res. Develop., Vol. 26, 286 (1982).
2. T. Nukii, H. Iwasaki, Y. Matsuda, H. Yoshida, S. Nakadu and K. Awane, Proc. Int. Microelectronics Symp., International Society for Hybrid Micro-electronics, 115 (1980).
3. C.A. Chang, J. Appl. Phys., 53, 7092 (1982).
4. D.Y. Shih and P.J. Ficalora, IEEE trans. Electron. Dev., ED-26, 27 (1979).
5. J.E.E. Baglin and J.M. Poate, "Thin Films - Interdiffusion and Reactions," Edited by J.M. Poate, K.N. Tu and J.W. Mayer, John Wiley & Sons, New York (1978), pp. 305-357.
6. A. Hiraki, M.A. Nicolet, and J.W. Mayer, Appl. Phys. Lett., 18, 178 (1971).
7. R.K. Lewis and S.K. Ray, to be published in Thin Solid Films.
8. D. Gupta, D. R. Campbell, P. S. Ho, "Thin Films - Interdiffusion and Reactions," Edited by J. M. Poate, K. N. Tu, J. W. Mayer, J. Wiley & Sons, New York (1978), pp. 163-165

ELASTIC RECOIL ANALYSIS OF HYDROGEN IN ION-IMPLANTED
MAGNETIC BUBBLE GARNETS USING 44 MeV CHLORINE IONS

A. LEIBERICH,[*] B. FLAUGHER,[*] and R. WOLFE[**]
[*]Nuclear Physics Lab., Rutgers University, Piscataway, NJ 08854
[**]AT & T Bell Laboratories, Murray Hill, NJ 07974

ABSTRACT

Elastic recoil analysis using 44 MeV chlorine ions is a precise and timesaving method of measuring hydrogen profiles of ion implanted magnetic bubble garnets. Neon, nitrogen and hydrogen ion-implanted modified yttrium iron garnets, which are used for magnetic bubble devices based on ion implanted propagation patterns, (I2P2), were studied with respect to their hydrogen concentration-depth distributions. A description and calibration of the elastic recoil analysis setup are presented along with the necessary simple algorithms to obtain the final hydrogen concentration-depth profiles from the measured energy spectra generated by recoiled protons. The garnet samples were annealed with or without a surface coating of SiO_2. At $250°C$, the hydrogen was found to diffuse throughout the damaged surface layer but not into the underlying undamaged material. At $350°C$, the hydrogen diffused out of the uncoated garnet, but was sealed into the implant damaged garnet layer by the thin oxide coating. Such hydrogen retention correlates with previously reported beneficial effects of an oxide layer deposited at a low temperature after implantation but before processing on ion implanted bubble devices.

INTRODUCTION

Hydrogen implantation of magnetic bubble garnets drastically changes their magnetic properties. The stress induced by the implant is the major contributing factor to the change in magnetic anisotropy of the surface layer [1], although the presence of the hydrogen in the garnet crystal in itself is also important [2,3]. We measured the hydrogen concentration-depth profiles of various garnet samples which had been predamaged with nitrogen and neon implants before implantation with hydrogen. This procedure emulated a processing step in the fabrication of ion implanted propagation patterns of I2P2 magnetic bubble memory devices [1].
Particle solid interaction physics offers several hydrogen profiling techniques. Both elastic and inelastic collisions of energetic ions with sample materials yield characteristic signatures that can be read by a variety of detection systems. For example, by using Rutherford backscattering (RBS) [4], the hydrogen can be detected by measuring differences in the backscattered ion energy spectra of samples which do and do not contain hydrogen. Such differences in yield arise from the perturbation of the beam ion stopping power due to the presence of the hydrogen in the sample material [4]. Nuclear reaction analysis (NRA) [5] makes use of the nuclear reactions induced by energetic isotope ions upon collision with the hydrogen located in sample materials. The resulting emission of reaction products allows for the determination of the hydrogen profiles.

The method used in our present study is elastic recoil analysis (ERA) [6]. It entails the detection of energetic sample protons which are recoil scattered elastically during sample irradiation with a heavy ion beam. In the literature, the technique is also referred to as elastic recoil detection (ERD) [7,8], elastic recoil detection analysis (ERDA) [9,10], elastic recoil scattering analysis [11], or high energy ion recoil analysis (HEIR) [12], the latter specifying ion energies of larger than 10 MeV.

Elastic recoil analysis of hydrogen using a 44 MeV chlorine ion beam provides the desired maximum profiling depth as well as the required sensitivity and depth resolution. The distinct advantage of elastic recoil analysis is the measurements of whole hydrogen profiles at fixed ion beam energy, resulting in reasonable analysis time per sample. This feature of being able to follow the progression of whole profiles during data taking allows the monitoring of the target samples with respect to possible damage created by the incident ions.

MAGNETIC BUBBLE GARNETS - I2P2 DEVICES

The importance of ion implantation as a processing step for magnetic bubble propagation devices has been extensively covered in the literature. To better understand the role of the hydrogen in I2P2 devices [1], samples of modified yttrium iron garnet films, 2 μm thick, were grown epitaxially onto non-magnetic gadoliniumm gallium garnet substrates. To emulate the processing step which forms the bubble propagation patterns, the garnet samples were triply implanted with neon, nitrogen and hydrogen in order of increasing depth [13]:

$$80 \text{ KeV} \quad 1.0 \cdot 10^{14} \text{ Ne}^+ \text{ per cm}^2$$
$$200 \text{ KeV} \quad 2.7 \cdot 10^{14} \text{ N}^+ \text{ per cm}^2$$
$$130 \text{ KeV} \quad 2.0 \cdot 10^{16} \text{ H}_2^+ \text{ per cm}^2$$

These implantations result in a fairly uniform damage profile extending from the sample surface up to a depth of 0.5 μm. The magnetic properties of the garnet in the region of the ion implants are altered due to the presence of the lattice expansion arising from the implant damage. The magnetic garnets used in this study, being of the type $(YSmLuCa)_3(FeGe)_5O_{12}$, are characterized by a negative magnetostriction coefficient. The compressive stress induced by the ion implants therefore results in planar magnetic anisotropy. For the I2P2 device fabrication, a photoresist mask of a 'contiguous disk' shape formed on the garnet surface before the above implantation treatment produces the magnetic bubble propagation rails. The unimplanted regions of the garnet remain characterized by an easy axis aligned perpendicular to the garnet surface. In this configuration the bubbles are confined to the unimplanted garnet material underlying the ion implanted layer. Clinging to the edges of the implant treated region, the magnetic bubbles can be propagated parallel to the surface of the film by applied external magnetic fields. Refs. [2,13] should be consulted for comprehensive reviews on this subject matter.

Proper annealing treatments have been shown to optimize the bubble propagation behavior [13]. A greatly reduced anisotropy change after annealing was observed for garnet wafers that were covered with a thin coating of silicon oxide, directing the attention towards the hydrogen, the most mobile constituent of this device configuration. The presence of hydrogen in magnetic garnets and its unusual magnetic effects have been discussed in recent reviews [2,3].

To study the effects of I2P2 device processing on the concentration-depth profiles of ion implanted hydrogen, garnet samples were prepared and furnace annealed with and without prior plasma deposition of a 0.15 μm SiO_2 surface coating. During deposition of the oxide the samples were kept at $150^\circ C$ to avoid pre-annealing.

EXPERIMENTAL

The elastic recoil analysis measurements were performed using the Rutgers 8 MV FN Tandem Van De Graaff Nuclear Accelerator. The chlorine ion beam is produced using Freon gas in a universal negative ion source (UNIS). Singly charged negative ions are injected from ground potential into the Tandem accelerator. Reaching the accelerator terminal which was kept at +5.5 MV, the chlorine ions are stripped of some of their electrons by transmission through a carbon foil of 10 $\mu g/cm^2$ thickness. The resulting positive chlorine ions are accelerated to ground potential and subsequently analysed by a 90° magnet. Selecting the Cl^{+7} ions results in the 44 MeV chlorine ion beam used for the hydrogen profiling measurements. The experimental setup parameters matched such prerequisites as a maximum profiling depth of 1.0 μm into the garnet material, a detection sensitivity of about $5 \cdot 10^{-4}$ hydrogen atoms per garnet formula unit to clearly resolve the $4 \cdot 10^{16}$ H/cm^2 implants as well as a reasonable profile measuring time of an average of one hour per spectrum.

The elastic recoil scattering events can be seen to occur in the following way. The incident chlorine ions penetrate the sample with a decreasing ion energy as a function of profiling depth. The elastic recoil scattering events occur at all depths throughout the sample, resulting in kinematic energy transfer to the hydrogen as well as to the other substrate atoms. The recoiled sample ions as well as the scattered incident chlorine ions emerge from the sample and are subsequently selectively absorbed in the 40 μm Mylar foil, which only allows the energetic protons to reach the detector.

As shown in fig.1, the recoiled protons were measured at a total scattering angle of $\Theta = 40^\circ$, with the sample tilted $\alpha = 25^\circ$ with respect to the beam axis. Beam dose integration and monitoring were provided by a mechanical beam chopper which scattered a fraction of the total number of incident chlorine ions into a separate solid state detector. The target material of the beam chopper consisted of a thin evaporated film of 0.1 μm gold deposited onto a silicon substrate. This assembly, consisting of the seperate solid state detector and the beam chopper, was mounted directly at the target chamber entrance after the final collimation aperture.

The proton detector, shown in fig.1, consisted of an energy sensitive solid state silicon detector of 475 μm thickness. Good hydrogen-concentration profiling sensitivity with reasonable average sample irradiation times of approximately one hour per profile, required a beam spot size of about 0.25 cm^2 on the target. This, in turn, demanded the installation of an aperture of 0.9 mm in diameter in front of the solid state detector. The aperture was located at a distance of 4.8 cm from the target beam spot. This experimental setup geometry resulted in a depth resolution of about 0.05 μm at the garnet surface [12].

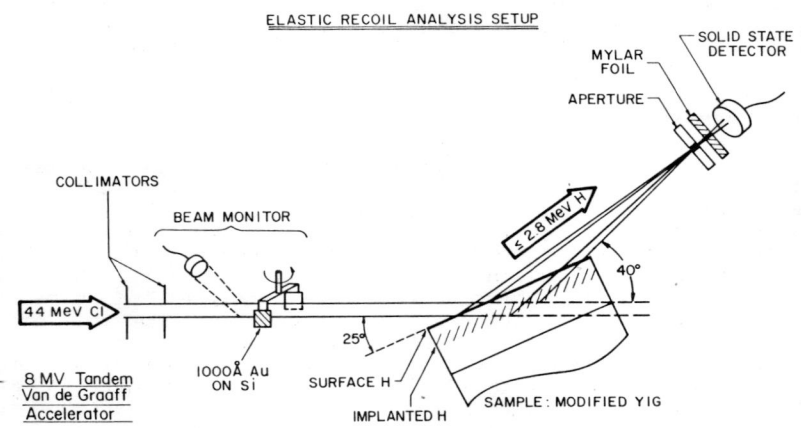

Fig.1. Experimental arrangment for hydrogen profiling by elastic recoil analysis.

FORMALISM

Inspecting Ziegler's ion stopping tables [14] shows that energetic chlorine ions experience maximum stopping in the yttrium iron garnet material at 44 MeV. The garnets used for our present study had a density of 5.18 g/cm^3 and were characterized by the following stoichiometric composition [13]:

$$(Y_{1.04}Sm_{0.51}Lu_{0.71}Ca_{0.74})(Fe_{1.97}Lu_{0.03})(Fe_{2.26}Ge_{0.74})O_{12} \tag{1}$$

Using the above and Bragg's Rule [14], the approximately constant chlorine ion stopping power is evaluated at S_{Cl} = 6.93 MeV/μm. The chlorine ion energy $E_{Cl}(j)$ inside the sample substrate as a function of the channel number j is given by

$$E_{Cl}(j) = E_{Cl}(j_o) - S_{Cl}x_{Cl}(j) \tag{2}$$

with $x_{Cl}(j)$ and $E_{Cl}(j_o)$ representing the chlorine ion depth and the incident chlorine ion energy of 44 MeV, respectively. The following expression published in the literature correlates the profiling depth perpendicular to the sample surface x(j) to the proton energy as measured by the detector $E_p(j)$ [12]:

$$x(j) = \frac{K \cdot E_{Cl}(j_o) - E_p(j) - S_M(j) \cdot x_M}{\left(\dfrac{K \cdot S_{Cl}}{\sin \alpha} + \dfrac{S_p(j)}{\sin (\theta - \alpha)} \right)} \qquad (3)$$

$S_M(j)$ and x_M are the average proton stopping power and thickness of the Mylar foil, respectively. $S_p(j)$ represents the average stopping power of the recoiled protons during their exit from the sample material. The parameter K is the kinematic factor which gives the energy transfer of the incident chlorine ions to the sample hydrogen atoms [9]:

$$K = \frac{E_p(j)}{E_{Cl}(j)} = \frac{4 m_p m_{Cl}}{(m_p + m_{Cl})^2} \cos^2\theta \qquad (4)$$

m_p and m_{Cl} represent the masses of the hydrogen and chlorine atoms, respectively, while the energy of the protons as measured by the solid state detector $E_p(j)$, is given by

$$E_p(j) = E_p(j_o) \cdot (j/j_o) \qquad (5)$$

with $E_p(j_o)$ the maximum proton energy, j the spectrum channel number and j_o the channel number corresponding to the maximum proton energy of hydrogen atoms that have been recoiled from the sample surface.

Since the parameters $S_M(j) \cdot x_M$ and $S_p(j)$ are functions of the proton energy, a suitable parameterization had to be established. Ref. [15] was used to obtain an expression for the energy dependence of the proton stopping power to fit the stopping power data given by Ziegler's tables [16]. Redefining the constants, the stopping power of a proton with energy E inside a substrate X is given by:

$$S_X(E) = \frac{m}{E} \left(\ln E + \frac{b}{m} \right) \qquad (6)$$

Here the energy dependence of the parameter b for small proton energies is neglected. The form of eq.(6) is convenient for it allows the fitting of the stopping power data using a linear least-squares fit. The parameters m and b represent the fitted slope and intercept, respectively, of data plotted as $S_X \cdot E$ versus $\ln(E)$. Such simplicity is desirable in view of the extensive usage of the stopping power functions during the calculations of the hydrogen concentration-depth profiles, which are described in the next sections. Fig.2 and table I display the fitted stopping powers data of magnetic bubble garnets, water and Mylar. The garnet and water data was obtained from ref. [16], while the Mylar data was taken from ref. [15]. For the setup calibration as well as for the determination of the hydrogen concentration-depth profiles, the proton energy loss in the various materials listed above had to be calculated. A proton incident onto a slab of material of thickness Δx with energy E_1, emerges from the slab with energy E_2 such that:

$$\Delta x = \int_{E_1}^{E_2} \frac{dE}{S_X(E)} \qquad (7)$$

Using the stopping power S_X as defined by eq.(6), the integral in eq.(7) can be solved using ref. [17]:

$$m\, e^{2b/m}\, \Delta x = \ln|B| - \ln|A| + 2(B-A) + \sum_{n=2}^{\infty} \frac{2^n}{n \cdot n!} (B^n - A^n) \qquad (8)$$

A and B are defined as $\ln(E_2)+(b/m)$ and $\ln(E_1)+(b/m)$, respectively. The following iterative algorithm was used to evaluate the infinite converging sum. All applications of eq.(8) required the calculation of the proton exit energy E_2 for given thickness Δx and incident proton energy E_1. Thus, the left hand side of eq.(8) and the part of the expression containing B were evaluated using the constants given in table I. The parameters E_L, E_H and E_M were then introduced, with E_L and E_H spanning the proton energy interval of interest and E_M representing the average of E_L and E_H. Defining the range of proton energies as: E_L = 0.1 MeV < E_M = E_2 < E_H = $K \cdot E_{Cl}(j_o)$, the right hand side of eq.(8) was evaluated truncating the infinite sum when the

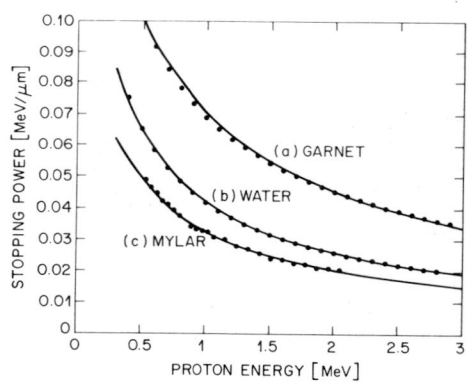

Fig.2. Fitted stopping power data from ref. [16], curves (a) and (b), and ref. [15], curve (c), using eq.(6).

higher term contributions were equal to less than 10^{-4} % of the total sum. If the right hand side of eq.(8) was found to be larger than the left hand side, E_2 was incremented to $E_M+(E_H-E_M)/2$ and E_L and E_H were set equal to E_M and E_H, respectively. If the right hand side was smaller than the left hand side, E_2 was decremented to $E_M-(E_M-E_L)/2$ and E_L and E_H were set equal to E_L and E_M, respectively. For new E_L, E_M and E_H the above algorithm was repeated until E_M converged to the proper E_2 within the specified error of 10^{-5} MeV.

Table I
Proton Stopping Powers.

Material	Symbol	m[MeV^2/μm]	b[MeV^2/μm]
Garnet	S_G	0.02913	0.07046
Mylar	S_M	0.01166	0.03263
Water	S_W	0.01401	0.04208

The above procedure defined the stopping power functions. The average proton energy loss in the magnetic bubble garnets $S_G(j)$, water $S_W(j)$ and Mylar $S_M(j)$ are thus defined as:

$$S_{G,W}(j) = \frac{K \cdot E_{Cl}(j) - E_p'(j)}{x_p(j)} = S_{G,W}[x_p(j), K \cdot E_{Cl}(j)] \qquad (9)$$

$$S_M(j) = \frac{E_p'(j) - E_p(j)}{x_M} = S_M[x_M, E_p'(j)] \qquad (10)$$

where $E_p'(j)$ is the exit energy of the protons after having traversed the distance $x_{p,M}$ in the sample material.

The calculation of hydrogen concentration profiles from the ERA spectrum yield $Y(j)$ requires the knowledge of the scattering cross-section. Here plain Coulomb scattering was assumed since the center of mass energy remains far below the Coulomb barrier for collisions of 44 MeV chlorine ions with stationary protons [9]. Combining all the constants of the expression given in in the ref. [9] into one calibration constant C, the spectrum yield $Y(j)$ was defined as:

$$Y(j) = \frac{C \cdot N(j) \cdot \delta x_{Cl}(j)}{[E_{Cl}(j)]^2} \qquad (11)$$

C is given in units of [MeV$^2 \cdot$cm^2/μm] to conform with the units chosen for the other constants in this study. $N(j)$ represents the target hydrogen density in units of [H/cm^3] while $\delta x_{Cl}(j)$ is given in [μm] by the following expression:

$$\delta x_{Cl}(j) = \frac{E_{Cl}(j+1) - E_{Cl}(j)}{S_{Cl}} \qquad (12)$$

$\delta x_{Cl}(j)$ is defined to be the interval in chlorine stopping depth $x_{Cl}(j)$ corresponding to the energy width $E_p(j_o)/j_o$ in the proton energy spectrum.

CALIBRATION

For precise calculation of the hydrogen concentration-depth profiles an independent determination of the calibration constant C used in eq.(11) was obtained. A convenient calibration substance is water ice, which was formed by freezing distilled water onto a target holder that was kept at liquid nitrogen temperatures. The large sputtering during irradiation with the high energy chlorine ions assures that the target remained stoichiometric H_2O during the calibration runs. Fig.3(b) displays a hydrogen elasic recoil analysis spectrum of water.

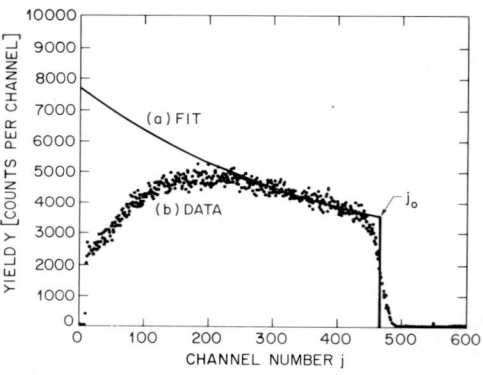

Fig.3. Recoil proton energy spectrum of water ice profiled at liquid nitrogen temperature using elastic recoil analysis. Curve (a) represents a fit to the experimental spectrum (b) using eq.(13); the fitted region spans channels $0.64 \cdot j_o$ to $0.90 \cdot j_o$. The divergence of the fit from the data for lower proton energies is due to the energy dependence of the proton stopping powers of water and Mylar.

Making the approximation suggested by ref. [8] of assuming a constant proton energy loss in the Mylar foil and neglecting the energy loss of the protons upon exit from the sample material, a curve was fitted to the data of fig.3(b). Fig.3(a) displays the fitted function which has the form

$$Y(j) = (m' \cdot j + b')^{-2} \qquad (13)$$

Eq.(13) was derived by combining eqs.(2), (3), (5) and (11) and applying the above assumptions. For channels $j > j_o$ the spectrum yield $Y(j)$ was set equal to zero and only the data points between $j_1 = 0.64 \cdot j_o$ and $j_2 = 0.90 \cdot j_o$ were used for the fitting. For $j < j_1$ the data differs from the fitted curve which reflects the proton energy dependence of the water and Mylar stopping powers. The stopping of 44 MeV chlorine ions in water was evaluated to be $S_W = 2.63$ MeV/μm using ref. [14] and the method used earlier to evaluate the chlorine ion stopping in the magnetic bubble garnets.

To fit the calibration constant C, the average proton stopping in the Mylar foil $S_M(j)$ was calculated as a function of the detected proton energy $E_p(j)$, directly using the water profile data displayed in figs.3(a) and 3(b). The motivation for this procedure was to obtain a smooth function from experiment that could be chi-square fitted to the stopping power table values of ref. [14].

To obtain the average Mylar stopping power $S_M = S_M(E_p(j))$ from experiment, the depth scale of the profiles had to be established. This was accomplished by recursively applying the combined eqs.(11) and (12), starting at channel number $j' = j_o-1$ using $E_{Cl}(j'+1) = E_{Cl}(j_o)$. $E_{Cl}(j')$ was then evaluated in recursive iteration of decreasing integral j'. The strong effect of the experimental resolution onto the spectrum yield in the region of $j' = j_o$ required the evaluation of $Y(j')$ using the data of fig.3(a) and 3(b) for $j > j_c = (0.64+0.90) \cdot j_o/2$ and $j < j_c$, respectively. The use of this algorithm and eq.(2) determined the parameter set $x_{Cl}(j)$. The depth scale perpendicular to the sample surface represented by $x(j)$ was established using the following expression obtained from inspection of fig.1:

$$\sin \alpha \cdot x_{Cl}(j) = x(j) = \sin(\theta-\alpha) \cdot x_p(j) \qquad (14)$$

After calculating the parameter set $E_p'(j)$ using eqs.(9) and (14), $E_p(j)$ was evaluated using eq.(5) and $E_p(j_o)$, the latter being determined by eq.(10) having set $E_p'(j_o^p) = K \cdot E_{Cl}(j_o)$. $S_M(j)$ was then evaluated by setting $S_M(j)^p = (E_p(j)-E_p(j))/x_M$. A plot of $S_M = S_M(E_p(j))$, which has thus been generated is shown in fig.4(b). Fig.4(a) displays the average stopping power of protons traversing a 40 µm Mylar foil, which has been calculated using eq.(7) and the data given in fig.2(c). The curve in fig.4(b) was calculated for a value of the calibration constant C, which had given the best chi-square fit to the curve shown in fig.4(a). The fitted region was restricted to proton energies larger than 1 MeV. The 6% discrepancy between the curves (a) and (b) in fig.4 for small proton energies was attributed to the energy dependence of the chlorine ion stopping power and energy straggling. Since the hydrogen profiles of the magnetic bubble garnets required about two thirds of the maximum profiling depth attainable by the present ERA setup, only the regime of good fit shown in fig.4 was needed to calculate the concentration-depth profiles.

It should be noticed that the data scatter of the hydrogen

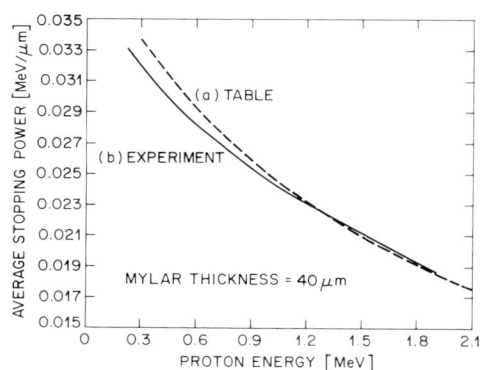

Fig.4. Average stopping power of protons traversing 40 µm of Mylar plotted as a function of the proton exit energy E_p. Curve (a) was calculated using the Mylar stopping data shown in fig.2(c). Curve (b) was generated using the spectrum of Fig.3(b). A chi-square fit for proton energies larger than 1 MeV, allows the evaluation of the ERA calibration constant.

recoil spectrum shown in fig.3(b) was compensated by variations in depth scale while generating curve 4(b). This feature is a consequence of the recursive nature of the algorithm. The calibration constant C for our setup was found to be $0.3020 \cdot 10^{-13}$ [MeV^{2}cm^2/µm] for a chlorine beam dose of approximately 50 µC of Cl^{+7} on target. Spectrum normalization was obtained by integrating chlorine ion backscattering events as detected by the beam chopper assembly shown in fig.1.

CONCENTRATION DEPTH PROFILES

Having determined the setup calibration constant, the hydrogen concentration-depth profiles shown in figs. 5 and 6 were extracted from the measured proton energy spectra of the garnet samples, which will be published elsewhere [18].
The procedure to calculate the profiles of the garnet samples differed from the one chosen to evaluate the calibration constant using the water spectrum. The hydrogen concentration profile $N(j)$ in the garnet was an unknown and eqs.(11) and (12) could be used to determine the profile depth scale.
The calculation of the hydrogen profiles in figs.5 and 6 entailed the following steps. The proton energy measured in the solid state detector as a function of channel number is given in ref.13:

$$E_p(j) = K \cdot E_{Cl}(j) - S_p(j) \cdot x_p(j) - S_M(j) \cdot x_M \qquad (15)$$

Combining eqs.(2), (5), (14) and (15), and evaluating $E_p(j_o)$ from eq.(5) using eq.(10), the dependence of the channel number j on the chlorine energy $E_{Cl}(j)$ was established to be:

$$\frac{j}{j_o} = \frac{K \cdot E_{Cl}(j) - \left(\frac{S_p(j) \cdot \sin \alpha}{S_{Cl} \cdot \sin(\theta - \alpha)}\right) [E_{Cl}(j_o) - E_{Cl}(j)] - S_M(j) \cdot x_M}{K \cdot E_{Cl}(j_o) - S_M(j_o) \cdot x_M} \qquad (16)$$

To obtain the hydrogen concentration-depth profiles, x_{Cl} was integrated from the sample surface at $j = j_o$ where $E_{Cl}(j) = E_{Cl}(j_o)$ and $x_{Cl}(j_o) = 0$. Choosing integration intervals of $dx_{Cl} = 0.001$ µm, eqs.(2) and (14) were used to evaluate $E_{Cl}(j)$. $S_M(j)$ and $S_p(j)$ were evaluated using eqs. (2), (9), (10) and (14). Increasing x_{Cl} by Δx_{Cl}, decreased j by Δj using eq.(16). Beginning with $j = j_o$, $E_{Cl}(j-1)$ was found by summing until $\Delta j = 1$. Then j was set equal to $j-1$ and the calculation was repeated. While summing between j and $j-1$ the parameter $y_i(j) = Y(j)/N(j)$ was evaluated using eq.(11) and (12). If I steps were required to complete an integral channel (ie. $\Delta j = 1$) then $y_I(j)$ was evaluated by summing over $y_i(j)$:

$$y_I(j) = \sum_{i=1}^{I} y_i(j) \qquad (17)$$

with i representing the differential steps in the summation for a single channel j. N(j) is defined as the average hydrogen concentration over the channel interval $\Delta j = 1$, since this interval just corresponds to the resolution of the digital-to-analog converter. Summing over the parameter $y_i(j)$ therefore was equivalent to summing over the number of events occuring at differential intervals inside the channnel j. For every completed channel j the hydrogen concentration N(j) was evaluated by setting:

$$N(j) = Y(j)/y_I(j) \qquad (18)$$

The choice of $dx_{Cl} = 0.001$ μm was proven reasonable for decrementing this interval has shown not to affect the calculated profiles.

RESULTS AND DISCUSSION

The hydrogen concentration-depth profiles of garnet samples without and with oxide coating are displayed in figs. 5 and 6, respectively. Fig.5(a) shows the profile of a garnet sample which had been ion implanted without further treatments. The approximately Gaussian shaped implant profile peaked at 0.36 μm with a straggling length of 0.07 μm, taking into account the experimental resolution of 0.05 μm, which has not been deconvoluted from the calculated profiles. This shape and depth are consistent with the earlier measurements [3].

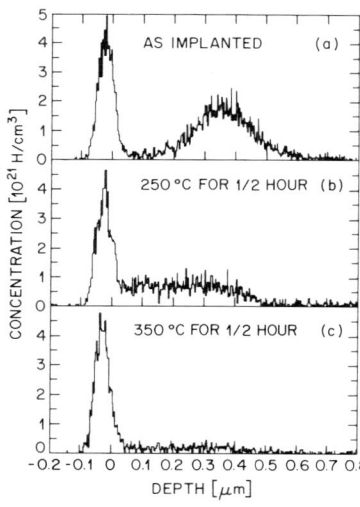

Fig.5. Hydrogen profiles of triply implanted garnet films, as implanted, (a), and after annealing, (b) and (c).

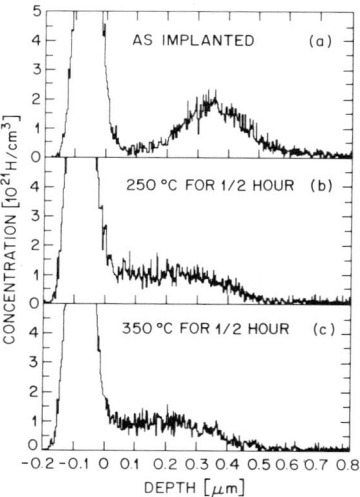

Fig.6. Same as fig.5, but with 0.15 μm of SiO_2 deposited on the garnet surface before annealing.

For quantitative analysis of the garnet profiles, two regions of interest were set. One region was placed across the hydrogen implant, while the second one was set at greater depth past the hydrogen implant region, allowing for monitoring of bulk hydrogen concentrations. The regions with their respective integrated hydrogen contents for the various samples are listed in table II.

Table II
Garnet Hydrogen Concentrations

Treatment	Implant Region (.075-.75µm)	Bulk Region (.65-.75µm)	Equivalent Dose
Without Oxide:			
as implanted	$4.51 \cdot 10^{16} H/cm^2$	$9.17 \cdot 10^{14} H/cm^2$	$3.89 \cdot 10^{16} H/cm^2$
250°C, 1/2 hr	$2.90 \cdot 10^{16} H/cm^2$	$7.24 \cdot 10^{14} H/cm^2$	$2.41 \cdot 10^{16} H/cm^2$
350°C, 1/2 hr	$0.87 \cdot 10^{16} H/cm^2$	$2.61 \cdot 10^{14} H/cm^2$	$0.69 \cdot 10^{16} H/cm^2$
With Oxide:			
as implanted	$5.09 \cdot 10^{16} H/cm^2$	$14.6 \cdot 10^{14} H/cm^2$	$4.10 \cdot 10^{16} H/cm^2$
250°C, 1/2 hr	$3.88 \cdot 10^{16} H/cm^2$	$8.79 \cdot 10^{14} H/cm^2$	$3.29 \cdot 10^{16} H/cm^2$
350°C, 1/2 hr	$3.43 \cdot 10^{16} H/cm^2$	$6.89 \cdot 10^{14} H/cm^2$	$2.96 \cdot 10^{16} H/cm^2$

Since the cross-section of the elastic recoil events is a function of depth, the placement of the parameter j_o (eg. $x(j_o) = 0$) was critical to obtain quantitative results. Two positions on the hydrogen profiles seemed reasonable. For the following analysis only the profiles in fig.5 qualify, for the profiles in fig.6 display a 9 atomic percent hydrogen concentration in the 0.15 µm oxide coating. This hydrogen was incorporated during the plasma deposition, a process which entails the reaction of silane (SiH_4) and oxygen gas. Placing j_o to the left or right with respect to the hydrogen surface peak, decides on whether or not this surface hydrogen was incorporated into the garnet lattice. Both positions of j_o resulted in hydrogen concentrations that were larger than the expected dose of $4 \cdot 10^{16}$ H/cm^2. However, placing j_o where shown in fig.5(a) resulted in the smaller calculated hydrogen dose.

Two things were derived from this result. First, the surface hydrogen stemmed from a contamination which we attributed to deposition of hydrocarbons present in the vacuum of the ion implantation accelerator. Secondly, there seemed to be excess hydrogen present in the implant region of the garnet which had not originated from the hydrogen implantation.

At this point a correction of the incident chlorine ion energy was made for both the hydrogen profiles in figs.5 and 6, taking into account the extra energy loss of the chlorine ions while traversing the surface hydrogen layer and the oxide coating, respectively. This correction was made by subtracting $\Delta E_{Cl} = \Delta E_p/K$ from the incident chlorine ion energy of 44 MeV. ΔE_{Cl} is evaluated by applying the approximation used for the generation of the curve in fig.3(a) to eq.(15) and differentiating the result. $\Delta E_p/K$ is obtained by differentiating eq.(5) and evaluating the result using the channel width of the surface hydrogen layer.

Setting the regions shown in table II, the hydrogen concentration was integrated both in the implant as well as the bulk region of the garnet samples. Subtracting the excess hydrogen from the implant profile by using the data in table II, resulted in agreement of the hydrogen implant doses of both figs.5(a) and 6(a) within 3% with the expected implant dose of $4 \cdot 10^{16}$ H/cm^2. The excess hydrogen in the sample without, (fig.5(a)), and with oxide coating, (fig.6(a)), amounted to respectively 3 and 5 times the hydrogen concentration expected for garnet samples for which the hydrogen implantation was omitted. Such concentrations were equivalent to about 4% of the implant dose when averaging this dose over the region spanned by the hydrogen implants.

To check wether the excess hydrogen had diffused into the garnet from the surface, a sample implanted with the standard neon and nitrogen doses but no hydrogen, was oxide coated and annealed at 350°C for 20 hours. No extra hydrogen was detected in the garnet and the hydrogen level in the oxide was unchanged. In addition by inspecting table II, a decrease in excess hydrogen concentration after annealing treatments suggested that the heat treatments only dilute the excess hydrogen concentrations.

The profile in fig.5(b) shows a triply implanted garnet sample after annealing in nitrogen atmosphere at 250°C for half an hour. The protons have diffused towards the surface through the layer damaged by the shallow neon and nitrogen implants, leaving 60% of the initial hydrogen implant dose in the garnet. No hydrogen diffusion deeper into the sample was detected. After annealing at 350°C for half an hour 80% of the implant hydrogen had diffused out of the garnet.

The oxide coated garnets in fig.6 displayed very similar features to the garnets in fig.5, with the exception of the large hydrogen concentrations present in the SiO$_2$ layer which covered the garnets of fig.6. The concentration scale in figs.5 and 6 is only meaningful for the hydrogen located in the garnet. The parts of the profiles corresponding to the surface hydrogen layers should be regarded as schematic indicating qualitatively the presence of hydrogen. After annealing at 250°C, fig.6(b), most of the hydrogen remained in the damaged garnet. However, after the oxide coated sample had been annealed at 350°C, fig.6(c), most of the hydrogen was retained in the damaged region of the garnet. The hydrogen concentrations after the last two anneals amounted to 82 and 74% of the original dose, respectively.

This result thus correlates the earlier discovered reduced change in magnetic anisotropy for samples that were coated with silicon oxide, to the actual presence of trapped hydrogen in the magnetic bubble garnets for the device configuration as shown in fig.6(c). The SiO$_2$ on the garnet surface reduced the total amount of hydrogen diffusing out of the garnet during annealing by a factor of more than 4. At room temperature the hydrogen profiles were very stable, resulting in devices with good longevity.

To determine the importance of the damage produced by the shallow implants, we profiled garnet samples which had been implanted only with hydrogen to the dose that is used for the triple implants. The hydrogen diffused toward the surface, but this time the profile retained more of its Gaussian character after a 250°C anneal. Even though the hydrogen implant produced a smaller level of damage near the surface most of the hydrogen was lost after the 350°C anneal.

Both the shallow implants and the SiO_2 coating are therefore important processing steps for I2P2 devices, contributing to the formation of a constant and stable hydrogen concentration within the first 0.4 μm of the implanted garnet films.

CONCLUSIONS

Elastic recoil analysis with 44 MeV chlorine ions as set up in our present study, represents an excellent method to study hydrogen in magnetic garnets and related materials. The method allows both precise and time saving analysis of magnetic bubble garnet samples which represent processing steps of I2P2 devices.

The central conclusion that was drawn from our measurments, is that the reduced change in magnetic anisotropy of the garnets, that have been annealed with a silicon oxide coating, is directly correlated to the presence of trapped hydrogen in the garnet film. The experiments have shown that hydrogen diffused rapidly in the ion implant damaged near-surface layer of the garnet in the temperature range of 250°C to 350°C. No hydrogen diffusion into the undamaged garnet was detected for any of the profiles. The nature of the damage profile affected the hydrogen diffusion. The hydrogen implantation in itself generated a significant damage in the garnet lattice allowing for hydrogen diffusion towards the surface.

An interesting observation was the effect of hydrogen diffusing out of the garnet film, but not diffusing back into the film from the oxide coating which in itself contained large amounts of hydrogen.

The role of the trapped hydrogen in improving the performance of bubble devices based on ion implanted propagation patterns has been well established, but the mechanisms by which it contributes to the magnetic anisotropy is not yet understood.

ACKNOWLEDGMENTS

We are grateful to T. W. Hou who provided the silicon dioxide layers, and to L. C. Feldman, C. J. Mogab and Prof. G. M. Temmer for valuable discussions. The research at Rutgers University was funded in part by the National Science Foundation.

REFERENCES

[1] R. Wolfe, J. C. North, W. A. Johnson, R. R. Spiwak, L. J. Varnerin and R. F. Fisher, AIP Conf. Proc. 10 (1973) 339.
[2] K. Yoshimi, in: Recent Magnetics for Electronics, JARECT, Vol. 10, ed. Y. Sakurai (North-Holland, Amsterdam, 1983) p.43.
[3] P. Gerard, Thin Solid Films 114 (1984) 3.
[4] W. K. Chu, J. W. Mayer and M. A. Nicolet, Backscattering Spectrometry (Academic Press, New York, 1978).
[5] W. A. Lanford, H. P. Trautvetter, J. F. Ziegler and J. Keller, Appl. Phys. Lett. 28 (1976) 566.
[6] P. J. Mills, P. F. Green, C. J. Palmstrom, J. W. Mayer and E. J. Kramer, Appl. Phys. Lett. 45 (1984) 957.
[7] B. L. Doyle and S. Peercy, Appl. Phys. Lett. 34 (1979)

811.
[8] A. Turos and O. Meyer, Nucl. Instr. and Meth. B4 (1984) 92.
[9] C. Nolscher, K. Brenner, R. Knauf and W. Schmidt, Nucl. Instr. and Meth. 218 (1983) 116.
[10] J. P. Thomas, M. Fallavier, D. Ramdane, N. Chevarier and A. Chevarier, Nucl. Instr. and Meth. 218 (1983) 125.
[11] C. C. P. Madiba, J. P. F. Sellschop, H. J. Annegarn and B. R. Appleton, Nucl. Instr. and Meth. 218 (1983) 409.
[12] C. Moreau, E. J. Knystautas, R. S. Timsit and R. Groleau, Nucl. Instr. and Meth. 218 (1983) 111.
[13] T. J. Nelson, J. E. Ballintine, L. A. Reith, B. J. Roman, S. E. G. Slusky and R. Wolfe, IEEE Trans. Mag. MAG-18 (1982) 1358.
[14] J. F. Ziegler, ed., Handbook of Stopping Cross - Sections for Energetic Ions in all Elements, Vol. 5 (Pergamon Press, New York, 1980).
[15] A. L'Hoir and D. Schmaus, Nucl. Instr. Meth., B4 (1984) 1.
[16] H. H. Anderson and J. F. Ziegler, eds., Hydrogen Stopping Powers and Ranges in all Elements (Pergamon Press, New York, 1980).
[17] H. B. Dwight, in: Tables of Integrals and Other Mathematical Data, 4th ed. (MacMillan Publishing, New York, 1961) p.143.
[18] A. Leiberich and R. Wolfe, Appl. Phys. Lett., to be published.

APPLICATIONS OF SURFACE ANALYSIS BY LASER IONIZATION
(SALI) TO INSULATORS AND II-VI COMPOUNDS

C. H. BECKER[*], C. M. STAHLE[**], and D. J. THOMSON[**]
[*]Chemical Physics Laboratory, SRI International, Menlo Park, CA 94025
[**]Stanford Electronics Laboratories, Stanford University, Stanford, CA 94305

ABSTRACT

Examples of the mass spectrometry of sputtered or evaporating neutral species obtained by SALI are presented for NBS Glass 610 (primarily a silicate), and an anodic oxide of HgCdTe. For the NBS glass, a SIMS spectra was recorded for comparison with SALI using the same apparatus. The raw SALI spectra of the glass is in semiquantitative accord with the known composition, in contrast to SIMS. Relative secondary ion yields can be determined for unknown complex materials by comparing SALI and SIMS spectra. Depth profiling measurements on the anodic oxides of $Hg_{1-x}Cd_xTe$ show a significant though depleted concentration of Hg in the oxide in contrast to numerous other analyses; this result is corroborated by RBS studies. Hg and also Te evaporation is monitored in real-time by SALI with large dynamic range capabilities.

INTRODUCTION

Surface analysis by laser ionization is an analytical tool that uses the mass spectrometry of neutral atomic and molecular species leaving a surface [1,2]. A high intensity, pulsed, focused, UV laser passes parallel and close (~1mm) to the surface and ionizes the species in an efficient and general fashion. The separation of the desorption or sputtering step from the ionization step means that major matrix effects are avoided with SALI, in contrast to SIMS, which is the other general analysis method of comparable sensitivity. SALI has been demonstrated to have quantitative ppm-level detection capabilities for all masses simultaneously while sampling submonolayer quantities [1-3]. Further improvements are anticipated. Studies using sputtering [1,2], laser desorption [3], and thermal evaporation [1] have been performed for metal and semiconductor (Si and GaAs) substrates. This paper reports on applications of SALI to an insulator and the oxide of a II-VI compound.

These two types of materials can be difficult to analyze and therefore they offer an opportunity to test SALI's versatility. The main difficulty with insulators is their propensity to charge-up. Difficulties encountered with the II-VI compounds, notably $Hg_{1-x}Cd_xTe$ and its oxides, are preferential sputtering, preferential diffusion and thermal evaporation, ion beam induced reactions and decomposition, and electron-stimulated desorption. The ability to analyze this difficult class of sensitive material as well as monitor multicomponent thermal evaporation down to very low rates will described.

ANALYSIS OF NBS GLASS 610

Charge build-up on insulators has been a concern to surface analysts for many years [4]. The usual approaches are the use of a charged particle flood beam of charge compensating that of the primary beam, deposition of a conducting overcoat, and the use of a small metallic aperture or metallic grids [4]. Compensating charged particle beams (such as an electron flood gun for positive ion bombardment) and conducting overcoats are not always satisfactory [4]. We have found that the use of a fine mesh of tungsten with apertures of a few hundred micrometers was a very suitable solution.

Using dc Ar^+ bombardment at 3 keV and approximately $50\,\mu A/cm^2$, and excimer laser ionization at 193 nm and 10^9 W/cm^2, the SALI mass spectrum shown in Fig. 1 was obtained in 100 laser pulses for NBS glass 610. Analysis of the spectrum and comparison to the NBS values of the composition of the matrix materials is given in Table 1. The low value for Na may be due to depletion of the neutral sputtering channel and/or migration of Na^+ away from a slightly charging surface. The particularly high sensitivity to Aℓ is expected because the laser photon energy overlaps a one-photon autoionizing resonance. With more extensive signal averaging (a few thousand laser shots) the analysis of impurities in the glass (around the 100 ppm atomic concentration level) gave better agreement between the raw data and the NBS values of the relative concentrations than shown in Table 1. Accuracy in determining concentrations can be improved by multiplying the raw data by previously determined relative detection sensitivies for the laser conditions.

By turning off the laser, adjusting an electrostatic potential on the reflecting time-of-flight (TOF) mass spectrometer, and pulsing the Ar^+ beam (about 100 ns pulse widths used), SIMS spectra was obtained with the same instrument. (Laser desorption mass spectrometry similarly can be obtained

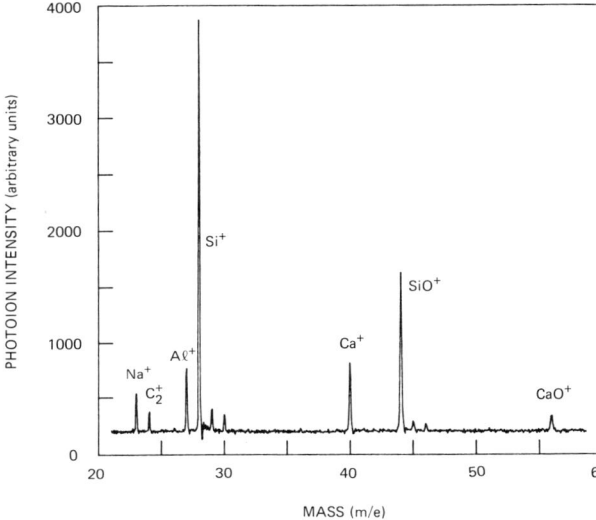

Figure 1. SALI TOF mass spectrum from NBS glass 610 by 3 keV Ar^+ sputtering with ejected neutrals ionized by 193 nm light at 10^9 W/cm^2. The spectrum was averaged over 100 laser pulses.

Table 1. Relative mass spectral intensities of species bearing Si, Ca, Aℓ, and Na sputtered from NBS glass 610 measured by SALI and SIMS.

	SALI	SIMS	NBS Values
Si	1.00	1.00	1.00
Ca	0.14	1.87	0.17
Aℓ	0.10	0.21	0.03
Na	0.06	8.82	0.37

with the instrument.) An example of SIMS spectra obtained from the glass is shown in Fig. 2 and the results of the analysis of this spectrum is given in Table 1 along with the SALI and NBS values. The very strong influence of the species' ionization potential on the positive ion SIMS sensitivities is clearly apparent. Another comparison between SALI and SIMS has been presented previously [3] for many species on a contaminated GaAs surface. Before and during the course of this analysis, with no dc sputtering, some build-up of hydrogen is apparent from residual water vapor (about 5×10^{-9} torr) released from the sample introduction probe. The width of the mass peaks is a result of the fairly long ion pulse length used.

It should be noted that, in general, charging will be a less severe problem for SALI than other methods where the surface potential defines the charged particle transmission through the instrument. This is because it is the electrostatic potential above the surface (typically about 1 mm) where laser ionization occurs that defines the instrument's transmission and because the reflector for TOF mass analysis can compensate (cause bunching) for spreads in ion energies of 10-20%. A cover plate with an aperture of 2-3mm diameter helps maintain a fairly uniform potential in the ionization volume of the laser. Thus as long as the charge build-up is not overly severe such as to completely deflect the primary beam, the mass spectra will be expected to be neither shifted nor broadened.

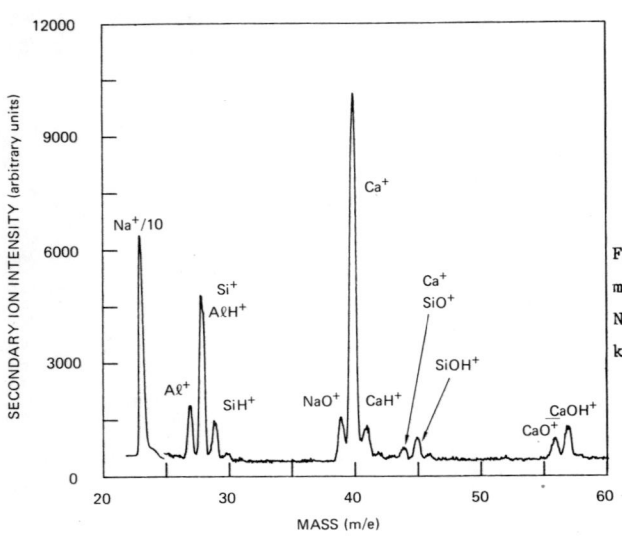

Figure 2. SIMS TOF mass spectrum from NBS glass 610 by 3 keV Ar^+ sputtering.

The relative intensities of masses in the SIMS and SALI spectra provide a direct measure of the relative secondary ion yields for a complex unknown material because the neutral yields measured by SALI generally are representative of the material's composition. However, the ratio of secondary ion to neutral yields may only be available for the sum of atomic and molecular species due to the possibility of photofragmentation. For example, one can compare the yields of $Si^+ + SiO^+$ to $Si + SiO$ but not Si^+ to Si or SiO^+ to SiO unless the SiO photofragmentation is known or can be determined in an independent study. While the SALI and SIMS data provide a measure of the relative yields of charged to neutral emission for the different species, the absolute secondary ion yields for complex materials can be obtained with the two spectra and the additional knowledge of total secondary ion and total sputtering yields. In this regard, surface composition determinations using SALI can be improved for species sputtered to a major extent as ions, such as may be the case for Na (Table 1), with supplemental SIMS, total ion yield, and total sputter yield data.

One final note regarding this glass analysis, that also applies to the $Hg_{1-x}Cd_xTe$ oxide analysis presented below, pertains to the low sensitivity to sputtered atomic oxygen. The mass 16 peak for this analysis was found to be approximately a factor of 10^3 below that expected from the materials stoichiometry. This sensitivity problem, which is significant under the present conditions for atomic N, O, F, He, Ne, and Ar, is characteristic of limited laser power. An increase of a factor of 10 to 100 in power _density_ is needed to approach saturation for these species, which should be achievable with commercial lasers. In this analysis however, the molecular species containing oxygen are still indicative of the oxide nature of the material though direct information on the oxygen content is not obtained.

ANODIC OXIDES GROWN ON $Hg_{1-x}Cd_xTe$

Determination of the composition of anodic oxides of $Hg_{1-x}Cd_xTe$ is of considerable interest [5-8] because this is important to understanding the chemical stability of the oxide of this infrared detector material. The glaring concern in most of the past analyses has been the measurement of Hg content. Frequently, very little or no Hg has been found in the oxide by standard analyses [5,6]. If the analysis method looks directly at the surface of the material (such as for AES and XPS), then preferential sputtering and preferential evaporation can deplete the surface in one or more components. Furthermore, the use of an electron beam can cause electron-stimulated desorption, again, in a preferential fashion. In a depth profiling situation, as long as a steady-state or quasi-steady-state sputtering condition is reached, the true material stoichiometry is removed; it is this removed component that is probed by SALI.

Anodic oxides grown on bulk solid state recrystallized $Hg_{0.78}Cd_{0.22}Te$ samples were used. The samples were mechanically polished and etched in a 1/8% bromine in methanol solution. The oxides were grown to thicknesses up to about 700Å in a KOH solution (0.1 N KOH, 90% ethylene glycol, 10% H_2O) at a current density of 0.3 mA/cm^2 [9]. Three keV Ar^+ was used for sputtering and the ionization was performed with 193 nm radiation focused to 10^8 W/cm^2, in 10 ns pulses. Experiments were conducted at room temperature.

Figure 3 shows a SALI spectrum taken from the middle of the anodic oxide. It represents an average of 50 laser pulses. The relative sensitivities (ionization efficiencies) for the elements were determined by steady-state sputtering of the bulk of a crystal of known composition. For these laser conditions the relative sensitivities for Te:Cd:Hg is 100:89:73. Figure 3 shows a significant though depleted Hg content relative to the bulk. A comparable depletion of Te leads to a concept discussed by Stahle et al. [10] that Hg and Te are dissolved in equal proportions in the electrolyte solution during oxide growth.

The Hg content in the oxide and bulk also was probed by Rutherford backscattering spectroscopy (RBS) measurements. The RBS results quantitatively corroborated the SALI work regarding the Hg reduction in the oxide from that of the bulk [10]. Unfortunately these RBS experiments could not separate the Te and Cd components or provide information on the oxygen content.

An important consideration in material performance for $Hg_{1-x}Cd_xTe$ is outdiffusion of Hg. Experiments are in progress using SALI to monitor in real-time the material loss in vacuum. Key parameters being explored are material history, surface pretreatment, temperature, and diffusion time.

Figure 4 displays an example of such data showing Hg and also Te evaporating from a bulk substrate shortly after a brief sputtering period. The spectra was recorded with 200 laser shots. The loss rate is actually quite low. Loss rates from Figure 4 are estimated to be 5×10^{10} Hg atoms/cm^2 s and 4×10^9 Te atoms/cm^2 s, corresponding to vapor densities equivalent to about 1×10^{-10} torr for Hg and 6×10^{-12} torr for Te. These

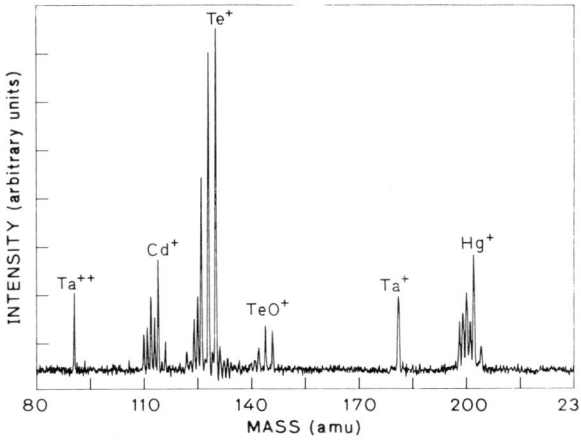

Figure 3. SALI spectrum taken from the central part of depth profiling through a 400Å thick anodic oxide grown on $Hg_{0.78}Cd_{0.22}Te$. The spectrum was averaged over 50 laser pulses.

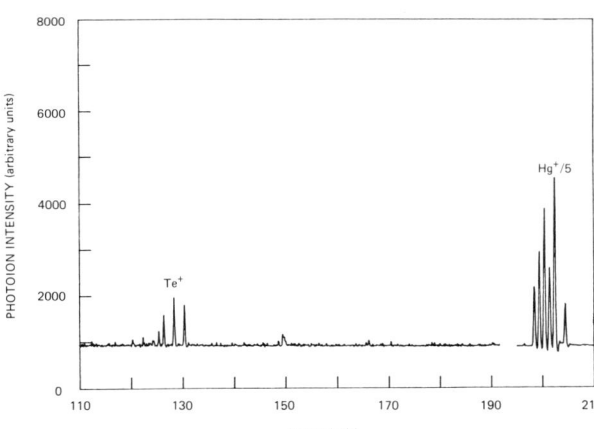

Figure 4. SALI spectrum taken of Te and Hg evaporating from the bulk of a $Hg_{0.78}Cd_{0.22}Te$ crystal recorded a few minutes after 1 minute of sputtering at room temperature. The spectrum represents an average of 200 laser pulses. The peak at m/e 149 is due to an organic contaminant.

low rates demonstrate SALI's large dynamic range because measurements can be made to ~10^{-4} torr, or higher pressures if differential pumping is employed.

SUMMARY

The method of surface analysis by laser ionization has been shown to be extremely useful for the analysis of insulators illustrated by the case of an NBS silicate glass, and for II-VI compounds illustrated by the case of the anodic oxides and bulk of $Hg_{1-x}Cd_xTe$. Charging of the glass by the

Ar^+ beam was minimized by a fine metal mesh. Comparison was made between SIMS and SALI analyses of the glass. SALI depth profiling through a $Hg_{0.78}Cd_{0.22}Te$ anodic oxide shows a Hg content reduced from that of the bulk, with similar Te depletion; the Hg content was corroborated by RBS studies. SALI has also been shown to be applicable to real-time analysis of material evaporation from $Hg_{1-x}Cd_xTe$ capable of covering a large dynamic range.

ACKNOWLEDGEMENTS

This work was supported in part by gift funds to Stanford from Texas Instruments. The surface analysis by laser ionization facility was made available by SRI Internal Research and Development Funds. This work has also benefited from facilities made available to Stanford University by the NSF-MRL Program through the Center for Materials Research at Stanford University. The authors thank Professors Bob Helms and Bill Spicer for stimulating discussions and their encouragement on II-VI compound research.

REFERENCES

1. C. H. Becker and K. T. Gillen, Appl. Phys. Lett. 45, 1063 (1984).
2. C. H. Becker and K. T. Gillen in "Laser Chemical Processing of Semiconductor Devices," eds. F. A. Houle, T. F. Deutsch, and R. M. Osgood, Jr., MRS Proceedings, Fall 1984 Meeting, p. 48.
3. C. H. Becker and K. T. Gillen, J. Vac. Sci. Technol. A 3, (in press).
4. H. W. Werner and A. E. Morgan, J. Appl. Phys. 47, 1232 (1976).
5. G. D. Davis, T. S. Sun, S. P. Buchner, and N. E. Byer, J. Vac. Sci. Technol. 19, 472 (1981).
6. P. Morgen, J. A. Silberman, I. Lindau, and W. E. Spicer, J. Vac. Sci. Technol. 21, 161 (1981).
7. M. Seelmann-Eggebert, G. Brandt, and H. J. Richter, J. Vac. Sci. Technol. A 2, 11 (1984).
8. U. Kaiser, P. Sander, O. Ganschow, and A. Benninghoven, Fresenius Z. Anal. Chem. 319, 877 (1984).
9. P. C. Catagnus and C. T. Baker, US Patent No. 3,997,018 (24 August 1976).
10. C. M. Stahle, D. J. Thomson, C. R. Helms, C. H. Becker, and A. Simmons, submitted to Appl. Phys. Lett.

INFLUENCE OF THE DENSITY OF OXIDE PARTICLES ON THE DIFFUSIONAL
BEHAVIOR OF OXYGEN IN INTERNALLY OXIDIZED, SILVER-BASED ALLOYS[+]

F.H.SANCHEZ[**], R.C.MERCADER[*], A.F.PASQUEVICH[*], A.G.BIBILONI[*] AND A.LOPEZ-GARCIA[*].
[**] Physics Department, The University of Connecticut, Storrs, CT 06268, U.S.A.
[*] Departamento de Fisica, Universidad Nacional de La Plata, 1900 La Plata, Argentina.

ABSTRACT

This paper presents strong evidence for the influence of the density of oxide particles on the internal oxidation kinetics of silver based alloys. Measurements performed by Mössbauer Spectroscopy, on 1 at% Sn in Ag alloys oxidized at temperatures between 523 and 823K, clearly indicate that the oxidation kinetics are described by a power law of the time with an exponent close to the unity for a high density of oxide particles (between 0.3 and 1.0×10^{-2} oxide particles per alloy atom). For low densities ($<10^{-4}$ oxide particles per alloy atom), the exponent is close to 0.5). Previous kinetics measurements in AgIn alloys are shown to be in general agreement with this rule. These results can be interpreted on the basis of the existence of strain fields around the oxide particles, which produce a network of channels for easy oxygen migration when the density of oxide particles is high enough.

INTRODUCTION

Internal oxidation of alloys has attracted attention some years ago because it makes materials harder and brittler, and increases their electric conductivity. We want to present here another consequence of this process: the modification of the oxygen diffusional behavior when the density of oxide particles becomes high.

Wagner's theory of internal oxidation kinetics(1) is known to fail at low temperatures. Although it is also known, both from theoretical(2) and experimental(3) work that the density of oxide particles is a decreasing function of the oxidation temperature, to our knowledge nobody has so far correlated these two facts. The recent stream of internal oxidation studies by TDPAC and Mössbauer techniques on silver alloys(4-8) has provided information about the onset of the process of agglomeration of the oxidized phase. The spectra from samples oxidized at low temperatures (473-573K) or from very diluted ones were interpreted as coming from probes in single-solute atom-oxide complexes isolated from each other in the silver matrix, while samples oxidized at higher temperatures showed characteristic spectra of stoichiometric bulk solute oxides. Here we present results on the internal oxidation of AgSn alloys (preliminarily reported(9)), obtained by Mossbauer Spectroscopy. These results will be analysed together with previous data(4,7) in order to establish and discuss the connection between the internal oxidation kinetics and the density of oxide particles.

[+] Work partially supported by Consejo Nacional de Investigaciones Cientificas y Tecnicas, Republica Argentina.

EXPERIMENTAL

Alloys with 1.0 ± 0.1 at % of Sn were made by fusion in quartz tubes under an Ar atmosphere in an electric oven using Ag and Sn with purities of 99.99 and 99.5 %, respectively.

The ingots were rolled down to the proper thicknesses for the Mössbauer measurements and three of them were annealed in an Ar atmosphere in order to get rid of the damage produced by the cold work. Carbon was used as a reducing agent. Annealing treatments were performed for about one hour at 1023K. The oxidizing treatments were performed in stages by heating the samples in open air at the specified temperatures (see table I).

The experimental device as well as details about data acquisition are described in reference 8.

RESULTS.

Table I contains the relevant information about the samples considered in this paper.

Table I.
Solute molar fractions C, thicknesses d and oxidation temperatures T. The experimental kinetics parameter m (see eq.(1)), the calculated mean number N (see eq.(2)) of solute atoms per oxide particle, the density ρ of oxide particles and the mean interparticle distance R are also listed. N was calculated using diffusion data from references 14 (In and Cu), 15 (Sn) and 16 (O). Asterisks indicate samples which were annealed prior to oxidation.

Alloy	Sample	C (at/at)	d (mm)	T (K)	m	N (at)	ρ	R (lat.par.)	Technique	Ref.
AgIn	1*	10^{-4}	0.290	573	0.61_2	1	10^{-4}	13.6	TDPAC	4
"	2	10^{-4}	0.165	573	0.88_2	1	10^{-4}	13.6	"	4
"	3	10^{-4}	1.000	673	0.53_4	1	10^{-4}	13.6	"	7
"	4	10^{-4}	1.000	773	0.49_4	1.7	5.8×10^{-5}	16.3	"	7
"	5*	10^{-2}	0.035	573	0.74_4	1	10^{-2}	2.9	"	4
"	6	10^{-2}	0.250	823	0.58_{11}^{12}	640	1.6×10^{-5}	25.2	"	4
AgSn	7*	10^{-2}	0.030	523	0.87_4	1	10^{-2}	2.9	Mossbauer	a
"	8*	10^{-2}	0.050	623	0.91_7	2	7×10^{-3}	3.7	"	a
"	9	10^{-2}	0.100	623	0.75_{18}	10	10^{-3}	6.3	"	a
"	10	10^{-2}	0.100	723	0.62_8	74	1.4×10^{-4}	12.3	"	a
"	11*	10^{-2}	0.125	823	0.60_9	670	1.5×10^{-5}	25.5	"	a
"	12*	10^{-2}	0.165	823	0.59_5	1000	10^{-5}	29.2	TDPAC	10

a: This work.

Figure 1 shows typical Mossbauer spectra obtained after cumulative oxidation treatments from one of the Ag with 1 at% Sn samples. The oxidation kinetics were obtained by measuring the percentage (F) of oxidized Sn atoms after an oxidizing treatment of duration t at the proper temperature. F was obtained from the reduction of the area of the nonoxidized Sn peak. Fig. 2 is a plot of lnF vs. lnt for the samples measured in the present work. It also includes TDPAC results from a 1 at% AgSn:In sample oxidized at 823K(10). The slopes m of the straight lines fitted to the data are listed in table I. Values of m obtained in a similar way from AgIn oxidized samples measured by TDPAC (4,7), can be also found in there.

Table II shows the measured isomer shifts of the oxidized Sn peak, referred to $CaSnO_3$ at room temperature, for the AgSn samples.

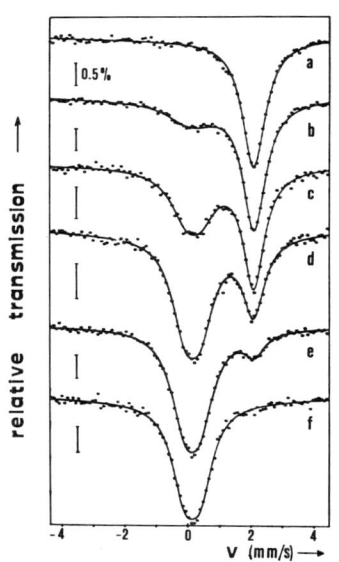

Figure 1.
Mössbauer spectra from sample 9 oxidized at 623K for: (b) 154, (c) 472, (d) 963, (e) 1629 and (f) 2259 minutes. Spectrum (a) corresponds to the non oxidized sample.

Table II.
Isomer shifts of the line corresponding to oxidized tin for the samples measured in this work. The values are referred to $CaSnO_3$ at room temperature.

Sample	Isomer Shift (mm/s)
7	0.30_1
8	0.22_1
9	0.16_1
10	0.08_1
11	0.01_1

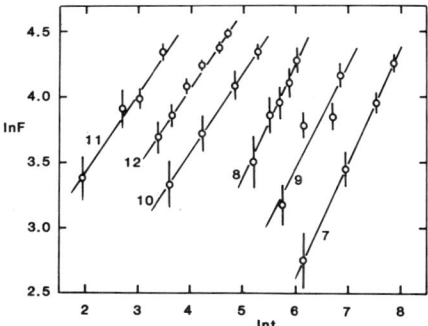

Figure 2.
Percentage F of oxidized Sn as a function of the oxidation time (in minutes) for the measured AgSn samples. Straight lines are least-squares fits to the experimental data. Sample numbers (see Table I) are indicated. Sample 12 from reference 10 has been also included.

DISCUSSION AND CONCLUSIONS.

The linear relationship between lnF and lnt evidenced in fig. 2 may by expressed by

$$F = k \cdot t^m, \qquad (1)$$

where k is a constant. According to Wagner's theory(1), if the oxidation process is controlled by diffusion, m=0.5. However, in most of the cases considered here, we found m>0.5. A quick look at table I reveals that m>0.7 for the more concentrated samples oxidized at the lower temperatures (with the exception of sample 2), i.e., when a higher density of small oxide particles inside the metal matrix can be expected.

Based on Ehrlich's results(2), it is possible to calculate the mean number N of solute atoms per oxide particle:

$$N(T) = K\{D_2(T) \cdot d \cdot (C_2)^{0.5}/(C_1(T) \cdot D_1(T))\}^{1.5}. \qquad (2)$$

C_1 is the oxygen concentration at the sample surface, C_2 is the initial solute concentration, D_1 and D_2 are the oxygen and solute diffusion coefficients in the alloy, d is the oxidized layer thickness, T the absolute oxidation temperature and K is a constant. As stated in references 4 and 8 for oxidations performed on 1 at% Sn or In silver based alloys at temperatures T < 573K, the TDPAC and Mössbauer results are coherent with the formation of single-solute atom-oxide complexes isolated from each other in the silver matrix. Hence, we have chosen a K value such that N=1 for C2=0.01 and T=573K. The N-values obtained in this way for the samples in table I are also listed there. Next, we calculated the mean density of oxide complexes as ρ =C/N. It's values, together with the mean separation between adjacent oxide particles can also be found in table I.

It should be mentioned that in the whole procedure, a homogeneous solute distribution has been assumed. This hypothesis is well supported by the Mössbauer and TDPAC spectra which never showed signals of solute clustering in the nonoxidized alloys(4-8). Nevertheless, in the case of Ag with 1 at% Sn alloys we pointed out(8) that an easier agglomeration of the solute oxide took place in the "as rolled" samples as compared to the ones subjected to a previous annealing treatment. Since this fact may account for a less homogeneous solute distribution in the "as rolled" samples, we obtained the N-values for these from their measured isomer shifts by using a N vs. isomer shift plot made with data from annealed specimens(8) (See fig. 3).

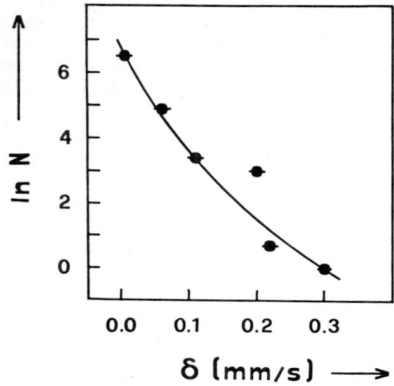

Figure 3.
The solid line shows the connection between the mean number N of tin atoms per oxide particle and the isomer shift (referred to $CaSnO_3$ at room temperature) of the oxidized Sn peak. (Data were taken from reference 8).

In the case of the diluted Ag:In alloys it was necesary to make some assumptions about the amount and nature of the existent impurities, in order to calculate N(T). As mentioned in reference 7, the In concentration in these alloys was lower than 5 ppm. However, the measured oxidation kinetics suggest an impurity concentration of about 100 ppm. According to the ASTM standards a 99.99% pure silver may have about 100 ppm of Cu and much lower amounts of other elements(11). Hence, we calculated N using this Cu concentration.

In fig. 4 we show the resulting relationship between the m-values and the density of oxide particles. It can be clearly seen that m decreases as soon as the distance between oxide complexes becomes longer than 3 or 4 lattice

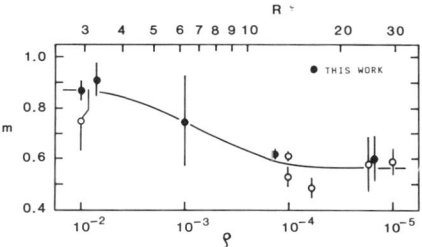

Figure 4.
Dependence of the oxidation kinetics parameter m (see eq.(1)) on the density ρ of oxide particles. R is the mean interparticle distance in lattice parameters.

parameters. The only exception to this rule is for sample 2 (not included in fig. 4), which will be commented on later. For even longer interparticle distances the m-value approaches the 0.5 one expected from the Wagner's theory. We interpret this situation as follows:

(i) When one-solute atom-complexes are formed, the oxygen incorporation in the neighborhood of the solute atoms may occur in a variable stoichiometry as evidenced by TDPAC measurements(7) and with an oxygen/solute atomic ratio higher than in normal oxides(12). The oxide complexes should then compress the lattice producing inhomogeneous elastic fields around them. When the complexes are 3 or 4 lattice parameters far from each other it is possible that a network of channels for easy oxygen migration be set up. The effect of these "channels" would be to allow oxygen to reach the nonoxidized alloy at nearly constant rates, giving rise to almost linear kinetics. It is worth mentioning that elastic inhomogeneous fields were observed and considered responsible for the anomalous O2 diffusion found in an oxidized Ag with 0.33 at% Sn alloy(12).

(ii) When the oxide is mainly in the form of precipitates or the alloys are very diluted, the oxide complexes are further apart from each other. Hence, no short circuit paths for oxygen diffusion can occur and the oxidation kinetics are closer to the parabolic one.

(iii) The consequence of cold work on the samples is the creation of a high density of dislocations, grain boundaries and other defects. It is known that grain boundary diffusion is particularly important at low temperatures. This may be the cause for the abnormally high kinetic rate observed for sample 2. One can wonder why grain boundary diffusion did not affect other "as rolled" samples. In first place, we can argue that samples oxidized at temperatures T>773K (T>0.6xMelting Temperature) recovered during the oxidation treatments. Secondly, it has been shown(13) that solute segregation to grain boundaries occurs in AgSn alloys at rather low temperatures, and a similar process may occur in the AgIn ones. Because the grain boundary solute concentration may be more than two orders of magnitude larger than the bulk one, one can expect the grain boundaries to become blocked with solute oxide for 1 at% alloys, and to no longer contribute appreciably to oxygen diffusion. Finally, the thickness reduction of sample 3 was probably too small to affect its oxidation kinetics in a measurable way.

These results confirm that it is possible to modify the internal oxidation kinetics, by controlling the density of oxide particles. Hence, they indicate that the oxygen diffusion process can be altered in that way. Alloys which undergo internal oxidation, will become largely stressed when the density of oxide particles is high. In this regard, special attention should be paid in alloys with more than 0.3 at% of highly immobile solutes, as they would produce a high density of one-solute atom-oxide complexes, turning the materials harder and brittler. We believe that it will be interesting to perform diffusion studies of other elements besides oxygen, to see how the strain fields created in internally oxidized alloys affect the process.

ACKNOWLEDGMENTS

We wish to thank Dr.J.I.Budnick and Dr.F.Namavar for their useful comments on the manuscript.

REFERENCES.

(1) C.Wagner, J.Electrochem.Soc. 63(1959)777.
(2) A.C.Ehrlich, J.Mat.Sci. 9(1974)1064.
(3) M.F.Ashby and G.C.Smith, J.Inst.Metals 91(1963)182.
(4) J.Desimoni, A.G.Bibiloni, L.Mendoza-Zelis, A.F.Pasquevich, F.H.Sanchez, and A.Lopez-Garcia, Phys.Rev. B28(1983)5739.
(5) A.F.Pasquevich, F.H.Sanchez, A.G.Bibiloni, C.P.Massolo, and A.Lopez-Garcia, in Nuclear and Electron Resonance Spectroscopies Applied to Materials Science, edited by E.N.Kaufmann and G.K.Shenoy,(NorthHolland, Amsterdam, 1981), p.435.
(6) A.F.Pasquevich, A.G.Bibiloni, C.P.Massolo, F.H.Sanchez, and A.Lopez-Garcia, Phys.Lett. 82A((1981)34.
(7) A.F.Pasquevich, F.H.Sanchez, A.G.Bibiloni, J.Desimoni, and A.Lopez-Garcia, Phys Rev. B27(1983)963.
(8) F.H.Sanchez, R.C.Mercader, A.F.Pasquevich, A.G.Bibiloni and A.Lopez-Garcia, Hyp.Interactions 20(1984)295.
(9) F.H.Sanchez, R.C.Mercader, A.F.Pasquevich, A.G.Bibiloni and A.Lopez-Garcia, to be published as a short note in Phys.Stat.Sol.(a)
(10) J.Desimoni, private communication.
(11) 1984 annual book of ASTM standards. Sect.2, Non Ferrous Metal Products. Vol.02.04, p B413. American Society for Testing Materials, 1984.
(12) G.P.Huffman and H.H.Podgurski, Acta Metall. 21(1973)449.
(13) J.Bernardini, P.Gas, E.D.Hondros, and M.P.Seah, Proc.Roy.Soc. London, A379(1982)159.
(14) John Askill. Tracer Diffusion Data for Metals, Alloys and Simple Oxides, IFI/Plenum. New York-Washington-London. 1970.
(15) P.Gas and J.Bernardini, Scripta Metallurgica 12(1978)367.
(16) W.Eichenauer and G.Muller, Z.Metallkd. 53(1962)321; 53(1962)700.

HIGH TEMPERATURE RAMAN STUDIES OF PHASE TRANSITIONS IN THIN FILM DIELECTRICS

GREGORY J. EXARHOS
Pacific Northwest Laboratory,* Richland, Washington 99352

ASTRACT
   Rapid and unambiguous characterization of crystalline phases in submicron sputter deposited TiO_2 films on silica substrates can be inferred from measured Raman spectra. Pure anatase and rutile, mixed phase, and amorphous films to thicknesses of several hundred Ångstroms and greater yield Raman spectra exhibiting little interference from the substrate when the appropriate component of the scattered light is analyzed. In situ Raman spectra were acquired as a function of temperature to 900°C using conventional radiant heating techniques and to temperatures near 2000°C using 10.6μ radiation from a CW CO_2 laser as a localized heating source. Pulsed Raman excitation/gated detection techniques were used to minimize blackbody radiation interference at these high temperatures. Anatase and amorphous TiO_2 films transform irreversibly to the rutile phase at temperatures below 900°C while rutile appears to be stable at much higher temperatures. Measurements performed on uncoated silica substrates at temperatures where the glass becomes fluid suggest that the strongly crosslinked glass has partially transformed into a chain-like structure.

INTRODUCTION
   The atomic composition of reactively sputtered thin dielectric films on silica substrates is readily discernible by a variety of surface spectroscopic methods, however, phase characterization usually requires diffraction techniques which probe localized and extended chemical bonding in the material. A primary structural probe of such films has been x-ray diffraction. For thin films of low Z material, signal acquisition times can be appreciable and in many cases, little structural information is afforded by these techniques. Vibrational spectroscopic probes can provide localized bonding information and unambiguously characterize particular crystalline and amorphous phases having identical atom compositions. The speed and adaptability of laser Raman measurements to thin film characterization make it an attractive probe for in situ film stability studies.
   This non-destructive measurement is well suited to thin film characterization where advantage may be taken of optical interference phenomena which can considerably enhance signal strengths.[1,2,3,4] Appropriate choice of scattering geometry can suppress interferences from the substrate allowing direct measurement of the deposited film.[5] Low Z materials such as TiO_2, SiC, BN, or Si_3N_4 have relatively large Raman cross sections making them good candidates for analysis by this method.
   Temperature induced structural changes in sputter deposited TiO_2 films have been observed by in situ Raman spectroscopic methods. The behavior of anatase, rutile, and amorphous TiO_2 to temperatures in excess of 1500°C has been characterized by Raman spectroscopic measurements. A two laser technique was used involving focused 10.6μ radiation from a CW CO_2 laser as the localized heating source and colinear pulsed .532μ radiation from a Nd:YAG laser as the Raman probe. Pulsed gated detection techniques were used to suppress blackbody radiation which otherwise would overwhelm the Raman signal at these temperatures. While rutile appeared to be stable to

---

*Pacific Northwest Laboratory is operated by Battelle Memorial Institute for the U.S. Department of Energy under contract DE-AC06-76RLO 1830.

temperatures in excess of 1500°C, the other phases transformed irreversibly to a rutile phase at much lower temperatures. High temperature vibrational measurements on bare silica substrates provide evidence for subtle structural changes as the glass melts.

EXPERIMENTAL

All dielectric coatings used in this investigation were prepared by reactively sputtering Ti in Ar/O_2 atmospheres in an rf diode system onto fused silica substrates. Experimental parameters used to control sample thickness and phase appear in the literature.[6,7] Samples investigated by Raman scattering techniques ranged in thickness from several hundred Angstroms to over 5μ.

Raman spectra were excited at normal incidence using the 180° backscattering geometry depicted in Figure 1. In most cases, the perpen-

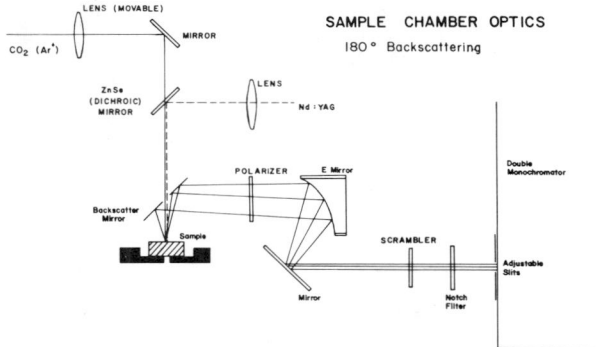

Figure 1. Raman scattering geometry.

dicular component, $Z(XY)\bar{Z}$, of the scattered light was analyzed in order to suppress Raman scattering from the silica substrate. 5] Scattered radiation was collected at f/1.4 and imaged onto the slits of a SPEX 0.85m double monochromator. Slit widths were maintained at 80μ. Conventional photon counting electronics were used for signal detection or a gated intensified diode array detector was used to record spectra in a time resolved mode. This required using a notch rejection filter centered at the Raman probe wavelength which also effectively rejected the first 350 cm^{-1} of the Raman spectrum.

Equilibrium measurements as a function of temperature below 1000°C were acquired with CW 488 nm Ar+ excitation of samples mounted in a resistively heated furnace. For measurements above 1000°C, 10.6μ CW radiation from a CO_2 laser (localized heating source) was combined with low energy pulsed 532 nm Raman probe radiation from a Nd:YAG laser. Sample temperatures over the heated area were determined from a two color optical pyrometer, and a gated detection scheme served to suppress blackbody emission from the Raman signal.

RAMAN CHARACTERIZATION OF THIN FILMS

Anatase and rutile phases of TiO_2 sputter deposited films exhibit vibrational features at 143, 395, 517, 636 cm^{-1} and 235, 440, 607 cm^{-1} respectively in good agreement with measured frequencies from bulk samples.[8] The perpendicular component of the scattered light is analyzed, particularly for the thinnest films where scattering from the silica substrate can be appreciable. Raman scattering from polycrystalline films is isotropic whereas scattering from the substrate (below 800 cm^{-1}) is highly polarized.[5] While vibrational frequencies identify particular phases, band intensities are proportional to the amount of material present.[8] Figure 2 shows the Raman band intensity dependence for pure

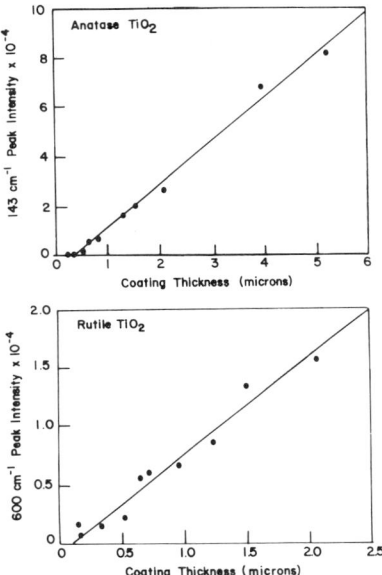

Figure 2. Thickness dependence of principal Raman peak intensities.

rutile and anatase phases as a function of film thickness determined from transmission measurements. For particular instrumental parameters, excellent linearity is observed over a wide dynamic range. (Under these conditions, the signal/noise ratio becomes small for coatings less than 2000Å thick.) The detection limit (500μ slits) is estimated to be a film having a thickness of 100Å. Observation of Raman scattering in much thinner films is possible using multilayer interference enhancement techniques.[1,2]

Mixed phase coatings have also been prepared and phase compositions can be discerned by reference to Figure 3 which shows marked spectral changes as a function of relative anatase/rutile content. Trace amounts of anatase in rutile films (.1 at %) can be determined by using the 143 cm^{-1} anatase feature which is about an order of magnitude stronger than other features in the anatase and rutile spectra.

A third TiO_2 phase was deposited on silica which gave a diffuse x-ray diffraction pattern. A 1μ thick coating yielded a weak Raman spectrum having two broad features at 440 and 600 cm^{-1}. The similarity of this spectrum to that of rutile suggests that both materials contain the same localized structural groups (near octahedral coordination of titanium with oxygen) but that long range order is absent in the amorphous phase. Oxygen deficient coatings have also been prepared and exhibit Raman features different from those observed for phases of TiO_2 stoichiometry. Broad weak features at 150, 420, and 610 cm^{-1} are observed. This phase apparently contains vestiges of both anatase and rutile phases in addition to other components such as TiO which have not as yet been identified.

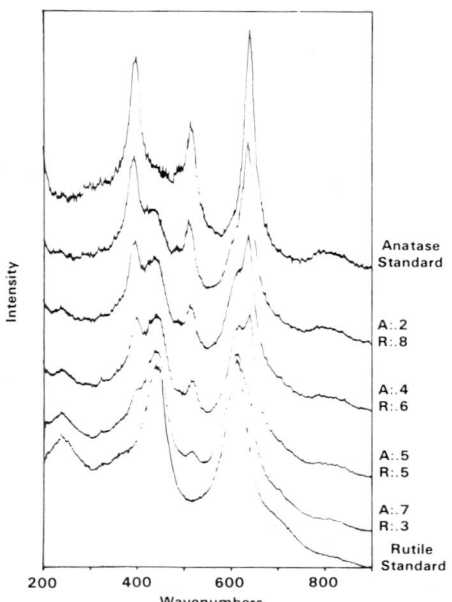

Figure 3. Raman spectra of mixed anatase-rutile phases.

HIGH TEMPERATURE RAMAN MEASUREMENTS

In situ vibrational measurements were obtained for several phases of TiO_2 deposited on silica as a function of sample temperature. Features observed in the amorphous TiO_2 coating broadened and increased in intensity as the temperature was raised as seen in Figure 4. At 800°C, two major features characteristic of the rutile phase at ca 440 and 600 cm^{-1} were observed which persisted as the sample was cooled to room temperature. After heat treatment the film remained intact, however, it appeared translucent following irreversible transformation to the rutile phase.

Similar irreversible phase transitions were observed in both single and multilayer anatase coatings as shown in Figure 5. Irreversible crystallization to the rutile phase initiated at ca 800°C and was complete at 900°C. The feature near 800 cm^{-1} is assigned to a vibrational mode of the silica substrate which becomes evident as fissures develop in the coating. Cracking results from the large volume change which accompanies the phase transformation. At higher temperatures (1560°C), the rutile lines shift to lower frequencies as discussed below.

Rutile coatings are stable with regard to phase transitions at temperatures approaching 1600°C. However, significant band shifts to lower frequency were measured. For instance, the a_{1g} mode at 610 cm^{-1} at room temperature shifts to 530 cm^{-1} at 1560°C and the e_g mode shifts from 440 to 410 cm^{-1}. Upon cooling to 25°C, the initial room temperature spectrum is recovered. Severe damage to the film is evident and could result from a significant amount of thermally induced stress or increased interaction with the heated substrate.

Raman spectra of a silica glass substrate have been acquired at temperatures in excess of 2000°C. Figure 6 compares the Bose-Einstein temperature corrected spectra at 25°C and 2050°C. To minimize volatilization and decomposition at these temperatures, the sample was flooded with oxygen. Raman features at 490 and 602 cm^{-1} show little

465

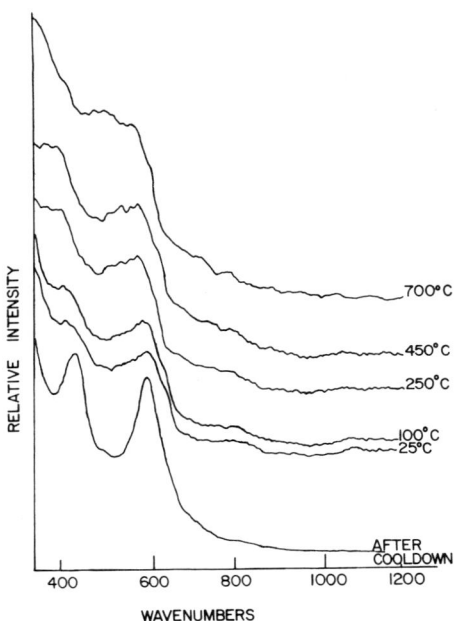

Figure 4. Raman spectra of an amorphous TiO_2 film as a function of temperature.

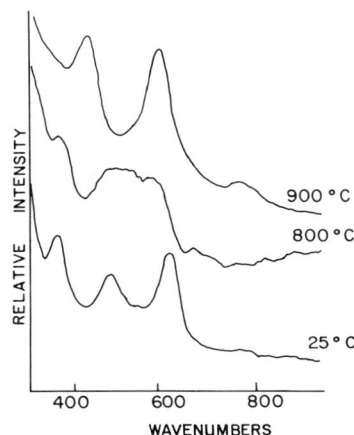

Figure 5. Raman spectra of an anatase multilayer coating on silica as a function of temperature.

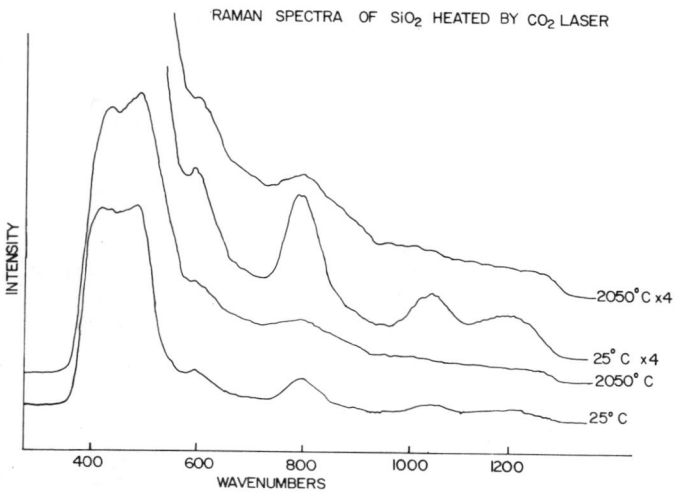

Figure 6. Raman spectra of a silica glass substrate at 25°C and 2050°C.

frequency shift with temperature although some band broadening is evident. The 800 cm^{-1} band broadens and shifts to lower frequency; a broad shoulder is evident in the high temperature spectrum at 880 cm^{-1}. The 1060 cm^{-1} band shifts to 990 cm^{-1} and broadens considerably while the 1200 cm^{-1} band broadens and shifts to higher frequency. The high temperature spectrum is representative of silica in the liquid state.

DISCUSSION

The relative instability of anatase films near 800°C is explained by a thermally induced irreversible phase transition to rutile which is in agreement with thermodynamic predictions. The higher density rutile phase stresses the thin film causing microcracking which leads to catastrophic failure of the film. Such effects are observed for single layer films as well as multilayer structures consisting of alternating anatase/silica layers.

Amorphous TiO_2 films crystallize into a rutile phase with increasing temperature. Figure 7 shows the intensity dependence of the 600 cm^{-1} band in the amorphous phase as a function of temperature. The intensity increase is indicative of recrystallization phenomena since Raman intensities normally decrease with increase in temperature due to Stokes, anti-Stokes partitioning.[9] The kinetics of the transformation is thermally controlled, however, the onset of crystallization appears to be near 200°C. The recrystallized film exhibits a two order intensity increase of the 600 cm^{-1} band over that from the amorphous film.

Rutile films undergo no detectable phase change at temperatures up to 1560°C. Significant Raman band shifts to lower frequency indicate a weakening of the bond force constants resulting from thermal excitation and associated bond length increases. At these temperatures, the silica substrate also begins to soften and flow. Following cooling, the rutile phase is maintained but the film has suffered damage and appears opaque probably due to partial dissolution of the coating in the high temperature silica substrate.

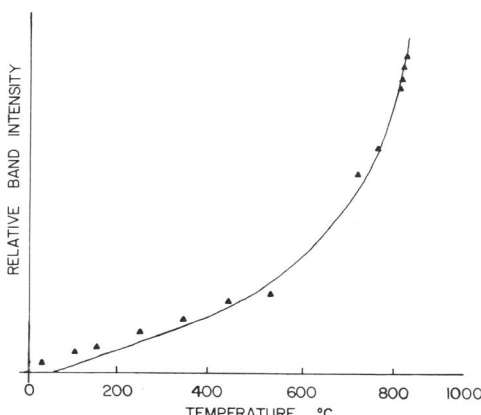

Figure 7. Raman intensity of the 600 cm^{-1} band of an amorphous TiO_2 film as a function of temperature.

Marked vibrational band changes observed for heated silica substrates when compared with the room temperature Raman spectrum suggest that the glass structure has been reorganized in the liquid state. The 490 and 602 cm^{-1} bands which have been assigned to three and four fold silica ring structures[10] persist at high temperatures suggesting that these structures are still present in the high temperature melt. Although vibrational modes in the 800-1200 cm^{-1} region are dominated by Si-O stretching[11], changes in the O-Si-O bond angle will act to perturb these frequencies as seen in radiation damaged SiO_2 glass.[12] A decrease in frequency corresponds to an increase in bond angle as observed for the 800 and 1060 cm^{-1} modes while the frequency increase of the 1200 cm^{-1} mode at 2050°C suggests a bond angle closing. These effects act to distort the regular tetrahedral geometry of individual SiO_4 units in the glass. In the extreme case, a linear structure could result. Appearance of a feature near 900 cm^{-1} in the 2050°C Raman spectrum of silica provides evidence for such linear structures since it is observed at room temperature in silicate glasses containing chain-like silicate anions.[11]

CONCLUSIONS

Raman spectroscopy is a sensitive non-destructive technique for phase characterization of thin dielectric films on silica under ambient conditions and during high temperature thermal treatment. Anatase and amorphous TiO_2 films are observed to undergo irreversible phase transformations to a rutile phase at temperatures below 800°C. At significantly higher temperatures, the substrate softens and exhibits marked Raman spectral changes which suggests that partial distortion of localized silicate tetrahedra is thermally induced to form chain-like structures.

Raman measurements can provide detailed information about the film/substate interface. Recent measurements on thick (8573Å) and thin (780Å) anatase coatings on silica reveal subtle changes in the Raman spectra. The 143 cm^{-1} feature increases in bandwidth by 40% for the thin coating. In addition, the 635 cm^{-1} band broadens to the low frequency side and the 395 cm^{-1} band broadens to the high frequency side. These observations suggest that traces of a rutile phase are present in the thin coating and this phase most likely is formed at the interface. For antase films on silica, then, a thin rutile layer acts to bond the film to the substrate.

Raman measurements for other sputter deposited thin films such as Si_3N_4 are currently in progress. Comparison between Raman spectra of an amorphous Si:N phase containing hydrogen and crystalline beta silicon nitride are shown in Figure 8. The broad feature at 200 cm^{-1} forms an

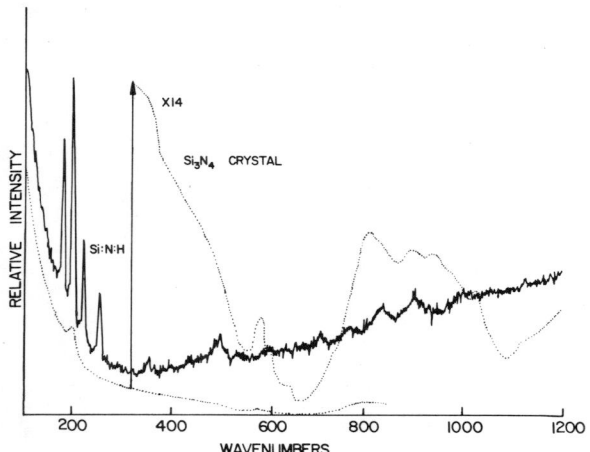

Figure 8. Raman spectrum of a sputter deposited Si:N:H film on silica compared with that from beta silicon nitride.

envelope around the crystalline silicon nitride lines at 183, 202, and 225 cm^{-1} suggesting lattice disorder and an amorphous structure. Additional broad features around 900 cm^{-1} have been observed in silicon oxynitride glasses and indicate a degree of oxygen contamination in the film.

The <u>in situ</u> capability of Raman spectroscopy to characterize thin sputter deposited films rapidly makes it an attractive probe for materials identification and reliability studies.

ACKNOWLEDGEMENTS

This work has been supported by the Air Force Weapons Laboratory under contract PO-84-004. Optical coatings have been supplied by Dr. W. T. Pawlewicz and Dr. P. M. Martin who are also thanked for helpful discussions regarding this work. The author also wishes to acknowledge Dr. M. R. Fischer, C. H. Nguyen, and P. F. Stevens for assistance with experimental measurements.

REFERENCES

1. R.J. Nemanich, G.A.N, Connell, T.M. Hays, and R.A. Street, Phys. Rev. B., <u>18</u>, 6900-6914 (1978).

2. R.J. Nemanich, C.C. Tsai, and G.A.N. Connell, Phys, Rev. Lett, <u>44</u>, 273-276 (1980).

3. W.T. Pawlewicz, G.J. Exarhos, and Conaway, W.E., Applied Optics, 22(12), 1837-1840 (1983).

4. G.J. Exarhos and W.T. Pawlewicz, Applied Optics 23(12), 1986-1988 (1984).

5. G.J. Exarhos, J. Chem. Phys., 81(11), 5211 (1984).

6. W.T. Pawlewicz, P.M. Martin, D.D. Hays, and I.B. Mann, Proc. Soc. Photo-Opt. Instrument, Eng., 325, 105-116 (1982).

7. W.T. Pawlewicz, D.D. Hays, and P.M. Martin, Thin Solid Films, 73, 169-175 (1980).

8. R.J. Capwell, K. Spagnolo, and M.A. DeSesa, Appl, Spectrosco., 26, 537-539 (1972).

9. G.J. Exarhos and W.M. Risen, J. Am. Ceram. Soc. 57(9), 401-408 (1974).

10. F.L. Galeener, J. Non-cryst. Sol., 49, 53-62 (1982).

11. P. McMillan, Am. Mineralogist, 69, 622-644 (1984).

12. G.J. Exarhos, Nucl. Ins. and Methods in Phys. Res. 299(B1)2,3, 498-502 (1984).

THE USE OF ENERGY LOSS STRUCTURES IN
XPS CHARACTERISATION OF SURFACES

J.E. CASTLE, I. ABU-TALIB AND S.A. RICHARDSON
The University of Surrey, Guildford, England.

ABSTRACT

This paper describes advances in the use of the energy loss background associated with individual photoelectron peaks. The subtraction of a Shirley-type background is now normal practice in quantitative XPS analysis. However, in the case of a composite peak containing features from differing depths the subtraction of a common background has a clear disadvantage: i.e. the proportion of background rise associated with each component should be different but is, in fact, fixed. A peak-fitting procedure is described which enables individual backgrounds to be used for each component. The method has been tested using evaporated overlayers and this enables a mean free path for electrons undergoing small energy losses (less than 10 eV) to be determined. The findings are in accord with those of Tougaard and Sigmund and suggest that the use of background intensities in conjunction with the peaks themselves enables the information depth of XPS to be extended by about 10%. A few observations on the behaviour and use in analysis of the large energy loss structure are made.

The use of the findings to aid in characterisation of the near surface distribution of elements and ions is described for the following systems: the distribution within oxide films on alloys; the locus of disbondment of organic films on metals; and the surface contamination of surfaces removed from aqueous media.

INTRODUCTION

A characteristic feature of X-Ray Photoelectron Spectroscopy (XPS) is the energy loss background associated with each individual photoelectron peak. The typical spectrum in Figure 1 illustrates the manner in which the background builds up as a series of steps or edges originating at each peak position. This background arises from those photoelectrons which have lost energy within the solid and are shifted to a lower kinetic energy in the spectrum. In some cases the energy is lost in discrete amounts such as by the excitation of plasmons or by shake-up processes and these cause a peak in the loss structure. This paper is not concerned with such characteristic losses, however, but with the smooth continuum on which the photopeaks sit. This continuum must be removed for quantitative analysis, sometimes by using a straight line to approximate the background rise across the peak but usually by the better approximation in which the background rise is assumed to be proportional to the peak intensity above it: the "Shirley-type" background[1].

In a recent paper[2] we pointed out that this procedure is inaccurate for a composite peak containing features from differing depths, e.g. for the native oxide on a metallic substrate. Figure 2 illustrates the situation for the oxide on aluminium: clearly the background associated with the chemically shifted aluminium ion is limited by its small thickness whereas that of the metal is not. The simple application of a Shirley background gives them in fact equal weighting.

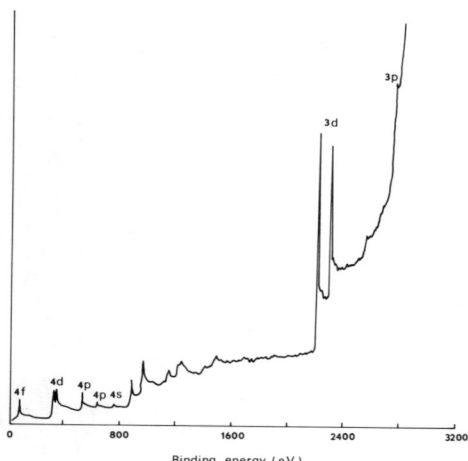
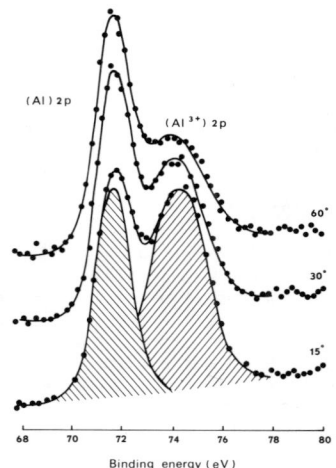

Figure 1. The Photo-Electron Spectrum of Gold (in silver L alpha radiation). Note the Step-Like Increase in Background

Figure 2. Spectrum of Native Oxide on Aluminium Foil. The use of a Standard 'Shirley' background would give too much weight to electrons scattered from the oxide.

   A peak-fitting procedure has been developed which enables individual backgrounds to be used for each component[2]. The effect of using this method on the complex chromium 2p peak is shown in Figure 3a. This best fit obtained by the computer iteration has ascribed a much smaller background "tail" to the oxide than to the metal. We find that the tail height associated with the clean chromium peak is 28% of the peak height whereas that associated with the oxide overlayer is only 4%. As the oxide increases in thickness the tail height increases but for thicknesses of oxide up to about 2nm the metal tail remains approximately constant at 30%. The resultant peak and background structure for chromium with a thicker oxide is shown in Figure 3b.

   The general dependence of the intensity of the energy loss structure for chromium has been examined using evaporated overlayers of titanium[2]. This enabled a mean free path for electrons undergoing small energy losses (less than 10 eV) to be determined. The findings are in accord with those of Tougaard and Sigmund[3]. They suggest that this energy loss mean free path is about ten percent greater than the normal inelastic mean free path. This order of difference between the attenuation of the photopeak itself and of the loss-electrons within about 10 eV of the peak is probably constant across the whole of the spectrum[4]. It is not very large: for example the change in inelastic mean free path consequent on changing between the two usual photon sources Mg and Al k-alpha would be greater than this for all peaks of greater binding energy than the carbon 1s level.

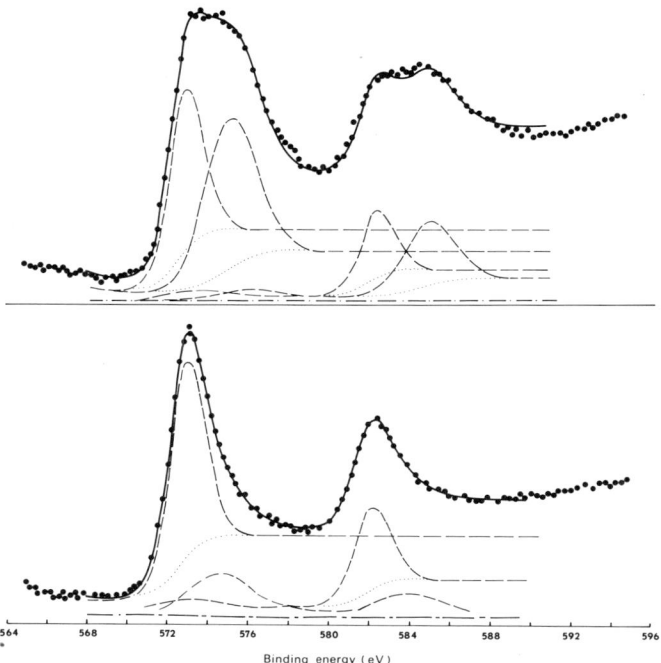

Figure 3a (lower) Slightly Oxidised Chromium. The rise in background intensity is associated only with the metal peak.

Figure 3b (upper) Oxidised Chromium. Both oxide and metal contribute to the rise in background although that on the metal is greater in proportion to its peak area.

The attenuation of loss-electrons does, however, spread across a significant part of the spectrum, a surface film of only 3 nm will reduce the electron signal over a range of 200 eV, and this effect is easily observable in the "wide" or survey scan frequently used by spectroscopists as has been pointed out by Proctor and Hercules[5]. In this paper we extend the investigation made previously to these post-peak loss structures (P-PS).

These structures provide a method to aid in characterisation of the near surface distribution of elements and ions which is particularly useful when ion-etching is undesirable. The effect will be described using the distribution within oxide films on alloys; the locus of disbondment of organic films on metals: and the surface contamination of surfaces.

SAMPLE PREPARATION

The detailed method of sample preparation has been described in a previous paper[2]. Briefly, a sample of pure chromium was etched by argon ions in the XPS spectrometer of an ESCA 3 instrument manufactured by VG Scientific Ltd. Overlayers of titanium were then produced by successive evaporations and the thickness estimated by means of the changes in relative intensities of the Ti2p and the Cr2p peaks respectively. The thicknesses used are given in Table 1. The spectra were recorded over regions of interest and were transferred by a direct interface to a Prime 750 computer which was used for background subtraction and for peak synthesis.

ELECTRON SPECTRA

The behaviour of the energy-loss spectrum over a range of 80eV beyond the peak is shown in Figure 4. Over this wider range the increase in background height associated with the near peak structure is seen to be far less important than the increase in slope at 50eV from the Cr2p peak (Table 1).

TABLE 1. The Values of the P-PS Slope in the Energy Loss Structure

Deposit Thickness	Increase at Peak (%)	Post-Peak Slope (%/eV)*	
		Cr	Ti
0	29	-0.14	-
1.85 (nm)	30	+0.22	-0.5
2.88 (nm)	34	+0.51	-0.5
3.34 (nm)	50	+1.04	-0.45

* The P-PS Values are measured as a % of the peak height: they have not been corrected for the transmission function of the instrument which is ca. $E^{-0.5}$.

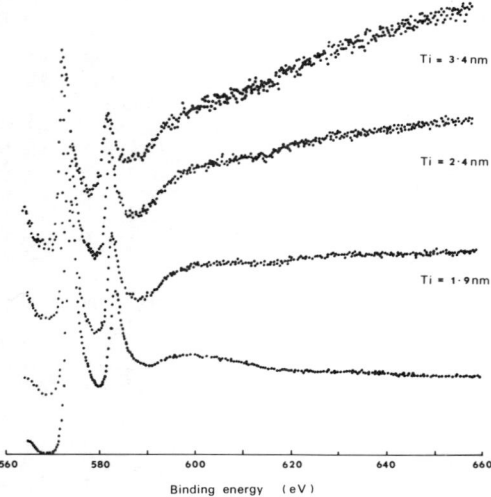

Figure 4. The Post-Peak Energy Loss Structure of Chromium 2p. There is a steady increase in slope with increase in overlayer thickness.

The peaks from the overlayers of titanium which are responsible for
the development of this P-PS structure on the chromium are shown in Figure 5.
These do not show any change in slope or background height with increase
in thickness. The energy loss structure is thus fully developed at a
thickness of approximately 2nm or twice the inelastic mean free path.
This difference in background slope between overlayer and substrate is
thus able to differentiate the relative positions in depth of individual
elements. An indication that this effect extends beyond the depth
normally accessible by XPS is given by the spectrum in Figure 6. This is
taken from a chromium surface covered with evaporated aluminium almost
sufficient to obscure the Cr2p peaks yet the abrupt change of slope at the
peak position stands out as a signal that chromium is present. Moreover,
whereas a direct analysis of the peak areas gives only the apparent atomic
ratio, Cr/Al = 1:32, it is the relative slopes that gives the information
that this is not an atomic mixture but a distinct layer sequence.

Further examples of the importance of this background structure to
the characterisation of the near-surface chemistry are given in Figures
7-10. Figure 7 shows the steady increase in slope associated with the
chromium peak as carbon contamination from an evaporation source builds up
on the surface. Notice that the carbon peak has the decreasing slope

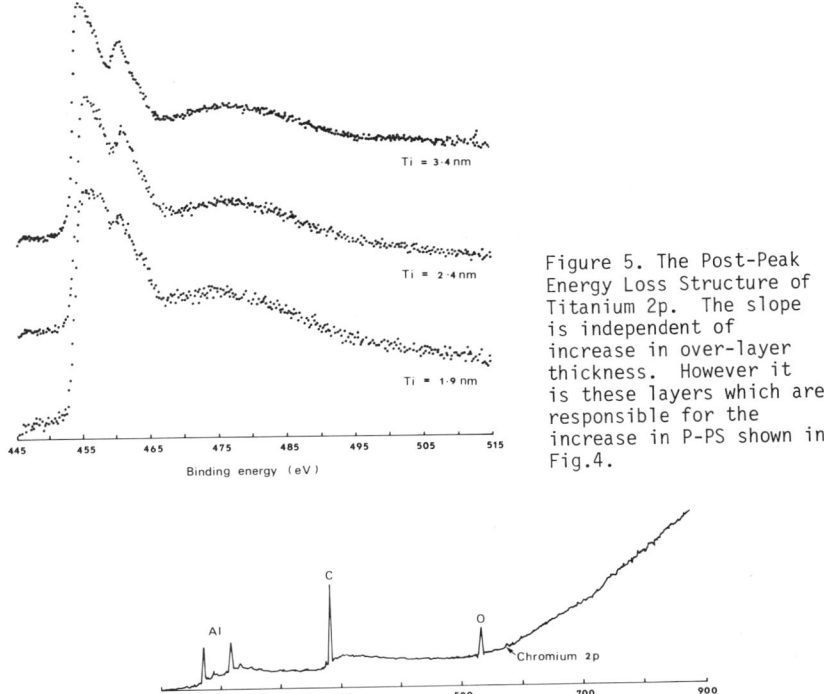

Figure 5. The Post-Peak Energy Loss Structure of Titanium 2p. The slope is independent of increase in over-layer thickness. However it is these layers which are responsible for the increase in P-PS shown in Fig.4.

Figure 6. A Spectrum of Chromium with a Heavy Over-layer of Aluminium. The position of the Cr2p peak is only made apparent by the change in slope.

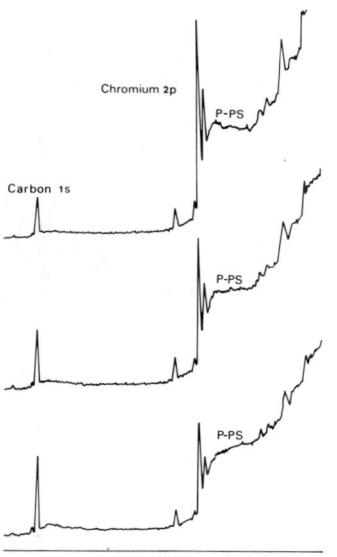

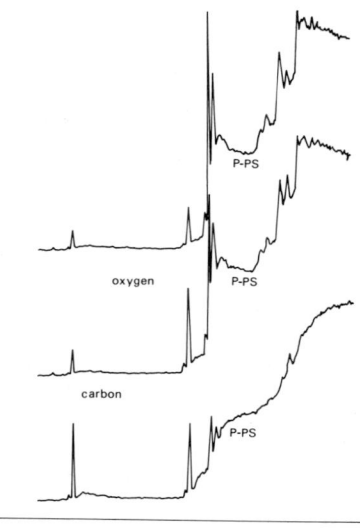

Figure 7. Spectra of Chromium. Note the change in the post-peak slope (P-PS) as carbon contamination accumulates on the surface.

Figure 8. Spectra of Chromium. Note the difference between reacted and unreacted overlayers: the oxygen uptake (centre) has changed the height and shape of the peak but has not altered the P-PS; Carbon deposition on top of oxide increases the P-PS of both the chromium and the oxygen peaks (lower spectrum).

which characterises the surface species. Figure 8 illustrates the difference between unreacted surface carbon and reacted surface oxygen: here the presence of a high surface concentration of oxygen (center) has not altered the post-peak slope of the clean chromium (upper) showing that chromium is included in the atomic mixture of the outer layers. By contrast the presence of carbon on the surface increases the post-peak slope of both oxygen and chromium. It is this sequence of slope changes that often characterises a dirty or contaminated surface but this information is also precisely that which is required when examining processed surfaces e.g. the locus of failure of a bonded joint. It can easily be lost by over-enthusiastic resort to the ion gun in order to "clean-up" the surface prior to examination.

In Figure 9 is shown the spectrum of a sample of a copper-nickel-iron alloy which has been held at a controlled electropotential in raw seawater at 40°C. There is some carbon up-take but this seems not to have displaced the alloying elements from the surface hierarchy: note that the post-peak slopes on each of the elements are approximately equal showing them to be present as an intimate mixture rather than as a layered sequence. The lower spectrum in Figure 9 is of a similar sample which received the same treatment but at the lower temperature of 20°C. In this case the rate of generation of electrochemical products has been lower and in consequence a surface film of contamination has formed and the post-peak slopes have increased.

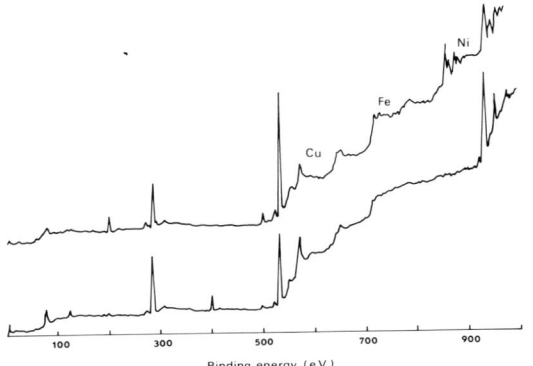

Figure 9. The practical use of P-PS in the corrosion of Cu-Ni-Fe alloy: The P-PS of each element is the same showing that each is present in the outer surface after exposure at 40°C (upper) but not at 20°C (lower spectrum)

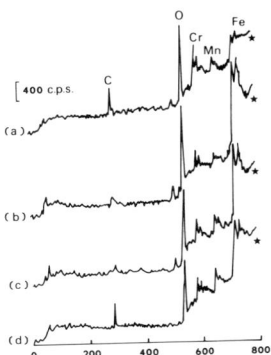

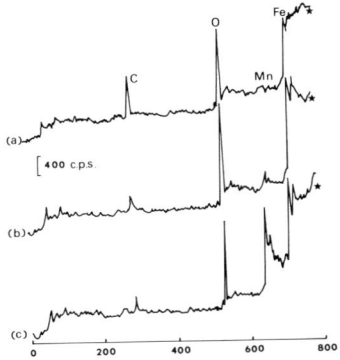

Figure 10(a) Nickel Steel. Initially (a) iron is covered with carbon contamination as shown by the increasing P-PS value (*). On oxidation (b) iron becomes the surface species but finally (c) on vacuum heat treatment is overgrown by manganese.

Figure 10(b) Chromium Steel. After vacuum heat-treatment iron is overgrown by chromium and manganese as shown by the P-PS (*) of curve (a). On oxidation iron moves into the surface (b) and is then stable on further oxidation (c) or heat-treatment (d).

Information on the layer sequence of metal oxides on alloy steel is readily obtained by interpretation of the post-peak slopes. An illustration is taken, Figure 10, from a paper on the oxidation of steels[6] in hydrocarbon-rich atmospheres. On the nickel steel the post-peak slopes show that, as prepared by abrasion, the steel has an initial covering of carbon and oxygen contamination. On heating in carbon dioxide, iron oxide forms as the surface phase and finally, on vaccum heat treatment, manganese oxide forms on the surface and overlays the iron substrate. The similar sequence on the chromium steel shows that after vacuum heat treatment both chromium and manganese oxides overlie iron oxide but on further oxidation iron oxide reforms leading to a reversal of the post-peak slopes of iron and chromium. Finally the spectra in Figure 11[7] distinguish between cohesive (upper) and adhesive failure of a polymer (lower). Note the presence of iron and oxygen in the surface of the adhesive failures whereas both are covered by residual polymer in the cohesive failure.

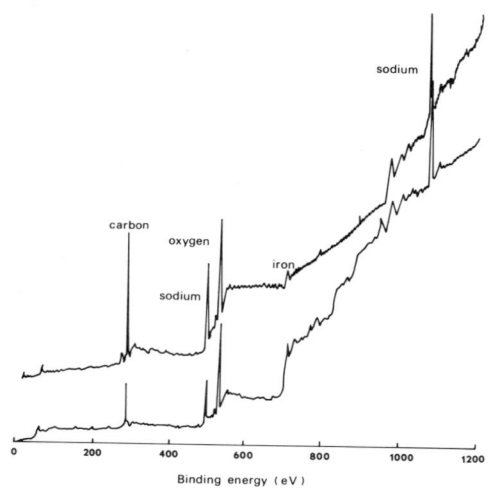

Figure 11. The Disbondment of Polymer Coatings: The presence of an overlayer of carbon in the locus of failure in the upper curve is shown by the increased P-PS on the Na, O, and Fe peaks; the truely adhesive failure is shown in the lower curve where all the P-PS values are lower.

WIDE SCAN BACKGROUND STRIPPING

The information in the background could perhaps be made more readily available by suitable manipulation of the spectrum and initial steps have been taken to achieve this. The Shirly background function used to strip the individual peaks can also be used across the entire spectrum (although the iterations converge more rapidly if this is done in at least two parts). In the final Figure (12) we return to the data obtained by evaporating titanium onto chromium. These show the combined peak and background for the clean chromium merged with the backgrounds of the subsequent substrates. This presentation focusses well on the increasing slope of the background and could be a useful method of enhancing information available from XPS. The stripped peaks also give a better focus on the elemental concentrations than is usually obtained from a set of wide scan spectra in which peak intensity and energy loss data overlap and sometimes interfere.

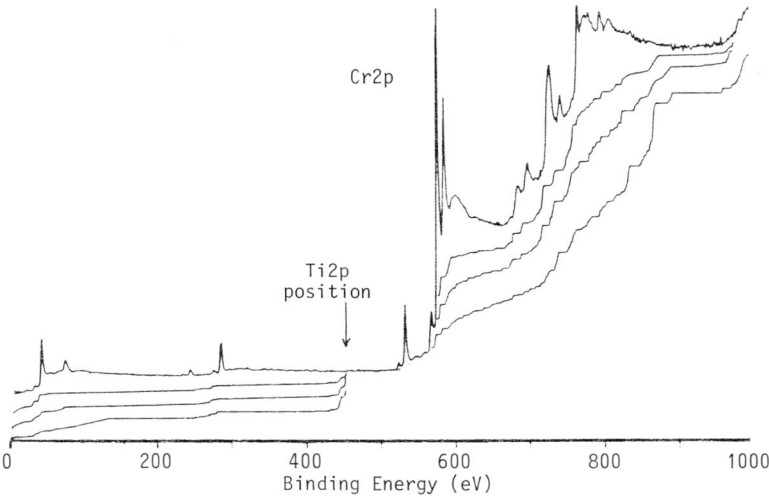

Figure 12. The total removal of peaks from the wide scan spectra of Figs.4 and 5 using a 'Shirley' function: curves all normalised at Ti2p position.

CONCLUSION

Specific attention to the general energy loss structure in XPS and Auger spectra can reveal information relating to the layer sequences of surface elements. The procedure can be formalised by use of the background slope as a function of peak height at a point about 50eV lower in kinetic energy than the peak. These post-peak slopes (P-PS) can be a reliable indicator of order in practical situations.

REFERENCES

1. P.M.A. Sherwood, Appendix 3, Practical Surface Analysis. Ed. D. Briggs and M.P. Seah, John Wiley & Sons Ltd. (1983).

2. J.E. Castle, I. Abu-Talib and S.A. Richardson, Paper to Quantitative Surface Analysis, NPL, Teddington, England. To be published, Surf. Interface Anal.

3. S. Tougaard and P. Sigmund, Phys. Rev.B, $\underline{25}$, (1982), 4452-4466.

4. S. Tougaard and A. Ignatiev, Surf. Sci, $\underline{129}$ (1983), 355-365.

5. A. Procter and D.M. Hercules, Applied Spectroscopy $\underline{26}$ (1), (1984), 46-51.

6. J.E. Castle and M.J. Durbin, Carbon $\underline{13}$ p.23-31 (1975).

7. J.F. Watts and J.E. Castle, J.Mat. Sci. $\underline{18}$, p.2982-3003.

Author Index

Abu-Talib, I., 471
Adams, F., 331
Alexopoulos, P., 117
Amarakoon, Vasantha R.W., 281, 299

Bayly, A.R., 319
Becker, C.H., 447
Beekman, D.W., 309
Benedict, J.P., 341
Bernasek, Steven L., 3
Bibiloni, A. G., 455
Bloch, J., 13
Botnick, Ephraim M., 229
Bowling, R.A., 215
Brown, R.A., 403

Capani, P.M., 203
Castle, J.E., 471
Chang, Jyh-Kao, 137
Chen, James T., 387
Chen, N.Q., 185
Cho, Chih-Chen, 3
Chow, T.P., 355
Crosson, C.A., 365

Davidson, R.D., 169
Degreve, F., 241
DeLuca, Alan, 79
Dowben, S.L., 355
Draper, Clifton, W., 3

Eastman, L.F., 203
Ehrlich, Gert, 47
Exarhos, Gregory J., 461

Fathers, D., 319
Figura, Paul M., 145
Flaugher, B., 431
Fluss, M.J., 419
Furman, B.K., 341

Geiss, R.H., 117
Gijbels, R., 331
Glass, H.L., 37
Golecki, I., 37
Granato, K.L., 341
Griem, H.T., 203
Gruzalski, G.R., 19

Habraken, F.H.P.M., 387, 395, 409
Hecq, M., 55
Heiblum, M., 13
Helms, Aubrey L., Jr., 3
Hill, C., 179
Hobbs, Linn W., 127
Howell, R.H., 419

Hua, Z.Y., 185
Hunt, J., 203

Issacson, Michael, 107

Johnson, Paul F., 281, 299
Jones, M.W., 179

Katz, W., 355
Kingham, David R., 319
Kiss, Klara, 145
Kohiki, Shigemi, 71
Kuiper, A.E.T., 387, 395

Lang, J.M., 241
Lapides, L.E., 365
Lareau, Richard T., 273
Lau, W.M., 263
Leiberich, A., 431
Lewis, N., 355
Lewis, R.K., 425
Lin, A.L, 37
Lin, Min-Shyong, 137
Ling, Peiching, 137
Lopez-Garcia, A., 455
Lou, Jen-Chung, 137

Maddox, R.L., 37
Magee, Charles W., 229
Manasevit, H.M., 37
Marsh, D.W., 355
Matteson, S., 215
McCallum, J.C., 403
McCune, Robert C., 27
McKenzie, C .D., 403
Mercader, R.C., 455
Meyer, P., 419
Mukherjee, S.D., 203

Naegele, Erich, 289
Nakazawa, Masatoshi, 85

Okamoto, H., 85
Ownby, G.W., 19

Parks, J.E., 309
Pasquevich, A.R., 455
Ploc, R.A., 169
Pollock, Gary A., 379
Prestipino, R.M., 341
Pruppers, M.J.M., 409

Rathbun, L., 203
Ratnam, P., 263
Ray, S.K., 425
Richardson, S.A., 471

Rosenberg, I.J., 419
Roy, J.A., 169

Salama, C.A.T., 263
Sanchez, F.H., 455
Sawhill, Howard T., 127
Schlenker, M., 117
Schmitt, H.W., 309
Schneider, Ulrich, 289
Scilla, Gerald J., 79
Seshan, K., 425
Shih, D.Y., 341
Siegel, Edward, 63
Skinner, D.K., 179
Smith, G.A., 355
Smith, S.R., 191
Solomon, J.S., 191
Spaar, M.T., 309
Stahle, C.M., 447

Taylor, Jenifer A.T., 281, 299
Thomas, D.R., 191
Thomson, D.J., 447

Van der Weg, W.F., 409
Verlinden, J., 331
Vlaeminck, R., 331
Vohralik, P., 319

Wang, Kuang, 91
Warburton, Michael J., 159
Waugh, A.R., 319
Whitney, R.L., 365
Wicks, G.W., 203
Wildi, Eve A., 79
Williams, J.S., 403
Williams, Peter, 273
Wolfe, R., 431
Wrigley, John D., 47

Zehner, D.M., 19
Zhang, Q.J., 185
Zhang, Wenqi, 91
Zhu, Rizhang, 91
Zhu, Yingyang, 91
Zijderhand, F., 109

Subject Index

Analytical applications, 241, 281, 289, 299, 309, 319, 331
Atomic interactions, 13, 37, 47, 63, 85, 91
Auger electron spectroscopy, 137, 145, 159, 169, 179, 185, 191, 203

Depth profiling analysis, 169, 179, 191, 203
Dopant distributions, 229, 263, 273, 355, 365

Energy loss spectroscopy, 19, 471

Gallium arsenide characterization, 137, 191, 203, 215

High energy interactions for materials characterization, 379, 395, 409, 419, 425, 431, 455
Hydrogen analysis, 379, 395, 431

Laser ionization/excitation, 447, 455, 461
Low energy electron diffraction, 3, 13, 19

Nuclear reaction analysis, 379, 387, 395, 409

Semiconductor device processing, 229, 263, 273, 341, 355, 365
Surface and in-depth analysis, 241, 263, 273, 281, 309, 319, 331
Surface composition, 3, 13, 27, 37, 47, 55, 71, 79, 85

Transmission electron microscopy, 107, 117, 127, 137

The Thirteenth Gate

Also by Tudor Parfitt

Operation Moses: The Story of the Exodus of the Falasha Jews from Ethiopia

The Thirteenth Gate

*Travels among the
Lost Tribes of Israel*

Tudor Parfitt

ADLER&ADLER

For my parents

Published in the United States in 1987 by
Adler & Adler, Publishers, Inc.
4550 Montgomery Avenue
Bethesda, Maryland 20814

First published in Great Britain by
Weidenfeld (Publishers) Limited

Copyright © 1987 by Tudor Parfitt
All rights reserved. No part of this book may be used or
reproduced in any manner whatsoever without written
permission except in the case of brief quotations embodied
in critical articles and reviews.

Library of Congress Cataloging-in-Publication Data

Parfitt, Tudor.
　　The thirteenth gate.

　　Includes index.
　　1. Jews—Asia.　2. Asia—Description and travel—
1951-　.　3. Parfitt, Tudor.　4. Asia—Ethnic
relations.　I. Title.
DS135.A85P37　　1987　　307.8'924'05　　87-1295

Printed in the United States
First U.S. Edition

Contents

Illustrations		vii
Acknowledgements		ix
Introduction		1
One	The Hostage Jews of Syria: The Forgotten Tribe	5
Two	Lost Tribes in India	36
Three	Singapore: A Dwindling Community	68
Four	The Secret of the Tribe of Zebulun	88
Five	The Deliverance of the Tribe of Dan	120
Six	The Prisoners of Venda	149
Glossary		165
Select Bibliography		167

Illustrations

Anti-Zionist posters at the entrance to the Jewish Quarter in Damascus
A Jew studying in a Damascus synagogue
The interior of a Damascus synagogue
The last Bene Israel inhabitant of Navagaon
A monument marking the place where the ancestors of the Bene Israel first landed in India
The last Jewish *teli*
A Bene Israel woman making *rotis*
The Sassoon mausoleum in Bombay
A Shinlung youth
Noa, a Japanese Jewess
A Falasha in Um Raquba, Sudan
A Lemba girl in Johannesburg wearing traditional dress
A Makuya candelabra
Master Kampo in his Kyoto synagogue (*by courtesy of the Japan Calligraphy Education Foundation*)
A Falasha feeding a child in Um Raquba camp
A Falasha woman inside a *tukul*
A Falasha girl
A Lemba boy in Venda
A Lemba woman
Solomon Sadiki, a Lemba from Soweto
A Lemba potter

(Unless otherwise stated, all the photographs were taken by the author)

Acknowledgements

My first thanks go to Lord Weidenfeld who suggested that I should write a book something like this one. I am extremely grateful to those who were generous with contacts and advice before I set off on my quest and to those who gave me kindness and hospitality on the way: particularly I would like to thank Jonathan Webber, David Patterson, Morris Marshall, Jeff Ryman, Julia Sherwood, Naim Gubbay, Alec Manasseh, Robert and Barbara Straus, Iris and Christopher Stockbridge, Professor and Mrs. Takao Ohkubo and Professor and Mrs. F. Nabarro. A few of the names used in the book are not the real names of the people concerned.

I am grateful to David Kessler for help and suggestions of various kinds and for permission to quote the passage on p. 128 from his book *The Falashas*. I am grateful for the help of many colleagues at the School of African and Oriental Studies, but particularly for that of Michael Cook, Joanna Firbank, David Hawkins, Erhardt Kratz, John Wansbrough and Sasha Piatigorsky. My thanks also go to the s.o.a.s. Research Committee which, in part, financed my visit to the Lemba, and to Glenda Abramson who first brought this strange tribe to my attention. I would like to thank my son Justin for persuading me to suppress one egregious infelicity of style and my father for spending most of Christmas 1986 reading and commenting on the manuscript. My friend Raffat Khan has been a great source of encouragement and helpful criticism. My greatest thanks go to Linda Osband, whose friendship and clear-minded editorial help have been invaluable throughout the writing of this book.

"Twelve of the thirteen gates of Jerusalem correspond to the twelve tribes, through which the prayers of each of them ascend to the heavens.... The thirteenth gate is for him that does not know which is his own tribe."

Dov Ber the *Maggid* of Mezericz,
Hasidic leader, d.1772

Introduction

The history of the northern kingdom of Israel came to a dramatic end in 721 B.C. when a considerable proportion of the population of the ten tribes was taken into captivity. The description of this calamitous event in the Book of Kings is terse: "The king of Assyria took Samaria, and carried Israel away into Assyria, and placed them in Halah and in Habor by the river of Gozan and in the city of the Medes." Of the subsequent fate of the ten tribes we know little. It seems probable that they assimilated to the cultures and religions of the people among whom they lived. But biblical prophecies maintained that one day the ten tribes would return to the Holy Land to be reunited with the tribes of Judah and Levi. Thus, the belief arose among the Jews that the tribes had not disappeared or assimilated but had been "lost" in perhaps some remote part of the world and would in time be restored to their land. In the Apocrypha, in the writings of Josephus and in many Talmudic passages, the continued existence of the ten tribes is taken as historical fact. Rabbinic tradition insisted that the tribes were to be found in three distinct areas: one group lived to one side of the Sambatyon, the fabled river which stopped flowing on the Sabbath; another group lived beyond the river, while a third group lived in the vicinity of Antioch. Fantastic details of the lost tribes were provided by a succession of Jewish writers and travellers, the most influential of whom was the ninth-century traveller, Eldad ha-Dani, who himself claimed to be of the tribe of Dan. Sightings of the lost tribes living under their tribal names in India, Arabia or Ethiopia were reported by a succession of Jewish writers until at least the middle of the nineteenth century, when a Jew from Palestine reported having found shepherds in northern Yemen who belonged to the tribe of Dan. Such travellers' tales certainly fed the myth of the lost tribes; in turn, the belief in the existence of an independent Jewish polity somewhere in the world kept alive Jewish hopes of restoration and Messianic expectation throughout the dark centuries of persecution.

Speculation about the lost tribes has not been confined to Jews. Indeed, there is scarcely a country in the world where the fate of the ten tribes has not been at some time the subject of concern. And there is scarcely a people in the world which has not at some time been identified as one of the lost tribes. This ancient myth has become almost a universal one. In the seventeenth century a particularly lively literature was devoted to the topic: some argued that the North American Indians were descended from the lost tribes; others preferred South American Indians, some of whom were allegedly found reciting Hebrew prayers. The most widespread and long-lasting of the seventeenth-century theories identified the British people with the lost tribes. This theory, which still has adherents, is explained in a recent publication of the British Israel World Federation, which is based in London but has branches throughout the English-speaking world: "In exile the ten tribes became divided into two groups in quite separate places. One near the Upper Euphrates, the other in Media. These two groups migrated by different routes to the British Isles, where they arrived, the first as the Celts or Ancient Britains, the second as the Anglo-Saxons and Normans. Here they fused once more into a single nation." Since the seventeenth century, groups in Georgia, Afghanistan, Kurdistan, Iraq, Persia, China, Japan, various parts of Africa and the aboriginal peoples of Australia either have claimed descent from the lost tribes or have been put forward by travellers as contenders for that distinction.

My journey to the contemporary lost tribes of Israel came about by chance rather than by design. Towards the end of 1984 I was asked by the Minority Rights Group to write a report on anti-Semitism throughout the world. A few days later I met Lord Weidenfeld at a dinner at St. Antony's College, Oxford. Hearing of the work I was about to undertake, he asked me to write a book, not about anti-Semitism, but on more general aspects of the Jewish diaspora. I agreed to do so. But what I found stirring my imagination as I travelled among far-flung communities were the groups, both Jewish and non-Jewish, who claimed descent from the lost tribes or who, like the Jews of Singapore or Syria, viewed themselves as "lost" in a rather different way. This journey to some of the lost tribes also seemed to me to be an exploration of the outer limits of Judaism in the contemporary world.

The journey described in this book follows the little-known route of the Jews towards the East and subsequently the spread of Jewish ideas into Africa. The journey was not continuous. The part of the journey which took in Syria, India, Singapore and Japan took place between June and October 1984. My visit to Sudan took place in November and December 1984 and my journey to the Lemba in South Africa took place in December 1986. The completion of this book was somewhat delayed by my decision to write a separate book about Israel's rescue of the Falashas from the Sudan, *Operation Moses*, which was published in 1985.

Throughout the book I have made little comment on the validity of the claims and aspirations of the people I encountered. This is, in part, because I believe that what a group thinks about itself is at least as important as what outsiders think about it. But a question which perhaps ought to be asked is: to what must Judaism be reduced before it stops being Judaism? In the case of a number of the groups described here, many of the normal attributes of Judaism are missing. They have no Jewish language; they have none of the texts of theology of modern Judaism; some have no Jewish blood. They do, however, have a sense of identification with the Jewish people and a sense of belonging to the religion of Israel. Should the "Jewishness" of such manifestations be judged by the austere criteria of orthodox Judaism? Or may they rather be taken as objective expressions of belief, important as such on their own merits?

To have encountered so many groups in quite different cultural situations throughout the world seeking membership of the community of Israel barely half a century after world Jewry was almost annihilated is in itself remarkable. One is almost tempted to view the phenomenon as some natural process of replenishment. For David Ben-Gurion, Israel's first and greatest prime minister, some of the developments described in this book might not have been unwelcome: after all, he believed that "a Jew is someone who believes himself to be one". But Israel's religious authorities, with their faces set firmly against religious pluralism, are unlikely to be any more enthusiastic about the Lemba and the Shinlung than they were in the past about the Falashas or the Bene Israel. And the Japanese sects, not least because of their audacious religious syncretism, will always be beyond the rabbinic pale. Some of the groups who appear in this book on the other hand are to be regarded as mainstream Jews. What then do those within the pale have in common with those beyond it? It is by virtue of not knowing where they belong or which tribe they belong to that the lost tribes, would-be lost tribes and forgotten and dying communities described here qualify for the Thirteenth Gate.

ONE

The Hostage Jews of Syria: The Forgotten Tribe

Jezebel's husband, King Ahab of the Northern Kingdom of Israel, waged two successful wars against Ben Hadad, king of Syria, eight hundred years before Christ. After his defeat, the Syrian king promised: "The cities which my father took from thy father, I will restore; and thou shalt make streets for thee in Damascus, as my father made in Samaria." Today there are still Jewish streets in Damascus.

A tradition of the Jews of Damascus has it that since the time of Ahab there have always been Jews in the city. Thus, they claim, they are one of the few Jewish communities in the world who can trace their descent back to the ten tribes who made up the population of the kingdom of Israel as it was before 721 B.C. The tradition of continuous Jewish settlement has some substance. Certainly, throughout Old Testament times Damascus had a large Jewish population and by the time of Herod had become an important centre of Jewish life. It was undoubtedly an attractive city for Jews to settle in not least because of its proximity to the Holy Land. In addition, the city had attractions in its own right: its position on the edge of the Arabian desert but surrounded by lush vegetation and gardens led one Talmudic writer to call it "the gateway to the garden of Eden". But perhaps most of all Jews were drawn to the city because it lay at the centre of the vital caravan routes which connected the Mediterranean, North Syria, Mesopotamia and Arabia.

After the Muslim conquest of A.D. 635 Jews continued to live in their traditional quarters in the south-east corner of Damascus and from time to time made notable contributions to the wider society. During the Islamic period the heyday of the community came in the tenth century when a Jew, al Qazzaz, became the head of the financial administration of the whole of Fatimid Syria. But when Benjamin of Tudela, the great Jewish traveller, visited Damascus two hundred years later, the Jews were still living comfortably. In the record of his travels Benjamin wrote:

> This city is very large and handsome. It is enclosed by a wall and surrounded by beautiful countryside which for a distance of fifteen miles all around boasts the most fertile orchards and gardens, in such numbers and of such beauty as to be without equal upon earth. . . . This city contains three thousand Jews, many of whom are learned and rich men.

The fortunes of the Jewish community took a turn for the worse in 1260 when the Egyptian Mamluks defeated the Mongols at the decisive battle of Ayn Jalut near Nablus and established their rule in Syria. The first reaction in Damascus to the defeat of the Mongols was a riot directed against the Jews and the Christians. The Mamluks, who were to rule from 1250 to 1517, soon introduced discriminatory anti-Jewish (and Christian) legislation. Jews were forbidden to ride horses or mules and were only permitted to ride donkeys sidesaddle. In order to comply with the Islamic ruling requiring non-Muslims to be dressed in such a way that they could be readily differentiated from Muslims, Jews were made to wear yellow turbans while Jewish women were required to wear one red and one black shoe. For similar reasons Jewish men had to blow on a whistle every time they entered a public *hamam* (Turkish bath) and inside the bath-house were obliged to wear little bells around their necks.

In the fifteenth century the Jewish community of Damascus was increased by the arrival of Jewish exiles from the Iberian Peninsula after the expulsion of the Jews from Spain (1492) and Portugal (1496–7). During the last years of Mamluk rule there was little persecution of the Jews. Indeed, after the arrival of the Spanish Jews the community went from strength to strength not least in the area of Jewish scholarship and within a few decades Damascus was to become one of the most important centres for the study of, among other things, the Jewish mystical tradition, the Kabbalah.

The Turkish conquest of Syria in 1516 marked the beginning

of an improvement in the position of the Jews. Whereas the Mamluks had tended to exclude Jews from public office, under the Turks the Jews were invited to be civil servants, financial advisers and linguists who played a vital role in maintaining relations between Turkey and Europe. In Constantinople the Jews had particularly close commercial links with the ultimate source of political power in Turkey – the Corps of Janissaries, the renowned military elite which had been founded in the fourteenth century. The importance of the Jews in the capital of the empire had some effect on the lot of their provincial brethren in Syria and elsewhere, at least as long as the empire exercised real control over Syria. But during certain periods, the Jews of Syria and Palestine were at the mercy of local rulers who for most internal purposes were independent of Ottoman authority.

The fortunes of the Jewish community of Damascus reached their lowest ebb during the second half of the eighteenth and the first half of the nineteenth century, a period during which Ottoman influence in the area was particularly feeble. It is a curious paradox that during these years the position of the Jews was precarious even at times when Jewish notables were helping to run the internal affairs of the city and province. Ordinary Jews were persecuted and humiliated as a matter of course: any Jew had to walk in the gutter if a Muslim approached him in the street. But even Jews who had amassed great fortunes or attained positions of authority were not immune to sudden reversals of fortune. Thus, in 1823 the intrepid traveller, Joseph Wolff, a Jewish convert to Christianity, noted that

> the high priest of the Jews, Joseph Abulafia and R. Farhi, Prime Minister to the Pasha ... were put in prison by express orders of the Sultan ... it is left to them to pay 40,000 bourses of piastres or to lose their heads. A renegade was appointed Prime Minister instead of the Jew and the Turks began to shout, saying with smiles, "Praise be to the Lord! A curse upon Raphael their Kakham! A curse upon all the Jews, their fathers, their mothers, grandfathers, grandmothers, their children and their children's children."

Three decades later, Porter, a British traveller, described the Jewish community of his day in similar terms:

> The Jews of Damascus are very influential on account of the vast wealth of some of the great families. They have been for many years

the bankers of the successive pashas and great merchants. Until the interference of European powers in the internal administration of affairs in Syria, the changes of fortune and circumstances through which some of these families passed were truly wonderful, and had more of the character of an Eastern romance than of stern and fearful reality. At one time the Jew would be the actual ruler of Syria, and then in a few weeks he would be stripped of fortune, and perhaps cruelly mutilated, or even murdered.

In 1840 the Jews of Damascus were charged with the ritual murder of a Capuchin friar, Father Thomas, and his servant Ibrahim. Many members of the community were arrested and tortured; a number died as a result of the tortures they had suffered. Jewish and liberal opinion in Europe was outraged at what was viewed as mediaeval barbarism and in what came to be known as the "Damascus Affair" European influence was effectively brought to bear on behalf of the local Jews. Thereafter, the position of the Jews improved radically. The Ottoman Empire introduced a series of reforms known as the *tanzimat*, which gradually gave a range of civil rights to the Jews as well as to Christians. Having achieved virtual equality before the law, throughout the second half of the nineteenth century the Syrian Jews were able to make considerable contributions to the commercial and intellectual life of the area. But despite the reforms and their improved position, by the end of the nineteenth century many Jews started leaving Damascus to settle elsewhere.

In 1947, at the time of the Israeli War of Independence, there were about 30,000 Jews in Syria. But a wave of hostility towards Israel culminating in officially orchestrated riots in Aleppo and Damascus left hundreds of Jewish homes and several ancient synagogues in ruins. Many Jews were killed. Of the rest, 15,000 fled the country, many of them making their way to Israel. A further 10,000 were able to leave between 1948 and the early 1960s during periods when restrictions on emigration were temporarily lifted. But since then there has been a continuous ban on Jewish emigration: the Syrians argue that they have no wish to supply the Zionist enemy with more potential troops. From time to time Jews are allowed to go abroad on brief visits for family, health or business reasons, but they are required to deposit large sums of money with the authorities and to leave behind immediate members of their family to guarantee their return. Thus, the majority of those who have left Syria over the last twenty years have done so illegally. According to some sources several dozen Jews a year manage to escape, usually over the border into Turkey. The dangers involved in emigrating in this way are obviously great and many have died in the attempt. None the less, entire families have escaped. Recently one such family left behind goods and property to the value of $3 million and arrived in New York a few weeks later with nothing. But naturally the majority of those who opt for this route are young men who leave alone or in small groups. Consequently, there are in Damascus several hundred more Jewish women of marriageable age than there are men. In recognition of this problem in 1977 President Hafez al-Assad permitted twelve unmarried Jewish women to leave Syria and emigrate to the United States, as a concession to the personal intervention of President Carter. Now there are only 5,500 Jews in Syria: 4,500 in Damascus, 1,000 in Aleppo and a few dozen families in Qamishli, a small town beyond the Euphrates on the border with Turkey.

The Jews are not the only oppressed group in Syria. Even the Sunni Muslim majority feel a sense of oppression because of the secular nature of the state. Syria and Lebanon are the only Arab states whose constitutions do not specify Islam as the state religion. The fact that the Syrian head of state, at least, is now required by the constitution to be a Muslim, is largely a consequence of the riots which struck the country in the early 1970s after an earlier draft constitution emphasizing the central position of the Baath party made no mention of Islam at all. In any case, the majority of Syria's Muslims do not regard Assad as a *real* Muslim. This is, in part, because he belongs to a "heretical" group known as the Alawites or Nusayris, a traditionally troublesome and turbulent sect, historically an offshoot of Shia Islam, which preaches the near-divinity of the Prophet's son-in-law Ali. The Alawites are now the ruling group in Syria. To make matters worse, Assad was responsible for suppressing the Arab world's most serious attempt at Islamic revolution, staged by the fundamentalist Muslim Brotherhood, in February 1982. More than 15,000 people are thought to have died in the suppression of the uprising. Most of those who were killed were Sunnis.

The chief minorities in Syria are Kurds and Armenians. The Armenians complain of harassment and attacks by members of the security forces and other forms of discrimination. Many of them have already left the country. The position of the Kurds is even worse. Since 1973 some 12,000 Kurds from Syria's northern border regions have been forcibly deported to other parts of Syria and their traditional homelands settled by ethnic Arabs. Many have been arrested and detained for years without trial. The smaller minorities fare no better. The internal situation in Syria is thus not a happy one. Many Syrians from all sects would like to emigrate. Indeed, since the mid 1950s, over a million have done so.

In many respects, however, the situation of the Jews is even worse than that of other groups. Many of the difficulties encountered by other sections of Syrian society are at least in part the result of Assad's heavy-handed attempts to create a modern, secular state in which traditional divisions will play no part. The Jews are excluded from this process. The authorities attempt to play down the differences between Kurd and Arab or between Shiite and Sunni, but there is no desire to play down the separateness of the Jew. The restrictions on them leaving the country are more rigorously enforced than is the case with others. Until recently the identity papers of Jews were stamped in red with the term *Musawi* (of the Mosaic faith). In the remote town of Qamishli "*Musawi*" is daubed in red on the outside of Jews' houses. In practice, as part of the campaign to minimize sectarian and ethnic differences, the religious affiliations of other Syrians are not noted on their identity papers, and certainly not in red, and in no cases on the outside of their homes. In 1977 somewhat half-hearted reforms were introduced whereby the designation "*Musawi*" in red ink on Jews' driving licences, identity cards, passports and bank documents was done away with, only to be replaced by a smaller entry in blue. Jews are not permitted to work in the government services, in any of the nationalized industries, the army or the police force. The access of young Jews to the universities is severely restricted and departments of science and technology are now more or less closed to them. A dozen or so Jews of an older generation are doctors and chemists, but more than 95 per cent of the male Jews in employment are tradesmen and artisans. As Jews are considered to be potential spies for Israel, they are not allowed to have contact with foreigners. There are *Muhabarat* (Syrian secret police) checkpoints controlling the areas where Jews live. The agents who patrol these neighbourhoods harass the Jews and allegedly engage in extortion, bribery and rape.

The Hostage Jews of Syria: The Forgotten Tribe

The morning after I arrived in Damascus I took a taxi to the British Embassy, an undistinguished building in the smart modern area which houses the President, the embassies and Damascus's small foreign community. The Embassy was closed for Ramadan, but I was able to telephone the duty officer at home. He had just finished an early morning game of tennis and sounded cross. "I am awfully sorry, I have rehearsals for *Travels with my Aunt* all weekend – the British community here is putting it on – and a picnic on Monday. I don't think it will be possible to see you." The Foreign and Commonwealth Office in London had told the Embassy that I was collecting material for a report for the Minority Rights Group. I had been promised a briefing on the subject of the Syrian Jews and their alleged persecution by the Syrian authorities. I could hardly say this over the phone. With considerable reluctance he finally agreed to come to the Embassy. Once I had told him what my business was in Damascus he recalled the correspondence and was anxious to be of help. We walked back to his flat. I asked him about the situation of the Jews in Damascus and about Syria generally. He was well informed about Crusader castles in Syria, but less knowledgeable about the topography of modern Damascus and had no idea where the Jewish Quarter was situated. As we entered his modern, ground-floor flat he called out and asked his wife if she knew the whereabouts of the Jewish Quarter. She was not quite sure either. They were, however, certain that there was no persecution of Jews in Syria. I remonstrated mildly: "But at the very least there appears to be a denial of basic rights."

"Yes – but that is not persecution. They are far better off than many others I could name." He laughed humourlessly. "There are always rumours here in Damascus. Rumours about anything and everything. There are sometimes rumours about persecution of the Jews but most of these start south of the border." He meant Israel.

I asked him about the case of a pregnant young Jewish woman, Lilly Abadi, of Aleppo, who had been killed with her two children some months before. The story was gruesome. Hayyim Abadi, a jeweller from Aleppo, had received a telephone call in his office and was told of terrible acts that had been perpetrated against his wife and children. When he returned home, he discovered that she had been shot and stabbed, her breasts cut, her stomach slit open and the unborn child killed. Her son was dead: his arms had been broken. The girl's body had been repeatedly slashed with a knife.

Subsequently other members of the Jewish community received threatening calls warning them that they would be next. "Ah yes," replied the diplomat, "the Abadi business. Well, you know, it was her husband who was basically at fault. He had been involved in various sorts of criminal activity and this murder was the result of family feuding."

"Whose family? His own?"

"No. Not necessarily."

He gave the impression of knowing far more about the Abadi case than he could reveal. He was silent for a moment and began to twitch with annoyance.

"You simply can't imagine the amount of pressure we have been under from Jewish groups. It's just a propaganda exercise for them down there." He nodded somewhere in the direction of the Golan Heights. His face puckered with what looked like distaste but as he spoke there was a suggestion of passion in his voice, too. The conflict between Jew and Arab in the Middle East rarely leaves foreigners neutral. As I left I remembered E. M. Forster's words: "It is impossible to regard a tragedy from two points of view."

Later that morning on the campus of the University of Damascus I was able to track down a Syrian scholar I had known years before in England. I mentioned the purpose of my trip and asked him about the Jewish Quarter and how best to get into it without being noticed by the *Muhabarat*.

"I really do not know, my friend. We never go there. It is too dangerous. No one goes into *Harat al-Yahud* any more and anyone who wants to be left alone keeps well away from the area. I'll tell you this though. Do not go in by the main streets. There are permanent *Muhabarat* check-points there and they will not allow you in." He pulled down a street map of Damascus from his shelves and showed me some less obvious routes into the Jewish Quarter in the old city.

It was Friday evening. Damascus was being slowly transformed by the setting sun from a neglected, litter-strewn Middle Eastern town into a rose-tinted monument to high mediaeval art. From Ibn Assaker Avenue I entered the old city through Bab Sharqi where centuries before a Jewish woman is said to have thrown down the rope which let in the besieging forces of Nur al-Din of Aleppo. Despite the failing light the craftsmen were still working in their booths; most of them were making the intricate mosaic boxes for which Damascus is famous. To get to the Jewish Quarter I turned off the main street into a warren of narrow alleyways, which finally led me to a wider street bordered by high stone houses with elaborate wooden balconies. From the *mezuzot* on the door-frames I realized I was in the Jewish Quarter.

Rounding a corner I came across a group of children playing together in a small group. They looked surprised to see me and fell silent. I greeted them in Arabic and ceremoniously shook hands with one of the smaller boys.

"Are you Syrian?" I asked.

"Yes. We're Syrians."

"Muslim? Christian?"

They looked at each other nervously. One of them giggled. The oldest child, a girl, said, "We're Jews."

"All of you?"

"Yes. All of us."

I spoke to them in Hebrew. They understood a little. Using a mixture of Hebrew and Arabic they warned me that there were police on the next street. The girl said, "We'll show you how to avoid the police." They disappeared noisily into a dark passageway and I followed. We scrambled together over a broken-down stone wall and through a series of yards full of rubble and collapsing stone houses. We emerged some minutes later in a square surrounded by neat whitewashed houses. A woman standing in front of one of the houses called to the children. I could not understand what she said. The children scattered in all directions. As they disappeared one of the smaller ones shouted *"Shalom"*. The sound reverberated in the empty square. I felt uneasy. A few minutes passed and I was joined in the square by a young man. He asked me for a cigarette.

"You know this is the Jewish Quarter?" he said.

"Are you Jewish?" I replied.

"The only people you find in the Jewish Quarter are Jews or *Muhabarat* agents." He laughed. Suddenly he asked me if I spoke Hebrew. I told him that I did. We spoke Hebrew for a few minutes. His Hebrew was rudimentary. He asked me where I was from and what I did, if I was married and how many children I had. Exhausted, apparently, by his efforts, he got up and walked away. I walked in the opposite direction and came across the same children I had met earlier. "Look!" they said, and showed me into a small courtyard with a well covered by a great vine supported by four metal posts. Pointing at a handsome stone building one of them said, "This is our synagogue." An old Jew was carefully sweeping the flagstones. "He speaks Hebrew," the girl said pointing at me. Indicating the old man, one of the boys shouted out raucously, "He's mad." The other children took up the refrain and started shouting and laughing. He lifted his broom menacingly. They ran off shrieking in mock terror. As I approached the man, I saw that he was not as old as I had thought. His eyes bulged and his mouth gaped open. "*Shalom,*" I said.

"Ah! You speak Hebrew. Good. I learned Hebrew when I was a boy and I have never forgotten it, thank God. Other things, many other things, I have forgotten. Hebrew I have not forgotten. I have a brain tumour." He touched his head with his hand. "It makes you forget. Sometimes it is no bad thing to forget. I used to be a carpenter, but now I am sick and just look after the synagogue. They'll be coming soon," he said, resuming his sweeping very slowly. His wife came into the yard with their six children. She was wearing a clean cotton dress with a floral design. She looked strong and patient. Their thirteen-year-old son said to me, "My father's crazy." He did not say it unkindly. The man said nothing, but put his broom to one side and sat down on a stone bench which ran the length of the synagogue. I sat next to him and we talked.

"The situation for us Jews is not good and not bad. We have food, no one starves, we have families, we have our synagogues. You'll see the synagogues are full. If we mind our own business they leave us alone. Sometimes they don't, but usually they do. We had a lot of trouble back in 1965 during the trial of the spy Eli Cohen and during the Six Day War. But generally they leave us alone."

"Whom do you mean by 'they'?"

He shrugged and said nothing.

"The real problem for us is the girls. We have far too many girls and not enough boys. They can't marry. That's no way for a Jewish girl. You must have a family, children. Some of them have married Muslim boys. Not the best Muslim families either. There's nothing else for them to do. They should let the girls go. They can go to New York. There are plenty of Syrian boys there. They won't go to the Land."

"You mean Israel?"

"We don't use the word. It is too dangerous. You never know

who is listening." He touched his head and grimaced with pain.

"Can't you get treatment?"

"I need an operation. It is a difficult operation – I'd have to go abroad. If I go I must first deposit $5,000 with the government. I could borrow the money, but if I died while I was out of the country my family would have to pay the $5,000. If you don't come back you forfeit the money even if you die abroad. I can't do that to my family." He paused and shouted at his children to go home and get ready for the evening service. "You have to be very careful what you talk about here. You can't trust everyone. There are even Jews working for the Syrians."

"Do you know who they are?"

"We've all got a good idea. We don't blame them. Some have a choice, some don't. The authorities have got ways of making you do what you don't want to do."

"Are you afraid of them?"

"Not like before. Assad looks after the Jews. If he goes and the Muslim Brotherhood takes over the government, they'll kill every Jew in Syria. Hafez is our only hope."

"Are the *Muhabarat* here to protect you?" He flinched at my use of the term.

"They used to say they were here to protect us from the Palestinians. A few years ago we used to have a lot of problems with the Palestinians. The young men and also the girls. They live near by. But now the Palestinians are forbidden to enter the ghetto and there's no trouble. Not for some years. But our friends are still protecting us." I saw the old man stiffen and I felt a light touch on my elbow. A short young man was standing there, wearing tight-fitting grey trousers and a shiny black shirt. He asked the old carpenter in Arabic, "What were you talking about?" He was still holding my elbow, very gently.

"The Englishman was asking about our synagogue. I told him. It's nothing, he just wandered in. A tourist." I confirmed this.

"You speak Arabic?"

I smiled and removed his hand from my elbow.

His voice dropped to a whisper and he asked me to go with him. I asked him who he was.

"Who I am, my friend, is none of your business."

"Are you a policeman?" He did not answer.

"If you are, show me your papers. Otherwise I'm staying here."

I tried to sound reasonable. Two more figures emerged from the shade offered by the vine. One of them put his hand on my shoulder and pushed me out of the courtyard. As I was led unceremoniously to the *Muhabarat* headquarters we came across small groups of Jews, dressed for the Sabbath, making their way to the synagogue. They avoided looking at us.

Opposite the concrete, two-storey police station was a sort of sentry box. Three *Muhabarat* officials armed with light machine-guns and pistols lounged outside. I was taken into a small room where ten men were smoking and drinking coffee. On one side of the room a wooden rack held rifles and sub-machine-guns. A heavily built man in civilian dress asked me for my passport.

"You want coffee?" he asked.

"No thank you."

"You English?"

"Welsh," I snapped.

"You've heard the news about your great prime minister, Mrs. Thatcher, what she's going to do?"

"No."

"She's going to send all the Jews from England to Israel to fight the Arabs. All the Jews must go. Women, children, old men, everyone. You'll have to go to Israel too." There were staccato bursts of laughter from around the room.

"You are a Jew?"

"No."

"You speak Hebrew. You spoke Hebrew to one of our friends."

"I speak Arabic too. Does that make me an Arab?" They laughed. "And by the way, your agent speaks very bad Hebrew." They laughed again, but with less enthusiasm.

The heavily built man started copying details from my passport into a ledger. He wrote laboriously. I was told to go and wait outside. The guards in the street gave me a stool and offered me a cigarette. I sat out in the courtyard for some time. Through the open window I could hear the commandant telephoning his superiors. He kept repeating some hopeless corruption of my name. Above me I could see a brilliant night sky. A muezzin wailed not far away. The guards baited me mildly: "England is very weak. Syria is very strong. England is very weak like a woman. Like Mrs. Thatcher!" I agreed wearily that this was so. I was called back into the small crowded room and interrogated. "No, I am not working for Israel." "Yes, I am interested in Jews." This they found difficult to understand. The interrogation dragged on. Eventually I was told that I could go.

"I came here to see the Jewish Quarter and I have seen nothing," I protested. "I want to see the synagogues. There are fine synagogues here – the most beautiful ones in the world." Seeing that this more or less untruthful appeal to the *Muhabarat*'s patriotic instincts was unlikely to succeed, I took out from my wallet a rather dog-eared letter from the Registrar at the School of Oriental and African Studies requesting that I should be given every assistance in my research. As the commandant could not read any English I can only assume that he was impressed by the School's embossed crest which features a camel. In any event, without taking his eyes off the document, he told me that I would be permitted to visit the synagogues. A *Muhabarat* agent, however, would have to accompany me. Under no circumstances was I to talk with Jews again.

"Why can't I talk to Jews?"

The commandant picked up a pistol which had been lying on a desk in front of him. He held it in both hands for a moment, then replaced it. "If you talk to Jews, my friend, I guarantee nothing." He look at me steadily. Everyone else was silent.

I was accompanied to the door by the young agent in the black shirt whose name was Ahmed. He took me to a magnificent synagogue not far from the police station. The empty courtyard resounded to the singing of the *hazzan*. "This musical tradition goes back long before the rise of Islam," I said, more to myself than to Ahmed. He swaggered into the building, pointing out some of its features in a proprietorial way. No one paid him much attention.

The Al-Iraki synagogue was so full it was difficult to get in the door. The prayers continued unabated. A number of men spoke to me in Hebrew as I passed. "Blessed be he that comes." "Are you a Jew?" "Are you from the Land?" I covered my head with the skullcap that was passed to me and greeted the small crowd that had gathered around with the Sabbath greeting, "*Shabat Shalom.*" I was handed a Hebrew prayer book, someone found the place for me, and the service continued as before. Ahmed soon grew bored and went out. The man next to me spoke insistently over the noise of the prayers. "You're a journalist? Things here are bad enough, but they could be worse. My son got out, thank God, and for us old ones it is not too bad. I won't leave, I want to be buried here with my ancestors. You get used to it. It's been thirty-five years. I can't really remember better times. And during the war it wasn't so good either. Over the last few years I've been beaten myself a couple of times. What are they going to get out of me?" Ahmed came back to the synagogue door, smoking a cigarette. He raised a finger to his lips to indicate that I should not speak.

I visited more synagogues with Ahmed. I gave him three packets of English cigarettes and, while I talked to people inside, he sat in the yard smoking moodily. I spoke to many people that night. In the end the individual stories they told seemed to merge into one: they were about beatings, escapes, extortion, theft and petty humiliations. But everyone said that the situation could be worse and that if ever the Islamic fundamentalists took over from Assad, it would be. I left the ghetto and walked back to my hotel. I slept uneasily that night.

The following morning I returned to the Jewish Quarter. The street which divides the Christian Quarter from the Jewish Quarter is a busy thoroughfare of the old city. A number of cars were trying to get past a barrage of donkeys and handcarts piled high with fruit and vegetables. While I waited for the traffic jam to clear, I went into a coffee shop. In one corner there was a group of youngsters, otherwise the place was empty. One of them, a serious young man wearing jeans and a checked shirt, spoke to me in English: "I saw you in the synagogue last night." I asked him to join me for a coffee. He accepted and we started to talk. He told me that he had just left school.

"What sort of school did you go to?"

"We have some schools in the Jewish Quarter. Some of the teachers are Jews although the directors are Muslims. Our Chief Rabbi has some control over things, but they aren't really Jewish schools. Exams were always on Saturday just to annoy us. The school I went to was not very good. The teaching in maths and languages was poor."

"But your English is good."

"I read a lot and listen to the radio. I like languages. I know Hebrew too from listening to the Israeli radio station *Kol Yisrael*. The worst thing about the school for us was the anti-Jewish propaganda. In some of the school-books Jews are shown as murderers and Zionist agents. We were told that Jews everywhere hate peace and that Israel must be completely destroyed.

"As far as our textbooks were concerned it all goes back to Mohammed's dislike of the Jews. It was confusing for us, but we listen to the Israeli Arabic broadcasts and that gives us some idea of what is happening in the outside world."

"What are your plans now that you have finished school?"

"I'll join my father's business. He does some foreign trading, so I may have the chance to travel when I'm older."

"Would you come back?"

"Yes, I would have to. My father has had a lot of problems. Because of his business contacts abroad he's been forced to take a Muslim partner. They don't get on well and we're fairly sure that he reports on us. My father has to get a special visa even to travel to Aleppo. If I went to the States and stayed there, he would lose the $5,000 deposit and he'd really get into trouble. I mean, he'd be tortured." He spoke English fast and with an American accent. "My father has already spent too much time in Rowda prison. That's where they put the Palestinians and the Jews. Either Rowda or al-Mezze prison. It's strange putting us together."

"What are you doing today?"

"I've been to synagogue. As you can see I'm not very observant. A friend of my brother's has his *Barmitzvah* today. That's it, there's not much to do here outside the synagogue. Even though we have a Jewish life, everything is dead!" Ironically, he quoted Jeremiah: "'Damascus is waxed feeble ... the city of my joy.'"

I left the coffee house and turned off into the Jewish Quarter. From the latticed window of an upstairs room I saw two plump, pretty Jewesses watching me. When I returned their smiles they blushed and disappeared from sight. Further on, a ruined arch partly obscured by a ramshackle door gave on to a courtyard shaded by a eucalyptus tree. Carefully tended flowers were growing in whitewashed tins beneath open windows. A small girl with blotches of iodine on her legs ran across the polished flagstones towards me. In one corner of the yard was a table piled up with food and a dozen Jews were eating their Sabbath morning meal. A patriarch beckoned from a couch which ran down one side of the table and I was summoned to join them in their meal of eggs, tomatoes, *kubbeh*, *tabbouleh* salad and aubergines. The convivial conversation was soon brought to an end by the unexpected arrival of Ahmed, still wearing his black shirt.

"Mr. Tudor. What you talking about?"

Ahmed sat down next to me and helped himself to a *kubbeh*. The conversation became stilted. One of the women asked me uncertainly if I would like to see the house. Two families shared the courtyard. One family lived in the ground-floor flat. The first-floor flat was reached by a stone stairway on one side of the courtyard. One of the rooms was kept for visitors. It was immaculately clean. There were old black and white tiles on the floor and in one corner stood a brass bed covered with a patterned white bedspread. Parts of the house were mediaeval; the walls boasted painted tiles and mosaics with gilded borders carved with *menorot* (the seven-branched candelabra) and other Jewish emblems. The room was dominated by a large framed *kame'a* or charm, consisting of the outline of twin *menorot* made up of Hebrew verses and acronyms. This was a charm against Lilith, the consort of Satan, the queen of demons, who is believed to kill new-born babies. They explained to me that this room was used for formal occasions, for visitors and also for childbirth. The decoration of the room was rich and ornate in contrast to the simplicity of the yard: traditionally the Jews of the Middle East went to pains to conceal their wealth, not wishing to arouse the envy of their Muslim and Christian neighbours. A carved painted door with intricate brass hinges led into a storeroom full of trays of drying peppers, courgettes and herbs. The family pressed me to be their guest and to stay in their beautiful guest room as long as I wished. They realized, of course, that this would be impossible, but the children took the suggestion seriously and ran off, reappearing minutes later with sheets and towels. Before Ahmed led me away I saw the rest of the house. Old photographs of earlier generations of Damascus Jews hung on many of the walls: most of them were pre-war. The men wore bristling moustaches and the fez. To accommodate the large number of children of this generation there were several beds, carefully made up, in each room.

Ahmed was bored and insisted that we report back to the *Muhabarat* headquarters. We passed some imposing residences which had once been the homes of wealthy Jews; now they appeared to be empty and neglected. Again, I was made to wait outside while the duty officer went through my papers. The agents in the small room seemed more tense than on the previous day. Cups of Turkish coffee were being handed around. As an act of calculated rudeness I was not offered one. Ahmed took me by the arm and we returned to the Jewish Quarter where I insisted on seeing the rest of the synagogues. Outside one of them, a *Barmitzvah* was in progress. The sight of Ahmed's well-known figure was sufficient to dispel the holiday atmosphere. The men became quiet and watchful although some of the older boys glowered bravely at us. The boy who was *Barmitzvah* was wearing a dark blue jacket, white shirt and black bow tie. He had a gold chain underneath his collar which hung down his shirt front; on his wrist he had a thick gold bangle. If his dress was more or less Western, the oriental Jewish rituals marking the beginning of his manhood seemed completely indigenous. I asked him to recite some of the biblical passage which he had learned for the morning's ceremony. Proudly he recited a few Hebrew verses. Everyone smiled and applauded, and even Ahmed was drawn for a moment into the spirit of festivity and threw his arms around the shoulder of a young man of his own age. The Jew shrugged him off and walked away without a backward glance. Ahmed's eyes flashed dangerously and, looking at me, he put two fingers to his lips indicating again that I should not speak.

Like the rest of the Jewish Quarter the area in front of the synagogue was full of children. Two Jewish boys ran out of the synagogue dressed up in miniature uniforms of the Syrian army complete with gold braid and campaign decorations. I supposed that this was the closest they would ever get to being in the Syrian armed forces. Ahmed walked away to talk to two men who were standing at the end of the alley watching us. I asked a Jew standing next to me why there were so many children in the ghetto. "There is nothing to do here. We go to synagogue, we make children and we hope. That's Jewish life!"

In a street leading off from the synagogue stood a group of young women wearing high-heeled leather shoes and frilly pink, light blue and red dresses with lace handkerchiefs tucked in their sleeves. Head held high and hips swaying, a stately young woman approached the group; she was dressed in a shiny black and silver dress and an elegant silver brocade jacket. She would not have looked amiss in the Crush Bar of the Royal Opera House. In this dusty alleyway, in the heat of the Syrian summer, she looked absurd. As I walked towards the group of girls, Ahmed hurried after me with the jauntiness of all short men. One of the Jews whispered to me in Hebrew, "These are our unmarried girls. Every

Sabbath they meet together. They have nothing else to do." A pall of heady Damascene perfume hung over them. One of the girls glanced at me. She looked accusing and ashamed at the same time. The other girls laughed too easily, like girls in a convent. As I wished them good morning, they giggled and said nothing.

We visited another synagogue. It looked as if it had been destroyed and recently reconstructed, in large part from concrete blocks. The morning service was over and the synagogue was empty except for one old Jew. His discarded jacket and prayer shawl lay next to him on the upholstered bench which skirted the room. Crosslegged and barefoot he sat reading from a rabbinic commentary. He sat by an open window which overlooked an orchard. Immersed in his study he barely spared us a glance.

Ahmed was perspiring from the heat. He said that we should return once again to the *Muhabarat* headquarters. We walked back slowly through the Jewish Quarter. Through an arch I saw the courtyard of another synagogue. It was covered with bougainvillaea and jasmine. Water gushed from a beautifully wrought brass faucet set into a marble fountain. Around its base stood blue-painted pots of yellow irises. Excited at seeing a stranger, a number of children ran towards me across the spotless marble slabs. When they saw Ahmed, they froze. He pulled my arm and we moved on.

The alley led into a decrepit square. One side was dominated by a wall painting of the Palestinian national flag, which marked the entrance to the small Palestinian quarter of the old city. A group of young Palestinians stood in front of their flag. Ahmed nodded at them and said, "You British gave their land to the Jews. You've heard of Balfour?" I nodded. One hears a great deal of Lord Balfour in the Middle East. I pointed a camera at the Palestinian flag, but Ahmed told me it was forbidden to take photographs. One of the Palestinians said, "Everything is forbidden here: it's worse than Israel." Ahmed gave the boy a hard look but said nothing.

We walked back to the police station, where again I was taken to see the commandant. "You've seen everything now. You can go. We don't want to see you here again." I remonstrated that I had not seen the Jewish schools. "The Jews' schools are closed today because it is the holy day of the Jews. They will be closed for the next few days because of the Muslim holidays. They have some Muslim students and teachers in their schools who cannot work on Muslim holidays."

"Palestinians?" I asked.

"You know a lot, my friend Mr. Tudor. We shall be looking out for you." He looked at me balefully and I left.

For several days I wandered around the rest of the old city talking to Christian and Muslim Arabs and to Armenians. It became clear that the dislike felt for the Jews was ancient and instinctive and only rarely connected with personal experience. Some of the people I spoke to reiterated the official anti-Jewish propaganda commonly heard in the Middle East which turns latent dislike into active hatred. Others evoked historical events such as King David's conquest of the city to explain their dislike of the Jews. One prosperous Christian merchant told me indignantly, "The Jews even drove St. Paul out of town. The poor fellow only escaped with his life because someone let him down in a basket from a window in the city wall," and he gestured vaguely towards a church next to one of the city's gates. Another Christian told me the story of the Capuchin Friar Thomas, who disappeared in 1840, and of the famous blood libel which was brought against the Jews who were accused of murdering him and using his blood for making passover *matzot*. "You see what sort of people the Jews are," he said. The Armenians I met were more even-handed. But one Armenian dentist told me that the Jews were treated far better than the Armenians because the Jews were exempt from army service while the Armenians were conscripted and frequently, according to him, selected for the most dangerous missions.

The old city of Damascus is crossed by its most famous street: the one known in the New Testament as the Street called Straight and in Arabic as Suq Madhat Basha. On the south side of this street in the direction of Bab Sharqi there are many Jewish antique and copperware shops. Strictly speaking these are not situated in the Jewish Quarter but in a common thoroughfare. I was able to spend a lot of time talking to these Jewish merchants. One day I saw Ahmed watching me from a distance, but he did not approach and I did not speak to him again.

The Jewish artisans pride themselves on gold and silver inlay work on copper and brass. For centuries the Jews of Damascus have been engraving on copper and wood: they produce good quality work which is exported throughout the Middle East. The majority of the designs are Islamic, with quotations from the Quran and other Islamic elements, but Jewish items are also to be found although they are not displayed prominently. These are family businesses with aunts, uncles, brothers and sisters all working together. They assured me that business was good and that in economic terms they had no complaints apart from the fact that they had difficulty travelling with their wares within Syria and had to rely on Muslim middlemen for much of the export work. "Sometimes they cheat you more, sometimes they cheat you less. No matter, thank God, we make a living. By the standards of Damascus we even make a good living."

They told me that there are a number of really wealthy Jewish merchants operating in Damascus.

"Would they leave if they could?"

"We would all leave if we had the chance – except a few old people who want to die here."

In one of these shops I found a craftsman who was putting the finishing touches to an ornate tray intended for the Gulf market. He spoke fluent colloquial Hebrew which he had picked up from Israeli radio. "I learnt Hebrew because I feel sure that sooner or later we shall be able to go there, so I thought I should prepare myself. We must be allowed to get out. We are not living in a prison here. We are living in a tomb. Even our Judaism is dead. It has no contact with the real world. There is no way in which it can change. We have no respect for the official rabbis; they work too closely with the Syrian authorities and they refuse to ask the government for the one thing we all want – the right to emigrate. All we have is a sense of belonging to something whose ritual gives some meaning to our lives. But I don't know what. The situation is not that bad: it is worse in Aleppo. If you want to see Jews suffering, you must go to Aleppo." I realized I had done all I could in Damascus and the following day I left my hotel and made my way to the central bus station.

There was an absolute absence of foreigners: no back packers, no package tours. Soldiers, *fellaheen* returning home after *Eid al-Fitr* (the feast which concludes the month of Ramadan), and businessmen made up the bulk of the passengers. The endless cadenzas of Um Qulthoum were punctuated from time to time by the honking of buses.

Next to me on the bus, a young soldier with the calloused hands of a *fellah* was leafing through a glossy Arabic journal. The lurid cover conveyed the Syrian obsession with the "Zionist American Imperialist" threat to Islam: a star of David inset with the stars and stripes superimposed on a view of the Dome of the Rock in Jerusalem. The soldier was reading an article entitled "Zionism against world peace".

The older suburbs of Damascus are indistinguishable from Tel Aviv or any other modern Mediterranean city. Small apartment buildings of worn concrete, palm trees and the occasional good modern building line the road. Aleppo lies about two hundred miles to the north. Cultivated land soon gives way to mountainous desert with occasional Bedouin settlements.

The desert before Aleppo is dotted with villages made up of unusual dome-shaped houses known in Arabic as *akwakh*. The bus pulled to a halt in front of one of these villages at a *Muhabarat* road block. Before boarding the bus I had been required to fill in my name, nationality and destination on a form which I assumed at the time would be passed on to the police. A plain-clothes official mounted the bus and made straight for me. He was a hard-looking man in his late twenties, his face barren of expression. He asked, in Arabic, for my passport. Pretending not to understand I smiled back at him. He moved to the back of the bus and started interrogating three *fellaheen* wearing *kaffiyyas*. They were taken off the bus and their bundles were searched by black-shirted officials of the *Muhabarat*. For no apparent reason they were beaten before being put back on the bus. Once again I was asked for my papers and again I smiled. This time they gave up. The soldier next to me said, "You are right not to talk to these people. In the army we hate them."

Shortly after I arrived in Aleppo I was approached in a tree-lined street near the famous Baron Hotel by a young man, skinny and diffident. He asked me, anxiously, if I was English. He shot out his hand and introduced himself as a first-year student of English at the University of Aleppo. Apparently, he had been waiting for a chance to practise English conversation with a native. We walked through a *suq* which is considered to be the finest in the Middle East and talked of literature. He had been studying Wordsworth, Keats and *A Passage to India*, and his English was derived almost entirely from these sources. He told me that he was of a religious inclination. As we sat in the courtyard of a twelfth-century mosque of great beauty and simplicity, he told me that this was the first time that he had entered a mosque other than the modern mosque next to his parents' home.

"Many of us believe that it is not progressive to maintain such ancient monuments. Surely they should be dismantled and replaced with edifices suitable to the age." He waved his hand in the direction of the minaret: "This should be exploded, all these old-fashioned buildings." Ibrahim was Sunni and told me that he did not know any Shia people although he knew people who did. I asked him, "Are the Shiites as Arab as you Sunnis?"

"Yes."

"Do you know any Christians?"

"Yes. They are Arabs too."

"Do you know Jews?"

"There are Jews. But they are not Arabs; they cannot be Arabs."

"But," I protested, "they speak Arabic and they have been living here for two thousand years or more. Are they not as Arab as the Christians?"

"No," he retorted fiercely, "they cannot be Arabs, they are blackguards and curs." He paused and asked me, "You know what it means, this word 'curs'?" I nodded and he continued: "I have never met one, but I know that they are curs. They are our foes, our eternal foes." I asked him where the Jews lived. "Just near where I met you. But I never go to the area. It is closed to us and also to you. You are partial to them, I believe?" As he asked the question, he looked hurt and perplexed.

That evening I was invited for drinks by an acquaintance living on the outskirts of the city. After a while I told him that I wanted to meet Jews. He looked horrified.

"I meet Jews from time to time. I have Jewish friends. But they would refuse to talk to you and would be very angry with me for putting them in a difficult position. What can they say? All I can do, perhaps, is to create a situation in which, by chance, you will meet Jews. Then it's up to you."

He did not elaborate and went off to make some telephone calls. He returned a few minutes later and said, "Tomorrow at 9.15 you will be at this address," and he handed me a piece of paper. "A Jew will arrive and he will try to sell you a carpet. He will probably succeed, he is a very good carpet seller." He laughed. "Some of the Jews here could sell pork to a rabbi."

I returned to the Baron Hotel where I spent some time talking with its venerable Armenian owner. His conversation was laced with heady names. He had known Philby, the English spy, and recalled him as "a charming man who stuttered"; T.E.Lawrence, who had spent a long time at Baron's, had been a friend of his father's. But most of his conversation was about Armenian affairs: the iniquities of the split patriarchate and the depravity of certain members of the Armenian community in Jerusalem. He was drinking mineral water from Soviet Armenia, but urged me to try the dark Armenian brandy which his health no longer permitted him to enjoy. Gradually I tried to turn the conversation to the Jews. I had no sooner mentioned the word than he rose slowly to his feet and put on an ancient recording of the Armenian *Sanctus*. "There are some things we do not talk about," he whispered, "not even here. But I can tell you this. Their position is no worse than ours. One of our Armenian priests was murdered recently and we are afraid. We fled here from the Turks. If necessary we'll flee again: some of us could go to Soviet Armenia. It's the closest thing we have to home."

I asked him about the murder of Lilly Abadi. "You know there are a lot of rumours. A hairdresser we know said it had something to do with an upstairs neighbour, others say it was connected with Abadi's business. Other people say other things." He shrugged and would say no more.

I dined alone that evening in the hotel dining-room wondering how I would be able to make contact with the Jewish community the following day and whether it was worth the risk to them and to me. As I passed through the lobby after dinner I noticed two armed men examining the hotel register. I went back into the restaurant and left the building by a side door. Behind Baron's there was what looked like a red-light district. As I walked through it shapes and shadows seemed to change places. But I was not sure if I was being followed by anything other than my own imagination.

In mediaeval times the Jews called Aleppo by its Hebrew name, Aram Zobah, the ancient Syrian kingdom which is mentioned in the Bible as one of the states which fought in coalition against King David. The more normal Hebrew form of Aleppo is Haleb, which Jewish tradition connects with Abraham who pastured his flocks on the hillsides around the town and freely gave the milk (Hebrew *ḥalav*) to the town's poor. Certainly Aleppo figures prominently in the traditions and the history of the Jews. The most ancient of the

Aleppo synagogues, the Mustaribah, is said to date from the fourth century A.D. Benjamin of Tudela found 1,500 Jews living in the city when he visited Syria in the twelfth century and the community expanded after the expulsion of the Jews from Spain, three hundred years later, in much the same way as did the Jewish community in Damascus.

Situated between the Euphrates and the Orontes, Aleppo was on the trade route to Baghdad and India. The Jews were active in trade and maintained close links with Jewish communities further east: for some four hundred years they had trade and religious links with the far-flung Cochin Jews of southern India. For the most part, Jews were able to prosper and live in harmony with their Muslim neighbours, but none the less they frequently suffered discrimination and persecution of various sorts. In the fourteenth century the Arab historian Ibn al-Fuwati wrote:

> In the province of Aleppo ... they are still bound to wear the badge: a yellow star of David. What is more, according to Islamic law, when the poll tax is to be paid, the person who delivers the sum must be standing, he who receives it must be seated. The former places it in the other's hand so that the Muslim receives it in the palm of his hand, the Muslim's hand being above and that of the Jew being below. The latter then stretches forth his beard and the Muslim strikes him on the cheek with the words: "Pay the dues of Allah, O enemy of Allah, O infidel!"

When Tamerlane, the lame warrior prince of Samarkand, conquered Syria in 1400, many Jews were slaughtered along with the rest of the non-Muslim population. None the less, the community survived and over the following centuries foreign Jews, particularly from France, settled in Aleppo. Known as Francos, they played an important role in the burgeoning trade between Europe, Persia and India.

As late as 1895 the Jews and Christians suffered from the religious fanaticism of local Muslims: indeed, wholesale slaughter was only prevented by the intervention of the Turkish military commander, Adham Pasha, who threatened to turn his cannons on the Muslim quarter of the old city unless the rioters dispersed. Blood libel accusations were levelled against the Jews on more than one occasion during the nineteenth century. Despite these setbacks the Jewish community continued to prosper and on the eve of the First World War there were 14,000 Jews in Aleppo, although in the inter-war years the community dwindled as a result of emigration, mainly to the West. During the riots of 1947 most of the Jews fled to Lebanon and Turkey, where some of them settled. The majority, however, moved on to the United States, Europe and Israel.

After the First World War the more affluent members of the Jewish community left Bahsita, the old Jewish walled ghetto in the north-west corner of Aleppo, and moved into a new suburb outside the walls known as Jamiliyya. Today almost all of Aleppo's Jews live there: *Muhabarat* surveillance of Jamiliyya is intense.

I made my way through the early morning bustle of the *suq* to the carpet store where I was to meet my Jewish contact. Carpets from Turkey, Iran and even the Far East were heaped on the pavement outside the shop along with saddle bags, cushions and prayer mats. Sure enough someone was waiting for me. He presented his card which described his profession as Interpreter and Corresponding Agent. Being fluent in a number of European languages he made a living by wandering around the *suq* helping the stall-holders to sell their wares to the infrequent foreign visitors. Throughout Ottoman times Jewish exiles from Europe had used their linguistic skills to good advantage and provided a necessary link between the Arab world and Europe. This man, Albert, was following an ancient tradition.

I told him that I would like to learn something of the Jewish community and, if possible, to meet some Jews. He looked aghast and hurriedly suggested we go for a walk. The Muslim owner of the carpet store followed us through the bazaar for a while imploring me to return. But Albert pushed his way through the crowd energetically until we came to a quiet square where he sat down and rested. "Jews," he said. "Who wants to know about Jews?" Having heard my reasons Albert looked very unhappy, but agreed to show me something of the Bahsita area and some of the synagogues if I would pay him the usual tourist guide rate. Albert set off again, this time less energetically.

"How many Jews are there in Aleppo?" I asked.

"Very few these days, my friend, very few indeed. Hardly any. But there are many carpets here."

"How would you describe the condition of the Jews at the moment?" I persisted.

"Not so good and not so bad. But the carpet business is truly awful. There is no one to buy and the prices are ridiculously low.

It's a good time to buy but a very bad time to sell. A man has to make a living. I have a wife and children to support." Firmly I explained to Albert that I was interested in Jews, not carpets. He laughed politely.

Albert took me for a brief visit to a synagogue which had been totally destroyed in the 1947 riots, but which had been restored largely through donations from the wealthy Syrian Jewish community in the United States. A few children were playing in the courtyard and in a corner of the beautifully restored building an old man was studying alone. But Albert was too nervous to want to linger and, looking constantly over his shoulder, he suggested that we return to the *suq*. "Our greatest synagogue was turned into a mosque five hundred years ago," he observed glumly.

As we walked back through Bahsita he began to talk more freely. In its day the old city must have been beautiful. Finely built fountains were now full of rubble; intricately carved arches were overgrown with skeins of rotting electric cables. Albert showed me the house where he had been born and where his family had lived for generations. "Everything has changed. It was clean, we all lived on top of each other but there was a real sense of community. It's not like that in Jamiliyya. No one trusts anyone else. I've been arrested by the *Muhabarat* seven times. Each time I was beaten. Each time they seemed to know more about my business than I do myself. They can always get informants. They threaten you with *Al-Abd al-Aswad* [the Black Slave] and you'll do anything." The Black Slave is a much feared device used by the *Muhabarat* into which the victim is strapped; when switched on it inserts a heated metal skewer into the anus. "I've never suffered that but I've been beaten often enough. Do you know what *Fallaqa* is? They tie you to a table and beat the soles of your feet. That's the worst."

I tried to commiserate with him. Nonchalantly he said, "Oh, that's quite normal here. I'm not worried by that. My main problem is that I'm not allowed anywhere near the embassies in Damascus. They've warned me away. I suppose they think I'm spying for Israel or something. The fact is a number of American diplomats owe me money for carpets. How am I going to get it?" and he raised his arms in a dramatic gesture of supplication. "But I still have some very fine carpets. Some beautiful ones. I'll show you something that no one else has seen. Something special."

Again I told Albert that I was not interested in carpets but that I wanted to hear more about the condition of the Jews. He smiled and led me into a narrow alley which forms part of the spice-sellers' quarter. He pointed out a short ginger-haired man of middle age with characteristic mongoloid features. "See that man – he's a Jew. He's also a cretin. To sell spices here you don't have to be much more than a cretin. He manages, with God's help. His father was killed in the riots and he inherited the business. But you wouldn't believe me if I told you that he has one of the most beautiful and intelligent wives of the Jewish community in Aleppo. He could not find a wife here so he tried to find one in Damascus, and there are so few young men there and so many unmarried girls that he came back with this beauty of his. He's not stopped smiling since." I asked him for some good quality henna and he served me slowly and deliberately from a sack in the corner of his shabby vaulted shop. His shop was indistinguishable from the other stores in the area. They all had the same shabby interiors and carried the same produce; according to Albert he was able to carry on his business without interference from anyone. "If he were a business genius he might have problems," he said caustically. "In general, although we have difficulties we are protected by Assad and this makes us unpopular with some of the other minorities. The Armenians in some ways have it worse. They have to serve in the army and there is no great pressure on the government to treat them well. But despite the great Soviet presence in Syria, the Americans still have a lot of power and they are pro-Israeli. If the American Ambassador wants to see Assad he can see him within minutes. At least that's what they say."

"Would you leave Aleppo if you could?" I asked.

"Immediately. I have sons in Paris. They run a gift shop and are doing well. But before I go I must sell off my stock of carpets. It's all I have." He led me back in the direction of the carpet shop. On the way we met a Jewish friend of his. A young man wearing a white shirt and American blue jeans, he could have walked off any Western university campus. He told me that he had trained as an engineer in Damascus, but was now prohibited from practising and helped to run his father's shop. He was not prepared to say much and kept looking around fearfully. He gave me a firm handshake, said "*Shalom*" and disappeared into the crowd. Albert took my arm and led me to the carpet shop, sat me down and brought me a cup of sweet Turkish coffee. "Now, my friend I have something for you. You are interested in Jews? Then let me tell you a story. A famous old Jewish family from Aleppo, great friends of mine, escaped over the Turkish frontier only last week. As far as I know they all made it. They left me to dispose of all their assets and I shall send the money as and when I am able. Their finest piece is a Jewish carpet which used to hang in one of the synagogues, but during the riots it was taken down because of its great value – there are very few Jewish carpets – and it has been with the family ever since. You can have it for next to nothing. A few hundred U.S. dollars. What's that to you?" I told Albert that it was a lot for me and that in any case I was on my way to India and had no desire to travel with a carpet.

"Ah, yes, there are Jews in India," he smiled enigmatically. "See the carpet anyway." By now, it was late afternoon and the only light in the *suq* was the setting sun filtering through a five-sided star carved into the stone vaulted ceiling. Albert laid the carpet out in a pool of dust-filled sunlight. I had never seen such a beautiful carpet: it was radiant with gold, red and the faded blue sometimes seen on old Chinese porcelain. "You see the stars of David?" whispered Albert. "And the *menorot* – you know the Jewish candlesticks. This, sir, is the only Jewish carpet left in Aleppo. In fact, it is against the law for me to have it here. These Jewish symbols are banned." And Albert hurriedly threw a simple Turkish kelim over it. "The last Jewish carpet in Aleppo," I reflected. I had to have it. I gave Albert the money and, as I left, he smiled radiantly at me and said, "The meaning of life lies in this carpet." It was only later that I realized what he meant.

I returned to Baron's with my purchase and immediately spread it out on the floor of my room. In the neon light of the hotel it looked very different. The greens and pinks were vulgar and modern; the *menorot* were not *menorot* at all and I remembered too late that the six-sided star is a common Islamic device. Albert, may he live to be a hundred and twenty, as the Hebrew blessing has it, had sold me a good story and an awful carpet, embroidered skilfully by his inventive imagination.

Later that evening I saw my old acquaintance out in the suburbs of Aleppo. I told him the story and he laughed: "I warned you. But you have to remember that the Jews and the other minorities here have had to live by their wits for centuries and it's still true in these difficult times and particularly here. We have a saying: 'It takes three Greeks to cheat a Syrian, three Syrians to cheat an Armenian, and three Armenians to cheat a Jew.' But it sounds as if our carpet friend is in a league of his own." He paused and lowered his voice: "By the way I should mention that some official gentlemen were around here today and were asking about you. I advise you to be very careful and, forgive me, but please don't visit me again."

The next morning I made my way furtively to the *suq* to meet Albert as planned, but of course he was not to be found.

None the less, over the next few days, using other contacts and some discretion, I succeeded in meeting a number of other Jews. I spoke to one relatively prosperous old man in his shop in the new city. He ushered me into a back room almost completely filled with an ancient desk and ledgers going back to the days of the French mandate. I asked him if he would like to leave Aleppo. "No, not really. I am too old to move anywhere. I would like to be buried in the Holy Land. Many of my family went to Jerusalem to die and their graves are on the Mount of Olives. That is where the Messiah will come: and they will be the first to greet him. But it will be impossible for me, impossible for most of us. But you know even Aleppo can be considered almost part of the Holy Land: in the Talmud it states that Aleppo is the most northerly point to which a Jew living in the Land of Israel can travel without being considered a 'traveller'. That is to say that in some way Haleb is within the periphery of the Holy Land. This is our home. We have been here since before the Muslims, before the Christians. And it's not true that they hate us exactly. They also are afraid of us. Let me tell you a story. My father, of blessed memory, used to travel around Syria on a horse selling goods in the small towns and villages. One day when he was travelling in the desert a young Bedouin set upon him: he wanted to take his horse and kill him. My father took out his little account book and started to write in it. The Bedouin, being superstitious and anyway believing that all Jews are magicians and sorcerers, asked, 'What are you doing?' Calmly, my father replied, 'I'm going to turn you into a donkey.' Of course the Bedu fled and my father died in his bed at an old age." The old man laughed. "The real point is," he continued, "if you are, like me, a religious Jew you can live anywhere, even in the bitter exile of Syria. We Jews have always suffered to some extent or other in every generation. We are used to it – that is what being a Jew is about. But our faith in redemption and our day-to-day rituals and prayers are sufficient to sustain us. That's what they are for. I have not long to live – and when the time comes I know I will not be buried on the Mount of Olives – but anyway Aleppo is almost part of the Holy Land." He did not look convinced.

Most of the other Jews I met wanted to talk about the Abadi case. Privately, I was told, even the Syrian authorities were convinced that the murders had a political or ideological motive even though their public position was that it was no more than a criminal act. A Syrian policeman I met at a bar on Baron Street told me after a few drinks that he was pretty sure that it was the work of the Muslim Brotherhood, but that it was certainly not the work of either the *Muhabarat* or the much feared forces of the Syrian President's brother Rifaat al-Assad – the so-called Defence Brigades. "You cannot separate the situation of the Jews here from what is happening throughout Aleppo all the time. Over the last few months there has been a great deal of violence. Nothing quite like the Abadi business perhaps. But a lot of murders and a lot of violence. A lot of it is political or religious if you like. The fear is that Aleppo will go the way of Hama, where so many thousands were killed. It could happen here. There's a lot of fanaticism in this town. It's under control at the moment but it could explode any time." He stared moodily into his beer. When I asked him how anyone could have got into the Jewish Quarter without the *Muhabarat* being aware of it, he had no answer.

All the Jews I met spoke of the murders with a fear born of the anxiety that it could happen again. For some time afterwards the Jewish school and the synagogues had all been closed and families moved in three to a room hoping that there would be some safety in numbers. One man I met who works near Abadi's shop in the old city told me, "Abadi himself has a good idea who did it but it's worth more than his life for him to say anything. He's just carrying on: business as usual."

The day before I left Aleppo I met Eli, a young Jew from Damascus, in a café near Aleppo's famous citadel. "I know the Abadis," he told me, "but here everyone knows everyone else. It's a very tiny community. Abadi is in direct descent from a famous Hebrew poet and composer of liturgical music. It's a famous name. But it's all in the past. I'll tell you something: if such a thing, God forbid, were to happen in Damascus, we'd fight. We are more numerous and far better organized. We don't have arms but we'd fight with our hands if necessary, even though we'd all be killed in the process." He spoke with such passion that I believed him. He told me that he was going to leave Syria very soon: "One day without telling anyone I'll just leave. We have a route across into Lebanon and from there it is not impossible to get to Cyprus and then on to Israel or the United States. My parents and perhaps my friends will be tortured if I disappear, but at least they'll have nothing to tell. My parents are quite old and they want me to get out at any price. They say that we Syrian Jews are descended from the lost tribes of Israel. But I say we are the forgotten tribe. And I don't want to be forgotten any longer."

Some months later while I was in the United States I visited Deal, a summer resort town fifty miles south of New York. Here around a thousand prosperous families of the New York Syrian Jewish community spend their summers. Magnificent mansions face the Atlantic and a handful of Syrian synagogues and Syrian grocery stores nestle together along the main street of what the casual passer-by might imagine to be a normal New Jersey township. From one member of this affluent community I heard that Eli had made it. He had crossed the mined border into Lebanon, had walked through what was then a war zone and made his way to America, like so many Jewish refugees before him. The prospects for those left behind are not good.

TWO

Lost Tribes in India

The trade routes which for centuries linked Damascus and Aleppo with the treasures of India and China used to traverse countless Jewish communities. But east of Syria today there are few signs of the historic settlements which used to play such an important role in the commercial life of the eastern Muslim world. The Jews of Mesopotamia – the "land between the two rivers" – have all but disappeared: 120,000 left for Israel in 1951 in the famous exodus code-named Operation Ezra and Nehemiah. The fate of the few who chose not to leave was highlighted by the execution of nine Jews in Baghdad's Liberation Square in 1969 on charges of spying for Israel. Further along the old trade routes, in Iran, there are about 30,000 Jews left trapped in the ambiguities of the "Government of Allah", which recognizes the Jews as a "protected" people free to practise their religion yet considers Jewish nationalism a crime punishable by death. The president of the Jewish community was executed in 1979 for "connections with Israel" and since then ten other Jews have been executed on similar charges. Since the outbreak of the on-going Iraq–Iran war there has been little contact between Iranian Jews and the outside world. Further east still, in Muslim Pakistan, a handful of old Jews in Karachi make every effort not to advertise their presence. But among what E.M. Forster refers to as the "hundred Indias" there are far-flung Jewish communities, most of them of great antiquity, living at peace with their Muslim and Hindu neighbours, but not, as I was about to discover, with each other.

There are three distinct Jewish communities in India: the Cochinis, the Bene Israel and the Baghdadis. Perhaps the best known is the Jewish community of Cochin in south India, which lays claim to having arrived in the sub-continent after the destruction of the first Temple. The Malayalam-speaking Cochin Jews have formed three separate endogamous groups: the White Jews, the Black Jews and the *Meshuhrarim* (freedmen). The fair-skinned White Jews represent a mixture of the indigenous Indian Jews and those European and Middle Eastern Jews who, like the Portuguese and the Dutch, found their way to the tip of the sub-continent over the centuries. The Black Jews are in most respects indistinguishable from local Indians and maintain separate synagogues and a separate social life. The *Meshuhrarim*, descendants of manumitted Indian slaves who had converted to the religion of their Jewish masters, were attached to both groups. Unlike the Bene Israel, the Cochini Jews never lost contact with mainstream Judaism, not least because of the importance of the ports of south India for the trade between East and West which assured a constant flow of visitors from Europe and the Middle East. Judaeo–Arabic documents found in the depository (*genizah*) of an eighth-century Cairo synagogue have shown that between the tenth and twelfth centuries strong commercial ties existed between the Cochin Jews and Mediterranean and Middle Eastern Jewish communities. Some of the great mediaeval travellers, Jews, Christians and Muslims, recorded the presence of Jews in the vicinity of Cochin: Al-Idrisi (1156), Benjamin of Tudela (1167), Qawzini (1280), Marco Polo (1293), Ibn Batuta (1342), as well as a number of others. The Cochin Jews have the earliest documentary evidence of the association of the Jews with India. A copper inscription still in the hands of the few remaining White Jews describes privileges granted to a certain Joseph Raban, thought to be the leader of the Jewish community of Cochin, by the Hindu ruler of Malabar around the end of the ninth century A.D. By 1948 there were 2,500 Cochin Jews of whom only a hundred were White Jews, but since then emigration mainly to Israel has reduced the community to no more than a handful.

The Bene Israel of West India believe that their ancestors left Palestine as a result of the persecutions of Antiochus Epiphanes (175–165 B.C.) although some of them claim that their forebears were one of the lost tribes of Israel. But even though the real origins of the Bene Israel are obscure, it seems clear that the community is an ancient one. According to tradition their ancestors arrived by sea "from somewhere in the north" and were shipwrecked near Navagaon, some twenty miles south of Bombay

Island. Seven women and seven men survived the shipwreck and established the community in the surrounding villages. For centuries they lived in isolation on the Konkan coastland, and the outside world was unaware of the presence of Jews in western India until the middle of the eighteenth century, by which time many of them had moved from the Konkan to Bombay. Thereafter, the Bene Israel were "discovered" by a "white" Cochini Jew, David Rahabi, a member of the eminent Rahabi family, and also by the American Missionary Society which was responsible for publishing among other things a Hebrew Grammar and the books of Genesis and Exodus in Marathi, the language of the Bene Israel.

At the time of their discovery, the only Hebrew of which they still had any knowledge was the first word of the elemental Hebrew prayer *Shema Israel* – "Hear O Israel, the Lord our God, the Lord is one." But Rahabi perceived many vestiges of ancient Jewish practices. The Bene Israel kept the Sabbath, maintained a number of festivals and fasts, and observed some of the Jewish dietary laws. For Rahabi the acid test of their Jewishness came when he gave the women some fish to cook including some which lacked scales and fins and which are, therefore, prohibited under the laws of *kashrut*; these the Bene Israel women refused to touch. Persuaded that they were indeed Jews, Rahabi suggested certain reforms to them and over the next decades they made great efforts to bring their customs into line with orthodox Jewish practice elsewhere.

Subsequently, a number of Bene Israel went with the British to Aden, which was ruled from Bombay, and there they created links with the Jews of Aden and the Yemen who taught them elements of their own observance. Some confusion has crept into the oral tradition of the Bene Israel who maintain that David Rahabi came from Egypt, not from Cochin, and undertook his mission in the eleventh rather than the eighteenth century! Although the Bene Israel are Indian in appearance, speak an Indian language and have been influenced by the surrounding culture, they have none the less over the centuries managed to maintain a quite separate existence from the other Indian groups of the area. The fascination of the Bene Israel lies in the fact that, despite their separation from world Jewry, they have managed to cling to at least some aspects of their ancient religion for perhaps as long as two thousand years. Today the Bene Israel are the only sizeable Jewish group in India.

Several thousand of them still live in and around Bombay, although thousands more have settled mainly in Israel but also in England and the United States.

The most recent of the Jewish communities in India is the group of Jews of Middle Eastern and mainly Iraqi origin, based chiefly in Bombay, known as the Baghdadis. The first Middle Eastern Jew to put down roots in Bombay and arguably therefore the founder of the Baghdadi community was Shalom ben Ovadiah ha-Cohen, a trader from Aleppo, who arrived in Bombay in 1790. In the early decades of the nineteenth century he was joined by other Arabic-speaking Jews from Basra, Aleppo and the Yemen and particularly from the great Jewish centre in Baghdad. But it was only with the arrival of the legendary David Sassoon in 1832 that the community fully established itself in West India.

David Sassoon's biography reads like an oriental romance. He was a remarkable man who among other things founded a great trading empire and created the fabulous wealth of a family which came to be known as "the Rothschilds of the East", although in view of the Sassoons' aristocratic lineage and their unlikely claim to be descended from King David, they may not have welcomed the comparison with the *nouveau-riche* Rothschilds. For even in the pre-modern period, the Sassoons of Baghdad were courtiers and merchant princes who served successive generations of Pashas as *sarraf* (chief financial adviser) and were frequently given the ancient title of *nasi* or president of the Jewish community of Baghdad. According to one eighteenth-century traveller: "The family flourished for centuries in Baghdad where their chieftains wore gold tissue to ride to the palace of the Pasha, those in the streets and outside the bazaar standing with bowed heads until he and his retinue had passed." But by the end of the nineteenth century Baghdad had fallen on hard times and with it the fortunes of its Jews. David Sassoon's father, Sheikh Sason ben Saleh, at one time *sarraf* to the Pasha and *nasi* of the community for almost four decades, fell from favour. According to a contemporary Jewish source, "The Sheikh Sason cannot show himself or go about free because of fear of the governor of the city, may his name be blotted out."

In 1828 David Sassoon, who was by then the unofficial head of the Jewish community, was gaoled at the orders of the city's governor, Daud Pasha. A considerable ransom was agreed between

Daud and Sassoon's father, but one of the conditions of his release was that he would be banished from Baghdad. Upon his release from prison, Sassoon chartered a ship which took him down the Euphrates to Basra, the city of his intended exile. According to family tradition he disguised himself in a long cloak in whose lining pearls had prudently been concealed. From Basra he sailed immediately to the Persian town of Bushire at the head of the Persian Gulf, fearing correctly that Daud Pasha would attempt to recapture him in order to extort more money from Sheikh Sason. Sassoon started trading in a modest way in his new home and within the next few months was able to create trading links with his elderly father, who soon found a way to escape from Baghdad and join his son – although he was obliged to leave the bulk of the family fortune in Baghdad.

By this time Bombay had become the obvious entrepôt for trade with the Far East, the Gulf ports and East Africa. Bombay was, of course, British. It had first come into British hands as part of the dowry of Catherine of Braganza on her marriage to Charles II, who in turn leased it to the East India Company for £10 a year. By the 1830s a new era of British mercantile tolerance seemed to offer boundless opportunities to the independent trader and already Parsees, Muslims, Jews and Hindus were trading peacefully and profitably in the benign atmosphere of the Bombay Presidency.

Sassoon's arrival in Bombay coincided providentially with the start of a commercial boom in the city which he was able to exploit to the full. He soon became involved in trade with the Gulf ports: in Persia there was a demand for manufactured cotton goods which British exporters were eager to exchange for silk, spices and pearls. Sassoon's offices in Tamarind Street, Bombay, became a crucial centre for this trade. At the same time, family tradition was maintained by re-establishing the name of Sassoon on the great overland silk route. But the area of trade which was to prove more profitable than any other was traffic in opium. By the time Sassoon arrived in Bombay, one third of the trade of the Bombay Presidency was given over to the endless cycle of profit represented by the drug. Opium was grown and refined at little cost in India before being shipped to China where it was sold for silver; the ships would return to Bombay with tea and silk which British ships bringing out manufactured goods for India and the Persian Gulf would take back to European ports. For some time the Sassoons almost held a monopoly in the opium trade. This led them eventually to follow the great British trading house, Jardine, Matheson & Co., and open general trading offices in Canton, Shanghai and Hong Kong and somewhat later in Japan.

Soon after his arrival in Bombay, David Sassoon started to acquire wharfage in the city. This policy was eventually to result in the construction of the giant Sassoon docks, which still stand as a memorial to his initiative and these days serve as Bombay's fish market. No less important in terms of the development of Bombay was Sassoon's decision to join a number of Parsee entrepreneurs and start constructing cotton mills in the city. By the middle of the nineteenth century Bombay had become the second largest city in Asia, after Tokyo, and the second largest in the British Empire, after London. Its economy was based largely on cotton. Sassoon's wealth grew with that of the city so that within twenty years of his arrival in Bombay he was a fabulously rich man by any standards. His descendants were to become British aristocrats with country seats, Park Lane houses and Scottish grouse moors. But David Sassoon remained an oriental and, until his death, wore the turban and flowing robes of a Jewish grandee from Baghdad.

As his enterprises developed Sassoon became aware of the need for a reliable and loyal workforce and it was not long before he started bringing out Jews from distant Baghdad to work for him. As the Sassoons and other Baghdadi families such as the Ezras, Ezekiels and Gubbays prospered, so did the Jewish community which depended upon them. By the end of the nineteenth century two thirds of the trade out of Bombay was in the hands of the Baghdadi Jews.

As white non-Indians, the Baghdadis enjoyed a special status under the Raj and, until the end of British rule in India, the community enjoyed considerable prosperity. But despite the fact that the Jews in India had never suffered from anti-Semitism and despite their material prosperity, after Indian independence in 1947, many of the community left. Although some were attracted to the new Jewish state, the majority went to Australia, the United States and Britain. The reasons that drew them away were many, but most felt that as whites they would have no place in the new Indianized India. New restrictive laws on the export of capital and import of luxury items made their traditional role as international traders difficult to maintain. Today there are no more than a few hundred Baghdadis left in India, almost all of them concentrated in Bombay.

The sight of Bombay from the air is not pleasant. One of the world's largest shanty towns stretches away from the airport in every direction. The monsoon had just arrived, covering the town in a layer of mud. According to the *Times of India*, thirty-seven people had died in the floods the previous day. The shacks which make up this metropolis of poverty are constructed with anything that comes to hand: more often than not cardboard. Here and there new tile roofs provided a permanent adjunct to sodden paper walls; the owners of the new tile roofs were evidently not planning to move. As my taxi took me to the centre of Bombay there was little change for the better. As Indian beggars clamoured around the cab, I wondered on what side of the bread-line the Jews would be and how, if at all, Judaism had adapted its ideas to cope with the sort of poverty that forces tens of thousands to live and die on the streets. Leo Baeck, the great German liberal rabbi who survived the Theresienstadt concentration camp, argued that Judaism's greatest distinguishing feature is its emphasis on practical moral activism. Could it be active here? Bombay seemed to be festering. Even the taxi, rotten with rust, stank damply of corrosion.

My first appointment was with the British Deputy High Commissioner in Bombay. I told him of my interest in the local Jews. He looked rather depressed. "It was once a very important community – under the Raj. But times have changed and many Jews have left. Until recently there was a very famous star of the Indian cinema known to everyone just as 'Uncle David'. He was a Jew – but he emigrated to Canada last year. There are others but I really don't know many of them. There is not much reason to. They have no political significance and in any case they are a very small community. And, of course, there is no anti-Semitism. If there were there might be some reason for Her Majesty's Government to get involved. But there are so many things to worry about: the poverty, the race riots, the problems in the Punjab, the never-ending demand for British passports." Staring out of the window in a distracted way he smiled wryly and said, "I shall be going home soon." He did not sound sorry. His secretary brought me some recent clippings from the Indian press detailing in terminal tones the disappearance of Indian Jewry. I began to wonder why I had come.

Lost Tribes in India

From the office of the Deputy High Commissioner I telephoned Naim Gubbay, an Indian Jew of the Baghdadi community, with whom I had made arrangements to stay while I was in Bombay. He offered to come and pick me up in his car. As I waited in the stifling heat I watched the passive line of Indians queuing outside the consular offices for the British passports which would enable them to leave Bombay for ever.

A car detached itself from the maelstrom of traffic and hurtled towards the queue. With a squeal of brakes it stopped inches short of a group of motionless children. No one protested. It was too hot. Naim Gubbay, the driver, hesitantly walked over to where I was standing and shook my hand. An heir to the glorious mercantile tradition of the Bombay Jews, Naim was a cadaverous man whose face was stamped with an expression of perpetual surprise. He was accompanied by a fellow Baghdadi, Ephraim, a middle-aged velvet merchant. As we drove back to Naim's apartment in the fashionable Fort area I told them what I had heard from the Deputy High Commissioner. "A disappearing community?" Naim laughed. "There are still about 300 of us who have not disappeared yet!" I knew that there were many more Jews in Bombay than this and I showed my surprise. They both looked a little uneasy and said that they meant Baghdadi Jews; they had no idea how many other Jews there were. "There *are* others," said Naim, "but they are Indian Jews – the Bene Israel – although some people say they are not Jews at all. They're black, more or less. If you want to find out more I'll put you in touch with an American Jewish woman called Carmel who lives in Bombay. She knows all about them." Half under his breath Ephraim muttered that I would get a biased account from the American. This was a foretaste of the suspicion with which the Bene Israel and the Baghdadis viewed each other.

Naim's luxurious apartment was on Nariman Point, just off fashionable Marine Drive which runs along the sea front near the area of central Bombay where the city's distinguished Gothic and Early English buildings are to be found. The apartment was decorated and furnished in an unmistakably Middle Eastern way. The walls were lined with mirrors, good Persian rugs covered the parquet floor, and in the centre of the room a large number of plastic-covered chairs were arranged around a coffee table. Naim told me that he was planning to leave India as soon as he could. One of his problems was that in Bombay there were no eligible young Baghdadi men for his daughter to marry. For Naim it was unthinkable that she should marry a Bene Israel. He had bought some property in the United States and his daughter was to go to an American college the following week. "It will be the end of the Gubbays in India. You know we were here even before the Sassoons. But before I close down my printing works I have somehow to get rid of 150 workers and it's bringing me close to a nervous breakdown." I asked him why he was going to the United States rather than to Israel, where so many of the Indian Jews had settled. "Actually, not so many of the Baghdadis from Bombay have gone to Israel. You see, the Bene Israel and the Baghdadis are two distinct communities. The Bene Israel are Indian and black. We are Iraqis and white. Most of *our* people have settled in the States, Britain and Australia. But in Israel, too often, we get lumped together with the Bene Israel which, of course, we find humiliating. Apart from anything else, who can make a living in Israel?" None the less, he had firm enough views on the management of the Jewish state: "The only way to treat the Arabs is to hit them hard. That's all they understand. And the Arabs are Israel's biggest problem. I'd get rid of every last one if it were up to me."

The following morning I joined Ephraim and Naim for the Sabbath morning prayers at the old Baghdadi synagogue in the Fort area. The stark stone building which dominated the street was celebrating its centenary. Like everything else belonging to the Baghdadi community the synagogue had been founded by the Sassoon family. At the top of the steps which led up to the synagogue two elderly Jews were talking quietly. Next to them I noticed an ancient wooden chest with a dozen or so slits carved into it with the names of various Jewish charities inscribed next to the openings. These time-honoured charities were mostly designed to care for the local poor but some, such as the one named in memory of Meir the Miracle Worker, had traditionally been intended for the support of the impoverished Jews of the Land of Israel. Near the chest an old naked Indian was asleep on the flagstones.

Inside the synagogue the silence was broken only by the endless whirring of the fans and the distant traffic sounds coming through the open windows. A small group of elderly Jews stood silently in one corner. Ephraim told me it was good that I had come as I would be able to make up the *minyan*. When I explained that I was not

Jewish, he shrugged his shoulders and said nothing. Later when someone asked him if I were Ashkenazi or Sephardi he replied, "Oh, don't worry. He's nothing." Once the morning *shaharit* service started, Ephraim started to talk volubly about the affairs of the community. He pointed out a massively built man whose baldness was accentuated by a tiny crocheted skullcap. He was leading the prayers with some energy. "He's our professional busybody," said Ephraim. "He knows everyone in the community and organizes everything even if no one wants it organized." He warned me in similar terms against the rest of the small congregation. After the service I was introduced to the congregants. Some invited me to visit them while others took me to one side to tell me the secrets of the community. On that Saturday morning the synagogue was seething with suspicion and intrigue. The community still has considerable assets controlled by the synagogue trustees along with trusts established decades before by the Sassoons. According to some, millions of dollars are involved. No one seemed to know what would happen to these funds, although a number of people expressed the fear that the Indian government would take them over because there are no beneficiaries. One Baghdadi said to me, "We could of course give it to the Bene Israel but I would rather the government had it. At least they might put it to good use." The trustees, I was told, speak to no one – they are a self-perpetuating body answerable only to themselves. Repeatedly I was told that it was the Sassoons' money that established the Baghdadis in Bombay. But now a cloud hangs over the last of the Sassoon fortune in India.

After the service a middle-aged Jewish woman, who had been standing discreetly at the back of the synagogue rather than in the women's gallery, gave me a lift in her car, built in India along the lines of some ancient British model. Cultivated and elegant, with the mildest of blue rinses, she spoke in a crisp and dismissive way about the rest of the congregation. "I imagine you were not very impressed by the service," she said. "These are poorly educated people you know: small businessmen and community busybodies. The synagogue and our community's affairs are all that they have. All the bickering does not add up to much. I'm always surprised anyway about the expectations people have of religion and religious people. You know what our great poet, Siegfried Sassoon, David Sassoon's great-grandson, had to say about religion?

'Religion beats me. I'm amazed at folk/Drinking the gospels in and never scratching/their heads for questions'." I smiled and told her that I had not had any expectations one way or the other. But I had had certain preconceptions and they had been disappointed. I had hoped to find some sort of cultural synthesis here. After all, the Baghdadi Jews had been in India for two hundred years and it seemed inconceivable that they should have failed to have taken anything from their environment.

Ephraim had invited me to join him for the meal which customarily follows the Sabbath morning prayers. Respecting the laws of the Sabbath, Ephraim had walked back from the synagogue and was waiting outside his home. His house had once been a cotton warehouse owned by the Sassoon family. A number of half-naked Indians were squatting in the communal entrance. A wooden staircase led up to a spacious apartment divided down the middle by a simple wooden screen. On one side of the screen there were beds, on the other a group of children who were watching *Blazing Saddles* on a video machine. Ephraim's efforts to persuade them to turn down the volume went unrewarded. A dozen or so adults sat round a large table at the other end of the room. The sounds of the video were punctuated from time to time by the shrill song of small birds which were flying around the rafters in complete freedom. The chanting of the Hebrew prayers was barely audible.

A very old Baghdadi, who spoke Arabic and very little English, presided over the *qidus*. As he started the Hebrew blessings over the chupatti and the "wine" (a disappointing liquid made by boiling up raisins with a little sugar), he implored the children in Arabic to turn the volume down. He, too, was ignored.

The table groaned under the food; and here at least was some sort of cultural synthesis. An Iraqi chicken and rice dish was served with an Indian green mint and coconut chutney; the traditional Iraqi Sabbath boiled egg (boiled slowly for hours with the outer skins of onions) was served alongside a local salad of tomatoes and grapefruit. The spaces between the dishes were filled with piles of nan and colourful mounds of mangoes.

Over the meal I asked questions about the community and its relations with the predominant cultures of India. "Well, there's no anti-Semitism here," said Ephraim. "The Jews have always been top people in Bombay. I'm talking about the Baghdadis, of course. You British did not like the Jews all that much; but they did not like anyone else very much either and we were always considered by the Indians to be closer to the British than anyone else. We spoke English, naturally, and in many ways we became British. And naturally when the British left so did many of us." Remembering the naked Indians at the bottom of the stairwell I asked Ephraim how he thought Judaism had been affected by the poverty of India. He was clearly puzzled by the question. "I do not know what you mean," he said. "We have a simple community here without any embellishments. We don't have big parties any more. We cannot afford a properly trained ritual slaughterer [*shohet*] so we have to eat chickens the whole time. When the Sassoons and the other great merchant families went we were left high and dry. I mean, we are a poor community, but despite the poverty we manage."

'But I'm talking about the poverty of the Indians!" I exclaimed.

"Oh – that. It's only newcomers like yourself who even bother to mention it. Quite frankly, it has nothing to do with us. We have enough of our own problems."

Later that day a young Baghdadi couple took me to their club – the India Cricket Club. They told me with some pride that it had cost them over £4,000 to join. In many respects they still identified with the former British rulers of India. They had attended English schools in Bombay and, with the exception of a smattering of Hindustani, spoke no other language than English. They stressed that they were viewed by the Indians as British. "The Indians treat us like white English sahibs. You think they can tell the difference?" But despite their relative prosperity they would be leaving India soon. They were deeply concerned for their children: for the moment there were no Jewish children for them to play with – in time there would be no Jews for them to marry. I pointed out that there were still several thousand Bene Israel in Bombay. They smiled politely and explained the differences between the communities at some length. Intermarriage, they stressed, was not encouraged between the Bene Israel and the Baghdadis.

From the India Cricket Club I was taken by my host, Naim Gubbay, to his club – the United Services Club, which is some way outside Bombay – past the former British military establishments whose impeccably polished brass plates still bear such names as Nelson House, Fleet House and Air Force House. Naim's club with its lush, manicured gardens and ocean frontage was an imposing enough place. But he seemed ill at ease there and his half-hearted attempts to introduce me to Indian acquaintances were embarrassing for all of us. We stood for a moment watching the sunset and the giant rollers pounding the beach. "It's a very fine club," said Naim sadly, "but I don't really fit in. There is no anti-Semitism here but still I don't feel Indian. I don't really fit in anywhere." Beyond the immaculate clubs that fringe the sea-front outside Bombay the poverty started again. Along one side of the road hundreds of people were sleeping out on the sodden ground. As we drove back into the city the night sky was lit up by their small fires.

Shortly after we arrived there was a knock at the door and Naim's servant let in an Indian woman who lived in the same apartment building. She was dressed in a sari and wore the red *tika* on her forehead. Naim and the woman chatted for a few minutes in Hindustani. After she left Naim said: "You saw that Indian woman? She's one of those Bene Israel. She's married to a Parsee who converted to Christianity but she still goes to her Bene Israel synagogue. We didn't really spell it out this morning, Ephraim and I, but perhaps you can now understand why we look down upon these people. You see they are just like any other Indian group. You can see what we are up against. They are Indian and we are not. You can't help but look down upon them."

The two of us dined in some style. Dish after dish was brought to the table: pilau rice, nan, chicken, boiled rice, vegetable curries, salads, freshly fried popadums, vegetable pakoras, grilled fish. We were served by Naim's very old and black Indian servant who was wearing a white singlet and dark shorts. He spent eleven months of the year working in Bombay for Naim and one month with his wife in far-off Madras. He slept on the floor in the kitchen. Until a month before he had had the assistance of an Indian cook, recently deceased, who had known how to cook all the Iraqi Jewish dishes. Naim enumerated them fondly.

After dinner Naim made me prostrate myself on a studded contraption, not unlike a bed of nails, which he called a yoga board, while he carried on talking. As I lay in considerable discomfort he told me more about the Bene Israel: "As a community we have done all we can for them. For instance, we have allowed them into our Jewish club – but what happens? They always sit apart in separate groups. They are caste-conscious – I suppose just like us – and they don't mix even when we are prepared to. Of course there is no intermarriage between us – except for one marriage which very soon failed. I daresay there are extremists on our side, too. It's not many years since there was a court case in Bombay where some of the Baghdadis tried to prove that the Bene Israel were not Jews at all. They made a very convincing case of it. You see, there are basic differences which are very wide. We come from different backgrounds. They have adopted Hindu names, customs and dress. Their women wear saris. We have never accepted the idea of being Indianized. They *are* Indian. They speak Marathi. I hardly know a word of Marathi. The only thing we have in common is the fact that we are all supposed to be Jews."

The Nariman Point house had always been Naim's home. He had been born there and most of his friends, mainly other Baghdadis, lived within walking distance. Over the next few days I visited many members of the Baghdadi community. They spoke of their business difficulties and their future plans: nearly all of them intended to leave Bombay sooner or later to join their families in Los Angeles, London or Sydney. Time and again the Sassoon money was mentioned. The Indian Jews, the Bene Israel, were frequently referred to, always with a mixture of embarrassment and contempt.

The first member of the Bene Israel community I met was a retired customs official who had been appointed to the Indian customs and excise service under the British Raj. He invited me to take tea with him in his home – a cramped ground-floor flat in a bleak apartment house called the New Ready Money Building. The space around the apartment house was covered with rubbish and burnt-out bonfires. Children were scavenging hopefully among the rubble and the sleeping bodies. When I arrived my host, Shelem, was walking along the low veranda which extended the length of the apartment block. He greeted me in a worried sort of way and led me by a circuitous route to the back of the building, where a door led directly off the wasteland into a neat and formal little room which was apparently used for visits such as this. I was sorry not to see the rest of the house. Shelem treated the questions I put to him about the Bene Israel with considerable suspicion. To divert me from the subject of his co-religionists he showed me the few Jewish items in the room. On one wall there was an elaborate Sephardi charm printed in New York, on another a calendar advertising an Israeli bank. I asked him if he had anything which could be described as both Indian and Jewish, in other words as Bene Israel. He first asked me why I wanted to know and then told me insistently and repeatedly that there was nothing "particular" about the Bene Israel: they were Jews like other Jews and there was no more to be said. Certainly, there was nothing very exotic about Shelem. The same, however, could not be said of his daughter, who arrived bearing three cups of sweet Indian tea. She was a young woman of spherical construction wrapped up in what appeared to be a number of lime green viscose saris. "This", said Shelem, "is my daughter, the Professor." Earnestly, the young woman introduced herself as Professor of Sociology at the University of Bombay with a special interest in the Bene Israel Jews. She looked rather young for such a position, but I was none the less delighted to have met someone who would be able to tell me something about this remote offshoot of Judaism. I asked her what she had been working on. Uncertainly she started talking about sociological concepts which, she said, helped explain the community. She seemed no more comfortable with her second-hand ideas than her father seemed in his badly cut Western suit. Defiantly she kept returning to the terms "lateral mobility" and "vertical solidarity". Mercifully her father soon tired of this and sent her off to make some more tea. While she was in the kitchen he proposed that we should go off and visit the local Bene Israel synagogue. I suggested that perhaps I ought to bid his daughter farewell, but he looked at me suspiciously and wanted to know why.

To reach the synagogue we had to pass through an extensive slum area. The cardboard lean-tos along the pavement had been soaked through by what Shelem said were the heaviest monsoon rains for a hundred years. The pavement dwellers were too exhausted and hopeless even to ask for money. The synagogue presented a contrast to the decay and misery outside. A number of trees shaded the well-swept courtyard and the building had the classical simplicity of a Middle Eastern prayer house. A group of Indian Jews had already gathered on the veranda at the front of the synagogue. The windows and shutters were being opened for the afternoon prayers. The interior was adorned with a variety of Jewish artifacts imported in the main from Israel. Shelem pointed out an ornate brass candelabra which he said was of Bene Israel design and manufacture, but as far as I could see this was the only item in the building that was made by the Bene Israel.

Founded in 1796, the synagogue had originally been known as the Samaji Hasaji, although significantly it now has a Hebrew name – *Sha'ar Rahamim*, the Gate of Mercy. The building had been endowed by Samaji Hasaji Divekar, who had been a *subahdar* major in the Bombay army. He had four brothers, all of whom had served as officers in the Bombay army and, according to the story which Shelem told me, the five men had been taken prisoner when the Bombay forces had surrendered to Tipu Sultan of Mysore at Bednur during the Second Mysore War. They were about to be executed but at the last moment the mother of the Sultan, hearing that the men were Bene Israel, pleaded for their lives on the grounds that the sons of Israel had been mentioned with favour in the Quran. Seeing the hand of God in this last-minute reprieve, Divekar vowed to build a synagogue to commemorate their salvation.

I asked Shelem about the dishevelled group of Indian Jews who were talking quietly among themselves on the synagogue veranda. "Oh – they are the Jewish poor," he replied. "There are many poor people in Bombay as you have seen, but we have our poor Jews too. They are most harassing. They are sleeping and living on the veranda and are being paid by the community to recite the morning and evening prayers and in general to make up the *minyan*. This place can truly be called the Gate of Mercy. We look after our poor," said Shelem glancing balefully at the Indian beggars waiting beyond the gate.

Shelem did not stay for the service. He walked stiffly back to the New Ready Money Building explaining as we went that there were no differences at all between the practices of the Indian Jews and those of the Baghdadis or indeed of Jews elsewhere. "The Baghdadis used to pray in this synagogue when they first came to Bombay. Now they don't accept us entirely, it is true, but this is because of race not religion. It must be admitted that we are a lot darker than they are," he said wistfully.

That evening Naim arranged for me to meet Carmel, the American Jewish woman who had been working for some years with the Bene Israel in Bombay. She described herself as a sculptor cum historian. According to her own account she had been working on behalf of the Bene Israel community and getting very little thanks for her labours either from the Bene Israel or from the Baghdadis. Recently she had managed to raise $10,000 for the relief of the Bene Israel poor; but somehow the future of the funds was in doubt and the money had become another source of contention in the community. The Bene Israel community, she stressed, was just as internally divided as the Jewish community as a whole and so far she had not been able to achieve any consensus on how the monies should be spent. She smiled wryly: "They are a funny lot. They always seem to be at each others' throats, but I guess that does not make them different from Jewish groups any other place in the world. But the really fascinating thing apart from their being one of the lost tribes of Israel is their religion – their Judaism. You see their religious state has much more to do with India than with ancient Israel or the Bronx. It is emotionally felt. Like all Indians their instinctual life is alive and connected deeply with their religious life. They make no attempt to analyse their religion and in religious terms they live completely outside the world of reason. But they have no ethical sense at all. The ethical sense is a Judaeo–Christian concept – not an Indian one. So the idea of doing anything to help the poor is quite foreign to them, although they go through the motions of charity because they know that this is what the Baghdadis do. In fact, you see, the Bene Israel own quite a lot of property and they are not without funds, but they don't like to waste it on the derelicts you see hanging around the synagogue." I asked Naim and Carmel if any of the Sassoon money was still being used for the support of the Bene Israel poor. Naim laughed hollowly: "Don't ask me about the Sassoon money. I don't know what they do with it. But I shall give you one word of advice – don't believe anything that the Bene Israel tell you about it. They'll weep on your shoulder." He roared with laughter. "You'll be soaked to the bone!" His ravaged, lopsided face beamed at me: "They'll tell you what bastards we Baghdadis are – and they're probably right!" Again he laughed.

Tactfully Carmel tried to change the subject. Smiling reproachfully at Naim she said, "Quite apart from all these intercommunal rivalries, one of the exciting things about the Jews here is the whole angle of the lost tribes. As you know, some of the Bene Israel believe that they can trace their lineage back to the lost tribes of Israel. But there is another lost tribe here as well. There are some so-called Jews out in Assam who believe that they are the lost tribe of Manasseh. It's quite a story. If you go out to the Bombay O.R.T. school you'll find a group of them there."

Little is known of the part of India from which this "lost tribe" comes. The north-east of India has been a sensitive area ever since India's war with China in 1960. Not only the border state of Arunachel Pradesh but also Manipur, Nagaland, Mizoram and Assam are virtually closed to the outside world. These self-proclaimed Jews belong to the Shinlung tribe, most of whom are to be found in the Chin state of Burma, while others live in Manipur, Tripura, Assam and Mizoram and the Chittagong hill districts of Bangladesh. The usual name for the Shinlung in India is Kuki and in Burma, Chin. But traditionally they have preferred to refer to themselves by the names of their specific tribes such as Thadou, Paite, Hmar and Simte. The name Shinlung has been adopted as a means for these separate tribes to achieve a degree of common identity. They all share the tradition that their remote ancestors came from "sinlung" – the Chin-Kuki word for a cave – although some of them explain that "sinlung" means "closed valley".

At the heart of the Shinlung's identification with Judaism lies the conviction that all the two million Chin-Kuki tribal people are descendants of one of the lost tribes of Israel. According to the beliefs of the Shinlung their common ancestor was Manmasi, whom they claim was none other than Manasseh, the son of Joseph. There are many Shinlung legends and songs about Manmasi; one song describes how he went into exile:

> Manmasi, you came crossing seas and rivers.
> You came through hills and mountains.
> You came through all this
> To the Land of Strangers.

Another describes how Manmasi brought his people across the Red Sea after the tribulations of the sons of Israel in Egypt:

> We observed the *Sipkui* feast
> Crossing over the Red Sea, running dry before us.
> And the walking enemies of mine,
> The riding foes of mine
> Were swallowed by the sea in their thousands.
> We were led by fire at night
> And by cloud in the day.

According to the Shinlung, the tribe of Manasseh settled in Persia from where they were driven to Afghanistan and finally to China. Around 600 A.D. religious persecution forced them to leave China and they settled in Vietnam. Some seven hundred years later they moved on to Aupatuang in Burma, from where they spilled over into north-east India.

An article in the September 1986 issue of *India Today*, an English-language Indian magazine, quoted L.S. Thangjom, the Director of Tourism in Manipur and a prominent member of the "Jewish" community, as saying: "If there can be black Jews in Ethiopia and Mongoloid Jews in China, what is so surprising about *our* ancestry?"

The initials O.R.T. stand for the Russian Obshchestvo Remeslenovo Truda – Society for the Encouragement of Handicraft – a Jewish aid organization founded in Russia in 1870, which maintains aid and development projects throughout the world. In Bombay O.R.T. runs the Kedourie School, situated beyond the Byculla Bridge, which is a vocational college largely for Bene Israel teenagers. The next morning I rode out in one of Bombay's yellow and black minicabs to the O.R.T. establishment. It is surrounded by a large unpaved area in which a dozen or so young Bene Israel were playing football. In one corner of the playground a group of paler-skinned, Mongol-looking youngsters looked on impassively. These, it seemed, were the vanguard of the tribe of Manasseh.

A good-looking young man left the group and greeted me in passable English. According to his account the Shinlung had been converted from their ancestral faith by Baptist missionaries one hundred and fifty years before. Until then they had maintained certain ancient Jewish practices: they had a supreme god called Pathien; believed in a sort of heaven and hell; they had practised animal sacrifice, draining the blood from the slaughtered beasts before they were consumed; a day of atonement had figured prominently in their calendar; a special blessing had been said over a male child on the eighth day – the day when Jewish boys are circumcised; and they had worn a variety of blue and white *tzitzit* – the fringed undergarment worn by male Jews. His parents, he told me, had realized that Christianity was not the "right way" about twenty years before when they heard the preaching of a local prophet called Tanruma of Manipur, who told them that the Shinlung would be destroyed if they did not revert to their ancient Jewish faith. He had told them they were all destined to go to the Land of Israel.

"Will you go?" I asked.

"Of course. And my parents and thousands of others. Israel is lovely – it is the land we love above all others."

The move to Judaism started in Churachandpur in south-western Manipur and has since spread throughout the area. Several thousand families have already accepted the teachings of Tanruma, I was told, and small synagogues have been established in the towns and villages around Imphal. The movement is still gaining momentum. For some years the younger members of the "sons of Manasseh" have been coming to Bombay where some of them have been circumcised and where they learn Hebrew and aspects of modern Jewish liturgy and ritual, which they then take back with them to their towns and villages.

"The valley where we live", the young man continued, "is the gateway to the Garden of Eden. Our forefathers left Babylon to find it and have been there ever since."

A young Bene Israel took me by the arm and led me away. "Don't listen to these Manipur people. They are not really Jews at all. Like all Indians they would like to leave India if they could and to go to Israel would be a good way for them to get out. And if Israel ever needs to increase its population these people are good workers – even if they are not Jews." As I left to return to the city centre the Shinlung boys looked at me anxiously. Optimistically one of them said, in Hebrew, "Next year in Jerusalem." The others smiled.

The only real meeting-ground for the Baghdadis and the Bene Israel in Bombay is the Jewish Club which is not far from the Oval – one of Bombay's best-known landmarks. One evening I was invited to a meeting of the grandly named Council of Indian Jews which was to be held in the club. The premises are run down and undistinguished. In one room a group of Bene Israel were playing a slow-moving game of cards. An elderly man sat alone at the bar. In the committee room nine or ten men sat around a table discussing the evening's agenda. The windows were open, letting in the overwhelming din of a Bombay night, but as everyone in the room was shouting it did not seem to matter. The following month there was to be a meeting of the Association of Asian and Pacific Jews in

Singapore and considerable effort was being devoted to the definition of the term "delegate".

"Why should a Baghdadi go and enjoy an expenses-paid trip?" asked one Bene Israel. "And why should it be a Bene Israel?" came the roar of protest from the Baghdadis. Eventually it was decided to send two delegates: one Baghdadi and one Bene Israel. But before this arrangement was agreed upon the arguments raged back and forth. A young Bene Israel whispered to me: "Small community – big problems," and laughed. Representation at the international Jewish Games, the Maccabiah, in Israel, was the next item on the agenda. The Bene Israel wanted to be represented at the games, but their suggestion of sending a cricket team to Tel Aviv had met with little official encouragement on the part of the games' organizers. "Why should they be discriminating against our national sport?" asked one of the young firebrands.

After the meeting I sat and talked with Nissim Ezekiel, perhaps the most renowned member of the Bene Israel community. He is a well-known English-language poet and critic and teaches American literature at the University of Bombay. I told him about the conversation I had had with Carmel. "Yes. Carmel knows a lot about the community but you can't really characterize the Judaism of the Bene Israel like that," said Nissim speaking in a slow, rather donnish way. "Visitors to India always seem to have some preconception about what the Jewish community is like or should be like. They assume that the Hindu influence upon the Bene Israel is likely to make them very religious, very mystical, very spiritual. How could Judaism have existed in India for so many hundreds of years and not be affected by the Indian spiritual tradition? Well, as a community we have always been afraid of spirituality precisely because it is so redolent of Hinduism, yoga, meditation. The idea of the snake crawling up through your body and exploding into enlightenment is not one that drew us. In the most fundamental sense, our success in keeping away from all that is why we are special in the context of India. Our way was always one of great simplicity: we followed the ten commandments, we had our simple ritual and simple prayer houses. We went about our lives. And, in fact, we were always completely cut off."

"Cut off from the Jewish world?" I asked.

"No," said Nissim intensely, "from India. We were kept inside our own limited world by the pressures of Indian society on the

Anti-Zionist posters at the entrance to the Jewish Quarter in Damascus

A Jew studying in a Damascus synagogue

The interior of a Damascus synagogue

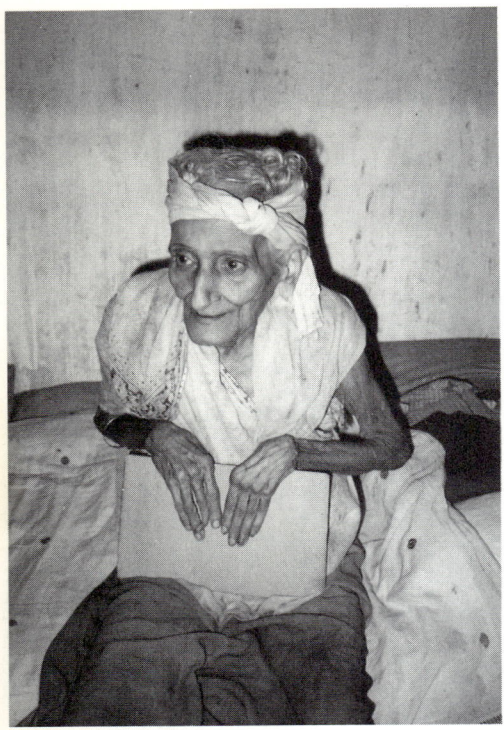
The last Bene Israel inhabitant of Navagaon

A monument marking the place where the ancestors of the Bene Israel first landed in India

The last Jewish *teli*

The Sassoon mausoleum in Bombay

A Bene Israel woman making *rotis*

A Shinlung youth

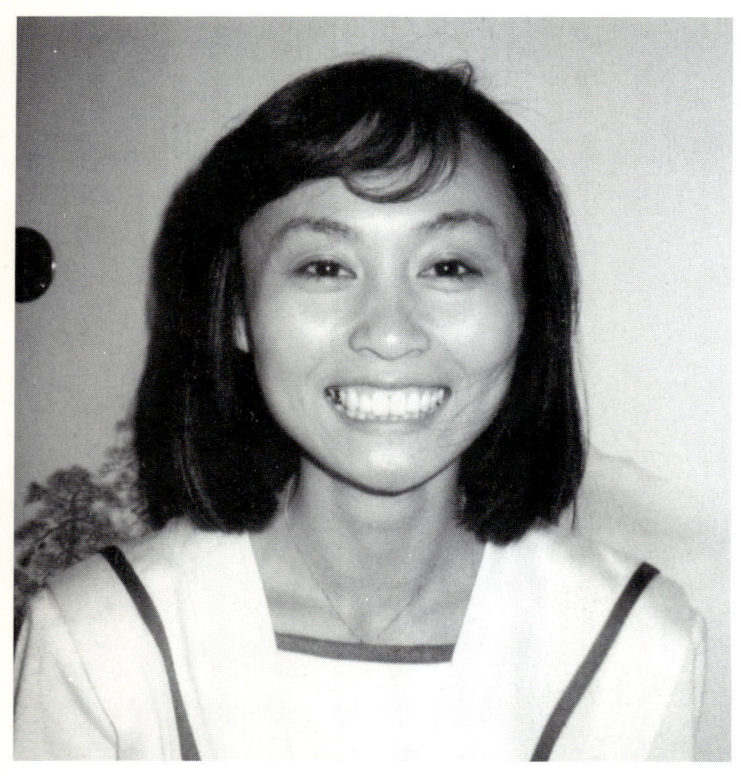

Noa, a Japanese Jewess

A Falasha in Um Raquba, Sudan

A Lemba girl in Johannesburg wearing traditional dress

A Makuya candelabra

Master Kampo in his Kyoto synagogue

Above left A Falasha feeding a child in Um Raquba camp; *above right* a Falasha woman inside a *tukul*

A Falasha girl

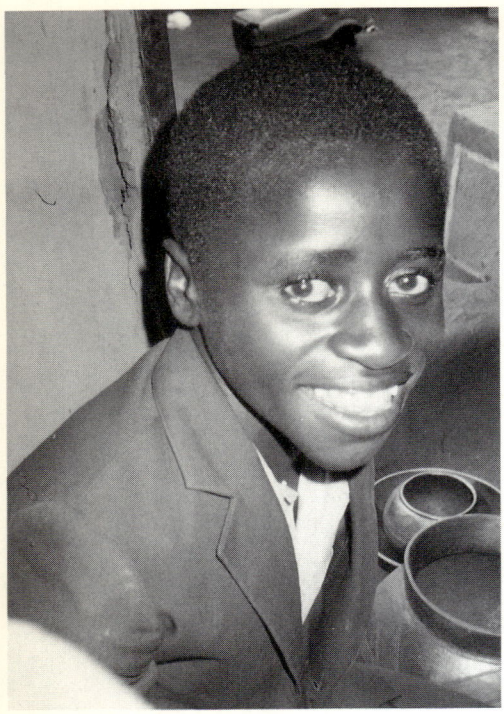
A Lemba boy in Venda

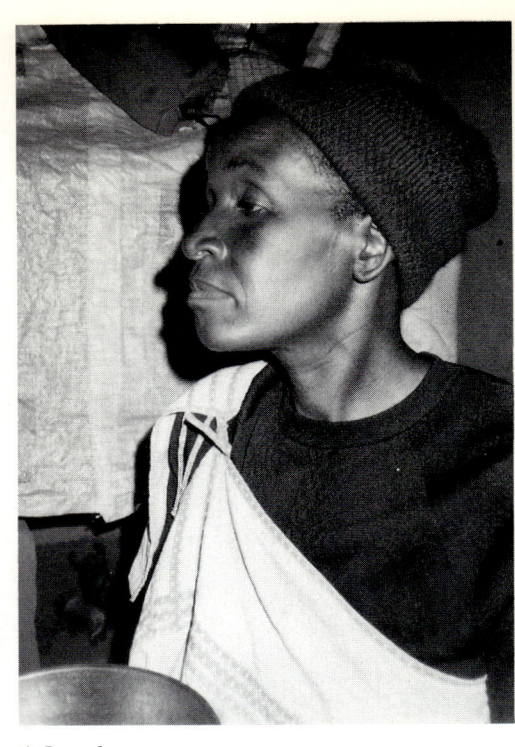
A Lemba woman

Solomon Sadiki, a Lemba from Soweto

A Lemba potter

outside. It was these pressures and particularly the Indian caste system which kept us separate from Indian religious influences for two thousand years. As far as India and Indians are concerned, the number of small groups is infinite. Each one of them has its place in the order of things and one more or less does not make any difference. Thus there were no pressures upon us to assimilate. All we had to do was stay as we were and not rock the boat. Well, as a result, we did stay separate and never did absorb very much, at least in religious terms, from Indian society." He paused and murmured: "Carmel, of course, has different views about us but her views have to be set in the context of her own origins. You see Carmel, a New York Jewess, wants to somehow validate her own Judaism through the supposed exoticism of Indian Judaism. But there is a more general problem as well. The trouble is that Westerners, sick and tired of their own cultures, come to India and see what they want to see. The reality is poverty, filth and degradation. Once I met an American visitor at the airport and as we drove back through the slums he noticed a barely covered Indian woman carrying a pitcher on her head. 'God!' he exclaimed, 'Indian women are so beautiful.' I told him I did not know what he was talking about. You see the reality was that the woman was dying of hunger. He could only see the beauty."

The Bene Israel lived mainly in villages of the Kulaba district on the Konkan coast. Part of this area had traditionally been ruled by the Maratha admirals, the Angria, part by a Maratha chieftain, the Pant Sachiv of Bhor, and part by the Muslim Nawab of Janjira. Over the last two hundred years or so the Bene Israel have distinguished themselves, out of all proportion to their numbers, as fighters in British "native regiments" and Indian armies. One Bene Israel medical officer was with Allenby's forces when they captured Jerusalem in 1917. But traditionally the Bene Israel were the oil pressers of the Konkan area where they were known as *shanwar teli* or Saturday oil pressers on the grounds that Saturday was the one day when they did *not* press oil.

During my talk with Nissim Ezekiel at the Jewish Club, he had spoken a good deal about the rural past of the Bene Israel. He told me that it was in the villages of the Konkan that the uniqueness of the community could still best be seen. So I decided to ask Shelem to accompany me there, where I could confront him with whatever particularity the Bene Israel still retained in their native habitat.

I arrived one morning in a hired car with a Hindu driver at the New Ready Money Building. Shelem was waiting for me on the veranda with a Bene Israel friend who had family business to attend to in one of the villages. As we left Bombay we passed the great docks erected in the last century by the Sassoons and still marked prominently with their name. A little further down the road my guides pointed out the extensive Baghdadi cemetery, dominated by the two great sepulchres of the Sassoon family. "Great men, the Sassoons," said Shelem, without conviction.

As we left Bombay I tried to get Shelem and his friend, Isaac, to talk about Judaism and the relations between the Bene Israel and the Baghdadis. But Shelem's suspicion of my questions made conversation difficult. They talked instead about money, inflation and prices. Persevering, I asked them how their customs differed from those of the villagers among whom they had lived for so many centuries. Shelem wagged his head in that peculiar Indian way which is half affirmation, half pleading. "We keep kosher," he said. His friend wagged his head in unison. They went on to mention some of the specifically Bene Israel foods which other Indians of the area did not eat: at thanksgiving celebrations a confection of raisins, rice, coconut and spices known in Marathi as *malida* was always served accompanied by two different varieties of fruit; at the end of the Day of Atonement special cakes made of coconut juice and honey were eaten. I mentioned tentatively that I thought that Hindus also ate *malida*. Solemnly they agreed that this was in fact so, but they assured me, in an apparently shamefaced way, that no other group ate the honey cakes. Subsequently neither Shelem nor Isaac would say anything and we drove through the beautiful rolling country of the Konkan in silence.

Eventually we reached the village of Penuel. Driving past the village pond and a colourful market the driver edged his car through a street crowded with tongas, animals and children which led through the Jewish quarter where several hundred Bene Israel had once lived. Most of them, however, had emigrated to Israel in the 1950s and the rest had moved into Bombay. Now there were no more than twenty or thirty in all. Penuel had once played an important role in the affairs of the community as it had been a centre for pilgrimage. At the synagogue vows and prayers had traditionally been made and barren women had come from miles around in the hope of being made fertile.

Removing my shoes I entered the *masjid* (the word the Bene Israel used to use to describe their place of worship notwithstanding its Arabic origin and Muslim association), while Shelem and Isaac wandered off in quest of the caretaker. The interior was cool and finely proportioned. When my guides returned they were arguing fiercely with the *shamash* about the price of coconut oil. Finally a price was agreed upon and they bought a tiny quantity which was poured into a glass with a wick. This was placed in an ornate brass holder which hung from the ceiling to one side of the ark. The *shamash* pulled a brass ring under the lamp, lowering it to shoulder height, and I was invited to participate in the ceremony by lighting the lamp. Once the lamp was lit the three of them, Shelem and Isaac in their dark suits, the *shamash* in a dirty white kaftan, stood together in sudden harmony, shaking their heads demurely.

Three Jewish families lived in simple rooms around the courtyard behind the synagogue. In one of them an old woman was frying *puris* on a small primus stove. I asked her if she minded being photographed and she fled shrieking into the interior. When I returned to the synagogue the community *hazzan*, who had been sent for, sang a hymn in Marathi, with a refrain in a sort of Hebrew, about the prophet Elijah. The Bene Israel believe that the prophet appeared to their forefathers on a hill near the village of Navagaon; he ascended to heaven in a chariot, but when he had gone wheel tracks and the imprint of a horse's hoof were found on the hill, where they can be seen to this day. When I asked Shelem about this legend he shifted uneasily and said nothing. Isaac vouchsafed that the return of Elijah would presage the end of the world and the coming of the Messiah. "The bear will fight the eagle," added Shelem portentously. I asked them about the history of the synagogue and the pilgrimage but they knew nothing, or if they did know they were keeping it to themselves.

For some time after leaving Penuel we drove in silence through the green countryside dotted with small villages. Shelem and Isaac wanted to visit distant relatives of theirs who owned a mill in a nearby village not far from the small town of Alibag. Shelem gave curt instructions to the driver to turn off the tarmac road down a muddy track. The mill was a considerable structure and, until the

1920s, had enjoyed a monopoly throughout the entire region. Since then Hindu mills had been established. Above the great wooden door of the mill a window in the form of a star of David had been set into the wall and underneath in Hebrew characters the word "Zion" proclaimed that this was a Jewish mill. Inside, on the walls, old copies of Victorian water-colours on biblical themes were interspersed with Hindu decorations. Shelem was quick to point out that the latter were for the benefit of the Hindu mill-workers. The miller asked me if I would like to see the synagogue in his village and we walked over to it. It was a well-built stone building which still served as a centre for the small groups of Bene Israel who live in a dozen or so villages within walking distance. "It's an impressive place," I said. He smiled, "It's an impressive place perhaps, but it is hardly being used any more. Years ago it was different. Now it is only used for the high holy days. Nearly all the Bene Israel have left Konkan. In another generation they will all have gone. You know we have been here for two thousand years. Some say longer."

We were joined by the *shamash* and his family who greeted us very deferentially and served tea in the synagogue. Perhaps in order to entertain us the *shamash* brought out the ram's horn, the *shofar*, which he blew expertly. His son meanwhile had gone to fetch me a book: a much-thumbed history of the Bene Israel by H. S. Kehimkar published in Tel Aviv in 1937. It was clear that it had not been thumbed by the *shamash* or his son for neither of them spoke or read a word of English. Through the mill owner they asked me if I had been to Israel. I replied that I had. "In that case," said the *shamash*, "you must accept this book as a gift. Because it too is from Israel." I would have liked to talk more with the *shamash* and his son but at this moment Shelem appeared, perspiring heavily from his walk, to say that we would have to go.

Not far from Alibag we came to Navagaon, a small village next to the sea where giant coconut palms towered above the simple thatched huts of the village. In his usual embarrassed way Shelem said, "We are proud of this place. This is where our ancestors landed so many thousand years back." He pointed over at a lush green field with slight undulations in it. "This is where the Bene Israel landed when they fled from Nebuchadnezzar," said Shelem. The undulations were graves, one or two of which were marked with dates from the late eighteenth century and inscribed in

Hebrew and Marathi. Most of the graves, however, were marked by smooth square tablets which lay unadorned in the grass. A hundred yards away the surf pounded on the beach. The only other sound was that of birdsong. The simplicity of the place and its evocation of the epic past of the Bene Israel were deeply moving and I wandered around for a while happily noting down some of the inscriptions.

In the village was an old Jewish woman who lived alone in a tiny hut at the end of a lane. She looked older than the eighty years she admitted to. Her limbs were pencil thin. Despite her fragility she was concerned about the future of the holy site and had fears that her Hindu neighbours, who were looking after her, had designs on the Bene Israel cemetery. I asked Shelem why it was that the Bene Israel community did not look after this old woman. Angrily, he insisted that they did, and to prove his point reached into his trouser pocket and pressed a handful of rupees into the frail hand of the last Jewish inhabitant of Navagaon.

The last point on Shelem's itinerary was Alibag, some miles south of Navagaon. Set in the middle of the village the Bene Israel synagogue is the dominant structure of the area. In the past Alibag had been the most important centre of the Indian Jews, although by now very few Bene Israel remain. Outside the synagogue there was a plaque in Marathi and English surrounded by finely painted flowers and birds. The English notice was headed, "Manners to be Observed in the Synagogue", and advised the visitor:

1. Please before entering the synagogue, boots, chappals, sleepers, footwear etc., should be taken off; and left them outside.
2. No noise except the praying should be made while the prayer is going on.
3. On every Shabbat or Eid festival at least one member of each family from the local resident of Alibag should attend the prayer.
4. Those against whom the balance of previous years bid is outstanding are prohibited to give further bids.

I asked about the last item and Shelem and Isaac explained eagerly, for this was something they really knew about, that the bids referred to were for the privilege of lighting the oil lamps on the high holy days and other feasts. On the Day of Atonement, they explained, the bids could run into tens of thousands of rupees. Once they were back to the subject of money they became animated and started arguing cheerfully in Marathi, using the occasional English word – "thousand", "ten thousand", "inflation", "exchange rate".

Since the original Bene Israel term for synagogue was *masjid*, it was not surprising to see that the word used for festival is *eid* – the Arabic word for a Muslim festival. I asked my guides about this. Immediately Shelem denied that the word was ever used by the Bene Israel. "Why should we use a Muslim word?" he asked indignantly. When I pointed out the word on the notice-board he showed intense surprise and examined it minutely. Isaac maintained that the word had somehow found its way onto the notice-board by mistake. I decided not to let this pass and when I pressed them they grudgingly admitted that the word *eid* was indeed the usual word for festival and that it had been borrowed, they supposed, from the local Muslims with whom they had always had far more in common than with the Hindus. That they were defensive on this issue is not surprising. For almost two hundred years they had been obliged to defend their claim to be Jewish in the teeth of the scepticism not only of their Baghdadi neighbours but also of Jewish communities outside India. In more recent times the Bene Israel have had the greatest difficulty in persuading the Israeli rabbinate of the legitimacy of their claim to be Jewish. Given the obscurity of their origins, any suggestion that their faith was dependent in any sense upon Islam was and is profoundly embarrassing to them.

The Bene Israel had been integrated into the Indian caste system to the extent that they had a caste profession which was traditionally that of oil presser. Throughout the nineteenth and twentieth centuries the Bene Israel diversified into a number of other occupations and by now, as a result of emigration and the flight from the countryside to Bombay, there is only one Bene Israel *teli* or oil presser left. He, I was proudly informed by Shelem, lived in Alibag. Shelem stepped jauntily ahead, his confusion over the *eid* issue apparently forgotten. At the side of the main street of Alibag he pointed out a stone house. Above the low lintel a brightly coloured Hindu decoration vied with a more simple testimony of faith: a sheet of paper against which had been pressed a man's palm dipped in blood. This ritual is performed every year in memory of the original Passover in Egypt.

The *teli* was not a rich man. His pale face was covered in light brown blotches, his hair a strange mat bronze. His eyes were expressionless. He looked like a product of generations of inbreeding. The last Jewish *teli* in India, the last of a caste, he brought me photographs of his children and grandchildren. They were all living in Israel. He pointed at the photographs dully without any sign of emotion or interest. Through Shelem I asked him how he pressed his oil and slowly he turned and took me back into the courtyard which was dominated by a segment of a tree trunk three or four feet high which had been hollowed out to serve as a mortar. A long and thick pole was connected to the yoke of a cow which, after some considerable time, was persuaded to walk around the hollowed log. Remarkably, the reddish beans which the *teli* had packed tightly around the pole were soon enough transformed by this process into a fragrant, gluey oil. These days, the oil presser explained, the product was difficult to sell because of the wide availability of cheap, commercially produced oil. He was standing in front of a row of large ceramic jars burnished by use to the colour and sheen of bronze. "When the cow dies," he said, "I shall stop. I am the last *shanwar teli*, the last one."

As we walked back to the car Isaac and Shelem made some jokes at the expense of the *teli*: "He's so poor he'll die before the cow! The cow will be the last *teli*!" When I did not join in their laughter they reverted to a sullen silence, but it was not long before their good humour returned. We were going back to Bombay and they were evidently glad to be leaving the countryside and its uncomfortable reminders of their past. Boisterously they started boasting to each other, for my benefit perhaps, of past exploits. "But Shelem, I had three hundred sepoys and one hundred and fifty officers under me." After a pause Shelem retorted that he had worked so hard that he had not been able to take a holiday in five years. Isaac soon capped that: "I, Shelem, never took leave in fifteen years." And so it went on.

What can be said of the Judaism of Shelem and Isaac? In their conversation their greatest efforts had been directed at demonstrating how Western they were; how removed from their country cousins and how different, indeed, they were from the rest of India. To some extent they had abandoned their past but their attempts to assimilate to the mores of the Baghdadis had failed, not least because the Baghdadis refused to accept them.

They seemed to be living in a cultural limbo. But despite all this they maintained a firm sense of identity with their Jewishness.

When my driver had picked me up that morning he had been friendly and alert. A copy of the *Times of India* and a flask of cold water had been put out for me on the back seat of the ageing Ambassador car. When I picked up Isaac and Shelem, a short salvo in Hindustani had silenced him for the rest of the day. I dropped off my Bene Israel guides at the New Ready Money Building and the driver took me back to Naim's apartment. He had been driving for fourteen hours. He turned to me and said, "You must be tired, sahib." I gave him a ten rupee tip and he said he was going off for a beer. I would have liked to join him.

Naim was waiting for me. "Did they weep on your shoulder?" he asked and, before waiting for a reply, went on: "Well, I expect you have taken their side now. They're right and we're wrong. All the foreigners do. I can tell you this, though. Without the Baghdadis the Jewish community in India would be a shambles!" A festive Iraqi dinner had been prepared for me, as I was leaving Bombay the following morning. With the smell of Arab food, the candles and the oriental carpets there was more of Baghdad in this room than there was of Bombay. And I was sorry to leave.

Later that evening I met Nissim Ezekiel, the poet, for the last time. We sat in the bar of India's most famous hotel, the Taj Mahal. I felt disappointed that I had learnt so little about the symbiosis between Judaism and India. Perhaps, after all, there was nothing to learn. Nissim was in a reflective mood. He spoke at length of the decline of the Bene Israel. In 1951, he told me, there had been 20,000 Bene Israel in India. Now there were no more than 5,000. Once it had been a really active and vital community, with many educators, journalists and other professionals. "We even had a Bene Israel Mayor of Bombay," said Nissim. "But now it's all over. Most of them have taken up Zionism and gone to Israel. When I was a child we never took Zionism very seriously: I remember joking about 'next year Jerusalem'. But after Israel's War of Independence things changed. Many people were very moved at the idea of a Jewish state and many more were talked into being moved by it. At the same time the upheavals after Indian independence made many think of leaving. In any case I suppose that one can imagine that by the end of the century there will be no Bene Israel community as such."

"It sounds as if in some way you resent the idea of Zionism."

"Not quite that. I just feel that it does not have a great deal to do with Judaism. But I would be the first to admit that my own experience of Judaism is rather strange. When I was eighteen I broke with religion entirely and remained a rationalist atheist until I was forty-one. I became more and more Indian and completely lost any sense of identity with Judaism. During India's struggle for independence I was fiercely pro-Indian unlike, for instance, the Baghdadis who were pro-British to a man. Gradually I was taken over by mysticism. I studied Zen, Christian mysticism and Kabbalah. But I was interested in it from a secular point of view: there was no faith involved. At this time I went to spend a year in New York and decided to try L.S.D., which was very fashionable at the time. I experienced the most extraordinary things. So much so that after my first trip I abandoned rationalist atheism and gradually returned to Judaism. Strange, isn't it? Thereafter I developed a close relationship with Judaism and Israel and now it would seem absurd for me to say I was not a Jew. You know, when I was in my non-Jewish phase I would show my poetry to foreign friends and they would invariably say – 'How Jewish!' But there was no reference to anything Jewish in my early poetry. I still don't know what they meant."

"What seems remarkable to me", I said to Nissim, "is that Judaism has survived at all in India, particularly as the Judaism of the Bene Israel seems to be such a feeble manifestation compared with the other great religions of India." Nissim smiled.

"But, you see, our survival has not so much to do with our watered-down Judaism. It has more to do with the fact that Hinduism is not a proselytizing religion. I think, too, that when the Bene Israel first arrived in India they must have realized within the first few days that they would only be able to survive by becoming a caste. Thus marriage outside the caste would be impossible. And that is what happened. A Jewish girl could not even talk to a Hindu boy. We became just another caste. If we imitated anyone it was the smaller religions – obviously Islam has influenced us in a number of ways. But all we accepted from the Hindus was their idea of the caste system. We were almost all *telis*. Which is not to say that the caste system follows professional lines: but most professions were filled by people from a certain caste. We did not suffer as a result, indeed we achieved some security and honourable obscurity from it. And as for our local Judaism being a feeble religious manifestation, I think that this is your disappointment talking. Like Carmel, you were looking for some exotic syncretism between the world of India and the world of Judaism. If it ever existed – and I do not really believe that it ever did, at least in the realm of ideas – it certainly exists no longer. From what you told me of your trip to the Konkan I presume that you think that the community is culturally shallow and unreal. But in a sense that is true of the whole of India: we all suffer from what has been called the 'crippled mind'. Our culture, or lack of it, or pretence at it, must be seen in the context of the whole of India. There has been a historical process which has immobilized millions of Indian minds: a slavishness, an inability to achieve even simple things. We are much the same. But for us it no longer matters. There is no future for us in India as a community. Of course some will stay – some of us have done very well by Indian standards – but the young will have to leave. Some of those who have decided to stay have taken Indian names and have completely assimilated. Those who have emigrated to Israel, with its advanced Western culture, have assimilated in a different way. Whatever there once was specific about us, specific and unique, is disappearing. It's good you came this year and not next."

Nissim was right. It is clear that for centuries the Bene Israel had formed an integral part of the "hundred Indias": each piece in the jigsaw had been held firmly in place by the other ninety-nine pieces. But now, with the gradual disintegration of the old Hindu order, the pressures were lessening. The Jewish piece of India looks as if it is going to drop out despite the hundreds of years of religious tolerance the Jews have enjoyed.

Nissim took a sip of cold Indian beer. His very Indian face beamed at me. "Poverty has been on your mind, hasn't it? You asked me last week if our Judaism here has managed in any sense to come to terms with the poverty of India. Well, the Bene Israel have produced no Jewish Mother Teresa! In part, this is because preoccupation with suffering is not very Indian. The Brahmin ideal after all is to see the world for what it is: an illusion – *maya*. With such a world view nothing much matters. But may I show you one of my poems?" Diffidently, he pulled out a volume of poetry from his battered briefcase: it was entitled

Latter Day Psalms. The poem which Nissim read slowly and simply was his reworking of the Twenty-Third Psalm:

> Is the Lord my Shepherd?
> Shall I not want?
>
> I lie down in green pastures, beside
> the still waters. Lead me
> away from these into thy
> work.
>
> When my soul is restored,
> I walk the path of self-
> righteousness.
>
> I *do* fear evil: thy rod
> and thy staff *do* comfort
> me.
>
> Let not my table be prepared
> in the presence of anybody.
> I do not need a cup that
> runneth over.
>
> I shall not expect goodness
> and mercy all the days of
> my life, even if I dwell
> in the house of the Lord.

This version of the biblical text was clearly one Indian Jew's reinterpretation of Judaism in the sub-continent. Nissim smiled and said, "I suppose this connects Judaism with the problem of India."

As I drove through the slums to the airport the following morning my attention was caught by the advertisements at the side of the road: "Bombay – Clean City – Green City"; "Wipe out Accidents!"; "Live without Fear – Oriental Insurance". Such messages seemed like ironical glosses on what I had seen of Bombay. And I wondered whether Nissim's poetry was any more than an ironical gloss on the problems of Jews and the problems of India. None the less, when I looked down over the slums as the plane rose from the airport towards the sea, I thought of some lines from Nissim's version of the Ninety-Fifth Psalm:

> The Sea is his; we may drown
> in it. He formed the dry land,
> on which many millions thirst
> to no end.

THREE

Singapore: A Dwindling Community

For hundreds of years the trading interests of Jews from Mesopotamia and other parts of the Middle East extended beyond India to Malaya, Siam and China. Partly through their efforts, coral, spices, indigo, sandalwood, elephant tusks and coconuts found their way to the bazaars of Baghdad and Damascus and the great cities of the Mediterranean. Although Singapura, the Lion City of the ancient Malays, had been destroyed in the fourteenth century and the island had remained virtually unpopulated for centuries, Jewish merchants may have suspected the commercial possibilities of Singapore with its strategic location on a major sea route and its proximity to the rich and barely tapped resources of the Malayan Peninsula and Indonesia. When Sir Thomas Stamford Raffles (1781–1826) of the East India Company acquired the right to establish a trading station on the island in 1819 and five years later bought Singapore and its offshore islands for 33,200 Spanish dollars, Jewish merchants were among the first to establish trading links with the new entrepôt. A Malay account of a journey from Singapore to Kelantan, *The Story of the Voyage of Abdullah Bin Abdul Kadir*, mentions that as early as 1838, "some twenty Chinese and Jewish (Singapore) merchants had dispatched merchandise in four Chinese junks to Pahang, Trengganu, Kelantan and Singgora". We are told little more of the Jewish merchants other than that their goods were impounded by the Rajah of Kelantan. From the first the Singapore settlement was richly cosmopolitan, with Europeans of every nationality, Chinese, Malays and Indians soon putting down roots. By the mid 1840s, of the mercantile establishments registered on the island twenty were British, five Chinese, five Arab, two Armenian, while no less than six were Jewish.

It was during this period that the Jews acquired property in the new city of Singapore for the construction of a synagogue and a cemetery. And for the next three or four decades the Baghdadi Jews, who made up the bulk of the community, chose to dwell in the vicinity of the synagogue which was situated at the heart of the commercial district of Boat Quay. The first synagogue, which was not markedly different from the Chinese shop houses which lined Synagogue Street, was sold off by the Jewish community towards the end of the nineteenth century. But it was still standing in the 1930s when it was inhabited by a Chinese family who were so in awe of the building and its sacred associations that they maintained the room which had formerly housed the Torah scrolls as their household shrine.

In 1842 a successful Baghdadi merchant by the name of Abraham Solomon settled permanently in Singapore. A ship owner, with British captains in his service, he was such a prominent figure that John Turnbull Thomson, who was appointed government surveyor to Singapore in 1841 and whose *Some Glimpses into Life in the Far East* describes the early days of the colony, devoted a chapter of his book to "the Jew of Bagdad". According to Thomson's account Solomon was

> a man of patriarchal appearance, and majestic mien – a man who might have sat to a sculptor as a model of the father of the faithful. He dressed in the flowing robes of the East, and a large turban covered his head. His complexion was fair, his eyes jet black and brilliant, his nose aquiline, his brow high; his hair, once black, was now grey, and his beard, the most capacious I had ever seen, flowed down over his breast. He was a man of large stature, and his arms and legs were proportionately stout, covered with black hair. Such was the personal appearance of Abraham Solomon, the Jew of Bagdad. . . .
>
> He was given to hospitality, and I believe a friend of mine and myself were the first Europeans that entered his house, and partook of his good things. . . .
>
> Our host could not speak English, so the Malay language had to do duty in our interchange of courtesies. It must be admitted that this language was a poor medium for so important an occasion; but good feelings and intentions made up all deficiencies. Our host first pressed us to fill our glasses, and quaff a bumper, before commencing on the solids, and this was readily acceded to by all of us. The

viands consisted of capons, mutton, and fowls, made up in various well-seasoned dishes, and our host was particular in explaining to us that all flesh had been most carefully killed by the priest, and that all the oil used was extracted from animals so killed. He dwelt much on this subject, which seemed to him a most important one, and he assured us that, during his life time, he had always been most rigid in observing the laws of Moses on these points, and in so doing, felt that his life had been holy, and his body clean. Meantime, the champagne flowed pretty freely, and our host's heart expanded. And as he chatted away, his memory reverted to Bagdad, and his younger days in that city. He spoke in ecstasies of the date, grape and the fig – not obtainable in Singapore. He then spoke of his father's house, and the oppressions thereof under a Mahomedan government. "At length," said he, "the oppressions became so grievous, that we had to flee the country. The soles of my feet were beaten till they were raw: for they wished to torture me into disclosing treasures that I had not."

We were charmed with the idea that we had sat, that night, with the true type of the contemporaries of those great kings of kings and lords of lords over Egypt and Assyria.

I do not mean to speak of our host in unmeasured terms of approbation. As an Oriental, he was narrow-minded, and probably fanatic; further, he was ostentatious and pharisaical.... For once, his company was interesting; but as it could not be repeated on equal terms, I only this time broke bread with Abraham Solomon the Jew of Bagdad.

This passage tells us something of the sort of Jew who settled in Singapore in the early years and something, too, of the attitude towards the Jews of the British, who were to be their masters on this tropical island for the next century and a quarter.

For several decades after Solomon's arrival in Singapore, Jews continued to leave the Middle East and seek their fortunes in India and the East, where colonial expansion was constantly opening up new areas to international commercial activity. Many of them chose to settle in Singapore, which had become the most important port in South-East Asia. Some members of the great merchant families of India such as the Sassoons, Gubbays and Ezras also moved to Singapore, where they maintained important trading links with their relatives in Bombay and Calcutta.

One of the earliest accounts of the community was written by a Jerusalem-based writer and traveller, Jacob Saphir (1822–85), who in 1857 visited Jewish groups in India, Singapore, Java and Australia with the intention of trying to raise money for the reconstruction of Jerusalem's Hurvah synagogue.

In Singapore, he wrote,

> I found about twenty households of our brothers, the Jews of Baghdad, who lived previously in Calcutta and, with the expansion of trade, arrived here, dealing in commerce and property, and some of them have been very successful. Among the community elders are Rabbi Abraham Shalom Solomon of Baghdad, Joseph Ambar of Damascus, and Shalom Gubbay, all well-known men. With them are their scribes and servants, ritual slaughterers and their children's teachers. They live in villas with gardens, overlooking the city and enjoying fresh air. Inside the city they have a beautiful synagogue, which has gathered cobwebs, because a *minyan* assembles there only on some Sabbaths or holidays, although they are observant, generous and wealthy. Here, as in Calcutta, their trade and properties are liable to vanish overnight especially since here most of their trade is in opium, with the cities of China. Their language, dress and customs here are as they were in their countries of origin; but they have learned the laws of the land, and are imitating European customs, as practised by the British.

By the 1870s there were a couple of hundred Jews living in Singapore, almost all of whom engaged in trade and many of whom had made their fortunes in opium. By this time most Jews had left the vicinity of Boat Quay. The old shop-house synagogue was sold off and in 1878 a rather more substantial synagogue, Magen Avot, was constructed on Waterloo Street, the main thoroughfare of the area to which the Jews had moved. Like all synagogues in the East its doors faced west – the direction of Jerusalem.

The wealth of Singapore made it one of the more spectacular gems of the British Empire. The city became a showpiece in which a number of superb buildings were constructed in the second half of the nineteenth century. The opening of the Suez Canal brought a greater number of shipping lines to the Far East and international trade out of Singapore increased dramatically. Tin mining and rubber planting in the Malay Peninsula had the effect of making the off-shore entrepôt even more important from an economic point of view so that by the end of the century Singapore had become a rich, teeming, cosmopolitan city. The Jews joined in the general prosperity. Their success owed something to the fact that from the first they had adopted English as their language of education and this facilitated their relations with the British. In addition their family and communal contacts throughout the Far East gave them access to a variety of overseas markets.

The most spectacularly successful Jewish merchant in Singapore in the nineteenth century was Sir Manasseh Meyer, reputedly the richest Jew in the East. Born in Baghdad in 1846, he came to Singapore in 1861 and by the 1870s had become the island's most important trader with India. Subsequently he concentrated on acquiring property on the island and did so with such success that, according to some reports, he acquired ownership of over half the real estate in Singapore. Among other buildings he owned the Adelphi Hotel, the Sea View Hotel, Meyer Mansions and Meyer Chambers, while his own mansion, Belle Vue (still lived in by his granddaughter), is one of the finest houses on the island. Towards the end of the nineteenth century, Meyer became a distinguished city elder and went on to make important donations to public institutions such as the University of Singapore (a wing of which still bears his name) and St. Andrews School which was attended by many Jewish children. He was knighted by the British for his philanthropic activities. The Adis family, too, were immensely wealthy and lived in a house which was often described as one of the most splendid mansions east of Suez, while their hotel, the Grand Hotel de l'Europe, was one of the most renowned in the British Empire. Somerset Maugham is reputed to have said, "You may stay at the Raffles, but dine at the Grand."

Jews continued to be enormously successful in economic terms into the twentieth century. By the 1930s, in a vastly expanded city, they still owned a great proportion of the city's real estate. According to the 1931 census the Jews, along with the Arabs, were the largest property owners in the island while a year later, according to another source, almost half the houses which were available to rent were owned by Jews. There are numerous streets and estates with Jewish names to this day: Synagogue Street, Solomon Street, Meyer Road, Nathan Street, Frankel Estate. And many buildings in the city still bear the star of David.

The wealth and opulent life-styles of some Jews in Singapore created certain divisions among the community. The wealthy Jews increasingly chose to live far away from the rather crowded conditions of the Jewish areas around Waterloo Street. There, small merchants, hawkers, bakers and servants, who had been brought out from Baghdad by wealthy Jewish families, eked out a precarious livelihood. Some of the more enterprising working-class Jews took household goods such as cotton and scissors to the mainland and peddled them in the Malay *kampong*, or even further afield to North Borneo and Sarawak. Others imported small items such as embroidered ladies' slippers from the Middle East. But there was a world of difference between these petty traders and the great Jewish merchants. As a symbol of the divide between rich and poor, in 1905 Sir Manasseh Meyer built an opulent new synagogue, Hesed El, at his own expense, on the high ground adjoining his family home. But for all its marble magnificence the new synagogue never attracted much support from the poorer members of the community. Even though local Jewish practice sanctioned travelling by rickshaw on the Sabbath, Hesed El was some distance away from the main Jewish area and most Jews continued to worship at Waterloo Street while the Meyers were obliged to pay indigent Jews to make up their synagogue's daily *minyan*, something which the humbler Magen Avot never had to do.

Although the bulk of the community consisted of Iraqi and other Middle Eastern Jews, there was a trickle of Ashkenazi immigration from the middle of the nineteenth century. The German–Jewish firm of Behn Meyer, which established itself in Singapore in the 1840s, was the first German company to have any success in trade with the East. One of the most important companies in Singapore, Katz Brothers, also an Ashkenazi concern, became another of the leading trading houses in the colony. By the twentieth century the few Europeans Jews in Singapore were beginning to make a considerable contribution to the life of the island. They owned some of the most luxurious shops and European bakeries. Iky's Bar, one of the more favoured drinking establishments in Singapore, was run and owned by a German Jew, while the city's most famous review was *Tingle Tangle* in the Burlington Hotel, which was owned by a German Jew called Puhlman. During the 1930s Jewish refugees from Europe started coming to Singapore in larger numbers, swelling the community so that by the outbreak of the Second World War the island's Jewish population had risen to a peak of around 2,000 souls.

The three years prior to the Japanese invasion of 1942 saw many Jews leave Singapore of their own accord for the safety of India, Australia and the United Kingdom, while some 250 were evacuated by the British. None the less, when the island fell to the Japanese many Jews were still there. For the first year of the occupation the Jews were allowed to carry on as normally as wartime conditions permitted. To start with there was no real discrimination against Jews; the only parallel with the situation of Jews in Nazi-occupied Europe was that those in Singapore were required to wear white and red armbands upon which were marked in Japanese their name, personal number and "race". But the Japanese showed no signs of anti-Semitism. One Jew wearing the armband was accosted by a Japanese official who asked, peering shortsightedly at the armband, "And what are you?" "I am a Jew," came the response. To which the Japanese replied: "Ah! Good! There are three great Js in the world. Japan, Julius Caesar, and Jews. Good!"

During this first period of the occupation the Japanese military administration asked the community for a written description of the Jewish religion and subsequently the Japanese Minister for Religious Affairs visited the Waterloo Street synagogue to observe a service at first hand. He assured the congregants that the Japanese would respect their religious practices and not interfere with the observance of their faith. But eventually, it is thought, the counsels of Japan's German allies prevailed and, in March 1942, over a hundred male Jews were interned, as Jews, in the Changi and Orchard Road internment camps. Before the end of the war the rest of the community, including women and children, were also interned. During the period of the Japanese occupation around 10 per cent of the Jewish population died, but by comparison with other groups the Jews were not treated badly. One Jewish former internee told me: "The Japanese felt somehow that we Jews were one of them. They treated us well. They used to talk to us about Judaism. We never really knew why. They seemed to like us and even be in awe of us. They certainly would never speak in such a friendly way with the other internees."

After the defeat of Japan many Jews who had spent the war elsewhere returned to Singapore. But not many of them were to stay for longer than a few years. The establishment of a communist regime in China (which among other things had the effect of destroying the largest Jewish community in the Far East in Shanghai) was seen by many Singapore Jews as marking the end of the stability which they viewed as the prerequisite for international trade. The general sense of insecurity was augmented by the break-up of old colonial régimes in the area which seemed to threaten the links between West and East which had been the *raison d'être* of the Jewish communities of the Far East in the first place. Today, forty years on, there are no more than two or three hundred Jews on the island and the community looks likely to disappear before the end of the century. The main reason for this continuing emigration is that for the younger members of the community here, as in India and Syria, there is not a large enough pool of potential marriage partners. So, one by one, the Singapore families are breaking up and moving to join relatives in Sydney, Perth, Los Angeles and London. None the less, many Jews are still active and influential in the glittering Singapore of the 1980s.

Changi Airport is the embodiment of the new Singapore: its shopping malls are temples to the cult of consumerism, the carpets of the reception areas are soft and elegant, the lighting is discreet. The area surrounding the airport is as clean and tidy as the interior. The grass is mown carefully. I glimpsed men picking up stray leaves from the ground and putting them into brightly coloured plastic bags. The pavements are immaculate. A Singapore Chinese girl had just told me that a government ordinance had recently banned chewing gum as an anti-social activity.

I was met at Changi by the President and the Secretary of the Jewish community of Singapore. The president, Alec Manasseh, was an elderly and portly Singapore-born Baghdadi. His father had come from Baghdad, his mother from the Baghdadi community in Bombay. Alec's father had traded in clothes – bringing made-up clothes from Zanzibar to Singapore and then shipping them on to Far Eastern markets. The secretary, Geoffrey Pinsler, was a young, English-educated lawyer, whose parents had come to Singapore before the war. He was driving a new BMW. Opening the boot, he pushed aside a pile of jogging and squash gear to make room for my small bag. Pointing at his trim waistline he told me that he had recently lost forty pounds. The Prime Minister, Lee Kuan Yew, had recently started waging a vendetta against what Geoffrey called "corporate paunches" and now everyone in

Singapore was jogging and trying to lose weight. Alec Manasseh eyed the mess of sporting equipment bleakly and snorted: "It's madness, mun, in this heat. Even the British didn't go running round the streets half-naked like this lot."

The previous evening, Alec told me in the characteristic Welsh-sounding Anglo–Indian accent of the Baghdadis, the Jews had met together to celebrate the birth of a new child to the community. "Oh! mun! We had a lovely time. Songs – you know – 'Yes! We have no bananas'; 'The saints are coming'; 'How much is the doggy?' and food! You should have seen the food. Yellow *basmati* rice fried in oil. I love that. I eat it five times a day. That's what we do in Singapore – eat. There is nothing much else to do. Chinese food. The best in the world. Indian spicy food. And Malayan food. At home we have a Malay cook. She used to cost me twenty dollars a month. Now she wants six hundred! And being a Jew doesn't come cheap here. You can buy a fresh chicken in the market for five dollars. A kosher one will cost you nine. And even so it's cleaned out by a Chinese butcher and not by the rabbi." Geoffrey interrupted to ask if Judaism was really about slaughtering chickens and if eating a kosher chicken made one more Jewish than eating a non-kosher one. As Alec had never in his life even considered buying a non-kosher chicken the question hung uneasily between them. Alec turned the conversation back to food and eating as we sat in the slow-moving, perfectly mannered traffic that was making its way to the centre of the city.

Alec's apartment, where I was to stay, had little of the Orient about it. A large modern flat in the fashionable Orchard Road area, it enjoyed panoramic views over a good deal of Singapore. The walls were decorated with a Passover plate from Israel and a large colour photograph of the Jerusalem skyline taken some years before. His Malay cook had a small room off the kitchen. In rapid Malay he shouted for the girl to bring me some tea and himself some yellow rice. Alec, now in his shorts and barefoot, eating his rice with grunts of pleasure, was of this place. "Singapore is home, mun. I know the people, know the languages. No one is more Singaporean than me!"

Late in life Alec had become a fairly successful stockbroker. Driving me to the centre of town in his chauffeur-driven Mercedes later that day, he talked about the depressed state of the market: "It's killing me, mun. I haven't sold anything for days!" Between

Singapore: A Dwindling Community

his complaints and imprecations he discussed the market and the business successes and failures of the Jewish community, and it became clear that there was very little he did not know about Singapore or about his co-religionists. I asked him if there were any intellectuals in the community. "Yes, we have two intellectuals. The rest of us are businessmen. We have David Marshall – the most famous Jew in Singapore – and we have Yahya Cohen – the second most famous Jew. Yahya Cohen used to be Surgeon General. When he was in private practice he used to charge $5,000 an operation and do twenty operations a day. I guess you'd call that an intellectual!"

Like the Baghdadis of Bombay, the Singapore Jews became thoroughly Anglicized under the Raj. None the less, although the wealthiest Jewish families were considered "Europeans" by the British, and their social events, for instance, were listed in the section of the local newspapers devoted to European rather than Asian events, the majority of Jews were always regarded as orientals or Asiatics. As such, not only were they banned from exclusively British institutions such as the Tanglin Club and the other European clubs, but they were also denied the right to serve in British armies in both world wars. As a member of the community put it to me: "There was a ruling class and there were the others. We were the others. Quite simply we were considered as natives. Even incredibly wealthy Jews who practically owned Singapore at the time never penetrated further than the periphery of British society. Obviously the fact that we are more or less white and after a while spoke English as our first language gave us more in common with the British than with the Chinese, Indians or Malays. And, of course, we thought of ourselves as being loyal British subjects. There was a picture of the King in every Jewish home. But the British did not want to have anything to do with us. The arrogance of the British became more and more irksome and we became increasingly aware of the anomalies and injustices which we had to endure. The best jobs always went to the British who were often less qualified than our people. The Jews have been called 'a justice-intoxicated people' and by some accident of history it was one of our 'justice-intoxicated' Singapore Jews who became so appalled by the arrogance of the British that he led the agitation which finally led to their withdrawal from the island. His name is David Marshall."

David Saul Marshall is the most prominent member of the Jewish community in Singapore. He is now serving as Ambassador to Paris and it was at his Embassy residence that I met him some months before I visited Singapore. A large man with impressively hooded eyes, he greeted me warmly and apologized for his dress. He was wearing an oriental silk dressing-gown which, notwithstanding his Iraqi origins, contrived to make him look distinctly Chinese. He spoke enthusiastically about the history of his community. His father, he told me, had not intended to leave Baghdad and settle in Singapore, but in 1900 doctors had advised him to go on a sea voyage to recuperate from an illness and this had taken him first to Bombay and then on to Singapore. The ship berthed at Singapore during the month of Ramadan. Laughing heartily, Marshall told me: "My father was horrified at the idea that the poor Singapore Muslims did not have any dates with which to break their Ramadan fast. In Iraq both Jews and Muslims always broke their fasts with dates. How anyone could break fast without dates was beyond his understanding. So he decided to stay in Singapore and make a living by importing dates from Basra for sale in Malaya and Singapore at the time of Ramadan. He did not make a fortune but he did well enough." He did not, however, do well enough to finance his son's law studies and David Marshall was obliged for a time to work as a car salesman in England in the 1930s before being called to the Bar in 1937. When he returned to Singapore he became the outstanding criminal lawyer of his generation. After the war Marshall, by now the leader of the Labour Front, led the agitation against the British which was to result in a thoroughgoing review of the constitutional status of the colony. When, in 1955, Singapore gained partial independence, Marshall led the United Labour Front coalition to a resounding political victory and became the island's first Chief Minister and Minister of Commerce. After this the days of the British Raj in Singapore were numbered.

Marshall's political activities were hardly more popular with the Jewish community than they were with the British. No matter how arrogant the British might have been, most Jews viewed the colonial presence on the island as a benign protectorate. As one old Jewess put it: "There was God above and the British below. Without the British we felt that we would be lost." Consequently, as Marshall told me, "The Jewish community damned me as a communist. They could not understand how I could attack the British who were so nice to us. But increasingly I felt that God had given me my voice to express the resentment of all Singaporeans against the miasma of racial discrimination and racial imperialism which was eroding our self-respect. I wanted justice. And the fact that I am a Jew", Marshall continued, "is not without significance. There is a Jewish cry for justice which echoes down the corridor of the centuries and that cry for justice has encouraged many social reforms in many parts of the world. Obviously Jews have a special relationship with injustice. As the archetypal minority they have been very close to it. You know, if you wear the shoe that pinches you know exactly where and why it hurts. We have worn the shoe of injustice for centuries. I developed an emotional resentment against being treated like a second-class human being. I began to resent the idea that God is white and that the whites' right to rule the world was God-ordained. I began to resent the idea that we non-whites should be humbly grateful for the opportunity to pick up the crumbs which fell from the white man's table." Marshall's voice had risen and he was speaking passionately. It was not difficult to visualize him haranguing groups of Malays and Chinese to throw off the yoke of colonial rule.

During his period in office Marshall was able to perform a very considerable service for the Jews of the East. In 1956 he was negotiating with the Chinese communist leader Chou En-lai in an attempt to persuade him to issue a communiqué encouraging Singapore Chinese to take up Singapore citizenship. As Marshall told me: "Seventy-five per cent of our population was Chinese and the British had made life impossible for them. Specifically they made naturalization acutely difficult by insisting that they take an oral examination in English as part of the process of acquiring citizenship. Every time they came along to take the test they slapped them down with words they did not understand. So their proud reaction was eventually to refuse to take up British Singapore citizenship. When I took office I wanted their loyalty to their country and I finally persuaded Chou En-lai to issue a communiqué urging the Chinese to give their loyalty to the country of their domicile. 'But', said Chou En-lai, 'there is the ancient tradition that when the Chinese are old and ready to die, they like to bury their bones in their ancestral country.' And he wanted a clause reflecting this inserted in the communiqué. I said in reply: 'I am moved, Mr. Prime Minister, to hear you voicing a tradition which is so close to that of the Jewish people who for millennia have wanted to go at the close of their life to be buried in the Land of Israel. And I am reminded that at this very moment there are five hundred tragic souls in Shanghai whose only wish is to be able to leave China, which they are being prevented from doing, to go and await their end in Israel. As it is they are not allowed to work, they are not allowed to earn, they are not allowed to leave.' Within a month", said Marshall smiling, "those Jews were allowed to emigrate. I'm very proud of that!"

Marshall's success in attracting the votes of a predominantly Chinese electorate must owe something to the way in which the Jews are perceived by the Chinese. There have been Jews living in China for a thousand years, and even before that there is evidence that Jewish merchants were not only heavily involved in the traffic of the overland silk route connecting the Far East with the Mediterranean but personally penetrated as far as what is today the Xinjiang Autonomous Region of China. Eventually Jews decided to settle and established themselves in Loyang, Beijing and Kaifeng, where the first synagogue in China was constructed in 1163. Chinese records of the Ming Dynasty (1368–1644), as well as the reports of mediaeval travellers such as Marco Polo and Andreas of Perugia, mention flourishing although increasingly acculturated Chinese Jewish communities. During the Ming period the Jews were so well regarded that they started playing a prominent role in the administration of China. As this tendency inevitably led to individuals being posted to administrative positions throughout the Empire, it can be seen as one of the factors which led to the dispersion and virtual disappearance of the community. During the Ch'ing period (1644–1911) the number of Jews in China dwindled and when, in the middle of the nineteenth century, China was rediscovered by the West, the only Jewish community to have survived was a small group in Kaifeng. For centuries the Kaifeng Jews have been physically indistinguishable from the other Chinese of the region. Today there are a handful of families left who, although not practising Jews, are still proud to claim Jewish ancestry. As one of them said to a recent visitor, "It puts us in direct family relationship with the great Karl Marx."

Few Singapore Chinese can be aware of the long history connecting the Jews with China. None the less, their attitude to the Jews is no less tolerant and respectful than that of their Chinese forebears.

One Jewish shopkeeper who had lived and worked among the Chinese all his life told me: "The Chinese respect the Jews because they consider that we have the same basic philosophy as them. We are keen to have our children well educated. Just like them. We honour our ancient traditions and ancestors just like them. And also they think that we are very intelligent and very able. And it is true that in the past the Jews were incredibly successful in Singapore compared with any other group. But now, although the Chinese have caught us up, they still think of us as a genius race and always believe that we'll outsmart them in the end!"

The Chinese attitude towards the local Jews has been influenced to some extent by the Singapore government's attitude towards Israel. Right from the beginnings of the independent state of Singapore its leaders have attempted to emulate the Jewish state. Singapore's authoritarian Prime Minister Lee Kuan Yew set himself the task of creating a garrison state on the Israeli model with a population as disciplined and as motivated as he believed the Israelis to be. The Singapore armed forces, which gradually took over the defence of the island republic from Britain, were modelled on the Israel Defence Force and have received considerable military assistance from Israel. Goh Keng Swee, thought by many to be the real architect of Singapore's economic success, said early in his career that there were two Asian countries that Singapore should try to emulate: one was in the extreme east – Japan; the other was in the extreme west – Israel. Albert Leila, a Baghdad-born Jew who for years has had a shop on Singapore's famous and colourful Change Alley, where Muslims from Sind, Tamils, Armenians and Chinese make a good living from the tourists, has had Chinese partners and worked with Chinese traders all his life. After the Six Day War, he told me, all the Chinese traders in the area came to congratulate him personally on the Israeli victory. In the euphoria that followed the brief war, many of the Jews started calling themselves "Israelis" rather than "Jews". But, as Albert Leila said, "That did not last long. The Yom Kippur War brought an end to all that and we're back to being just ordinary Jews again."

Many of the Singapore Chinese I met spoke with admiration of the enormous success of the local Jews, and the island's newspapers report on this subject regularly. According to Alec Manasseh the reason for this is straightforward: "The Chinese have always thought the Jews to be superior. Whatever they have done they have done well. They are good lawyers, good surgeons, good opticians and most of all good businessmen. Naturally the Chinese are impressed." But the fact is that the Chinese have taken their veneration for the local Jewish community a great deal further than their undoubted successes would justify. When I met Yahya Cohen, the community's most articulate member, he told me a revealing story. Some years before he had been waiting with a Chinese surgeon for a patient to arrive at the operating theatre. Their conversation turned to the various minorities in Singapore. Cohen asked his colleague: "How many Malays would you say there are in Singapore?"

"Oh, perhaps 15 per cent."

"Quite right," agreed Cohen.

"And Bengalis?"

"Perhaps a few hundred," said the Chinese surgeon.

"And Parsees?"

"Even fewer. No more than twenty or thirty."

"Absolutely right."

"And how about Jews?"

"I don't know exactly," said the Chinese cautiously, "but I should say there are about 200,000."

"But who are these 200,000 Jews?" asked a bemused Cohen.

"Well. There's David Marshall. There's you. And I'm afraid I don't know the rest!"

At the time there were no more than four hundred Jews in Singapore. Now there are even fewer, but it is still common to find Chinese who believe that the Jewish community is one of the most populous in the republic.

Like the Baghdadi Jews of India the Jews of Singapore have tended to resist eastern influences. As one Jew said to me, "Chinese culture has barely dented the surface of our way of life." This is partly because in Singapore the various nationalities keep firmly to their own cultural traditions. It could be that eventually there will be a homogeneous Singapore culture, which would inevitably be largely influenced by Chinese culture and in which all the minorities would participate. But that moment is far away, and by the time it arrives it is probable that the last Jew will have left Singapore.

Despite their cultural separateness many of the Jews have considerable respect for the local Chinese and for the way Singapore is run. "I am convinced", Yahya Cohen told me, "that if Jews were running Singapore instead of the Chinese we would have an internationally known philharmonic orchestra by now as well as five universities, ten political parties and twenty newspapers. But we would almost certainly be living in a filthy, dirty city where there was scant regard for law and order and we would have to make do without any real government at all. What the Chinese have done here is not to everyone's liking. But we have firm and stable government, the cleanest city in the world, and the highest per capita income in Asia after Japan."

Many of the younger members of the Jewish community take a less sanguine view of the "firm government" of Lee Kuan Yew. One boy who had just left school told me that more than anything else he wanted to live in a more open society than that afforded by today's Singapore. "I have no particular desire to chew gum, read *Playboy* magazine or smoke joints, but the fact that doing these things is against the law of the country I find ridiculous and oppressive. If I stay here I shall have to serve in the army of a country whose values increasingly I despise." He had decided to go to university in Los Angeles where already there are more Singapore Jews than there are in Singapore. He did not think he would be coming back.

Yahya Cohen agreed that the Jewish youth of the island find Lee Kuan Yew's obsession with discipline irksome. "Of course young people kick against a regime which makes littering, jay-walking and spitting in public fineable offences. But this is not really oppression. The self-discipline of Singapore is self-imposed. What is more troubling perhaps, and the Jews are conscious of it, is that dissent of any form has become almost unheard of in Singapore in recent years. People seem to believe that silence is better than protest. I sometimes think that the national motto ought to be 'no comment'. Well," Cohen smiled, "that's not a very Jewish attitude."

The one aspect of life in Singapore which few Jews will be happy to give up is the local Chinese and Malay cuisine. Albert Leila, the old-style trader of Change Alley, told me that after spending much of his childhood in Singapore he was taken back to Iraq in the 1920s by a father who had tired of selling small haberdashery items in the tropical heat of the Malay jungle. Just before the Second World War when the position of the Jews in Baghdad was beginning to deteriorate sharply, Leila received an anonymous letter containing two bullets. He was warned that if he failed to deposit U.S.$100 at a certain place a third bullet "would find its way to his black Jewish heart". Leila decided to leave Baghdad and return to Singapore. "But to tell you the truth," he said, "the reason I left was not because of the threats. I did not take them very seriously. The real thing was that the taste of the food of Singapore was in my mouth. Once you have eaten here you never want to eat anywhere else."

Jews of all classes have appropriated the local food. I was invited to a party by a young Jew who had recently made a fortune importing watches. His Malay cooks had prepared a meal which would have done credit to a Rajah's table. It was the classic fare of the Malaysian *haute cuisine*. When I complimented him, he smiled disarmingly, and said, quite innocently, "This is just our local Jewish food." An old Indonesian-born Jewess whom I met in the Jewish old folks' home, where around a dozen poor old people live on the charity of the community, was anxious to tell me exactly what the local Jewish food consisted of: "O Gosh," she said, "Jewish food here is so good. Chutneys, curries, pilau rice, *kitchiri*, noodles," and she worked her way nostalgically through the cuisines of most of the Asian communities of Singapore.

Jonel Craiu is one of the very few Ashkenazi Jews still living in Singapore. Born and brought up in Romania, he had little if any contact with oriental Jews before his arrival in the East some twenty years before. "But they are Jews like us," he said, "even though when I first came here I did not think so. The Ashkenazi Jews here laugh about their obsession with rice. Every year we celebrate the second Passover *seder* all together as a community. But what these Iraqis put on amounts to a rice feast. It's incredible for us. If ever you want to tell the difference between an Ashkenazi Jew and a Sephardi Jew in Singapore all you have to do is listen to them talk for five minutes. You can be sure that the Sephardi will tell you how many times he has eaten rice during the day. And what they call Jewish food is *biryani*, curry, local pickles. They have this breakfast, lunch and dinner and they think it's Jewish!"

A few evenings after my arrival in Singapore I was taken by Alec Manasseh and his wife to the Tanglin Club for dinner. This bastion of British exclusiveness long ago opened its doors to anyone able to afford the membership fees. In the course of the evening I became aware of another passion the Jews of Singapore have in common with the local Chinese: gambling. Alec's wife disappeared soon after dinner to try her luck on the fruit machines. As we waited for her to return Alec said: "We might have a long wait, mun! Gambling in one form or another is the greatest pastime of the Jews here. One of the richest Jews on the island is Jacob Ballas. A millionaire several times over. He's such a big gambler that he is consulted by many of the greatest casinos in the world. But all the Jews go to the horse races, both here and up-country. For years the rabbi has been trying to get people to stop going to the races on Saturday and go to the synagogue instead. But I ask you. What a hope, mun! Even the poorest of the poor gamble, and if they haven't got a stake they'll borrow it. If they are not betting on the races they are playing mah-jong, poker or a Jewish Iraqi card game called *yaka*. During the Japanese occupation of Singapore when the Jews were interned, gambling on cards became a real problem. I was in charge of one of the huts and tried to put an end to it. In theory people were losing millions. We did not have a table so they used to play on the earth floor, which I used to soak with water every morning to stop them playing on it. But what a hope, mun! Trying to stop a Singapore Jew from gambling is like trying to stop a British sailor from going to a Chinese whorehouse."

I met Jacob Ballas in his office in one of the new high-rise blocks that make up the modern city of Singapore. An aggressively jocular man, he is regarded by the other Jews as the *enfant terrible* of the community. He has had a spectacularly successful business career, but his real interest is gambling; and it was horses, jockeys and casinos that he wanted to discuss with me. It was with difficulty that I was able to arouse his interest in anything else. Ballas has a high regard for the local Chinese. "Most of them", he told me, "have the idea that the Jews are outstanding businessmen. That we are a genius race, as they always put it. But this is rubbish. The fact is that the Chinese are the most gifted businessmen in the world. They never talk or think about anything except money. The only reason I made a fortune is because I spent my life with the Chinese and I may have picked up a thing or two. I was practically brought up with Lee Kuan Yew's brothers and sisters. They always tell me that I am a Chinese Jew!" And, indeed, notwithstanding his Iraqi origins, with his yellowing complexion, small round eyes and cropped hair, he looked more Chinese than Jewish. Even his disconcertingly extended smile looked Chinese. Ballas came from a poor family and started in business by selling insurance. From there he went on to become the first president of the joint Singapore/Malaysian Stock Exchange. Although he is now a very wealthy man with homes in London, Israel and the United States, he is very bitter about the poverty of his childhood. "There were some very rich Jews and some very poor Jews. *We* were very poor Jews. I have always resented the fact that the Meyer family, which was fabulously wealthy at the time, could have used their money to ensure that no Jew in Singapore was on the breadline. But they didn't. The Meyers would not even have noticed it. If they had used their money properly we could have had a Jewish community here of twenty or thirty thousand people. We could have saved Jews from Europe perhaps and could have been one of the great Jewish communities of the world. But instead they used to make us crawl up the hill from Waterloo Street to their mansion to accept the two-dollar charity they doled out once a year. My father was a poor baker. We needed that money as badly as anyone. But he would never let us go up the hill and take Meyer's handouts." By his expression I could see that this was one subject closer to his heart than gambling.

Some days before I left Singapore I went with Alec Manasseh to see the old cemetery of the Jewish community which was being dug up to make way for Singapore's ambitious new underground railway system. The cemetery was on an elevated piece of ground fringed with palm trees and was divided into two distinct plots. The higher of the two plots, dominated by two great mausoleums, was reserved for members of the Meyer family. The lower plot provided humbler accommodation for the rest of the community. The Meyers had kept themselves separate in life and in death, but this was about to be ended. A team of three professional disinterrers from Israel was supervising the transfer of Meyer bones from their old graves into black plastic bags. The remains were to be buried in a new plot which was to be organized on altogether more democratic lines. "The Meyers won't like it," mumbled Alec, "but I think the fact that they are now being lumped together with everyone else gives the rest of us the last laugh."

David Marshall had spoken with pride of the massive contribution the island's Jews had made to the development of

Singapore over a period of a century and a half. But he realizes that the days of this community are numbered. "We have been abandoned," he told me, "both by world Jewry and also by Israel. No one, apparently, has any interest in perpetuating Jewish life in this tropical island. The Israelis are happy enough to receive our financial contributions, but they treat us merely like milch cows. And despite the fact that our spiritual life here is faltering we have not been offered the spiritual sustenance we need. The Iraqi Jews think of themselves as a lost tribe. Perhaps they are. We Singapore Jews left one exile for another and now this remnant of a lost tribe is disappearing. We are a vanishing community. And yet we did as much as anyone to make Singapore what it is today." Staring distractedly into the distance he concluded gloomily, "I think the problem is a spiritual one."

I was anxious to hear more of the spiritual problems of the community. Early one morning, having breakfasted with Alec on yellow rice, I went to Waterloo Street. At one end of the street the air was fragrant with the smell of Chinese food being served at one wok stall and in dimly lit shop-houses. The synagogue stood at the end of a row of dilapidated colonial-style houses. As I entered the synagogue precinct I could hear the sound of prayers through the open windows. Outside, in the aggressively lush garden, in the temporary shade offered by a slender coconut palm, a white parrot in a wooden cage chanted along in unison. "Wait a moment, *Tuan*," said the Malay caretaker, smiling sheepishly and rubbing his eyes, "the rabbi will soon be out," and he made room for me on the bench where he had been taking an early morning nap. A few minutes later the rabbi, a small, soft-faced Moroccan, walked briskly down the steps. "Good morning, Rabbi," I said. "I wonder if you could tell me something about the Jewish community."

"What do you want to know?" he snapped. "And why? You're not Jewish, are you? It's just a Jewish community. We have two synagogues, a school and a poorhouse. That's it. Nothing else." He clapped his hands together dismissively and went off to slaughter chickens.

FOUR
The Secret of the Tribe of Zebulun

In early mediaeval times Jews from Europe and the Middle East may have been involved in trade with Japan through their connection with the silk route. Later, during Japan's so-called "Christian Century" (1542–1639), some Jews participated in the limited trade initiated by the Portuguese and the Dutch. But it was not until after 1853, when Commodore Perry of the United States Navy arrived in Japan and initiated the process which was to open up Japan to outside influences, that Jews started to settle in the country. During the next few decades Jewish communities established themselves in Yokohama and Nagasaki, where they were primarily involved in the import–export trade, and subsequently Jews settled in Kobe and Tokyo. After the Russia revolutions of 1905 and 1917 many Jews made for the West, where they found havens particularly in Great Britain and the United States. But some took the long overland route eastwards and found sanctuary in Manchuria, China and Japan. During the 1930s and the early years of the Second World War, notwithstanding the close ties Japan enjoyed with Hitler's Germany, thousands of Jewish refugees from Nazism found temporary shelter in Japan before being relocated. This was partly the result of Japan's "Jewish policy", which was an important element in Japanese foreign policy considerations before and during the war. Today there are five or six hundred Jews in Japan, including a number of Japanese converts to Judaism. The backbone of the permanent Jewish community still consists of Russian *émigrés*, but the majority of the Jews now living in the country are temporarily based American businessmen and their families.

However, Judaism has far more remarkable expressions in Japan than this small Jewish community of exiles. The Japanese are known to be the most voracious consumers of all things Western and there are very few things more basic to Western civilization than Judaism. It is none the less surprising to realize that for the last half century the Japanese have been fascinated by Jews, Israel and Judaism. The most extreme form of this fascination is the view held by some Japanese that the Japanese themselves, or at any rate elements within the Japanese nation, are descended from the lost tribes of Israel. One of the first to put forward this remarkable theory, and perhaps its originator, was a Scottish missionary, N. McLeod, whose work *Epitome of the Ancient History of Japan* (Tokyo, 1873) enjoyed something of a vogue in the last decades of the nineteenth century. By and large his arguments are not compelling, but this "common origin" theory has adherents to this day. A more chauvinistic view, known as the "reverse theory of common origin", has it that Moses did not receive the Torah at Mount Sinai but travelled to Japan where he was instructed in the Shinto faith which he brought back in a modified form to the people of Israel. At the more sober end of the spectrum of interest is the Japanese Association of Jewish Studies, which encourages research in Jewish studies and produces a scholarly journal entitled *Studies in Jewish Life and Culture*, naturally written in Japanese. Perhaps the most striking outcome of the Japanese fascination with Judaism has been the establishment of religious sects, with tens of thousands of adherents, notably the Makuya and Beit Shalom, whose rituals and beliefs have a preponderance of Jewish and Israeli characteristics. That there is a widespread Japanese interest in the Jews can be adduced from the fact that a book comparing the Japanese with the Jews by one Isaiah ben Dasan (an otherwise unknown written whose identity is something of a mystery), entitled *Nihonjin to Yudayajin* (The Japanese and the Jews) (Tokyo, 1970), won one of Japan's most coveted literary prizes and has sold well over a million copies.

Jewish communal activity in Japan is based on the Jewish Community Centre in the Shibuya area of Tokyo. Shortly after the Second World War the community bought a stylish old house on the site of which the Centre now stands. With the escalation in land

and property values in Tokyo it was decided to pull down the old building, sell off some of the land and rebuild. With the $3 million profit it was possible to construct an elegant modern complex, which includes a restaurant, swimming-pool, library and synagogue.

The temperature outside the Jewish Community Centre was 36°C when I arrived there. Everyone had been complaining of the unusual heat. Inside the Centre, however, it was cool, almost cold, and the smells wafting up from the kitchen were not of Japan but of Central European cuisine: chicken broth, dumplings, goulash. I glanced into the restaurant: a Japanese waiter was playfully reprimanding a bearded European diner: "Alex! How many *piroshki* have you eaten today?" A bottle of chilled Stolichnaya vodka stood at Alex's elbow. A large group of Americans were talking volubly together.

Michael Schudrich, the young bearded American rabbi of the Jewish community in Tokyo, received me in his tastefully furnished apartment overlooking the Community Centre's pool. When I asked him why there were so many American Jews in Tokyo, he replied cheerfully, "If a Jew wants to be poor he can do it in New York or Jerusalem. Those that come here do so because of the fantastic business opportunities. So most of them are businessmen of one sort or another." The rabbi told me that at any one time there would be a few dozen Jews in Japan studying the martial arts or some aspect of Japanese language or culture. But great as the Jewish interest in Japan was, he was quite convinced that the Japanese were more interested in the Jews and in Israel than the Jews were interested in the Japanese. The primary reason he put forward for the Japanese fascination with Israel was the "incredible military success" of Israel in its wars against the Arab states. This the Japanese admired. The second was what the Japanese perceive as being the overwhelming economic success of the Jews. "In fact," he said, "we have something of a problem with that," and he reached down a couple of English–Japanese dictionaries. "You see the definition of 'Jew' here relies pretty heavily upon the idea that the Jews are rich." He showed me the entry in Sanseido's *New Crown English–Japanese Dictionary* (revised edition, 1964): "Jew (dzu) n. Jew: Jews covet money – consequently there are many Jewish millionaires. The word can be used in lieu of the following: 'avaricious', 'miser' and 'rich'." The rabbi sighed: "Entries like this are pretty alarming and we are trying to do something about it." Occasionally he had to respond to anti-Semitic articles in the press. Only the week before there had been one originally written by an American anti-Semite describing the Jewish domination of the Western press. Schudrich had shown this to some of the local Jews and they told him not to worry because the only impact that such an article was likely to have on the Japanese was to make them want to emulate the Jews, not despise or fear them! Recently a Japanese man had come to see him and told him he wanted to convert to Judaism. He asked him why he should want to do so and he replied: "The Jews are so clever. Look, they have produced Freud, Einstein and Barbara Streisand!" The Japanese was convinced that the act of conversion in itself would make him cleverer and more effective in business. One of the factors which further complicates the Japanese view of the Jews is the idea that the Japanese *are* Jews. "Up north there is even a tombstone thought to be the burial place of Moses. Moshe Rabeinu in Japan!" the rabbi expostulated in a bemused way.

"I get letters almost every day from Japanese who are interested in Judaism." He showed me one that he had received that morning from Yamato city in Kanagawa prefecture. The letter concluded: "I am afraid of the God of Abraham. I am sure that only the God of Abraham is God. I think that Jews are lucky because they are descendants of Abraham. I heard that Anti-Messiah may come out of the tribe of Dan. I want to know the reason. Would you tell me the reason? 60 yen stamp is put on this letter." The rabbi laughed apologetically. "But the most amazing thing is the phenomenon of the two big Judaizing sects, the Makuya and the Beit Shalom. Both of the sects have prophets – *neviim* – and, of course, both of the *neviim* had visions! The Beit Shalom are really Christians with a lot of Jewish colouring. But the Makuya seem to be giving up on the Christian elements of their religion and are getting more and more Jewish by the day. I should just mention that they believe that they are one of the lost tribes of Israel. But the most colourful 'lost triber' is Master Kampo Harada. He's Japan's most famous calligrapher and lives half way up a mountain near Kyoto where he keeps what is probably the best collection of Jewish books in the Far East, which he can't read. He also believes that he knows where the original tablets containing the ten commandments are hidden." It was clear that the laconic Rabbi Schudrich did not take Master Kampo Harada very seriously.

The rabbi had invited me to take a swim before I went. I walked across to the pool where I found a young Japanese woman swimming lazily backwards and forwards. She greeted me with a demure *"Shalom"*. I said *"Shalom"* in return and then asked her, in English, if she spoke Hebrew, assuming that she did not. Rather less demurely and in extremely good English she asked: "Am I to assume that as you reply in English, *you* do not speak Hebrew?" Defensively, I replied: "As a matter of fact I do." "It's strange," she said smiling, "you don't look Jewish." I was soon to discover that she spoke Hebrew fluently.

Noa Kayoko Uno was a convert to Judaism. She had first become interested in Jews, she told me, when she was ten years old and had read a Japanese translation of *Anne Frank's Diary* which her father had given her as a birthday present. A little later, like many other Japanese schoolchildren, she read *The Merchant of Venice* in English. As her knowledge of the Jews was based exclusively on these works she became increasingly puzzled about the nature of Judaism and why six million Jews had been murdered by Japan's German allies in the war. Shortly after she left school her best friend married an Israeli who had been studying the martial arts in Japan. When the couple moved to Israel, Noa was invited to visit them. She was to spend three and a half years in Israel.

On one occasion she was travelling on a crowded Israeli bus full of Jews from all over the world and found herself seated next to an elderly Polish woman who had an Auschwitz number tattooed on her wrist. Noa was stunned and afterwards read everything that she could get hold of on the Holocaust. Some weeks later she went to visit the old woman. They had no real language in common, nor did they need one. For a long time they simply held each other and wept. When she left the old woman's room she had decided to become a Jewess.

For a year she studied with an old Russian *rebbetzin*, mastered the laws of *kashrut*, continued with her Hebrew studies, went to the synagogue regularly, and kept the Jewish feasts and fasts. In due course she took an orthodox conversion. But the death of her father a year later obliged her to return to Japan and now she had the responsibility of looking after her mother. For the time being she would not be able to return to Jerusalem. "But that", she said, "is where I belong."

Talking to Noa in Hebrew was a strange experience. Her Hebrew was strongly accented but had a peculiar sweetness. "You must have charmed the rabbis in Jerusalem, Noa," I said. "Hardly that," she replied. "For some time they would not accept me. One rabbi came all the way from Safed to tell me that I could *not* be Jewish. I recited to him the passage from the Book of Ruth, in Hebrew of course: 'Entreat me not to leave you or to return from following you; for where you go I will go, and where you lodge, I will lodge; your people shall be my people, and your God my God; where you die I will die and there will I be buried.' Then, I must admit," Noa continued, "tears came into the rabbi's eyes and, like Naomi, he said no more."

For some time Noa spoke to me of her love for Israel and Judaism. She had found in her new faith a liberation from what she called the oppression of Japanese society and the tyranny of her father's Samurai family and traditions. In many respects she now found herself outside the Japanese system and that allowed her the freedom to do as she wished. "But my mother treats me somehow as a traitor," she said sadly.

I asked her if by any chance she was one of those Japanese who believed themselves to be descended from the lost tribes of Israel. Noa was not. But she was not completely dismissive of the idea either. "You have to understand that no one really knows where the Japanese came from in the first place and therefore any theory which helps throw light on our origins is always going to find followers. All we know is that the Japanese are a mixture of various groups who migrated to Japan from Asia and became more or less assimilated by about the fourth century under the leadership of the Yamato clan. I know that some people believe that the Yamato were Jewish. I am not sure about that," she said thoughtfully. "But if you like I'll take you to meet some Japanese people who believe that they *are* descended from Jews." She offered to take me to meet the Makuya.

Late the following afternoon a Makuya chauffeur was sent to pick us up. A haze of heat and dust largely obscured Tokyo, and a large blue sticker featuring a star of David obscured a good deal of the car's rear window. The driver exchanged a brief greeting with Noa and drove us in silence through a fashionable suburb of Tokyo until we reached a large, sombre house surrounded by a well-kept garden. This was the Makuya reception house. A Japanese girl greeted us in Hebrew and showed us into an anteroom, which was decorated in the Japanese style with *tatami* mats on the floor and paper screens around the walls. The girl murmured shyly that the sect's leader was expecting us. Looking around the room two objects caught my eye: a beautiful Japanese ceramic bowl and a substantial gold *menorah*.

With the death of Makuya's founder, "Professor" Avraham Ikuro Teshima, the spiritual guidance of the movement fell to Dr. Akiva Jindo. Jindo was an engaging young man with a quick smile. He entered the room quietly, bowing deeply to Noa and shaking hands with me, firmly, in the manner of evangelical Christians. After completing his Japanese High School, Jindo had been awarded a scholarship from the Grew Foundation which enabled him to spend four years at Swarthmore College, Pennsylvania, a Quaker foundation, where he studied chemistry. Thereafter he took a Ph.D. at the University of California at Berkeley. In 1966, in Japan for the summer vacation, he was invited to attend a Makuya meeting in Tokyo. During the prayer-meeting, Teshima was approached by a stream of sick and crippled people. "The Professor", in Jindo's words, "grabbed a polio child by the limp legs and stared up to heaven with eyes closed. During that breath-taking moment all eyes were on him. And, look! The crooked legs became straight. The power had entered into the numb legs and hands. And then a blind man was restored his sight. After each miracle hand-clapping burst from the audience and the voice of prayer became more intense." Immediately Jindo became an enthusiastic member of Makuya. After completing his doctoral work he returned to Japan, where he started working in the research division of the Hitachi company. But his real interests lay with Makuya and, when Hitachi refused to give him seventeen days' leave to join the Seventh Makuya Pilgrimage trip to Israel, he decided to incur the wrath of his family and friends and do something which is quite unheard of in Japan – leave the company.

Although Makuya is a Christian sect accepting the divinity of Christ, its Jewishness is very visible. Over the years thousands of Makuya have gone to Israel where many of them have learned Hebrew. Indeed, as I was to discover, it is possible to travel throughout Japan and find Hebrew-speaking members of the movement in most of the major towns. The importance of Hebrew for the Makuya can be judged by the fact that they have recently brought out a beautifully produced Japanese–Hebrew dictionary. Whenever they get together they sing secular and religious Hebrew songs, many of them the songs of modern Israel. They take Hebrew names, observe the Sabbath and keep a form of *kashrut*: they light candles on Friday evening, break *hallah* and read from the Jewish prayer book – the *siddur*. Their view of the world is informed by a profound admiration for Israel and the Jewish people. To some extent this sentiment derives from the Christian part of their ideology. But, in addition, it springs from the *national* nature of Judaism – the idea that Judaism is the religion of the Jewish people. The Makuya are intensely nationalistic and, in some ways, are looking towards the redemption of the Japanese nation.

Teshima met Martin Buber, the great Jewish philosopher, on a number of occasions. According to Teshima's written account Buber told him: "The religion of the Bible not only saves individuals but also redeems whole nations: it has never been an individualistic religion. If ever the Japanese mind is poisoned by so-called Western Christianity, which lays its basis on the individual's expression of faith, then I am afraid that the Japanese nation will collapse. God's concern is with the redemption of a people." Teshima's fear was that the Japanese, unlike the Jews who never lost their sense of being the people of Israel, had begun to view themselves increasingly as individuals. Teshima argued: "In postwar Japan, individualism and egoism, resulting from the prevailing, misinterpreted ideas of democracy, are gnawing at all aspects of national life.... What has made Japan such a pitiful country? It is the result of the wrong school education under the control of the communistic Nikkyoso [Japanese Schoolteachers' Union] which regards patriotism as the residue of the pre-war indoctrination of the militaristic totalitarianism and hence shuns teaching 'the pride of a nation' such as expounded in the Bible." Whereas he believed that, because the Jews had the Bible, they could bring about what he regarded as the greatest miracle of the twentieth century – the re-establishment of the Jewish state.

Jindo and I sat cross-legged on the *tatami* mats around a low table. As tea and rice cakes were served by another young Hebrew-speaking and kimono-clad Japanese woman, I asked him why there was no cross in the building, whereas it was full of Jewish emblems. Gently he replied: "Makuya is a Japanese Jewish religion. Most of our members were formerly Buddhist or Shinto and now they have become firm believers in the God of the Bible. But we are not following a Western religion and we do not need Western symbols. For us the cross is associated with 'Western' Christianity. We feel that we are members of a Jewish faith, but we still retain our Japanese customs. We go to the Buddhist temple to visit the graves of our ancestors and parents and for national reasons. Like the Zen master we meditate upon the *tatami* mat," and he pointed at the mats around us. "Indeed, we do not regard the Jewish symbols as being foreign. Our Japanese background is part of our Jewishness because our culture owes a lot to Judaism. The structure of the Shinto shrine, for example, is very much like the structure of the Second Temple. Shinto also has the idea of the invisible godhead. And once a year the Emperor, dressed in white garments, went into the holy of holies, in his capacity as High Priest, and asked forgiveness for the sins of his people. Very much like the Day of Atonement. As you know, there are various theories about the origin of the Japanese people. It is said that some came from South-East Asian islands and some from northern Asia. But the most important element, we believe, consisted of groups of Semitic people, particularly the tribe of Hada, whose origins are directly traceable to the Land of Israel. Their original worship was the religion of Israel. There are traces of their ancient Hebrew songs in a number of folk songs sung in Japan to this day. In the area of the great Shinto shrine of Ise people sing a song called *Ise Ondo*." He sang some unintelligible words. "It's the song of Miriam from Exodus 15. Can't you recognize it?" he asked excitedly. I had to confess that I could not. "It was brought to Japan by the tribe of Hada, and their descendants have kept the tradition for around 1,500 years." Jindo talked earnestly for some time about the tribe of Hada. According to his version of Japan's early history this tribe was to be identified with the lost tribe of Zebulun. The family crests of the Hada tribe showed sailboats similar to depictions found at archaeological sites in Israel. The Hada brought with them the knowledge of silk weaving and they soon became the industrial leaders of the country, while their sound financial instincts enabled them to serve the imperial court as economic advisers. Zebulun/Hada called their god Ogami or Yahada (which Jindo explained as the God of Hada) and their faith was gradually taken over by Japan's warrior class – the Samurai. The fact that their god Yahada was adopted by the royal family would suggest a special relationship between the Jews and the royal family. Jindo explained that Prince Mikasa, the younger brother of the ageing emperor of Japan, had made a special study of the golden mirror which forms part of the royal regalia of the imperial family. According to Japanese legend the sun goddess Amaterasu Ōmikami had been frightened by her brother Susano No Mikoto and had taken shelter in a cave, thereby plunging the world into total darkness. In order to entice the sun goddess out of the cave, the god Amano Koyane No Mikoto hung a golden mirror encrusted with gems outside the goddess's hiding-place and started to recite a specially composed prayer. Amaterasu finally emerged and light was returned to the world. "When Mikasa examined the mirror," said Jindo solemnly, "he found a Hebrew letter in each corner which together make up the divine name of Jehovah. This has been called the secret of the tribe of Zebulun." Jindo paused and looked at me intently. "If you look closely at the features of the imperial family, you can see that they look quite different from the ordinary Japanese. The late Premier of Japan, Prince Ayamaro Konoe, looked exactly like a European Jew and the same can be said of Prince Mikasa. It is quite likely that the royal family themselves are descended from the lost tribe of Zebulun!

"So you see, it is not all that surprising that we Japanese Makuya are interested in our Jewish roots. Through the line of Zebulun we *are* Jewish. And that means that we are passionately interested in Israel and support the state as much as we can."

I was later to hear from Israeli diplomats something of the effectiveness of this Makuya support. In the wake of the Yom Kippur War the Japanese government yielded to Arab pressure and was on the verge of severing diplomatic relations with Israel and imposing economic sanctions. Teshima at the time was a sick man. Despite doctors' warnings he led more than three thousand Makuya through the centre of Tokyo carrying placards in Hebrew, Japanese and Greek announcing that as Israel was "in the hands of God", it should not be abandoned by Japan. Teshima died a few weeks later. In November 1975, when the U.N. General Assembly condemned Zionism as racism, Makuya sent a petition of protest to the Secretary General, Kurt Waldheim, with 37,000 signatures.

During all the Israel–Arab wars Makuya has tried to send volunteers and other forms of aid to the Jewish state, while every year delegates are sent to participate in Israel's Independence Day celebrations. As one diplomat said to me: "There is no non-Jewish group in the world which gives such unequivocal support to Israel as do the Makuya."

"What people do not quite understand", said Jindo, "is that our support for Israel is not really a political expression. We see the creation of the state of Israel and the restoration of Jerusalem to the Jews as the fulfilment of biblical prophecy: as something *ayom*." The Hebrew word means "awesome"; Jindo gave it particular stress as if it had special significance for him. "Teshima's chief aim was to *protect* this fulfilment of prophecy. I mean to say that our support for Israel is an important part of our religious expression. We are trying to further God's work in the world by supporting the earthly Israel and by seeking to understand what theologians refer to as the mystery of Israel. But, in fact, the mystery of Israel is also the mystery of Japan." Jindo smiled enigmatically. "But to get back to your point about Jewish and Christian symbols. Even the Jewish star of David is fully part of our culture. The crest on the great shrine of Ise, Shinto's most sacred shrine, is the star of David. It also used to be the custom in Japan to present a kimono to a family celebrating the birth of a child and on the back of the kimono the star was sewn. Look!" and Jindo passed me a book with a photograph inside showing two young Japanese children wearing kimonos embroidered with the six-sided star. The book was entitled *The Ancient Jewish Diaspora in Japan: The Tribe of Hada – Their Religious and Cultural Influence*, by Ikuro Teshima. Glancing through the book I noticed a photograph of the Great Wall of China. The legend beneath the photograph read: "Architectural masterpiece by the hands of the Jewish Diaspora." It is possible that at this point my face started to betray signs of scepticism. In any event Jindo stopped smiling and rather abruptly told me that I clearly did not understand all this "with my soul". Pressing a button somewhere he turned on a video film of a Makuya conference which had been held the year before on the higher slopes of Mount Myoko. It was extraordinary.

The first part of the film showed Makuya members, some of them in their teens, undergoing *Hiwatari*: the ritual of walking barefoot over beds of burning coals. "This", said Jindo, "is one of the Japanese sides of our faith. It comes from Buddhism, of course. It is the expression of our total dedication to the God of the Bible. Of course, those who believe they will not get burned, by the power of faith do not get burned." The young initiates approached the fire fearfully; many of them completed the ordeal sobbing with pain. Obviously they had not believed enough. There were others who crossed in a sort of ecstasy, smiling and singing Hebrew Palmach songs from the Israeli War of Independence as they kept their eyes fixed on a giant *menorah* which had been placed on a prominent mound near the site of *Hiwatari*. I felt sure that the Israelites of old would have felt comfortable here on this high place with the smell of burning flesh in their nostrils. "Does this not remind you of the story of my namesake Rabbi Akiva?" asked Jindo. "When he was being tortured to death by Roman soldiers he started to laugh. When they asked him what he was laughing about, he replied that he had never fully understood what it meant 'to love God with all your heart and with all your soul', but now that he was able to praise God even during this torture he finally understood the biblical text." Jindo explained that the reason the Makuya were singing Palmach songs during their ordeal was because *Hiwatari* was testing their mettle as "soldiers of Israel".

Later in the film, crowds of Makuya could be seen whirling in ecstasy carrying velvet-covered Torah scrolls high above their heads in imitation of the Jewish feast of *Simhat Torah* which marks the completion of the year's synagogue reading of the Pentateuch. The scrolls had been donated to the Makuya by the Israeli Egged bus company after four hundred Makuya members had toured Israel in Egged buses after attending the twenty-fifth anniversary celebrations of the establishment of the State of Israel. This part of the ceremony, Jindo told me, was called *Simhat Makuya*. The film went on to show Makuya members saying fervent Hebrew prayers standing under freezing waterfalls, following the Japanese tradition of *misogi*.

Jindo could perhaps see that I was affected by this film and said, "Now I would like to show you something a little gentler, a little lighter," and he led Noa and me out of the building to a waiting car. We drove through the dark streets for a few minutes until Jindo stopped the car and pointed up at a high roof at the end of the street. A huge neon *menorah* shone against the night sky. This was the centre of Makuya activity in Tokyo. We entered a hall in which a dozen young people were preparing for an evening prayer meeting. A man was playing a Hasidic folk song on a great Yamaha organ. Around the walls were posters of the Wailing Wall and the Old City of Jerusalem with the inscription: 1967 – Reunification of a City. The Makuya greeted Jindo reverently. He said something to one of the girls in Japanese. The organist struck up and the sound of *Yerushalayim shel Zahav* (Jerusalem the Golden), the song always associated with the Six Day War, filled the hall. The girl stood simply, her hands behind her back, and sang the song in flawless Hebrew: "Jerusalem of gold, of bronze, of light,/Let me be the harp for all your songs." I could see that the Makuya were deeply moved. And so, in a strange way, was I.

Before I left, Akiva Jindo gave me an essay he had written explaining the centrality of Israel in his faith. The instrument of his "conversion to Israel" was Professor Bergman, the Israeli philosopher and first rector of the Hebrew University of Jerusalem. In Jindo's words: "I introduced myself to the benign professor and explained how I had come to Israel by giving up my scientific career. Nodding and approving, he listened to me and said: 'You have chosen the better. I studied with Einstein and wanted to devote my life to the study of mathematics and physics. But one day I was struck by the *Orot ha-Kodesh* [the holy light]. I was changed into a man who aspired to the way of religion. Formerly I had been living in a small world to please my ego, but then I was changed to live for a greater purpose – the future of our country and people and the eternal peace of mankind,' and he shook my hand. Then, all of a sudden, I felt as if boiling water was poured over my head; a shiver ran through my entire body, and I sensed clearly that the Holy Power that was pulsating within the professor was pouring into me. Encounter – this mysterious experience in the Holy Land of Israel had engraved a deep impression in my heart." As I read this I wondered whether Bergman had had any idea of the effect he was having on his affable Japanese guest.

As we drove back to the centre of Tokyo, Noa told me diffidently (for she considered my visit to the Makuya to be a failure and, as it was her idea, felt ashamed) that I ought to see Masayuki Kobayashi, a professor at the renowned Waseda University in Tokyo. He knew more about the connections between Judaism and Japan than anyone else and his views, unlike those of Dr. Jindo, had some scholarly and objective basis.

The Secret of the Tribe of Zebulun

Some days later I visited Professor Kobayashi. His wife opened the door of their pretty house and invited me in. I removed my shoes and entered a book-lined study almost filled with a large desk and two comfortable leather chairs. His shelves were full of the same books I had in my study at home. One wall was dominated by a relief map of Israel. In front of it on a small table stood a small Buddha; to one side a photograph of an orthodox Jewish sage. Through the window I could see a trellis covered with vines and beyond that an uncompromisingly Japanese garden. From time to time the house shook faintly from the passing of the trains on the nearby Yamate subway line.

Kobayashi was an elderly man with the slightly vain mannerisms adopted by many scholars. His wife brought us in Japanese tea and some coffee-flavoured blancmanges and then left us to our conversation. We soon discovered we had a friend in common: a Japanese called Avraham Ovadiah Hiroshi Okamoto, who had converted to Judaism after the Second World War. Okamoto had had the idea that the Japanese should be brought to the Jewish faith and to this end had gone to the United States where he studied for the rabbinate at the Hebrew Union College in Cincinnati. He returned to Japan a fully fledged rabbi, but had little success in attracting his compatriots to his adopted faith; he decided to increase his qualifications and perhaps his appeal by taking a higher degree at Oxford, where he did postgraduate work in Hebrew at the Oriental Institute. Kobayashi now told me that his Oxford career had had no perceptible influence on the way he was regarded in Japan. The foreign Jewish community still refused to accept him as a rabbi and, despite the fact that there had been a good deal of speculation about a mass conversion of the Japanese to Judaism, it finally transpired that the Japanese were not interested in converting. Finally he was forced to take a job in a university in Florida where he died a young and disappointed man, "with few students" Kobayashi added, with a hint of *Schadenfreude*. After the war a number of Japanese, like Okamoto, had become interested in Judaism, partly, Kobayashi suggested, out of a desire to atone for the crimes committed against the Jews by their German allies. Japan's most famous convert to Judaism was Setsuzo Kotsuji, a Hebrew scholar from Kyoto, whose family claimed descent from a long line of Shinto priests. At the beginning of the Second World War he had played a remarkable role in helping Jewish refugees who at that time were concentrated in Kobe, and he was used by the government to help implement their "Jewish policy". Taking the name Avraham, he converted to Judaism in 1959 and was buried in Jerusalem in 1973. His funeral was attended by many members of the Mir Yeshivah, who had escaped from Europe and had been helped by him years before in Kobe. "But of all the Judaizers," said Kobayashi thoughtfully, "Okamoto was the most serious and the most committed. Scandalously, the Jews here never accepted him and irresponsible people in the United States pushed him into his 'mission' to Japan. But none the less he was the most impressive example of the Judaizing principle."

Kobayashi was the chairman of the Japanese Association of Jewish Studies. Most of the Association's members had become interested in Judaism because of what had happened to the Jews during the war and also because of the great impact that *Anne Frank's Diary* had had on the Japanese imagination. But Kobayashi went to some pains to stress that although there was considerable interest in Judaism in Japan, it was an interest which had to be seen in the broad context of Japanese society. Almost every Japanese, he told me, is a member of some club or society devoted to the study of a foreign culture, religion or language. There are Norwegian, Welsh, Cornish and Finnish societies with enthusiastic memberships. That there should be a nucleus of people devoted to the study of Jewish affairs was entirely to be expected. "Of course," said Kobayashi, "there are aspects of the Japanese connection with Judaism and Israel which give it a certain piquancy. Something", he added dryly, "which might be absent in, let us say, the study of the Japanese interest in Cornwall." One of the key factors in keeping alive the Japanese interest in Judaism has been Japan's imperial family. Prince Mikasa, the younger brother of the Emperor, has had a scholarly interest in Hebrew and Jewish studies. But, in addition, there has been the fairly widespread idea that the royal family is itself Jewish. This is largely based on the notion first put to me by Jindo that the mirror of the royal regalia has Hebrew writing on it. But Kobayashi insisted that, as the mirror was "secret", it would be impossible for any scholar to verify or disprove the theory. "Of course," he said, "His Imperial Highness is a member of my Association. But it would be incorrect of me to ask him questions about such a matter."

The Secret of the Tribe of Zebulun

Kobayashi told me a number of other prevalent rumours. One such maintained that it was not Jesus who died on the cross but his brother – while Jesus left Palestine and made his way by boat to Japan, landing at a port on the Pacific in the northern prefecture of Almori where he married into the Sawada family and died peacefully at a ripe old age. It is said that to this day the family members bear a striking resemblance to Jesus. Kobayashi had checked the origins of this story and claimed that in 1937 the theory was first advanced as a way of attracting sightseers to the coastal village where Jesus was supposed to have landed. The idea was given to the villagers by one of the new breed of pro-Nazi "Jewish experts". As he told me this, Kobayashi laughed discreetly, in the Japanese manner.

"The idea of the Japanese being the lost tribe of Israel is older than that," Kobayashi continued. "Indeed, one of my main objectives in setting up the Association was to combat the common origin theory. One reason being that it is wrong. But there is another reason too," and he paused and looked more serious. "After the war a wave of philo-Semitism crossed Japan and many people embraced the common origin theory. But a lot of those were former anti-Semites who had got their ideas from the Nazis. I suppose they thought that their best cover was to become philo-Semites. And, of course, these days the Makuya have adopted the common origin theory too; but in their case the idea is to provide cover for their real purpose – to convert the Jews to some Japanese version of Christianity!" Of the founder of Makuya, Ikuro Teshima, Kobayashi spoke disparagingly: he had no right to call himself "Professor" (he was a businessman by training) and his religious eclecticism was the product of a disorderly mind.

Kobayashi had travelled widely in the West and had visited Jewish communities throughout the world. Jews, wherever they were to be found, were his absorbing passion. He had been brought up as a Methodist in a remote village in the Japanese alps, where he had enjoyed little social or intellectual activity outside the church. He got to know the Old Testament intimately. At the same time he got to know some of Japan's Koreans – the country's most pronounced ethnic minority and one that had never been allowed to assimilate. The Koreans triggered off in him an interest in the problem of minorities and this in turn led him to study the Jews – the archetypal minority. This interest first blossomed at the time of the Second Sino–Japanese War in 1937 when news of the Nazi persecution of the Jews started filtering into Japan. As early as 1905 Kobayashi's teacher of history at the University of Tokyo had himself written an article on Jewish history entitled "Anti-Semitism and Zionism", and this encouraged Kobayashi to pursue his scholarly investigations. His first article on a Jewish topic was published in 1939 at a time when what he called "a very strong stream of thought against the Jewish people" was becoming evident in Japan.

Japanese anti-Semitism is a remarkable and little studied phenomenon. One of the keys to understanding it is a strange story from the Russo–Japanese War of 1904–5. After enthusiastically engaging the Russian troops, Japan rapidly got to the point where it could no longer sustain the financial burden of fighting such a powerful opponent. The vice-governor of the Bank of Japan was sent to London to raise a loan. He had little initial success until, by chance, he met up with the American–Jewish financier, Jacob Schiff. In the wake of the infamous Kishinev Pogrom of 1904 Schiff's hatred of Tsar Nicholas II's anti-Semitic regime had reached boiling-point. Eager to damage Russia in whatever way he could, he raised a bond issue of $200 million which enabled Japan to go on to win the war. Subsequently Schiff became a national hero in Japan and the Emperor Meiji paid him the singular honour for a foreigner of inviting him to the Imperial Palace for lunch. This incident established in the Japanese mind the idea that Jews had unlimited access to money. Such a view appeared to receive confirmation in the notorious *Protocols of the Learned Elders of Zion*, an anti-Semitic forgery of the Tsarist secret police, the *Okhrana*, which came to prominence as a tool of White Russian propaganda after the Russian Revolution of 1917. It was translated from Russian by Norihiro Yasue, a Japanese officer attached to the staff of the White Russian general, Gregori Semenov, during Japan's ill-fated Siberian Campaign (1918–22). Semenov's proud boast had it that every soldier under his command had a copy of the *Protocols* in his knapsack. The book had a considerable effect. After Japan's defeat in Siberia many Japanese attributed their failure to an international Jewish conspiracy and this in turn led to a spate of anti-Semitic works being published in Japan during the 1920s and 1930s. During the Second World War the influence of the *Protocols* was plain to see. Kobayashi told me of one prominent Japanese,

General Nobutake Shioden, elected to the Diet in 1942, who until his death in 1962 continued to believe that not only had the international Jewish conspiracy brought about Japan's defeat in the Second World War but also that the plane which dropped the atomic bomb on Hiroshima had Hebrew writing on the fuselage!

Captain Yasue, the translator of the *Protocols*, went on to become one of the most prominent of the "Jewish experts" in Japan and perhaps the first to perceive the potential usefulness of the Jews for the Japanese cause. In 1926 he was sent to Palestine to study the Jews in their new "National Home" and nothing he saw there disabused him of his views: the kibbutz, he thought, was clearly the means by which the Jews would colonize the world once they had achieved the world domination prophesied by the *Protocols*.

"I have a good collection of *antisemitica*," said Kobayashi, and went to a pile of books standing on the floor. He passed me five volumes covered in very old brown paper jackets. They were consecutive issues of the journal of the Research Society of International Politics and Economics, *Studies of International Secret Power*. They appeared in the years before the war and were the work of the "Jewish experts" who had set themselves the task of researching the Jewish conspiracy and who became particularly important after Japan's Tripartite Pact with Germany and Italy in 1940. "It is very strange", Kobayashi continued, "that the anti-Semitic campaign was as successful as it was given that there were hardly any Jews here. Perhaps that is why it was so potent. We knew no Jews and could not distinguish between them and other *gaijin* [foreigners]. Thus, they were camouflaged and dangerous." Kobayashi shrugged apologetically. "But paradoxically it was this fear of Jews which led Japan to elaborate what was in effect a rather pro-Jewish policy initiative: the *Fugu* Plan."

The *Fugu* Plan was a scheme which emerged from the frequent deliberations of the "Jewish experts": Japan would offer the persecuted Jews of Europe a haven – "an Israel in Asia" – in their newly acquired territories in Manchuria. Few Japanese were inclined to colonize these vast and uncongenial areas, so Jewish settlers with the necessary technical, managerial and industrial skills would settle them instead. In gratitude for Japan's uniquely humane treatment of the Jews, Jewish "money power" would provide massive financial investment in Japan's Greater East Asia Co-Prosperity Sphere and enable Japan to develop properly her new dominions. At the same time the "Jewish newspaper, radio and film monopoly" would see to it that the world viewed Japan as a humane and civilized nation which should be accorded the respect of the international community. In 1938 this scheme was debated in Tokyo at what came to be known as the Five Ministers' Conference which was attended by the Japanese Prime Minister, Prince Konoye, the Foreign Minister Hachiro Arita, the Minister for the Army, the Minister for the Navy and the Minister responsible for Commerce, Industry and Finance. They decided that Japan's close diplomatic links with the anti-Semitic Axis powers meant that the policy could not be publicly endorsed but that it should be covertly pursued none the less. Even after the Tripartite Alliance Japan never abandoned the basic premise of the plan, which was that Jewish power in the world could be manipulated in Japan's favour and that persecution of the Jews would not serve that end. A clear lack of enthusiasm towards the plan on the part of American Jewish leaders meant that it was never fully implemented.

The *Fugu* Plan had two important effects. Thousands of European Jews granted Japanese transit visas by a soft-hearted Japanese consul in Kovno in Lithuania were allowed entry into Japan and were thus saved from almost certain death. Included among them were the three hundred members of Europe's most famous Jewish academy – the Mir Yeshivah. It was the only *yeshivah* to survive the war intact. But the plan had even greater consequences in Shanghai where tens of thousands of Jews were concentrated during the war years. Joseph Meisinger, the so-called "Butcher of Warsaw", was sent by the Gestapo to bring the final solution to the "Jewish problem" of the Far East. But the Japanese remained firm: they had promised Hitler a military alliance but this in no sense obliged them to be anti-Semitic. Meisinger's ambitions were frustrated and finally unequivocally blocked.

"It's not without piquancy, do you think, this Japanese–Jewish connection?" asked Kobayashi, smiling, as he guided me through the piles of books that lay between his study and the outside world.

That evening I had arranged to dine with Alex Tribuyov at the Jewish Community Centre – the same Alex who had been eating *piroshki* with such enjoyment when I had first arrived at the Centre. The Japanese waiters served us with cold sturgeon; cut-glass decanters of chilled vodka were placed in front of us. "I hear that you are interested in Japan's Jews," said Alex with mock severity, "but that you are spending all your time with the Japanese meshuganners like Jindo, Kobayashi and co. Oh yes, and with the beautiful Noa – not that she's meshuggane. She's O.K. But don't forget us Russians! I could tell you a tale or two you could use in your book. You ever hear of the *Fugu* Plan? Fugu – it's the Japanese globe fish. To cook it you have to have a special licence. You cook it well, it's delicious. Nothing like it. You make one mistake – you're dead," and he prodded me hard in the chest. "The Japs had the idea that they could harness the great Jewish power (we should be so lucky!), but they were afraid that if things went wrong for them and the Jews turned against them, that would be it. The war would be over and the Japanese would be finished."

Tribuyov has an almost legendary status among the Jewish community in Japan. He had studied law at the University of Kiev after the Russian Revolution but because his father had been a small factory owner he was classified as an "exploiter of the people" and had been forbidden to practise. In 1926 he decided to leave Russia, but this too was forbidden. None the less, he took the Trans-Siberian railway to Vladivostok. Arriving there one evening without any papers he could not risk staying at a hotel, so instead he found a prostitute and rented her room for the night. With the help of a local accomplice who had contacts among the Chinese smugglers who at that time were running perfumes and silk stockings over the border, Tribuyov managed to get into Manchuria where he made for Harbin. Harbin was then the capital of Heilung Kiang province in northern Manchuria and since 1898, when Russia had gained the concession to build the Chinese Eastern Railway, a Russified Harbin had become the administrative centre of the scheme. By the time Alex got to Harbin there were around 15,000 Jews there and the community boasted schools, its own banks, hospitals and a number of newspapers. In 1926 it was very much like a pre-revolutionary Russian town: Alex's first memory of the place was looking out of the window the first morning after his arrival and seeing schoolchildren with blue uniforms and epaulettes such as had been worn in Tsarist days. In time Alex's father's textile business expanded and Alex was sent to Japan to run the Japanese side of the business. Shortly before the war he was arrested by the Japanese police. In contravention of the currency regulations of the time he had thousands of U.S. dollars in his safe. Before the police had time to search his office his secretary hid the money and, upon his release, she returned it to him. The Japanese woman's loyalty so moved him that he married her and she subsequently converted to Judaism. "She's the only Japanese Jewish woman who speaks English with a Russian accent," said Alex. During the war he was one of the most active members of the small group which for two years laboured to assist the thousands of Jewish refugees who passed through Japan. Alex talked late into the night about his sixty years in the Far East. "The Japanese and the Jews? The important thing to remember is that the Japanese have never understood us. The *Fugu* Plan and all Japanese anti-Semitism were based on fallacies. Basically the Japanese dislike all *gaijin*. The Jews are *gaijin* – no more, no less. But if people are going to dislike you – at least let them do it politely – like the Japanese!" Alex told me to come and see him the following day. Apologetically I explained that I was taking the night train to Kyoto and that I would not be back for some days. "Kyoto?" he asked indignantly. "What you going to find in Kyoto? The Beit Shalom prophet and the guy who's building the *schul*? O.K. You want stories? You stick with us old timers!" He glared at me and threw back a glass of Russian vodka.

Beit Shalom is centred in Kyoto, the old imperial capital of Japan. Situated on the edge of the town in full view of the imposing mountains which surround the city, the movement's headquarters could easily pass for an Israeli educational establishment. From an open window I could hear a Hebrew lesson in progress. In front of the building stood a little statue of Anne Frank.

I was shown into a small reception room which forms part of a hostel built expressly for accommodating and entertaining Jewish and Israeli visitors to Japan. The son of the movement's "prophet" was waiting to greet me. In fluent English he described to me the origins of the movement. On 9 January 1938 his father, Takeji Otsuki, had a vision in Manchuria. "One evening he was praying and suddenly felt the presence of God. And God spoke to him: 'I am the God of Abraham, Isaac and Jacob.'" God then told Otsuki four things: that Israel would achieve its independence and that he should pray for the rebirth of the state; that he should pray for the peace of Jerusalem upon which depended the peace of the world; that the Jewish people were to become a holy people, "a priest for the world"; and that Otsuki should pray for the coming of the Messiah which was imminent. Alarmed by the increasingly close relationship between Japan and anti-Semitic Germany, Otsuki warned that the consequences of any Japanese alliance with the Nazis would be a humiliating defeat for Japan. During the war Christians in Japan were persecuted as a matter of course. Many of them gave up their faith or at least compromised with the authorities to the extent of worshipping the Emperor and going to the Shinto shrines to pray for a national victory. The Holiness Group, the evangelical church of which Otsuki was a member, was banned because it refused to make these compromises. Most of the churches were effectively closed. Some of the ministers died in prison. But Otsuki managed to evade imprisonment, to maintain his ministry and to propagate his message: that to harm the people of God would have unimaginable consequences for Japan. Some of the Jewish refugees from Hitler's Europe who flooded into Japan between 1939 and 1941 tell touching stories of small kindnesses shown them by members of the Holiness Group. With Japan's unconditional surrender in 1945 Christians regained their religious freedom, and in January 1946 Otsuki formed the Japanese Friends of Israel, later to be known as Beit Shalom. At first it had little following, but once the State of Israel had achieved its independence in 1948 disciples started to attach themselves to Otsuki. "Now," said the prophet's son, "we have perhaps twenty thousand followers throughout Japan, and also in Korea, Taiwan and the Philippines." With that he led me outside for a tour of the premises. Pausing in front of the statue of Anne Frank, he said: "By now you know that Anne Frank is the best-known Jew in Japan because of the great popularity of her diary. In 1971 we took the Beit Shalom choir on tour to Israel and, while we were having some tea in a restaurant in Netanya, we met, not by chance but by divine providence, Anne Frank's father, Otto Frank. From then, until his death in 1980, we were in frequent correspondence with Otto and now we keep the letters from Anne's father in Anne's Rose Church, the memorial we have built for Anne in Nishinomiya, overlooking Osaka Bay." The prophet's son showed me an album with illustrations of the Rose Church. One photograph entitled "Mementoes of Anne" showed a tie donated by Otto, Anne's silver spoon and shoe horn; another showed "the altar of Anne's Chapel" with a golden *menorah*, donated by Otto, flanked by photographs of Anne and her father. In the base of the *menorah* were stones from the Wailing Wall and a piece of stone from Bergen Belsen where she had died. A part of the church built around these relics is given over to a holocaust memorial museum which is visited by school trips from all over Japan, and outside the church is a rose garden called "Souvenir of Anne Frank", where yellow roses sent by Otto as a gift in 1972 serve as a further reminder of the life and death of his daughter. The building of the church was begun in 1979 to commemorate the fiftieth anniversary of Anne's birth, while the words recorded in her diary, "I shall work in the world and for mankind," have become the church's motto. "Thousands of schoolchildren come here," said Otsuki. "You see, *Anne Frank's Diary* was not only a best-seller here but has become a compulsory text in junior high school."

The prophet's son asked me what I did for a living. When I told him that I taught Hebrew at an English university, he exclaimed, "A Hebrew Professor!" and his face took on an altogether unexpected radiance. "Then I must go and find my father," he said excitedly. I wandered into the classroom where I had heard the Hebrew lesson in progress. A rather beautiful Israeli girl, heavily pregnant, was teaching a dozen or so Beit Shalom children modern Hebrew conversation. From this first-floor room I could see the tiled roofs, burnished by time, and beyond, the mountains which ring the city. The girls who sat together on one side of the room wore prim black skirts and white blouses, the boys black trousers and white shirts. The metal balustrade on the balcony outside the room was fashioned from interconnecting *menorot*. I asked the girl why the children wanted to learn Hebrew.

"Quite simple. For them it is the holy language. And what is more important, it will enable them to talk to Jews from Israel. They love Jews. They really love Jews." She shrugged helplessly.

In the courtyard below I could see an elderly Japanese man, wearing khaki trousers, a white shirt and a black skullcap, looking up at the classroom. "That's the prophet," said the Israeli. "He's waiting for you."

Using his son as interpreter, Takeji Otsuki told me about the movement he had founded and his love for Israel and the Jewish people. "God teaches us to rejoice with Jerusalem and mourn with her. And the entire Bible testifies that God loves Israel. Should we not therefore love the Jews? And should we not therefore love their country? Jesus was, after all, a Jew, and we stress all the Jewish elements. You remember when St. Paul was asked, 'What advantage hath the Jew?', he replied, 'Much more in every way.' Our families keep Shabbat, light the candles on a Friday night and say the *havdalah* and *kiddush* prayers. We do not eat pork and we use Hebrew prayers and songs. Our Hebrew choir is famous throughout the world. And, of course, Anne Frank is very important to us," and he looked devoutly towards the statue of the little girl. "I shall work in the world and for mankind," he said, his kind, intelligent face showing signs of deep emotion. Of all the things that had happened to the little Dutch girl, both before and after her death, I thought, the fact that she had become the saint of a remote Japanese Christian, Judaizing sect was perhaps the strangest of all.

Some days later I was visited by a delegation of Kyoto members of Makuya. They were accompanied by a sad Israeli woman called Nehamah who was from the Israeli kibbutz of Heftzibah which has a close relationship with the Makuya sect. For years Makuya members have gone to Heftzibah to learn Hebrew and have recently constructed an impressive Japanese garden in the kibbutz. Nehamah's mother had first started teaching Hebrew to the young Japanese visitors over twenty years before. Now, Nehamah told me, she was teaching their children. Nehamah's husband had recently died of cancer. As soon as they had heard the news the Makuya sent her a plane ticket and were now helping her to recover by giving her an extended holiday in Japan. "These people are my family," she said. "They are more Jewish than most Jews."

Among the Makuya were a young married couple: Malakhi and Hamato Matsui. She had studied singing with an operatic maestro called Ricci in Tokyo and was now putting her considerable talent to use as soloist in the Makuya Hebrew choir. Later that afternoon while we sat sheltering from a light summer rain in the garden of the Nimomaru Palace, built by Ieyasu Tokugawa, the first Tokugawa Shogun, in 1603, Hamato entertained us with arias from Mozart and Puccini and with the haunting Jewish melodies taken to Israel from the *shtetlach* of Eastern Europe.

The sites we visited were not selected at random. One of Kyoto's more memorable buildings is the Koryuji Temple, otherwise known as Uzamasadera, which belongs to the Shingon sect of Buddhism. Constructed by Prince Shotuku in 603 A.D., it now contains some of Japan's most revered treasures: a wooden statue of Prince Shotuku carved by himself in 606 and a statue of Miroku Bosatu dating from the Asuka period (552–645). But for the Makuya the temple and its statues had a specific meaning: they believed the temple had been constructed in honour of the messianic figure of King David and as proof of this they showed me a stone pillar in front of one of the shrines on which was inscribed: "To the God of weaving, musical instruments and dancing." "King David played the harp and danced before the Ark, did he not?" asked Hamato gaily. From the Koryuji temple they took me to what they regarded as the most important "Jewish" site in Kyoto: "the Israel well". They seemed content to believe that its Japanese name – Isriya – signified Israel and they approached it with reverence. Nehamah dropped a stone in it which hit the water level disappointingly soon. "Jacob's well in Israel is deeper," she said.

Hamato spoke operatic Italian and a little Hebrew. During the day she made a number of obscure references to Susanna. I imagined that at some time she had played Susanna in Mozart's *Marriage of Figaro*, but she kept returning to it and finally asked me if I had visited Susanna's grave. She explained that whenever the Makuya passed through London on the way to Israel they made a special pilgrimage to the grave of Susanna, the mother of John Wesley, the founder of the Methodist movement: "John loved God and loved Israel. He is very precious to us. But Susanna was a scholar, a saint, and a beautiful woman. She inspired him. If she were alive now she would be a Makuya." They all laughed happily.

Before leaving Tokyo I had telephoned Master Kampo Harada, calligrapher, multi-millionaire, mystic and, apparently, Jew. I spoke to his secretary explaining that I was a Hebraist from the University of London and would like to meet Master Kampo. He telephoned back a few minutes later to ask me if I was an *important* professor of Hebrew. I murmured inconclusively, but he interrupted my protestations to invite me to be Kampo's guest for as long as I wished to stay in Kyoto.

Master Kampo Harada has an impressive house in Kyoto complete with offices, a vast library and a museum of his collected treasures, a little to the east of the majestic Heian shrine, brilliant in its lacquered vermilion. His library contains around 300,000 works. One part of the complex is a gallery of Kampo's work along with superb examples of Chinese calligraphy. During China's Great Cultural Revolution Kampo persuaded the Chinese authorities to

The Secret of the Tribe of Zebulun

permit him to export thousands of ancient works which would otherwise have been lost. A side gallery contained a prized collection of *suzuri* – the trays on which the calligrapher's ink is prepared. According to Kampo's business manager, Tadashi Kimura, some of the *suzuri* were worth more than a million yen each. Other galleries contained a selection of objects from all over the world: ancient Greek vases, ugly British Victorian furniture, a stuffed lion. Kimura said: "After Nelson Rockefeller, Master Kampo has the most extensive collection of antiques in the world."

When I arrived at Kampo's house a young Israeli was waiting to meet me. David had been travelling the world for years. He had worked as a deep sea diver in Singapore and on the North Sea oil rigs. Now he was doing a Ph.D. thesis on an aspect of Zen Buddhism. Kampo had meanwhile employed him to catalogue and organize the part of his library devoted to Hebrew and Jewish books. Among the 3,000 printed volumes there were numerous old editions of the Talmud and Mishnah as well as hundreds of mainly rabbinic volumes going back to the seventeenth century. In addition, the library contained a dozen Torah scrolls, many of them housed in a specially made ark. Kampo was in the process of constructing a Jewish wing in his museum where he planned to include an authentic Ashkenazi synagogue complete with a women's gallery. Over a number of years he had collected *menorot*, yarmulkas, rabbis' vestments, prayer shawls, charms and divorce deeds. It was extraordinary to reflect that here in the shadow of a famous Shinto shrine was the largest collection of Jewish books and artefacts in the Far East.

"What do you think of all this, David?" I asked the young Israeli.

"I think he should sell the lot and give the money to refugee children in the Lebanon – or maybe to the Jews of Damascus," he said bitterly. "But don't ask me to be objective about all these Jewish things. On the one hand I dislike Judaism because it's clannish and self-righteous. But I can't stop being a Jew even if I want to. I could stop being an Israeli but I can't stop being a Jew. As a result of Jewish suffering and the holocaust I'm bound to Jews I have nothing in common with. It's sad and it's sick." The handsome, gruff Israeli was too moved by the subject to want to say any more and he walked away.

Tadashi Kimura, dressed in a sober business suit, walked formally through the Hebrew section of the Kampo Kaikan library and bowing deeply informed me that Master Kampo would be pleased to grant me an audience in his country retreat at Sumera, a remote place two hours north of Kyoto. A luxurious black Toyota Senator sedan was waiting outside complete with chauffeur and a pretty Japanese interpreter. Half an hour before we arrived at the house Kimura said that we were already traversing Kampo's land. The thick pine forests through which the road was now rising steeply all belonged to Kampo. Kampo's house, a largely wooden structure, was something like a vast ski lodge. It stood at the head of a valley with panoramic views in every direction. Two hundred yards up the side of a hill was another large structure which served as a dormitory for young girls, aspiring calligraphers, who would come here occasionally to study with the master. A group of them were busy scything grass at the side of the road: "You must be healthy in body and mind to be a calligrapher," said Kimura.

Kampo wore a not immaculate kaftan of prophetic rather than strictly Japanese cut and a flowing white beard which reached the middle of his chest. His appearance conveyed the message that here was an oriental sage and mystic. His opening words, "God told me that we would meet this afternoon," did little to dispel this idea.

We sat in Kampo's study, on *tatami* mats, around a low table. On one wall was an example of his own calligraphy, on another a small Indian goddess. A stack of the latest Japanese stereo, television and video equipment stood in a corner. The sun setting over the pine forests cast a pink light on to the paper screen walls.

We were still exchanging elaborate courtesies when a young and beautiful Chinese woman in her mid thirties entered the room. She had very red lips, a perfect body and the coldest eyes I had ever seen. "This is Kampo-san's secretary," said the interpreter, and giggled decorously. Kampo, an alert octogenarian, put his arm around the Chinese woman and laughed infectiously. While Kampo spoke, his secretary served us with fruit, cakes and tea.

Kampo talked for several hours, frequently pulling out maps from an adjacent map drawer to illustrate his theories. He believed that he was a Japanese Jew of Jewish descent. His family name, Harada, was originally Hada, he thought, and his ancestors had been members of the tribe of Hada of which Akiva Jindo had spoken. At one point in their history the Hada, notwithstanding the contributions they had made to the development of Japanese civilization, had made enemies and had withdrawn from the imperial court to Kyushu where they had built their own fortified castles. Kampo had recently bought hundreds of acres of land in Kyushu where he had constructed an enormous complex called the World Study Museum. He gave me a stack of brochures advertising the museum. With its red-tiled roof and black-and-white timbered walls it was somewhere between a Dutch barn and a Tudor mansion. As he explained the purpose of this museum, which was to bring Western enlightenment to the Japanese, Kampo's voice took on a musical tone which had been absent before: "It is fitting", he chanted, "that enlightenment should radiate out from the place where we Jews first established ourselves in Japan." Kampo was optimistic that the tablets of the ten commandments would soon grace his museum. His hopes were based on the fact that on Mount Tsurugi, on Shikoku Island 6,000 feet above sea level, recently excavated tombs had revealed what he considered to be incontrovertible evidence of early Jewish settlement in Japan. Kampo is convinced that the tablets are buried here and the following year he was hoping to lead an archaeological expedition to find them. "But I shall have to do it secretly. The site is near a Shinto shrine. I cannot tell you which one. But there are secrets within it which will disprove the idea that the blood of the royal family is purely Japanese and they will revolutionize our understanding of Japan, Shintoism – and Judaism. Prince Mikasa knows the secret because it is engraved on the imperial mirror, but he says nothing." I looked around the table at the faces of Kampo's attendants. The interpreter had been relaying Kampo's words enthusiastically and with conviction. Kimura sat as if in the presence of a minor deity. But the Chinese woman looked at me conspiratorially and, without moving a muscle of her beautiful face, almost winked.

Shortly after the war Kampo had had a vision, the first of many. In the sky he saw the Chinese character *"fude"* which, he said, means both "brush" and "the word of God". From this he understood that he should devote his life to the preaching of God's word using the calligrapher's brush. As the owner of a number of schools and kindergartens he was a very wealthy man at a time when most Japanese were desperately poor. But he gave the schools away and devoted himself to calligraphy. His friends and family considered him to be mad and he was forced to flee his home town and hide in the mountains. The vision came to him again. Almost immediately he started to teach calligraphy and to this end in 1953 founded the Nippon Shuji Educational Federation. By the time I met him he had had rather more than six million students – most of them enrolled on one of his correspondence courses. He had two thousand girls working for him correcting his students' exercises. His own work fetches incredibly high prices. A kimono incorporating a simple piece of Kampo's calligraphy will fetch over U.S. $1,000. "The fact that I have been so successful", said Kampo, "is the final proof of the validity of my vision. If I had been a commercial failure I would have put the vision down to some form of temporary dementia." He rose from the floor and looked out over his forest. "But I have everything a man could possibly want. God has rewarded me well. I have had further visions and many of them I have described in *haiku*," and he passed me a large printed volume of his collected poetry. "My most important vision is that there can be no peace in the world until the Japanese and the Jews become true friends and allies. Eventually the Japanese and the Jews will rule the world – and then there will be a peace which will last for a thousand years. You see, the Japanese and the Jews are so clever, so clever. The *haiku* contain these secrets. But Prince Mikasa already knows the secrets. Prince Mikasa knows all these secrets," he repeated gravely.

"I shall ask him when I see him," I replied.

There was silence.

"You *know* His Imperial Highness?"

"He was a student at the School of Oriental and African Studies where I teach. I'm seeing him tomorrow."

Kampo rose to his feet and came around the table to where I was kneeling. He squatted next to me and held my hand.

"You can tell the Prince everything I have told you. Most of it he already knows. And tell him about the synagogue and the Hebrew books. Perhaps he would be willing to be the patron of my World Museum in Kyushu?" The beautiful Chinese woman had also knelt beside me and was holding my other hand. It was definitely time to go.

As I left to return to Kyoto, Kampo Harada loaded me with presents, embraced me fondly and exclaimed, "God sent you to us in order for you to tell the world of my revelations." As the great car purred away into the forest he was still talking animatedly.

During the drive back Kimura told me of the fabulous riches of his master and that the calligraphy which he had given me was worth a small fortune. That night I spent in the Kampo Pavilion next to the Heian Shrine. Before I slept I reflected that over the previous few hours in Kyoto I had met two men who had had recent conversations with the living God.

His Imperial Highness Prince Mikasa No Miya Takahito met me at the Institute of Middle Eastern Studies of which he is the founder and director. An alert, quick-smiling man, he was dressed in a dark blue pin-striped suit. His sombre tie was secured by a tie-pin. He greeted me in the corridor outside his room with an unceremonious handshake and ushered me into his study. Showing me to a comfortable leather chair the Prince offered me a cigarette. It was stamped with a gold chrysanthemum – the emblem of the imperial household. Cautiously I asked him about the alleged connections between Japan's imperial family and the Jews. He laughed, his eyes crinkling in real amusement. "In the first place let me assure you that as far as the imperial insignia and the mirror are concerned – nothing can be proven one way or the other because they have never been seen by me or by anyone else. They form the core of the great shrine of Ise. The rest is a mixture of conjecture and fantasy. There have been dozens of people over the last forty or so years who have been convinced of the connection between the two peoples. There was a certain Japanese scholar who taught at an American university. He had this bee in his bonnet that anything that could not be otherwise explained in Japanese was Hebrew. He found a Hebrew basis for instance for many of our old Japanese songs and explained the Japanese *sumu sumu* [which you use to introduce an explanation] as the Hebrew *shem'u shem'u* [listen!], or the expression used when you are pushing or pulling something heavy, *en yakura*, as having something to do with *Yahweh* [Jehovah]. I can remember that he came to lecture here on this some fifteen years ago. At the end of the lecture everyone was too embarrassed to say very much – but he took the fact that there were no questions to mean that everyone agreed with him and he went on to write a two-volume book on the subject which he dedicated to me! And, of course, there are people who believe that Uzumasa in Kyoto is a Jewish shrine or that the Uzukini of ancient Japanese history was the Prince of Judah on the grounds that Uzu equals Yehudah." The Prince paused and, after some thought, said, "The proliferation of these ideas came about because before the war there were groups of people who did not like or even hated the Jewish people. Some of them began to study Jewish history from this point of view. During the war some of them changed their minds and started to believe that the Jewish people could be utilized to win the war against the Allies. So they began to study the Jews from this point of view. Then after the war there was a further shift in opinion. It was known that the Jews were good businessmen who excelled in international trade. So again the Jews were studied, this time from this point of view. I think the idea was that if it could be shown that we had the same origins or the same basic culture it would be very convenient; we would become intimate friends and we would share their trading successes. There are two tendencies, are there not? The one is anti-Semitic, the other is philo-Semitic. But they are both based on emotional rather than factual criteria. After the Second World War many Japanese people became disenchanted with traditional Japanese religions and started looking outwards for other faiths to replace them. Judaism was one of the religions that for a while people became interested in. I too became very interested in Hebrew and Jewish history but my studies soon led me to the conclusion that the Japanese and the Jews are quite different and that our religious systems are quite different. It seems to me that the Jewish religion is very severe, very strict, while Shintoism is emotional and not at all strict. I do not believe that the Japanese people could tolerate such a strict regime as Judaism!"

The Prince spoke engagingly for some time about the people who had contributed to the extraordinary and little-known overlapping of Japanese culture and Judaism. "It was a moment in history. I think it may now be over," he said. Fresh from Kyoto, I was not so sure.

I had arranged to meet Noa at the Middle East Institute later that afternoon. As Prince Mikasa was showing me around the exhibition halls I saw her in the distance and asked the Prince if he would like to meet a Japanese Jewess. He was amused at the idea. Quickly I introduced them. Noa was embarrassed and blushed deeply. Perhaps to put her at ease the Prince spoke a few words of Hebrew to her. They spoke for some minutes. As Noa and I left the building she said: "You cannot possibly imagine what this means to me. My mother still believes deep down in the divinity of the imperial family. That is why she thinks of me as a traitor. But I have spoken to the Emperor's brother about my Jewishness. What did he do? He smiled and asked me questions about the conversion and spoke to me in Hebrew!" Noa told me later that as soon as she heard the news of her meeting with his Imperial Highness, her mother was swiftly reconciled to her daughter's adopted faith.

The following evening I went with Jindo to a mass meeting of the Makuya. When I got there the ecstasy level was already high. An insistent high-pitched chanting filled the room. A man led them in prayer. Every time he paused for breath the congregation shouted fervent "amens". On the left the men knelt in the Japanese style – on the right the women. The impromptu prayers sounded hysterical; Jindo said they were speaking in tongues. Just near me a woman supported a spastic child on her knees, her hands raised in entreaty. To my right a man, carried away by the fervour of his prayers, was caught in spasm. Suddenly the cacophony stopped and a huge organ broke into the stirring strains of the *Hymn for the Makuya Elders* by Abraham Ikuro Teshima. "High up the pine clad hills/Down the mountain sides/the breeze is blowing softly/In the praise of God." But, the hymn went on, the Makuya were sad to have lost their leaders: one had introduced the techniques of modern photography to Japan; another had been the head of Mitsubishi steel; another was the president of the World Rotary Club. Behind the podium the entire wall was covered by the white background and red sun of the Japanese national flag. To my right Jindo whispered to me: "You know that Japan is called the 'Land of the Rising Sun', which is the meaning of the word *Nihon*. But for us the word has another meaning: the Hebrew root *nahah* means 'to follow'; and the word *'hon'* in Japanese means 'book'; so the word *Nihon* means 'the followers of the book', which is to say the Jews. So the flag of Japan is also the flag of Makuya." The speaker led the Makuya in prayer and then read, first in Hebrew and then in Japanese, from Psalm 137: "By the rivers of Babylon, there we sat down, yea, we wept, when we remembered Zion." In the background the Yamaha organ softly played *Jerusalem the Golden* and around me, the men and women of Makuya, the exiled tribe of Zebulun, wept in longing for their lost land.

FIVE

The Deliverance of the Tribe of Dan

The Falashas, the black Jews of Ethiopia, were cut off from all other Jewish communities for the best part of two thousand years. When they had their first encounter with white Jews in the middle of the nineteenth century, they were amazed. They thought they were the last Jews left in the world. For centuries they had been living in remote and scattered mountain villages in the heart of Christian Ethiopia which itself was largely cut off from the outside world. "Encompassed on all sides by the enemies of their religion," wrote Gibbon in the *Decline and Fall of the Roman Empire*, "the Ethiopians slept nearly one thousand years forgetful of the world by whom they were forgotten."

Greek sources show that Jews were living in Ethiopia as early as the second century B.C., but the first hint to the world at large that a sizeable Jewish community may have been flourishing in Ethiopia was given in a Hebrew book, *Sefer Eldad*, written by a ninth-century Jewish traveller, Eldad ha-Dani. He claimed that members of the lost tribes of Israel were living peacefully in an independent Jewish state in Africa called Sambatyon. Eldad's account of this remarkable country intrigued generations of Jews throughout the world. Among other things he maintained that Sambatyon was surrounded by a torrential and uncrossable river of stones which followed the biblical injunction and stopped flowing on the Sabbath. A further hint at the existence of a Jewish polity in Africa was provided by the fifteenth-century writer, Eli of Ferrara, who wrote of an encounter he had had in Jerusalem with an Ethiopian Jew who told him that for centuries the Jews of Ethiopia had been fighting wars to maintain their political independence. There were occasional references to the Falashas over the next few centuries: in the sixteenth century Isaac ibn Akrish noted that in Ethiopia a Jewish monarch had allied himself with Muslims in his struggles against Christian Ethiopia, and in the seventeenth century Moses de Rossi wrote of Jews living in mountainous parts of Ethiopia which he called "the mountains of the moon". But it was not until the publication in 1790 of *Travels to Discover the Source of the Nile* by the Scottish explorer, James Bruce, the Laird of Kinnaird, that any extensive or reliable information about the Falashas reached the outside world. And by and large the idea that there were black Jews practising an archaic form of Judaism in Ethiopia was considered so fantastic that for some time Bruce's account was either disbelieved or ignored. Certainly it was not until the middle of the nineteenth century when news of Protestant missionary activity among the Falashas was reported in the West that European Jews started taking any great interest in the Jews of Ethiopia.

The traditions of ancient Israel are woven tightly into the fabric of Ethiopian society. The story of King Solomon and the Queen of Sheba which links the histories of Israel and Ethiopia is the tradition revered above all others. According to the *Kebra Negast*, the Ethiopian national epic, when the Queen visited King Solomon in Jerusalem she was seduced by her royal host. None the less, she was sufficiently impressed by what she had seen at the Jerusalem court to convert to Judaism. When she returned to Ethiopia the Queen gave birth to a son, the future King Menelik. As soon as he came of age Menelik returned to Jerusalem where, perhaps to avenge his mother, he stole the sacred ark of the covenant which he secretly took back with him to Ethiopia. Thereafter the mantle of the "chosen people" fell upon the Ethiopians, and their capital, Axum, became the New Jerusalem.

The Falashas believe that their own origins can be traced back to the same events. But whereas the Ethiopian royal house claims to be descended from Solomon, the Falashas believe they are descended from the first-born sons of Israelite notables sent by Solomon as a guard of honour to accompany Menelik back to Ethiopia. In the 1840s a Falasha religious leader gave this emphatic but confused account of his people's arrival in Ethiopia to the French explorer, Antoine d'Abbadie: "We came with Solomon.... We came after Jeremiah. We came under Solomon;

we came by Sennar and from there to Axum. Undoubtedly we came under Solomon."

There are a number of other theories about the origins of the Falashas. Some argue that when the children of Israel left Egypt at the time of the Exodus, a group broke off from the rest and instead of crossing the Red Sea made their way down the coast of Africa to Ethiopia. Others believe that the Falashas are descended from groups of Jews who fled Palestine after the destruction of the First or Second Temple. But the most popular and persistent theory, perhaps based on the writing of Eldad, is that the Falashas are the lost tribe of Dan. Today, many orthodox Jews including Israel's Chief Rabbis subscribe to this view even though it has no historical substance. The Falashas, too, are growing to prefer this explanation of their past to the one favoured by their own traditions.

The ancient Ethiopian kingdom of Axum was founded in the first century A.D. Until the conversion to Christianity of King Exana, four hundred years later, the culture of Axum had been greatly influenced by the Semitic and Judaic concepts which had taken root in South Arabia. There was continuous two-way traffic between South Arabia and Ethiopia, which are separated only by the narrow straits of Bab el Mandeb. It is usually assumed that the Falashas are the descendants of those Judaized and Semitized elements within the Axumite kingdom which rejected Christianity. Included among these may have been Jews, Arabians or Ethiopians who had converted to Judaism and were reluctant to exchange their Jewish beliefs for Christian ones.

At the beginning of the Christian era, Judaism, in one form or another, had millions of adherents mainly around the Mediterranean, only a minority of whom were from Palestine. There were Jewish converts in South Arabia as well as Palestinian Jews who had migrated there before the destruction of Jerusalem by Titus in 70 A.D. The name Falasha, which perhaps means *"émigré"*, could originally have referred to Arabian Jews or Arabian converts to Judaism who had left Arabia and settled in Ethiopia. It can be presumed that this migration took place before the third century A.D. when the Jewish Oral Law was codified, because there are no elements in the Falasha religion which reflect any awareness of rabbinic Judaism. If there was any admixture of Jewish blood two thousand years ago, there is certainly no sign of it today. The

Falashas are indistinguishable, physically, from any other Ethiopians of their region.

None the less, throughout the ages the Falashas managed to maintain a separate identity from their Christian neighbours. In *The Blue Nile* Alan Moorehead wrote: "The Ethiopians were eaters of raw meat, heavy drinkers, uncouth in their manners and given to wild and primitive passions ... the Ethiopians on their icy heights huddled together with their cattle at night and seldom washed." The Falashas, on the other hand, cooked their meat, were sober by inclination and were so relentlessly hygienic that the Amhara accused them of "smelling of water".

Like the Bene Israel the Falashas have no knowledge of Hebrew: their sacred texts are written in Ge'ez, as are the scriptures of Ethiopian Christians. Under the probable influence of Christian Ethiopia they have adopted a monastic tradition which is extremely unusual, although not entirely unheard of, in Judaism. But the chief peculiarity of the Falashas' Judaism is that they have no knowledge of the rabbinic texts or of the later refinements of biblical law (such as the regulations prohibiting the mixing of meat and milk based on the biblical text "thou shalt not seethe a kid in his mother's milk"), or of the post-Exilic feasts such as Purim or Hanukkah. The Judaism of the Falashas is Old Testament Judaism, based on a literal obedience to the Pentateuch. One of the striking features of the faith of the Falashas is their strong attachment to the Land of Israel. Whenever an Ethiopian Jew prayed, he would first turn in the direction of Jerusalem. Falasha literature and prayers deal constantly with such themes as the return to Zion and the re-establishment of priestly worship in a rebuilt temple. The Falashas' love of Zion is no different, in essence, from that of any other Jewish group throughout the Diaspora, and considerably more fervent than some.

After the rise of Christianity in the fourth century A.D., those Ethiopians who had accepted Judaism and refused to embrace Christianity were persecuted and in consequence withdrew from the coastal areas to establish their community in the Gondar region. Bruce recorded that Falashas took "possession of the rugged, and almost inaccessible rocks, in that high ridge called the Mountains of Samen. One of these, which nature seems to have formed for a fortress, they chose for their metropolis, and it was ever after called the Jews' Rock." They may have been reinforced by Jews who were brought as captives from South Arabia, which by then was partly Judaized, by Kaleb, the *negus* (king) of Ethiopia, after his successful campaign of 525 against the king of Himyar. In the tenth century the Falashas, along with other Agau tribes, rose against the ancient Ethiopian dynasty of Axum. Ethiopian tradition maintains that they were led by a Jewish queen, Judith, who overthrew the *negus* and went on a wild rampage throughout Ethiopia, destroying churches and pillaging monasteries. According to Bruce, Judith decided "to establish her religion by the extirpation of the race of Solomon", by which he meant the Ethiopians. This may be pure legend, but it is a legend with wide currency in Ethiopia and it has some bearing on the subsequent history of the Falashas. Emperor Haile Selassie in his autobiography spoke of "the extermination of Christians when Yodit of the Falasha tribe reigned", and after an attack on a Falasha village in Wollo in 1972, during which around thirty Falashas were killed, the murderers claimed that they were taking revenge on the descendants of the Jewish queen, "the whore of Axum", who had almost destroyed Ethiopian Christianity nearly a thousand years before.

Between 1137 and 1270 the Zagwe dynasty, of Agau stock, ruled Ethiopia. This line of kings has been resented by Ethiopian tradition on the grounds that they were not of the Solomonic line and therefore not "Israelites" as the Amhara emperors claimed to be. The Zagwe kings seem to have left their Falasha Agau cousins in peace. But in 1270 one of the descendants of the ancient Amhara line of Axum, *Negus* Yekuno Amlak, who claimed direct descent from Solomon and Sheba, was restored to the throne. He determined to destroy the power of the Falashas, perhaps because they could not be relied upon in the critical wars against the Muslim kingdoms of the south which for hundreds of years were to threaten the existence of Christianity in the Horn of Africa.

For a period the Falashas held the balance of power between the Christians and the Muslims. During the reign of Amda Seyon (1314–44), the Falasha ruler Gideon decided to join forces with the Muslims against their ancestral Christian enemies. But when the Muslim forces laid waste the Falasha territories north of Lake Tana, Gideon changed allegiance and, together, the Christian and Falasha armies were able to halt the Muslim invasion. Amda Seyon's gratitude was not long-lasting and, within a short while, his forces were instructed to bring the Falashas to heel and to convert them to the religion of Christ. Under the emperors Dawit and Yeshaq, the Falashas lost even more ground. A decree issued by Yeshaq (1413–29) declared that, "He who is baptized in the Christian religion may inherit the land of his father. Otherwise let him be a Falasha!" At Kosage, north of Gondar, the Falashas were defeated in what proved to be a decisive battle. Internal Falasha divisions were exploited by the Amharas and by the middle of the fifteenth century the Falashas, now without a king, were only semi-autonomous.

Towards the end of the fifteenth century Judaism in some form started attracting tribes in the Gondar region and thus the Falashas, who were no longer so much a military threat, became a religious threat to the Amhara rulers. *Negus* Zara Yakob (1434–68), known in the Ethiopian chronicles as "the exterminator of the Jews", did what he could to undermine the attraction of Judaism. In his work, the *Book of Light*, he echoed mediaeval European belief by accusing the Jews of eating children. Under Zara Yakob and his successor, Baeda Mariam (1468–78), there were further massacres of the Falashas and many were forced to convert or, like the *marranos* of Spain, to profess Christianity in public while secretly clinging to their Jewish faith. But there were still some pockets of Falasha resistance in the Semien mountains.

In the reign of Lebna Dengel (1508–40) the balance of power shifted again and the Falasha leader Gideon and his queen, Judith, in an attempt to reassert their independence, joined forces first with the *negus* and then with the invading Muslim leader Ahmad ibn Ibrahim. The devastation that followed the Muslim conquest had grievous consequences for Ethiopian culture in general. It is considered possible that the lost literary monuments of the Falashas, which were frequently described by them in subsequent centuries, were in fact destroyed during this period when so much of the wider Ethiopian culture was lost. Led by "Radaet the Jew", to use Bruce's phrase, the Falashas had some temporary successes against *Negus* Minas and were able to expand their territory. But by the 1580s the dogged campaign waged by the new *negus*, Sarsa Dengel, forced them back into the most remote areas of the Semien mountains, where they made a last stand. The Amharas had great numerical superiority and one advantage besides: the Falashas were still armed with shields and spears, but by now the Amharas had managed to acquire guns. When Sarsa Dengel had cannons brought up into the mountains, the Falashas had little hope of success. But they fought to the last and burned the last of their crops so they would not fall into enemy hands. The Christian Ethiopian chronicles make much of the Falashas' bravery: some women preferred to throw themselves to their death from "the Jews' Rock" than to submit to Christian rule. "Radaet the Jew" was taken prisoner and promised his life if he would bow to the Virgin Mary. He replied: "It is forbidden to utter the name of Mary! Hurry! If I am to die, I depart from a world of lies to a world of justice, from darkness to light. Kill me!"

The spirit of the Falashas was not yet broken. During the reign of *Negus* Susenyos (1607–23), the oppressed Agau tribes rose in revolt, joined by the Falashas led by another Gideon. As Bruce relates it:

> The constant success of the king, and the bloody manner in which he pursued his victory, began to alarm Gideon, lest the end should be the extirpation of his whole nation ... the King gave orders to extirpate all the Falasha that were in Foggora, Janfakara, and Bagenarwe, to the borders of Samen: also all that were in Bagla, and in all the districts under their command, wherever they could find them.

The allies of the Falashas were defeated early on in the campaign and they were left to carry on alone. Their remaining strongholds were destroyed and many of the Falashas were killed. The rest were promised that if they laid down their arms they could return to their villages. But the promise was broken and the remnant were offered the choice faced so often by Jews through the centuries: conversion to Christianity or death. As Bruce wrote: "Many of them were baptized accordingly and they were all ordered to plow and harrow on the Sabbath day." But like the Jewish martyrs of the Inquisition many refused to convert and in 1616, during the slaughter that ensued, their leader Gideon was killed. Those that remained were sold into slavery. This last battle marks the end of Falasha autonomy, although Western travellers wrote of "Jewish kings" in Ethiopia until the end of the eighteenth century.

When he was in Ethiopia James Bruce was told that one hundred and fifty years before, in the early seventeenth century, Falashas could muster at least half a million people, which, considering the trouble they had given the Amhara emperors for centuries, seems possible. By the middle of the nineteenth century the Falashas were estimated to number between 150,000 and 200,000. Partly as a result of the efforts of Protestant missionaries, by 1900 there were less than 50,000. Today there are no more than 28,000 Falashas, only about 8,000 of whom are still in Ethiopia. Whatever their importance in mediaeval Ethiopia might have been, most Ethiopians outside Addis Ababa and the Gondar region have never heard of the Falashas. They are one tiny section of the forty million Christians, Muslims and pagans who make up the second largest black population in Africa.

Until the 1974 Marxist revolution in Ethiopia, the Falashas were not permitted to own land. They were, however, able to rent land and they eked out a precarious existence growing *teff* in the fields and raising sheep and hump-backed cattle on the dry grassland dotted with eucalyptus and acacia trees beyond the confines of their villages. But because of the restrictions on land ownership the Falashas developed other occupations. Over the centuries they became the artisan class of their part of Ethiopia, specializing in metalwork, pottery, weaving and the building trades. This did nothing for their standing in the community, for such work was despised by the Amhara farmers. None the less, in the past they had been responsible for some of the finest buildings in Ethiopia and, in Bruce's day, "carried the art of pottery to a degree of perfection scarcely to be imagined". To the present day Falasha men have continued to work as craftsmen while their women have looked after the children and engaged in spinning, weaving and pottery.

Through their occupation as blacksmiths the Falashas long ago acquired a reputation as being possessed by *buda*, supernatural powers associated with sorcerers and bearers of the evil eye. They were believed by their neighbours to have the power to turn into hyenas and in this guise to prey upon Christian children. One traveller reported that Ethiopian Christians considered the Falashas "the sort of mortals that spit fire and were bred up in hell". The crucifixion of Christ was attributed to *buda*; it was craftsmen after all who performed the deed: they fashioned the cross and the nails and hammered them through Christ's hands and feet.

In the first half of the nineteenth century Protestant missionaries, ablaze with the idea of converting Jewish souls in every corner of the globe and thereby bringing closer the return of the Messiah and his thousand-year reign, discovered the Falashas, many of whom they converted to Christianity. Seeing the missionaries succeed where a thousand years of Amhara oppression had failed, many Falashas lost heart. In 1862 some of them, seeing no other solution, began to trek to Jerusalem. This was partly out of fear of the missionaries and partly because they thought the Messiah had come (there was some confusion in their minds between the Messiah, whom they called Theodore, and the Emperor Theodore). In full expectation of some miraculous delivery, they left their mountain homes totally unprepared for a migration to the Holy Land. They got no further than the neighbouring province of Tigre, where many of them died. In the same year a Falasha called Abba Sagga wrote a letter in Ge'ez to the "Chief Priest of the Jews in Jerusalem" asking if the time had yet come for the Jews of Ethiopia to return to the promised land:

> Has the time arrived that we should return to you, our city, the holy city of Jerusalem? For we are a poor people and have neither prince nor prophet and if the time has arrived send us a letter which will reach us.... They say that the time has arrived; the men of our country say: "Separate yourselves from the Christians and go to your country, Jerusalem, and reunite yourselves with your brothers."

Abba Sagga received no reply to his letter. The Falashas were to wait one hundred and twenty-two years before the time was considered right.

Although a number of Jewish individuals interested themselves in the Falashas, Jewry as a whole had some difficulty in accepting that these black *montagnards* were really Jews. The establishment of the State of Israel did little to change the situation. David Ben-Gurion, the founding father of the Jewish state, and several of his ministers listened with attention to accounts of the Falashas brought back by Jewish travellers. Ben-Gurion held the non-halakhic view that a Jew is someone who believes himself to be a Jew and he might have been persuaded to encourage Falasha immigration. His advisers, however, were opposed to the idea. Emperor Haile Selassie, the Conquering Lion of the Tribe of Judah, the Elect of God, was a friend of Israel, but was opposed to Falasha emigration on the grounds that it might encourage other tribal minorities to demand autonomy or special treatment. In addition, there were now a number of scholars of international standing who insisted that the Falashas were not Jews at all.

During the 1950s and 1960s the argument that the Falashas are not Jews was prevalent in Israel. It was admitted that they respected the Sabbath and maintained some of the *kashrut* regulations; but they were illiterate and even their priests did not know a word of Hebrew or anything at all of rabbinic Judaism. Even their supposedly Jewish traits were shared by Christian Ethiopians and by certain non-Christian groups such as the Qemant, while even the Emperor regarded himself as a direct descendant of Solomon and Sheba and thought of himself as an "Israelite".

Whether they were officially regarded as Jews or not, as soon as the Falashas heard of the establishment of the Jewish state they started to trickle into Israel. The Falasha areas were hundreds of miles from the coast. But some walked to the coastal ports where they sometimes were able to get jobs on freighters and other ships. When they got to Eilat they would jump ship. Others managed to save the money for an air ticket from El Al, the Israeli airline, which in 1970 established a regular service from Addis Ababa to Lod. But the Israelis did not make it easy. One woman whose son had made his way to Israel applied to the Israeli Embassy in Addis Ababa for a visa, explaining that she wished to join him. She was told that she could not visit the country unless she had $600 in travellers' cheques. Three weeks later she returned to the Embassy with a cross hung around her neck and explained to another official that she was a Christian wanting to go on pilgrimage to Israel. She was given a visa on the spot.

Meanwhile the small band of Falashas who managed to make their way to Israel formed themselves into the Israel Falasha Committee. It was led by Professor Tartakower, a sociologist and demographer whose academic life had started in his native Poland, and Ovadia Hazzi, an Ethiopian Jew of Yemenite extraction, who had become the Israeli army's senior sergeant-major. The committee headed by this strangely matched pair played a major part in persuading the religious authorities that the Falashas were indeed Jews.

In 1973 the Sephardi Chief Rabbi of Israel, Ovadia Yosef, basing himself on earlier rabbinic rulings, most notably by the sixteenth-century Chief Rabbi of Egypt, David ben Solomon Ibn Avi Zimrah, and invoking the views of former Chief Rabbi Kook, declared that the Falashas were "Jews who must be saved from absorption and assimilation. It is our duty to speed up their immigration into Israel . . . for whoever saves a single soul in Israel, it is as though he had saved the whole world."

In the mountainous region of Gondar it was believed that if a Jew crossed on to Christian land the grass would never grow again. It was this sort of local prejudice rather than overt state discrimination which made life difficult for the Falashas. During the reign of Haile Selassie (1930–74), peasants, whether Christians, Muslims or Jews, were treated with indifference and little that happened in Addis Ababa had much bearing on the remote mountain areas.

In the Gondar region, the Falashas were the only religious group not to own its land. The Christian peasants had to suffer the burden of mediaevally rapacious land-taxes, but the Falashas had to pay the land-tax in addition to a high rent which was often as much as 50 per cent of an annual crop.

These economic pressures were augmented by social ones. The Falashas were scorned by the Amhara because many of them worked as artisans and feared because they were thought to be the killers of Christ. It was commonly believed in the villages that the Falashas brought Christian corpses back to life to work their fields for them at night. The Falashas were blamed by their neighbours for every misfortune which befell the areas in which they lived. In the mixed villages Falashas were required to apologize for local setbacks to the village leaders and, after the revolution, to the peasant associations.

For a Falasha peasant, farming land that was often adjacent to Christian land, there were constant difficulties. The Christians in some villages, claiming that the water belonged exclusively to them, objected to the Falashas using the streams and rivers for their cattle. The Falashas were considered to have no rights. They were the untouchables of Ethiopia. When travelling from one village to another the Falashas would often try to pass themselves off as Christians. Hoping to avoid molestation, many Falasha women tattooed a cross on their foreheads.

In 1974 Haile Selassie, the Emperor of Ethiopia, was deposed. The first act of the new revolutionary government which replaced him was to destroy the power of the landed classes and the church by redistributing their land to the peasantry. The revolution gave the Falashas the right to own their land for the first time since they had lost their political independence centuries before. But the political turbulence caused by the revolution more than offset this benefit. In the immediate aftermath of the revolution, right-wing counter-revolutionary forces of the Ethiopian Democratic Union as well as the forces of the extreme left-wing Ethiopian People's Revolutionary Party (who were opposed to the central government because in their view the revolution had not been taken far enough) resisted the forces of the Dergue and brought terror to the areas inhabited by the Falashas. Attacks from the two sides were called respectively the "red" and the "white" terror. Many were killed and many more fled from their villages in the outlying areas to join larger concentrations of Falashas in the villages closer to the regional capital of Gondar.

The policy of the revolution was "that no nationality will dominate another", and thus the Falashas along with the other minorities were invited to participate in the new society being created by the revolution. But in time it was decided that the imposition of uniformity was a revolutionary necessity and that the many minority groups in Ethiopia would have to assimilate to the majority. Amharic was given pre-eminence over other languages and minority religions were discouraged in favour of the two officially recognized religions – Coptic Christianity and Islam. This new policy was to have dire consequences for the Falashas.

In 1981 the tenuous links between the Falashas and Jews elsewhere were severed when O.R.T., the Jewish aid organization, was forced to leave Ethiopia. Subsequently Jewish schools and synagogues were closed down, Hebrew books were publicly burned, and Falashas were forced to work on the Sabbath. Religious leaders were harassed: one Falasha priest, for instance, was imprisoned for fifteen months for spreading "black magic". Any contact between the Falashas and the outside world was actively discouraged: Falashas caught talking to the occasional tourist were questioned and sometimes imprisoned.

The years of terror and persecution led the Falashas to leave Ethiopia in increasing numbers in an attempt to get to what they saw as the Promised Land. They knew that Falashas had already succeeded in getting to Israel and they knew that once they were there they would receive material assistance and automatic citizenship. In addition, they realized that their future as Jews lay not in Ethiopia but in Israel. Thus from early 1977 small numbers of Ethiopian Jews made their way to refugee camps in neighbouring Sudan close to the Ethiopian border. But getting out of Ethiopia was not easy. The distances involved were great and the only way to travel was on foot. The terrain they had to cross was wild and mountainous. Often the refugees were attacked by rebel or army units, bandits or other, Christian, refugees. From the early 1980s the Ethiopian authorities made their departure as difficult as possible: compulsory travel passes were introduced and any Falasha found outside his village without a pass signed by the chairman of the peasants' association was liable to imprisonment. Those caught far from their villages and obviously on their way to the Sudan were imprisoned for long periods and were frequently tortured before being taken back to their villages. Despite the difficulties, between 1977 and 1984 many thousands of Falashas managed to leave Ethiopia.

The story of how the Falashas managed to get from the Sudan, a fervently Muslim, Arab League state, officially at war with the Jewish state, to Israel, is an extraordinary one. Even before the Falashas left Ethiopia many of them had been helped by Israel's Intelligence Service, the Mossad, which established a number of support stations along the mountain paths taken by the Ethiopian Jews. Once they were in the Sudan, the Mossad arranged for some of them to leave by boat from the port of Suakin, near Port Sudan, to Eilat, while others were taken through Kenya where safe houses had been arranged for them in Nairobi. Others were flown out by civilian airlines from Khartoum's International Airport to various European cities and then transported on to Tel Aviv, frequently by El Al. From 1977 to 1984 thousands of Falashas were spirited out of the Sudan in this way without anyone in the outside world being any the wiser.

When I arrived in the Sudan I knew nothing of this major exodus which had been going on for years. Nor did I know that the day of my arrival in the country coincided with the beginning of a major Israeli initiative which had been designed to facilitate the emigration of the 12,000 Falashas who had fled their homeland over the previous few months not only because of the political situation in Ethiopia but also because of the drought and famine which were beginning to take their toll in the parts of the country where they lived. The Israeli initiative was to be called Operation Moses.

My first inkling that something might be afoot came the morning of my arrival at Khartoum Airport, when I happened to meet a British student working in the Sudan for Voluntary Service Overseas. Indignantly he told me that a month or so before, the Israelis had landed two Hercules aircraft near Gedaref, a small town in eastern Sudan, and had airlifted hundreds of Falashas to Israel. "I know that the refugees were starving," he said, "but what they did was illegal! Sudan is officially at war with Israel. It's appalling that Israel should be able to do things like that and get away with it."

The city of Khartoum, wedged between the Blue Nile and the White Nile, still retains something of its colonial past. The Presidential Palace, the National Museum, many of the ministries and the Grand Hotel were all built by the British. Along the banks of the Nile the trunks of the great, shady trees are painted white, in military style, to a height of two metres. But beyond the cordon of old colonial buildings the town soon turns into a depressing network of grey concrete structures; and beyond this on all sides are the shanty towns which were swollen by refugees from the great famine.

The great concentration of refugees from Ethiopia, among them the Falashas, lay to the east of the country in the vicinity of Gedaref and Kassala. The bus that was to take me to Gedaref was an imposing sight. It had come into service only the previous day and was the product of Italian design and German engineering. The driver stood by the gleaming bonnet and proudly yelled out, "The new bus to Gedaref! Take the new bus!" The drivers of the other buses at Suq al Shab, the bus station, looked at the new bus wistfully. Their own patched and frequently repainted vehicles looked as if they, too, were relics of British times. The road to Gedaref is an almost dead straight ribbon of tarmac which crosses an absolutely flat plain. Parts of it are always desert. This year, because of the drought, there was nothing green to relieve the grey dust which stretched into the distance where it was transformed into an endless mirage of water broken only by the occasional dried-out thorn tree. For much of the route the road follows the Blue Nile. It flowed sluggish and low; its mud flats merged into the grey of the surrounding countryside.

"No rain," I said, conversationally, to my neighbour.

"It is the will of Allah," he replied smoothly. He was a country-born Muslim who was now serving as a *Sharia* (Islamic religious law) magistrate in western Sudan. *Sharia*, he told me, had been in force in Sudan since September 1983. "So far", he said, "it has been successful. A man who loses his right hand is unlikely to repeat his crime. By the way, all *I* do is decide upon the punishment. It is the soldiers who carry it out." I was reminded of an entry I had seen that morning in the Sudan News Agency English language bulletin:

> Two porters and a butcher were convicted of causing injury to the plaintiff. One of the two porters was sentenced to twenty-five lashes, twenty Sudan pounds in fine and to be kicked in his head and bitten in his back. The other received similar sentences of flogging and fine beside a kick-in-the-belly sentence. A verdict of twenty-five lashes and twenty Sudan pounds was passed against the butcher. He was also sentenced to have his right hand broken and his forehead kicked.

Henry Stern, a Jewish convert to Christianity, whose *Wanderings among the Falashas in Abyssinia* was an account of a journey undertaken in 1860, also passed through Gedaref on the way to his encounter with the Falashas. Of the town he wrote: "Kedaref, which unites in itself the repulsive vices of Mohammedanism, with all the revolting pollutions of Paganism, was, at the time of our arrival, in a state of great consternation." There was no sign of consternation today. At three o'clock in the afternoon Gedaref is about as dead and hopeless a place as one can imagine. The heat was intense. A pall of dust hung over the market-place. Those with half-successful stores had closed them for the day. Those who had sold nothing in the morning, or who had nothing to sell, sat among the goats and debris and hoped. Just near where the bus had set down its passengers six knife-sellers sat next to each other waiting for a sale. The brightly coloured knives were identical. So were the prices they asked. Behind the knife-sellers I found a small dejected sidewalk café. Three Ethiopian refugees were sharing a Coke. One of them approached me jauntily. Dawit was an Eritrean: he spoke quick, flashy English. "Jesus Christ, man, what you wanna come here for? You come from England to a dump like this? You crazy?" I told him that I wanted to see the Ethiopian refugee camps. "Refugee camps? They're bloody concentration camps. We'll never get out of this fucking place. Nobody wants us."

"Will you take me to one of the camps?" I asked.

"What's in it for me, man?" he snapped.

I gave him a few dollars and we walked across to an open patch of ground where peasants from the surrounding countryside were selling fruit to the refugees returning to their camps. There were small, grey oranges and psychedelic pink grapefruits, puffy from exposure to the sun. "No one buys them. No one has *birr*," said

The Deliverance of the Tribe of Dan

Dawit rubbing his thumb and third finger together in the international sign for money.

The rutted track out of Gedaref was baked black and hard by the sun. The box-car carrying Dawit and me and a dozen refugees passed a pile of burning refuse which smoked against the afternoon sky. Storks picked through the rubble. Was there anything edible in Gedaref's rubbish tips, I wondered. A thin boy watched the birds intently.

The truck hit a pot-hole and swayed dangerously. One of the men fell against me. He seemed to weigh nothing. Everyone was silent. We reached the top of a small hill and I could see the once fertile plain stretching into the distance: now it was desert littered with stones and torn plastic bags. Tewawa refugee camp overlooks this sea of desolation.

In Tewawa the Ethiopians used to make a living working for local Sudanese farmers. Then came the drought. Now 20,000 people live here without work. Tewawa has become a permanent township of hundreds of *tukuls* (dome-shaped grass huts), each one surrounded by thorn hedges and bare patches of land which once were vegetable patches. Smoke rose from cooking fires, but there was no smell of food.

Dawit told me we had arrived. "It is a dangerous place, man. As many people are knifed every night as die of hunger." The central square of the camp was teeming with people. A hundred or so men stood in one corner selling personal possessions. One had a shirt, another a pair of trousers, another a skirt. Those with nothing to sell pushed their way feverishly through the crowd. "Most of the fights are over women," Dawit continued. "There are many prostitutes here."

We passed a few drunken men standing outside a couple of *tukuls* that served as a brothel. From the door of one of the huts two girls of striking beauty watched us walk by. On the other side of the brothel we saw a man lying on the ground, blood pouring out of a deep gash on his leg. He was singing. "He has been drinking *tedj*. The war widows make it and sell it, but the Sudanese police punish them if they find them out. They get thirty lashes and three months in prison." Since the introduction of *Sharia* law in 1983, alcohol has been banned in Sudan. "The old women make *tedj* and the young women are whores. They have to make a living."

We were walking down a wide path bordered by thorn hedges.

The *tukuls* on both sides were adorned with crosses. I asked Dawit if all the refugees were Christians. "I am a Christian," he said, pulling out a brass cross from beneath his shirt. "Most of the refugees here are Christians but there are some Muslims too. We also have some Israelites." He stumbled over the word. "Did you know that there are Jews in Ethiopia? We call them Falashas but they call themselves Beta Israel – the House of Israel. We do not like these people. They do very bad things to Christians. They brought the famine to Ethiopia. They killed Jesus." Dawit scowled piously towards the only brick building in the camp – a whitewashed Ethiopian church surrounded by a luxuriant garden. He was joined by some friends. "That is the city of Jews," one of them said, pointing at a group of *tukuls* even more pitiful and derelict than the ones I had seen in the Christian quarter. I asked them why these huts were worse than those further up the hill. "They don't fix the *tukuls*. They are not staying here. They are going to Port Sudan and from there, by German boat, to the Jewish country, Israel. Two nights ago some went. Tonight some will go. The Falashas have got Israel. What have we Christians got?" He gestured hopelessly at the surrounding camp.

Dawit and I walked on towards the Falasha section of Tewawa. In the main alley there were a few groups of Falashas who looked at us suspiciously. No one spoke. For the most part they were dressed in cast-offs: once fashionable bell-bottomed trousers and long-collared shirts collected by relief organizations in the affluent West. The older men wore the Ethiopian cloak, the *shamma*. The men looked hostile and worried, but the children seemed excited as if they were waiting for something. The women were sitting in groups in front of their *tukuls*. Because of the Jewish Sabbath the cooking fires were not burning.

I decided to send Dawit back to the main square to wait for me there. He did not seem reluctant to go. A large crowd of children gathered round smiling and pulling at my shoulder-bag. The Sabbath, which finishes at sunset, was drawing to a close. I greeted them with the traditional Falasha blessing, "*Sanbat Salam*" – "Sabbath Peace". The men looked at me stonily, but one or two of the boys laughed and replied in Hebrew, "*Shalom.*"

"Are you Jewish?" asked one man. "Yes," I replied. "Are you?"

"No," he said. "I am a Christian." We both lied. I said a few words in Hebrew to the group which had gathered around me.

A tall teenage boy took me to one side and whispered to me, in Hebrew, pointing at Dawit's retreating back: "Be careful. That man is bad. You cannot trust him. Hundreds of us are going to Israel, but you must not tell anyone. We are leaving tonight."

How could the Falashas be going there from a rigidly Islamic, Arab League state? How was it that Sudan's State Security Force, the *AmnulDawla*, reputedly the best secret service in black Africa, was unaware of what was happening? Was this a clandestine Israeli operation like the Entebbe mission or were the Sudanese themselves involved in some way? If the story got out, how would the other Arab states react? And if the Sudanese and the *Amnul-Dawla* were involved, why were there such poor security arrangements in Tewawa? If I could uncover this story, so could the dozens of Western journalists who were in Sudan reporting on the famine. At this point, the sight of two armed men at the end of the dusty track led me to believe that the security arrangements were perhaps better than I thought. Ducking behind a *tukul* I made my way back to where Dawit was waiting for me in Tewawa's main square. We jumped into a box-car and returned to Gedaref. Before we got out he said, "You know, man, I've been here for three years. Unlike the Jews, I'll never get out." He took the money I had given him and thrust it under my nose. "This isn't enough. Give me more, man, you've got it." I gave him some more money, the equivalent of a week's wages in Gedaref, so he said, but he did not look very happy. Nor did I expect him to.

I realized that I would have to go back to Khartoum to find out more about this seemingly incredible operation. In London I had been given a number of contact telephone numbers in Sudan of people working for aid organizations. The first person I telephoned when I got back to Khartoum was "Joshua". I mentioned various people we knew in common and he invited me to meet him that evening on the terrace of the Grand Hotel overlooking the Blue Nile. I briefly mentioned the interest I had in the Falashas and what I had discovered while I was in Gedaref. He looked shocked. Speaking slowly, in heavily accented English, he said: "I am very sorry you found out. This is a top-secret operation. The *Amnul-Dawla* agents in Gedaref are supposed to have the whole place sealed off. If just one word of this gets to the press the operation is finished. These people will be left to rot in the camps. Hundreds have already died. Hundreds, perhaps thousands, will certainly die if we don't get them out. If you are wondering what the difference is between the Falashas and the thousands of other Ethiopians who are dying in the camps in the east, I'll tell you. Like me, the Falashas are Jews. But that is not the main difference. The Christian and Muslim refugees from Ethiopia cannot escape the famine and the camps because they have nowhere to go. No country will take them. During my lifetime, Jews have not always had the chance to escape when faced with destruction. This time, because Israel is prepared to take them, these black Jews do have a chance. If this story gets out the Falashas lose that chance."

Joshua's real name cannot be told. Although I was later to learn that he was a remarkably courageous man, on this occasion he was nervous: partly on his own account and partly on account of the 12,000 people who for the next few weeks would be his responsibility. Joshua was the chief co-ordinator in Khartoum of Operation Moses.

He told me that it would be impossible for me to make any contact with the Falashas. Their camps were completely sealed off by the *AmnulDawla*. There were road blocks on all the tracks leading to the camps and there were Sudanese troops patrolling the country around them. The most extraordinary thing to learn was that the Sudanese secret service was organizing the transport of the Falashas from the Gedaref area to Khartoum Airport, from where they were being flown to Israel. And this was the reason for the great secrecy. If it were to be discovered that Sudan was co-operating with Israel, the repercussions in the Arab world would be enormous.

The next day, I returned to Gedaref. A police road block which I had not seen the previous day waved us on. In Gedaref I found a lorry which was taking thirty or forty Ethiopian refugees across the desert to the small town of Doka and on to Um Raquba (the Arabic means Mother of Shelter), the camp, where, according to Joshua, almost 10,000 Falashas were waiting to be taken to Israel. As the driver waited for the truck to fill up, a crowd of hawkers selling wares from deep trays balanced on their heads, surrounded the decrepit vehicle. A thin Ethiopian girl next to me picked up a cheap glass necklace and asked its price. "Fifty piastres only. It will bring you luck, if Allah wills." She put it back sadly. Just before the truck left Gedaref I bought it for her and for the next few hours she rewarded me with shy smiles of gratitude. It was an unfair transaction.

Soon the country on both sides of the rough track started to undulate. The land still had the appearance of desert, but occasionally we passed dusty Sudanese villages where dozens of men would be sitting idly in the coffee shop on the main square. With nothing growing, there was nothing for them to do. At one village we all got out to stretch our legs. I sat on the ground and smoked a cigarette. The villagers gathered around me curiously; a child touched my hair and wrinkled up its nose in amused disgust. Just past the village another road block waved us on.

Um Raquba is about twenty miles away from the border with Ethiopia. From a distance it looks like any other Sudanese village: a tightly packed collection of *tukuls*. The plain between the hilly border country and the refugee camp was a stony desert which most of the refugees had crossed by foot. The camp had been established in 1976 for 2,000 Christian Ethiopians from Gondar. Now it was a small township of 20,000 people. Um Raquba was divided into two sections. On a slight incline stood the "settled" camp where there were 10,000 Christians, most of whom had been there for some years. The settled camp consisted of *tukuls* as well as more solidly constructed buildings, such as the clinic, school and church. Half a mile away down a dirt road was the second part of the camp – the Reception Centre – which was an area of scores of *tukuls* and canvas tents radiating out from a simple field hospital built of wood and straw. There was no sign of any security guards.

The whole camp was under the joint aegis of the Sudan Commission for Refugees and the Sudan Council of Churches (s.c.c.). The Commission was represented by the camp's director and the s.c.c. by two Swedish relief workers on loan from the Swedish Council of Churches. In practice, the care and management of the newly arrived refugees was in the hands of the two Swedes, Peter, the camp administrator, and Elizabeth, the camp nurse. The Commission and the s.c.c. were dependent for supplies on the u.n.h.c.r. (United Nations High Commission for Refugees), which among other things was responsible for administering the u.n. World Food Programme.

As soon as I arrived I was taken to see the Swedes. They were evangelical Christians and, as they put it, "vocationally missionaries". Peter said to me jovially: "We are Christians, I mean committed Christians, so obviously we would like these ten thousand Jews to stay with us here in the camp and perhaps in time become followers of Jesus. But they are leaving and they do have a wonderful chance to escape and get away." He said it with a certain regret.

When they first arrived at the camp, Peter told me, the Falashas were hungry, in most cases starving, thirsty and ill. But after their arduous march from Ethiopia they were happy to see the camp. When some of the Falashas from the more remote villages arrived they thought that they had reached the Promised Land and knelt to kiss the barren soil in front of the field hospital. When they discovered that this was not Jerusalem they wept.

In fact, Um Raquba hardly proved to be much of a refuge for the Falashas. The physical condition of most of them deteriorated while they were there. Many of them died. Shortly after my arrival I walked with Peter to a low hill outside the camp and there I counted about 1,600 crudely built stone cairns, which served as the final resting-place for the Falasha dead. According to the camp records kept by the sanitation officer 1,939 Falashas perished between April and November 1984: 1,202 of these were under fifteen.

When we got back to the camp Elizabeth told me to be careful. There were killings here every night as there were in Tewawa. "Don't spend the night with the refugees," she said primly. "The only place they could put you up is in one of the whore houses; that's also where they drink *tedj* and that's where the stabbings take place." We were sitting in her neat, carefully swept compound. On one wall were colourful icons and other examples of Ethiopian folk art. Elizabeth was pressing a grey-blue cotton skirt with an old-fashioned iron filled with glowing charcoal. "You must keep up appearances," she said with a wry smile.

She sent me to report to the Sudanese camp director who told me that I would be accommodated in the sanitation officer's house – a low concrete building on the edge of the camp overlooking the desert. Inside the building there was only one room which was empty except for two neatly made beds. The sanitation officer, a lugubrious Sudanese, was sitting on one of them carefully tracing English words onto a piece of lined paper. "There is No Life Without Wife," he wrote, and began illuminating the head letters with coloured pencils. When I entered the room he greeted me sadly

The Deliverance of the Tribe of Dan

and from time to time lifted his head from his drawing to complain of the expense of getting married. He was twenty-six years old but now that he was sanitation officer he would have to wait for many years before he could afford the 10,000 Sudanese pounds necessary to throw a wedding party befitting his newly acquired rank. I asked him guarded questions about the refugees.

Straightforwardly he replied: "Most of the newcomers are Jews. No one is supposed to know they are here. Every morning for the last few days trucks have been coming to take them away. I supposed that you must have something to do with it. Otherwise they would not have let you in. Anyway, they are being taken to Israel. And that's a good thing. If they stay here they'll starve."

Following my host's directions I walked through the settled camp to the Reception Centre. Men and children stood around in groups eyeing me suspiciously. It was getting dark and the last rays of the sun were gilding the tents of this lost tribe of Israel as they prepared themselves for the journey that would finally take them to Zion. But any exhilaration I felt at the imminent deliverance of these courageous people was dashed by what I saw in the hospital. Dozens of grey-faced Falashas covered in little more than rags were lying on makeshift wooden beds. Relatives standing with them seemed completely at a loss. They looked at me blankly. One woman died as I stood next to her. Her husband, who had carried her on his back for the last week of their journey through the mountains to the border, was weeping quietly at her side. The next morning I looked for him but I was told that he was among those taken off by the lorries during the night. Most of the patients were older women, women with babies at their breasts, and children. These groups had also made up the bulk of the estimated 1,500 Falashas who had died on the long trek from their home villages to Um Raquba and other camps.

As far as I could see there were no orderlies or nurses on duty at the field hospital. Indeed, I was to discover that apart from Elizabeth, the Swedish nurse, there were no trained medical personnel in that camp of 20,000 refugees and frequently Elizabeth had been obliged to undertake complicated surgical operations for which she had had no training herself.

As I walked around the beds I was aware that a young Ethiopian was following me. I smiled at him and eagerly he told me, "I learned English at school. Are you English?"

"Yes," I said. "Are you Jewish?"

He nodded.

"They tell me that you are all going to Israel. I used to live there. It is a beautiful country."

"You are a Jew too?" he said. I nodded. Looking around carefully he said, "Come with me. I want you to meet my friends." We walked through the dark alleys of the camp until we reached a group of men sitting around a fire in front of a *tukul*. They were talking quietly together. From my bag I brought out tinned sardines, biscuits and sweets. For months, perhaps longer, they had been on a starvation diet: they were all emaciated. Yet they ate the food slowly, and with a sort of aristocratic restraint. Nothing was said. As they were eating, my guide Ya'kob said proprietorially, "He is a Jew from England." They moved closer and one of them touched my hand in a gesture of solidarity.

Crouched around the fire, their *shammas* drawn tightly around them, they smoked my cigarettes and sucked contentedly at the sweets. Gradually they started to talk. Most of them had been in the camp for six months or more and they told me about the grim losses their community had suffered during this period. Many had died of disease: the malaria against which, as highland people, they had no immunity, tuberculosis, hepatitis, measles, meningitis, dysentery and cholera. But this is not how they explained their losses: the water had been poisoned, they said; the Christians had taken their share of the food; there was not enough food anyway – sometimes a whole week had passed without them receiving anything. I told them I had counted the Falasha graves in the cemetery outside the camp and they retorted that there were more dead than had been buried in the cemetery: many dead Falashas had been dragged away at night and thrown into ditches around the camp by the camp workers. *Very* many had died, they assured me gravely. At one point their people were dying at the rate of fifteen to twenty a day and nothing, it seemed, could be done to prevent it. So the Falasha elders got together and after much discussion decided on a course of action. Half a mile away, between the camp and the cemetery, stood a hill on top of which an Ethiopian Muslim "wizard" had constructed a stone-built compound. Christians and Muslims from the camp visited the wizard and received cures and advice on personal matters. The Falasha elders visited him and were advised to sacrifice two cows, one red, one black, and a white goat. Money was collected throughout the community and the beasts were purchased from a neighbouring village and were sacrificed according to Falasha custom. But it did not help. Uneasily, Ya'kob said, "It is like what happened to the Israelites in Egypt before they went to the Land of Israel. They had many illnesses before they were saved. It made the people pure for the Holy Land." The men stared fixedly into the flames of the fire.

Yeshaq, one of the older men in the group, told me, in Amharic, how he had left his village and got to Um Raquba. Ya'kob translated: "One day in 1979 a rumour started going around that there was a road through the desert which would take us to Jerusalem. We came here to Sudan because we want to go to Israel. We had always dreamed of Jerusalem. First of all I sent my son from the mountains to see if there was a way to the Holy Land. He went away and six months later he sent me a letter. He told us that he had reached Jerusalem and that if we crossed the mountains and came down into the Sudan we would be able to go to Jerusalem too. I took my wife and sons and daughters and I came to this place. Now I am waiting to go to Jerusalem." He spoke these words calmly and patiently. "Did you not come because of the famine?" I asked. He smiled: "In my life I have seen a lot of famine. Some years there is food, some years there is none. Also there is no food here." He gestured at the camp around us. "No, we came here because the time has come for us to join our people in Israel."

I asked another older man how he had heard of Israel. He did not like the question and replied, aggressively: "It is written in the Torah that Israel is our home. We have *always* known about Israel. We are the descendants of Abraham, Isaac, Jacob, Moses and Solomon. Like the other Jews we have been waiting for many years. For hundreds of years. A few years ago we heard that Israel had become independent and that the Jewish state was fighting wars and defeating its enemies just like the Beta Israel in Ethiopia used to fight wars and defeat their enemies. Finally we left the place to which we were exiled and like the other Jews, like the white Jews, we are going home."

But there were other reasons for the Falashas' decision to leave their homeland and risk their lives in the process. One of the group told me he had been waiting in and around Gedaref for almost two years and thus far had never been included in a transport to

Khartoum. "If the trucks were going from Tewawa, I would be in Gedaref. If the trucks were going from Gedaref I would be in Tewawa." His reasons for leaving were more prosaic and he told his story in a flat matter-of-fact way: "When the revolution came they gave us land. We were happy to have the land because we had never owned our land in the past. But the drought came to the mountains and there was no rain. For a whole year there was no rain. Our lands were no good to us. That is why we wanted to leave our village and go to Israel."

I heard many stories of courage and endurance as I sat with the Falashas. But perhaps the most remarkable story was one I was to hear a few weeks later in Israel from a young woman who had been in Um Raquba that night and who had been rescued a few days later by Operation Moses: "When I first left my village in order to come to Jerusalem," she told me, "I was fifteen. A Falasha had come to our village and told the elders that we could walk to Jerusalem. We had a meeting and the elders told us what they had heard. They said that it was written in our holy books that one day we should go to Israel and that the time had now come. I set out with my brothers and sisters and some friends. My brother had brought malaria pills from the clinic in Gondar. We had water in plastic carriers, a supply of *injera* [flat porous bread], and the clothes we had were the *shamma* we were wearing. At one o'clock in the morning we left our village in the Semien mountains and crept out like thieves. For two nights we walked and in the day we slept. On the second morning soldiers found us and took us by truck to their army camp. We were questioned by an officer and my brothers were beaten. The soldiers wanted to have me, but the officer would not let them. We were accused of being Zionist spies and leaving for Israel. We told them that we were travelling to look for work, but they did not believe us. We were put in prison and stayed there for six months. We were all tortured, but it was worse for my brothers. When we were let out of prison we were taken back to our village. We stayed there for three weeks, but then we decided to try to leave again. This time we made our plans even more carefully. Our parents wanted to come with us, but they had been warned that the journey to Sudan was very difficult and that many had died. They decided to stay behind, but told us that we must go to Jerusalem. They gave us their blessing and told us that God would protect us. When we left there was no moon. My father

The Deliverance of the Tribe of Dan

walked with us as far as the next village. Then he returned home. This time our group was larger because all the younger ones in the village had decided to go. There were some older ones too. When we started off from the village we were twenty-eight people. We travelled for three nights and then rested in a Falasha village. Most of the young people in that village had already left and the older ones had decided that they would leave the following week. They killed chickens and made some good food for us. The following day we were captured by soldiers, who beat the men and did other things to two of the girls." As she said this she hung her head in shame and it was some time before she was able to carry on with her story.

"That night we escaped. We walked for about a week. For some of the way we had a Christian guide who showed us the path over the mountains. But he left us. The next day we were captured by the *shiftas* [bandits]. The men wanted to fight, but they had no guns. The *shiftas* threatened to kill us, but we pleaded with them and they took our food, our ornaments, our clothes and also the malaria pills which my brother had brought for us. They took the amulets against the evil eye which the children wore around their necks. They did not take our water carriers. We walked for another ten days. The older ones were unable to carry on and we had to leave them. They said they would try to return to the village. I left my uncle and aunt by the side of the track. We all cried. Some of the children died on the way. My sister had a baby. It was born early and was very small. The men carried her for a week on a stretcher until she was strong enough to walk. We hardly had any water and we had no *injera*. We ate berries that we found along the way. We were walking barefoot. Some of us had set off with shoes and the *shiftas* had stolen them. Our feet were bleeding from the stones and thorns. We walked at night and could not see the way. Often we lost the path and had to turn back. We had long sticks which we used so that we would not stumble. My brother and his friend became ill and died. I think they had malaria. But my brother had been badly beaten and was very weak. We had to leave them there, we were too weak to bury them. We piled some stones and branches on the bodies but we were not able to cover them properly. We were starving.

"One night we were resting by a big rock and a man came up to us. He was a Falasha, but not from our village. He was a tall, strong man. He was not starving. He gave us food and water and some of us medicine. He showed us the way to go and told us that people were waiting for us and would take us to Jerusalem. When we were on the right path he left us. He was going up the mountain to help other Falashas leave. We walked for two more days and then we came to the border with Sudan. The soldiers were waiting for us. They took us on a lorry to Um Raquba, where we found many Falashas from other villages. Of the twenty-eight who left, eight died on the way and four turned back. Every night as we walked we prayed that God would bring us to Jerusalem."

The next morning the sanitation officer woke me early and gave me a cup of sweet cinnamon-flavoured coffee. "The trucks have come and gone," he said. "The Jews are going to their holy city of Jerusalem. Also *our* holy city of Jerusalem," he said glumly. I made my way down to the Reception Centre but failed to find my friends of the night before. The Falashas kept their distance but other refugees were anxious to talk to me. With one group of refugee camp workers I sat and chatted about camp conditions and the political situation in Ethiopia. In the middle of this one of the group pointed at some Falashas who were watching us from the other side of the hospital compound. "Those people are Jews," he said. "They killed Christ." His eyes flashed as he said it and it seemed to me that he held this wretched group of Falashas personally responsible for deicide. Another refugee joined in: "Some people believe that they caused the famine in Ethiopia and that is why they have been driven out by the army; they have brought disease with them to this camp and many of us have died."

Later I was told that the Christians were often very afraid of the Falashas. Some of the recent Christian refugees in Um Raquba refused to go to the clinic in the Reception Centre as there were always Falashas in the vicinity. Aid workers in the camp told me that they had heard rumours that the Falashas eat people, particularly Christians, and cause women to have miscarriages. Falashas who were particularly active or energetic were believed to have the evil eye: a number were pointed out to me, rather discreetly, by other refugees. Falashas with the evil eye were attacked in the camps and from more than one source I heard the story of a young Falasha who, only a mile from the Sudanese border and what he believed to be safety, was torn apart by a group of Ethiopian refugees who feared he had the evil eye.

The Deliverance of the Tribe of Dan

Later that day I returned to the Falasha camp. Clouds of purple smoke hung over the *tukuls*. Within minutes I was besieged by a crowd of children. They were all thin; some of them had dreadfully distended stomachs. Shamefacedly I handed out the last of my English sweets. A youngish Falasha woman walked up to me in a slightly brazen way. Mockingly, she held out her hand and patted her pregnant stomach. "One for mother, one for baby and two for children," she demanded in halting Arabic. She smiled showing her gold teeth. Feeling embarrassed I gave her the sweets. "That's all?" she said. "How about some money for me and my children?" I told her that I only had travellers' cheques.

"Never mind – they'll do," she said, too quickly.

"Was it difficult to get here?" I asked.

"Many died. My husband died when we arrived. He's buried over there," and she pointed towards the cemetery. "My brothers died on the way. A child died."

"You are a Falasha?" I asked.

"No," she said mockingly. "I am a Christian, like you."

"What are your plans?" I asked.

"I am going to America," she half sang, ironically, elongating the vowels, and swinging her hips walked back to her *tukul* without a backward glance.

Over the next few days hundreds of Falashas left every night and the atmosphere among them became increasingly tense. They were well aware that the operation could be curtailed or cancelled at any moment. I kept away from them and spent my time interviewing the camp workers and the aid officials in order to gain a clearer picture of what had happened to the Falashas since they had been in the Sudan. One day the camp director came to my room early one morning and told me nervously: "Forgive me. There has been a mistake. You should not be here. I thought you had permission. You must leave."

Back in Khartoum I went to see Joshua. He was even more nervous than he had been a few days before. The airlift was working smoothly but he was living with the daily fear that one of the journalists covering the Ethiopian famine would stumble over the operation as I had done. His chief worry was that Senator Edward Kennedy was planning to visit Gedaref over the Christmas period and that the inevitable retinue of journalists would hardly be able to avoid seeing what was happening to the Falashas. "We need four weeks," he said, "and we can get every single Falasha out of the Sudan. The Sudanese are committed to helping us, and I think that many of them want to help, but if the story gets out they'll shut down the operation without a second thought." Had he known it, the story was already out. By the end of November, Jewish newspapers in the United States were circulating details of Operation Moses, and in early December the *New York Times* and the *Boston Globe* carried articles about the airlift. But for some reason the operation was able to continue for another month before a leak, emanating ironically from Israel, finally brought the airlift to a halt, stranding hundreds of Falashas in the Sudan.

The day before I left Khartoum, Joshua said, "You've got a good story. But don't publish anything until the operation is over. And just in case you are ever tempted, let me give you this." And he gave me a folded piece of paper torn out of a school exercise-book. On it was written: "A Falasha Prayer: 'Deliver me and put me with Thy people Israel, for thou art just, O Lord.'"

SIX

The Prisoners of Venda

Venda is a rugged territory situated between the southern border of Zimbabwe and the north-eastern corner of the Transvaal; much of it is covered by the Soutpansberg Mountains and their foothills. The majority of the population of Venda are VhaVenda, black tribesmen originally from Central Africa whose migration southwards extended over several centuries. In the nineteenth century the VhaVenda, under their greatest chief Thohoyandou, began to prosper and for a while controlled a vast area, from the Olifants river in the south to the Zambezi river in the north. The death of Thohoyandou led to a long struggle for the succession which was exploited both by whites and neighbouring black tribes. Paul Kruger, the president of the Boer Republic from 1883, tried unsuccessfully to bring Venda under the domination of the Boers. The country was finally conquered in 1899 and placed under white administration while a great deal of its territory was demarcated for white settlement. In 1979 2,500 square miles of the original land of the VhaVenda was designated the "independent" Republic of Venda by the South African Government.

The Lemba people, the so-called Black Jews of Southern Africa, live alongside the VhaVenda in and around Venda and also in parts of southern Zimbabwe. Paul Kruger was one of the first to perceive that the Lemba had a number of apparently Semitic characteristics and for this reason they are sometimes referred to as "Kruger's Jews". Although white missionaries and others have always shown a great interest in the "Jewishness" of the Lemba, there has still been no definitive study of the subject, and the claim that the Lemba are of Jewish faith or ancestry has still to be proved or disproved.

In recent years economic circumstances have forced many Lemba south to Johannesburg in search of work. Of the estimated total of 70–80,000 Lemba in South Africa, several thousand live in the sprawling black township of Soweto, most of them in the areas of Meadowlands, Dube and Tshiwelo. The state of emergency in December 1986 prevented me from visiting Soweto. Instead, I arranged to meet a representative of the Soweto community of Lemba in the relatively liberal atmosphere of the University of the Witswatersrand.

Witswatersrand University has a distinguished record of opposition to apartheid policies. None the less, it is still a predominantly white university. As I waited for my Lemba contact I wondered if he would feel uncomfortable here. Perhaps, after all, "Wits" was not the most neutral place for black to meet white. But in Johannesburg I could think of nothing better.

A plump, shabbily dressed black man was waiting for me. Near by a group of white students were exuberantly discussing their final examination results. The black was standing to one side, uncomfortable and self-conscious. Phillemon Matsherry was studying law at night school. During the day he worked as a clerk in an office not far from the university. Like almost all the blacks who work in Johannesburg he travelled from Soweto early in the morning and returned there at night. Matsherry was the executive secretary of the Lemba Youth Cultural Association.

"I am a Jew," he told me, "but unfortunately I have a Venda nose – not a Lemba one," and he rubbed his broad, flat nose ruefully. "All my family have fine long noses. If you can see my brother, you can like him. I am the only one with this!" He was born in Tshidele in Venda, a country village where two hundred Lembas lived together. "The important thing about the Jews and the important thing about the Lembas is cleanliness and purity. If a man is clean he will like himself. The name of my village means 'The Clean Place', probably because so many Lembas live there."

"How do you know you are Jews?" I asked.

"The most important point is cleanliness," replied Phillemon sternly. "All the rules of purity in the Bible we obey. We do not eat pork, unlike the VhaVenda. We have a Jewish language which our priests and traditional witch doctors know. A professor from Bar Ilan University in Israel has proved that there are many Hebrew words in this old Shona language of ours. I am busy learning this language. The young have forgotten it: only the learned elders like Professor Mathiva still know it. Then we have something like the Jewish *shofar* to call the Lemba to their assemblies: we use the rhinoceros horn. And for us, as for other Jews, circumcision is a most important matter. In fact, if a man is not circumcised he is not a man at all: he is regarded as a woman. We used to circumcise at seven or eight days like other Jews. But for hundreds of years we have been living among other nations and we have forgotten certain things: now we circumcise boys when they are nine or ten years and girls are circumcised by the Lemba women when they are thirteen. And we keep *kashrut*," said Phillemon, emphasizing his words by slapping his hand on the desk in the university classroom we were occupying. "We eat no self-dead animals. They must be slaughtered with a knife and have cloven hoofs. The priest used to slaughter – but now any Lemba male can do it."

Phillemon talked for a couple of hours as I took notes. Enthusiastically and perhaps selectively he led me through the lore of the Lemba, pointing out affinities with the ancient faith of Israel. He saw that his enthusiasm might not have entirely convinced me of the Lembas' claims and he exclaimed: "Many Jews accept us as Jews. Dr. Nabarro believes that we are. A few of the Jews of Johannesburg believe in us. The Rabbi Blumenthal who died last month was most believing. When we went to his funeral service there were no secrets from the Lemba: we were accepted and we saw everything. It was more than good!" he said excitedly. "It reminded me of Senna. But I was sad. Rabbi Blumenthal was a son of the Lemba: he was trying to bring black and white Jews together. He died at the height of his power. We shall not see his power again. His death reminded the Lemba that we are indeed a lost nation. A remnant of Israel."

"Would you like to go to Israel?" I asked.

"Even tomorrow", he said, his voice rising, "I would go to Israel! The Jewish and Lemba people should get together. It will bring great benefits to us both. I shall jump for joy if someone says 'Here's a ticket.'" Not for the first time during our conversation Phillemon looked at me hopefully.

Dr. Margaret Nabarro is an ethno-musicologist who is currently engaged in compiling material for a book on the Lemba; for many years she has been persuaded of the authenticity of the Lembas' claims. "Of course they are Jews," she said, her Lancashire accent in no way modified by decades in South Africa. "Very few of the local Jews take them seriously but facts are facts: they keep kosher, circumcise their young, separate milk and meat, have the same funeral arrangements and purity laws. There's no doubt about it at all! Do you know they have a musical instrument called a bush piano which, unlike those used by the VhaVenda, is played inside a gourd decorated with shells. The only other people who have an instrument like that are the Falashas. I believe that the Lemba and the Falashas are related and perhaps formed part of the same original migration. Some stayed on in Ethiopia as Falashas, the rest made their way down through Africa until eventually they came to Zimbabwe. If you want any more evidence – their tribal symbol is the star of David with an elephant inside it," and proudly she showed me the programme of a recent Lemba Cultural Association conference which had as its logo an elephant with raised trunk inside a star of David.

As further evidence Dr. Nabarro showed me her collection of Lemba pots: "Like the Ethiopian Jews," she said, "the Lemba are great artisans: potters, carpenters and, in days gone by, iron and copper workers. In fact, they introduced the VhaVenda to these skills and many more besides." The pots were formed by hand, not thrown on a wheel, and were decorated with geometrical patterns in ochre and silver-black. "Be careful with the black," she said, "they do it with boot polish and it comes off on your hands. You see the similarity to the Falasha pots? Not surprising: they both come from Senna."

The Lemba talk of Senna a great deal. According to their traditions the Lemba originate from the city of Senna "in the north"; then they "crossed the water" and created "Senna Two"; after this they migrated south and settled in southern Zimbabwe where, according to them, they were the chief architects and builders of Great Zimbabwe, whose ruins they still venerate. Some people have identified Senna with Saana in the Yemen; others with Sena on the Zambezi river. As Dr. Nabarro was talking I was reminded that when the Falashas were first encountered by Europeans they said they had come from "Sennar". This has been identified with Sennar on the Nile and has been used to support the theory that the Falashas, or the religion of the Falashas, came not from Arabia, but from the north and specifically from the Jewish colony at Elephantine. But perhaps "Sennar" meant Saana? Perhaps, indeed, Sennar and Senna are one and the same. This flight of speculation took me unawares as I looked at the Lemba pots in Dr. Nabarro's study. Her conviction, I could see, was contagious.

Missionaries, anthropologists and others have been interested in the Lemba for a long time. In 1875 A. Merensky in *Beitrage zur Kenntnis Sud-Afrikas* was perhaps the first to note the apparently significant and intriguing phenomenon that the Lemba closed their prayers with the word "Amen" – a word whose meaning they did not understand. Generations of missionaries were to take a special interest in the Lemba as a result. In 1908, R. Wessmann, in the *Bawenda of the Spelonken*, concluded,

> one cannot avoid the often striking similarity between the African and the Jewish types. Again and again we find laws and customs amongst the African which force the impression that there has been at some former times some kind of connection between blacks and the ancient Hebrews.... In the Lemba tribe, one may find, especially, distinct traces of such contact or connection.

Although it was mainly the beliefs and religious practices of the Lemba which led observers to the conclusion that they were of Semitic origin, their crafts and particularly their skills with metals also seemed to some to be proof of the fact that the Lemba came from somewhere else. In *Twenty-five Years in a Wagon: The Gold Regions of Africa*, Andrew Anderson wrote:

> The natives state that gold was worked, and the forts were built by men who once occupied the country, whom they called Abbelamba, and there is every appearance that it is so, for I am quite of the opinion that no African race of these parts ever built these strongholds or took the trouble to make such extensive excavations in the earth as we find all over the country.

The most important work on Venda is *The BaVenda* by H.A. Stayt, first published in 1931. Stayt devoted little space to the Lemba but his conclusions carry a certain authority. He wrote:

> The BaLemba are found scattered in small villages or single kraals throughout Vendaland, more particularly in the west, without chiefs and without any political bond of union. They possess marked Semitic characteristics and are the metalworkers, traders and businessmen among their Bantu brethren, wandering among the BaVenda, indeed anywhere among the tribes from the Zambezi to Pietersburg, absorbing the language of the people among whom they

have settled and maintaining by endogamy (at any rate until recently) the purity of their race. The life and customs of this peculiar people are strangely reminiscent of the wandering Jews of mediaeval times. They are rapidly losing their individuality since European traders have robbed them of their heritage.

Similar conclusions are still drawn. In a recent doctoral dissertation on the relationship between the languages of Venda and Shona, P.J. Wentzel states with some certainty:

> The Lemba as we have come to know them in Zimbabwe and the Transvaal cannot be seen as a tribe or even a sect. They are a cult community. This community is built up of at least three elements. The first is a Semitic element with its origin in Ethiopia with a group of people called Falashas fleeing to the south after upheavals in that country. This Semitic element was much smaller than the Mahommedan element which met and merged with this first group in the vicinity of the Lower Zambezi.... Apart from these two groups there was a third which came from the east via Kitevhe's Kingdom into Zimbabwe before the end of the sixteenth century.

Not everyone, however, is so convinced. John Blacking, an eminent ethno-musicologist, has explained the interest of outsiders in the Lemba in the following way:

> Many European missionaries and administrators were evidently attracted by the idea of a minority who inherited intellectual faculties "superior" to those of the Venda majority.... It happens that some of the Lemba taboos and rites, such as abstaining from pork and eating only kosher-killed meat, coincide with Jewish practices. On the minimum of scientific evidence and the maximum of romantic conjecture, numerous theories about the Jewish origin of the Lemba have been put forward, even the ideas that they are a lost tribe of Israel, or the builders of Zimbabwe. These theories may be correct, but as yet there is no proof that is even moderately convincing.

Margaret Nabarro had suggested that I should meet Professor Mathiva, the first Lemba professor, and a former student of the School of Oriental and African Studies. He now teaches at the University of the North, a black university in the "National State" of Lebowa in the northern Transvaal. But Professor Mathiva told me on the telephone that he was leaving that very day for the Republic of Venda. We arranged to meet two days later in Thohoyandou, the capital of the republic, in front of the Venda Sun

The Prisoners of Venda

Hotel Casino – the country's chief source of revenue. "Oh, by the way," said Mathiva, "I shall be driving a Mercedes. A white Mercedes. And what are you driving?"

"I'll be renting a car."

"Never mind, old boy," he said commiseratingly.

The three hundred miles which separate Thohoyandou from Johannesburg cross the Boer heartlands. Expanses of uncultivated high veld gave way every now and then to carefully tended fields and boring, respectable little townships. I stopped for petrol in one town which, according to the map, was dead on the Tropic of Capricorn. It was hot and bright. A few blacks were sitting outside the liquor store. I left the car to be filled up and went to what I took to be the men's lavatory. "That's for blacks," snarled the Boer proprietor and showed me to the one marked "Blanken" – whites. The last Boer town before the turn-off for Thohoyandou is Louis Trichardt: Venda lies to the north-east of it. The tourist agency in Johannesburg had given me a pictographic map produced by the Venda Ministry of Tourism which presents the republic as the "Land of Legend": it shows "kokwane" – prehistoric footprints – Mumwadi Hot Springs, the Sacred Lake Fundudzi, the Holy Forest and, up towards the Limpopo river which separates South Africa from Zimbabwe, an enticingly drawn Big Baobab Tree. As it is against the policy of the Republic of Venda to distinguish between Venda and Lemba, there was no mention of the extraordinary past of Kruger's Jews.

Venda is spectacularly beautiful. The wooded slopes rising to the Soutpansberg Mountains are crossed with rivers and streams; the land, where it has been cleared, is red and fertile. The thatched conical huts, not unlike the *tukuls* of Sudan and Ethiopia, climb the hills and blend with the woods which cover all but the highest mountain. There had been rains and the fields surrounding the villages were bright green slashed with scars of livid red.

The Venda Sun Hotel and Casino is a place to avoid. I went in for a coffee while I was waiting for the Professor: it was eight o'clock in the morning and the dining-room smelled overpoweringly of stale cigarette smoke. A red-eyed unshaven white sat smoking moodily, his coffee untouched. Perhaps he had lost heavily the previous night. Outside on the red-brick surround of the hotel swimming-pool a few South African Asian families sheltered from the morning sun in thatched booths, such as the ones used to advertise holidays in the Caribbean. They did not look as if they were enjoying themselves.

Mathiva was punctual. A tall, awkwardly built man, he was dressed, befitting his profession, in a severe grey suit of impressively antique cut. His head had recently been shaved; the slightest grey stubble was beginning to show through his scalp. The Lemba tradition, now not widely observed, requires both males and females to shave their head at the new moon for reasons of ritual purity. We sat down, but he seemed uneasy. Hoping to break the ice I mentioned colleagues of mine in the Linguistics Department at S.O.A.S. where Mathiva had studied. His face contorted; with memories, I supposed, of past difficulties. Quickly changing the subject I showed him the Minority Rights Group Report I had co-written on the Falashas. His eyes lit up.

"Falashas?" he enquired, eagerly.

'Yes. I have also written this book about the Falashas of Ethiopia." And I showed him *Operation Moses*, which features a colour photograph of Falashas on the cover.

Slowly and emphatically he said: "We Lemba are Falashas. We travelled from Ethiopia to Zimbabwe, and from Zimbabwe some of us came here. Even now, when we bury our dead, they are laid on a shelf in the grave with their heads pointing north, towards Senna. When a Lemba dies he returns to Senna. The Falashas also left Senna but they stayed in the north. We are blood relatives of the Falashas. They are the tribe of Dan and we are another tribe of Israel." For some time he studied the book intently.

"Is it available here in South Africa?"

I told him I thought it was.

"The Lemba will buy many copies of it."

"I shall let my publishers know."

Mathiva studied the photographs carefully, paying particular attention to a young Falasha girl I had photographed in Jerusalem.

"She is so beautiful. And can't you see she is a Lemba?" he said excitedly. "Most certainly she is Lemba – you see this nose?"

Now Mathiva was prepared to talk. The Lemba, he told me, were Jews. They had once had a book, perhaps the book of the Jews, he said, but it had been lost along with the great Lemba drum. They worshipped God whom they call Mwali. Mwali had occasionally revealed himself to the Lemba, particularly on mountain tops. The most sacred place in Lemba tradition, with the exception of Senna whose location is unknown, is a mountain in southern Zimbabwe called Mount Belengwa. This mountain is protected by a hereditary Lemba custodian who must give his consent to the entry of the aspiring pilgrim. "Even a good Lemba who has failed to receive the permission of this guardian", said Mathiva, "will not return alive from the mountain. And the man does not let every fellow in. If you have eaten pork you can never go there. We used to go together every year in a bus belonging to the Lemba. But now the bus is broken. The only way to go now is by private car."

"But would you not like to make a pilgrimage to Israel if you are a Jew?"

"I would. I would. Most certainly and with the greatest happiness. But it's a question of ways and means. I also want to visit Senna," he said, as if to emphasize the impracticality of going to Israel.

"The Jews do not recognize the Lemba," I pointed out.

"I believe the rabbis are very harsh. Every Jew must practise his religion if he wants to be taken for a Jew. We Lemba must prove through our deeds that we are Jews. Obviously, all human life is influenced by the places an individual inhabits. It is possible to lose your own cultural and religious heritage. It has happened in the past and it is happening today. That is why the Lemba Cultural Association is so important. We Lembas are indeed losing our culture. But historically there is proof that we are Jews. And if our book were ever to be discovered there would be final proof. And though there may be some little differences here and there, the similarities are even greater!"

Later that day I met a young educated Lemba whom I shall call Daniel. Having had difficulties with the South African authorities he was lying low in Thohoyandou where I met him by chance. He was less anxious than Mathiva to prolong the separate existence of the Lemba. "Divide and rule is the policy of Pretoria. Why should we serve the purposes of white supremacy? I am South African and my heritage is South Africa. At the same time I am a Lemba. But the distinction between Lemba and Venda these days is very slight. We all speak Venda. My father and grandfather speak the Lemba language but we young people do not. We can marry whom we want: all my aunts and many of my grandfathers are Venda. What happened was that because we were such good potters and traders the Venda kings gave those of us who had come south from Zimbabwe certain privileges: we were the bearers of the royal drum and carriers of the royal baggage. They gave us a small amount of land and we settled, and in time we multiplied. In many respects we can now be considered as Venda. As far as white South Africa is concerned we can be considered as black South Africans." The last sentence had come out as something of an outburst and he smiled apologetically. "But there is no question but that the Lemba are Jews; there is research going on now at Witswatersrand University that proves it. I am definitely a Jew. A South African Jew."

"Daniel," I asked, "how do you imagine Jerusalem?"

"Well," he said, "I don't know, but in my imagination it is like Senna. It is a big city but the people maintain their ancient cultures and traditions. Their houses are not similar to these" – he pointed at the thatched houses up on the hillside – "but rather they are tall and shining, cream, no, white. White. I am sure that eventually the Israelis will accept us as Jews. After all, if we can be accommodated by the Jews of South Africa, we can be accommodated by the Israelis, too. But for the moment we are prisoners in our own land. We cannot leave Venda for Israel. I dare not leave Venda at all."

"But where do you see your people's redemption? In creating the new order in South Africa you have been telling me about, or in hoping to be able to go to Israel?"

"Ah! That's a very modern question!" he laughed. "You see, in traditional Lemba society we did not think of redemption. We believed in Mwali and we believed in our ancestors. We did not need redemption. From what? We had our ancient religion and culture; we had rain; we had food. Mwali provided everything. Today it is different. Contact with Israel will help us to recreate our past; revolution in South Africa will help us recreate our present. But I'm not waiting to be taken to Jerusalem on an eagle! Look at me!" and he pointed at his substantial belly. "I would need a fleet of eagles!" He roared with laughter and using his joke as an excuse to bring our conversation to an end he left the hut where we had been sitting and went off to a "meeting" – a word which he managed, more or less, to invest with revolutionary authority.

The religious identity, or perhaps identities, of the Lemba have undergone many changes over the last century. Christian missionaries first started working in Venda in 1872. Since then many Lemba have been exposed to the Old Testament and, of course, to Christianity. Their identification with Jews is probably one which was suggested to them by missionaries at this time. Many Lembas converted to Christianity and, as Professor Mathiva put it to me, became "excellent Christians". But some became disillusioned. Phillemon Matsherry had said: "The missionaries got on very well with the Lemba but when the Lemba discovered that the missionaries ate pork and self-dead animals despite what it says in the Bible we lost respect. Education was a priority, also then, and we went to the mission schools and accepted their teaching. I am sorry to say that the Lemba are very clever people and we wanted to gain the learning of the white man. But we remained culturally Lemba even against the opposition of the missionaries. The missionaries were very much against the circumcision of women, for instance; but when the Lemba girls came to puberty they would slip out of the mission schools and go off to the initiation ceremonies, where the older women look after things, and do what their mothers had done before them." Another influence upon the Lemba may have been the small Jewish community in Louis Trichardt: some Jews had Lemba servants who had every opportunity of learning about Jewish dietary and other customs. One Lemba, Mtenda Mbelengwa, had worked for a Jewish family in the town: it is perhaps significant that he became the first president of the Lemba Cultural Association, which was founded in 1945, and it was he who introduced the star of David as the logo of the Lembas in 1947, combining it with the elephant which had been their traditional totem.

But perhaps the greatest change over this period has come about as a result of assimilation to Venda culture. Professor Mathiva took me to meet one of his nephews, Crause Mabudhafasi, an extremely well-educated young man, who was working in the clinic of one of Mathiva's sons (two of whom were doctors). In the late afternoon we drove around the villages together, Crause acting as my guide and, as need arose, interpreter. "It really is becoming a thing of the past," he told me. "Lemba, not Lemba. We have been living among Vendas so long that we are almost the same. In fact, they have become like us: they circumcise now, which they learnt from us; they are almost as skilled as us now as artisans, which they learnt from us; the only major difference these days is that they eat pork and keep pigs and we don't." Crause took me to meet some of his family. His grandmother and aunt lived in beautifully painted huts on the slope of a valley not far from Thohoyandou. The walls of the huts are made from hand-made bricks which are plastered over with mud. The floors inside the huts and in the walled yard are made of cow dung mixed with mud which, when baked by the sun, makes a smooth and durable surface. Crause, who had been named in honour of some Scandinavian teacher of his father, introduced me to his relatives: "This is my grandmother: she is one hundred and two. I have many grandmothers – most of them are over ninety; and this is my aunt, my grandfather's wife, she is eighty-two." It should be noted that the Lemba traditionally have tended to marry their cousins, and certainly within the clan, of which there are twelve among the Lemba people (some of them like Hamisi and Sadiki have distinctly Semitic names): this can help to account for the fact that in English the Lembas use relationship terms like grandfather loosely: Crause told me of six of his "grandfathers". Both of the women confirmed the fact that the Lembas eschew pork, will not mix meat and milk and would refuse to use cooking vessels belonging to non-Lemba. But on other details of Lemba practice they gave rather heterodox information. They claimed, for instance, that Lemba women were not circumcised but that only Venda women were; their denial might be explained, of course, by embarrassment in front of a white male. Neither of them knew what a "Jew" is (the Venda word for Jew is "*Yahud*"); neither had heard of Israel; and neither had heard of Senna. Crause seemed mildly surprised by their ignorance of the traditions of his people but explained it easily enough. "These women are not inquisitive like young people today. Traditionally they ate apart from the men who kept the tradition. They know about cooking and childbirth, pot-making and weaving. What else could they know?"

There was nothing dismissive about Crause's question. He had questioned his elderly, dirty and traditionally dressed relatives with elaborate courtesy and patience. Respect for the elderly and particularly veneration of the ancestors are claimed as specific Lemba virtues. Indeed, ancestor-worship, particularly of ancestral Lemba leaders such as Mbelengwa and Baramina (about whom little is known) used to be part of the Lemba cult.

Samuel Moeti, the Director General of the Foreign Relations Department of the Republic of Venda, is one of the acknowledged leaders of the Lemba today. I telephoned him from my hotel near Louis Trichardt and we arranged a meeting-place. "I drive a white Mercedes," he said, "with a licence plate vvii." Moeti took me to a large modern house on top of the hill that looks over Thohoyandou. It was next to the only building of real substance in the republic (apart from the Venda Sun Casino), the South African Embassy. There was a synthetic peach-coloured carpet on the floor with plastic strips protecting the main thoroughfare; in one corner stood a three-tiered object made of white plaster. It was a sort of Baroque fountain with basins of decreasing size at each level. The rim of the top basin supported a naked woman carrying an urn and a small bird, alarmingly out of scale. Everything in the room, with its magnificent view over the veld, was Western and hideous. Perhaps the only item in the entire house which had something to do with Lemba culture was a tape of a programme Margaret Nabarro had done for South African Radio on Lemba culture. Mr. Moeti was cautious and diplomatic in dealing with my questions about the status of Venda. It is well known that all the "independent" homelands are hopelessly corrupt. But when I asked him about corruption in the republic he answered blankly, "No corruption. No corruption at all in Venda," and he beamed at me. But his eyes were twinkling. He was more forthcoming on the Lemba.

According to him there are half a million Lemba in the whole of Africa: 250,000 in South Africa and the same number in Zimbabwe. He spoke generally for some time about the cultural and religious tradition of the Lemba more or less voicing what I had already heard.

"Do you really consider yourself to be a Jew, Mr. Moeti?" I asked.

"Of course I do. And it is causing a lot of problems for me and for the rest of us. The Venda are jealous of our association with Israel."

'But what association do you have with Israel?"

"You know, we have Israeli visitors sometimes, like Professor Sharon and Professor Sharvit. And some of our younger ones have been paired off with pen pals among the Falashas in Israel. Johannesburg Jews, white Jews, arranged that."

"But, Mr. Moeti, would you actually want to go and live in Israel? Here you are an important man, with a fine house and a white Mercedes. What sort of a future could you expect in Israel?"

"My dear friend – of course I would most dearly love to go there.

The Thirteenth Gate

I would of course have to make sacrifices," and he gestured in the direction of the white plaster object, "but my identity matters most," and he gave me another of his overwhelming smiles.

Moeti's greatest concern, as he talked on into the evening, was assimilation. In former times Lemba women were absolutely forbidden to marry a non-Lemba; non-Lemba women could be brought into the tribe, but only after a rigorous ceremony of purification which involved shaving off all body hair, immersion in water, having to lie under a wooden pyre which was then set alight and being made to crawl naked through a small hole in an ant hill. "Of course," said Moeti, "the ants bite and the woman bleeds. The idea was to bleed that pig blood off." Now, Moeti confessed, Lembas were intermarrying freely with Vendas. "Assimilation and anti-Semitism are our main worries," he continued.

"Anti-Semitism?" I asked. "What anti-Semitism?"

"Oh, just general," he said, and he waved an arm vaguely towards the grass huts that dotted the landscape beyond his picture window.

A few days later, back in Johannesburg, a fearless Margaret Nabarro and her husband, an eminent scientist, risked official "investigation" by inviting half a dozen Soweto Lembas to their home to meet me. When I arrived, the room in which the small reception was being held was in darkness and the Lembas and the Nabarros were watching slides of the annual Lemba' Cultural Association Rally and other Lemba events. The conversation was easy and relaxed as if the Nabarros and the Lemba were related and were watching family slides. In a sense they were. Margaret Nabarro told me that not long before she had been made an honorary Lemba – the first non-Lemba to be so honoured. "I am sorry we are so few," said Solomon Sadiki, a middle-aged man who acted as the group's spokesman. "Soweto is burning tonight. It was difficult to leave and we could not bring too many here for fear of attracting attention to the Nabarros," and he smiled gently at his hosts. "The Nabarros are Lembas," he said. "And they are our friends."

Solomon had all the appurtenances of a Semite. He could have passed easily for a dark Yemenite Jew. When I told him about my interest in the Falashas, Solomon became animated. "My grandfather used to tell me that we Lemba are Falashas. Once in 1956 I went with my boss on behalf of the Lloyds Insurance Company to Ethiopia. We stayed in Addis Ababa for some time. Well, of course, as soon as I got there I looked up the Falashas who were living in the city. Their customs are just the same as ours and I felt comfortable with them. Naturally they accepted me as one of them. I would have been happy to stay with them, but work and family brought me back to South Africa. I have no doubt that the Falashas and the Lemba are the same people." As he spoke Margaret Nabarro was showing slides of "typical" Lemba faces – men and women with lean Semitic faces, thin prominent noses and almond shaped eyes. Each photograph seemed to confirm what all of the people there believed: that the Lembas were Jews with a history as long and as proud as that of Jews anywhere else in the world.

As we sat eating Margaret Nabarro's mince pies one of the Lemba said: "They are calling this Christmas 'Black Christmas'. The revolution could happen any day. We are all being asked not to use the electricity and to show solidarity by lighting a candle and putting it in the window. Houses will be burned tonight. People will be killed. We had better go."

At the door, Solomon Sadiki turned to me and said, "Israel takes good care of its lost tribes, is it not so?"

"I believe it is so, Solomon," I replied.

"They rescued the tribe of Dan from Ethiopia, didn't they?"

"They did, Solomon."

"They might rescue us too."

"Yes, they might."

"But our trouble, you know, is that we have forgotten which tribe we are."

And with that Solomon and his friends disappeared into the night in the direction of Soweto.

Glossary

AmnulDawla (Ar.)	Sudanese security forces
Barmitzvah (Heb.)	The ceremony by which a thirteen-year-old boy becomes an adult Jew for ritual purposes
Eid (Ar.)	Muslim festival
Eid al-Fitr (Ar.)	The festival which concludes the month of Ramadan
Fellah, fellaheen (pl.) (Ar.)	Farmer or peasant
Gaijin (Jap.)	Foreigner
Genizah (Heb.)	A store for worn-out sacred books and scrolls
Hallah	Loaf used for the Sabbath
Harat al-Yahud (Ar.)	Jewish Quarter
Havdalah (Heb.)	Prayer indicating the distinction between the holy day which has ended and the normal day which is to follow
Hazzan	Cantor
Injera (Amharic)	Bread
Kaffiyya (Ar.)	Arab headwear
Kame'a (Heb.)	Amulet or charm
Kashrut (Heb.)	Jewish dietary laws
Kiddush (Heb.)	Prayer proclaiming the sanctity of the Sabbath or a holy day

Glossary

Masjid (Ar.)	Mosque
Matzah, matzot (pl.) (Heb.)	Unleavened bread
Menorah, Menorot (pl.) (Heb.)	Seven-branched candelabra
Mezuzah, mezuzot (pl.) (Heb.)	Biblical verse written on parchment kept in a special case affixed to the doorpost of a Jewish home
Minyan (Heb.)	Minimum quorum of ten adult males needed for corporate worship, e.g. in synagogue
Muhabarat (Ar.)	Syrian secret police
Musawi (Ar.)	Of the Mosaic faith
Navi, neviim (pl.) (Heb.)	Prophet
Negus (Amharic)	King
Rebbetzin (Yiddish)	Rabbi's wife
Shamash (Heb.)	Synagogue servant, sexton
Sharia (Ar.)	Islamic religious law
Shifta (Amharic)	Bandit
Shofar (Heb.)	Ram's horn blown on ceremonial occasions
Shohet (Heb.)	Ritual slaughterer
Shtetl, shtetlach (pl.) (Yiddish)	Small village or town in Central and Eastern Europe inhabited by Jews
Siddur (Heb.)	Prayer book containing synagogue liturgy
Tukul (Amharic)	Dome-shaped grass hut

Select Bibliography

S. D. Goitein, *Jews and Arabs* (New York, 1964)
B. J. Israel, *The Bene Israel of India* (London, 1984)
D. Kessler, *The Falashas: The Forgotten Jews of Ethiopia* (new edition, New York, 1985)
B. Lewis, *The Jews of Islam* (London, 1984)
E. Nathan, *The History of Jews in Singapore 1830–1945* (Singapore, 1986)
T. Parfitt, *Operation Moses: The Story of the Exodus of the Falasha Jews from Ethiopia* (London, 1985)
N. Rejwan, *The Jews of Iraq* (London, 1985)
H. A. Stayt, *The Bavenda* (reprinted London, 1968)
M. Tokayer and M. Swartz, *The Fugu Plan* (New York, 1975)